LEMON-AID
USED CARS
2000

LEMON-AID USED CARS 2000

PHIL EDMONSTON

Stoddart

Published in 1999 by
Stoddart Publishing Co. Limited
34 Lesmill Road
Toronto, Ontario
M3B 2T6

Canadian Cataloguing in Publication Data

The National Library of Canada has catalogued this publication as follows:

Edmonston, Louis-Philippe, 1944–
Lemon-aid used cars

Annual.
1999–
Continues: Edmonston, Louis-Philippe, 1944–.
Lemon-aid used car guide, ISSN 0714-5861.
ISSN 1485-1121
ISBN 0-7737-6046-6 (2000)

1. Used cars – Purchasing – Periodicals. I. Title.

TL162.E3398 629.222 C98-300799-3

Cover design: Bill Douglas @ The Bang
Typesetting and text design: Wordstyle Productions
Editorial services: Greg Ioannou, Colborne Communications Centre

Canada

The Canada Council for the Arts since 1957 | Le Conseil des Arts du Canada depuis 1957

We acknowledge for their financial support of our publishing program the Canada Council, the Ontario Arts Council, and the Government of Canada through the Book Publishing Industry Development Program (BPIDP).

Printed and bound in Canada

CONTENTS

MEDIUM CARS
206

LARGE CARS/WAGONS
305

LUXURY CARS
333

SPORTS CARS
402

MINIVANS
451

Appendix I/25 BEST INTERNET GRIPE SITES
497

Appendix II/SURVEY AND BULLETIN SEARCH
501

Key Documents

The following documents, memos, court decisions, and service bulletins aren't published in any other consumer guide. They have been put in the following index so that you can easily find and photocopy whichever document will prove helpful in your dealings with automakers, government agencies, dealers, or service managers. Most of the service bulletins outline repairs or replacements that should be done for free.

Part One

Part Two

Part Three

Bulletins Showing Factory-Related Defects

Introduction
LEMON-AID DOES MORE THAN COMPLAIN!

The 27th edition of *Lemon-Aid Used Cars* is unlike any other auto book on the market. Its main objective, to inform and protect consumers in an industry known for its dishonesty and exaggerated claims, remains unchanged. However, this guide also focuses on secret warranties and confidential service bulletins that automakers swear don't exist. That's why you'll be interested in the "Key Documents" list found on the previous pages. There you'll find the exact bulletin, memo, or news clipping reproduced from the original so neither the dealer nor the automaker can weasel out of its obligations.

The book's information is culled from mostly U.S. and Canadian sources and is gathered throughout the year from owner complaints, whistle blowers, lawsuits, and judgments, as well as from confidential manufacturer service bulletins.

Lemon-Aid does more than complain, however. It not only targets abusive auto industry practices, it also lobbies automakers for changes. For example, two years ago we blew the whistle on Chrysler Canada's failure-prone automatic transmissions and brakes, and defective paint. Following *Lemon-Aid*'s downgrading of Chrysler's products, company officials met with me and set up a Warranty Review Committee to pay claims previously rejected. After a similar downgrading last year of products by Ford of Canada, its officials arranged to see me in February of this year and began giving refunds to Taurus, Sable, and Windstar owners with engine and automatic transmission claims, even after their warranty had expired or they had been refused a refund in the past (see Part Two, "Invest in Protest").

Each year, *Lemon-Aid* announces its auto/automaker ratings. This guide makes a critical comparison of 1990–98 cars and minivans, and safer, cheaper, and more reliable alternatives are given for each vehicle.

Lemon-Aid combines test results with owner complaints, Internet postings, and surveys to determine its ratings. That's where you, dear reader, can help. Please fill out the questionnaire found at the end of this guide and list the degree of satisfaction or dissatisfaction you've experienced with your vehicle. Your information will be compiled and used in the next edition of *Lemon-Aid Used Cars.*

I wouldn't dare keep it a secret.

Phil Edmonston
www.lemonaidcars.com
June 1999

Part One
STEALS AND DEALS

"In the early '60s boomers grew up believing the Establishment, including the U.S. car industry, was no good, and in the '70s, we proved them right, we taught them that we build junk."

James C. Bulin
Ford Motor Company
Generational Studies Manager

"Out of 100 vehicles, we're apt to build 10 that are as good as any that Toyota has ever built, 80 that are okay and 10 that cause repeated problems for our customers."

Robert Lutz, President
Chrysler U.S.
Chrysler Times, 17 July 1995

Right on, Robert, and it's *Lemon Aid*'s job to warn buyers to steer clear of that last 10 percent and even reconsider that 80 percent of your production you consider "okay."

Thirty years of victories and defeats

Okay, so I may be a car curmudgeon, but I've earned the title over the past 30 years battling automakers who routinely cheat customers and treat their dealers as serfs. When I founded the Automobile Protection Association (APA) in Montreal in the fall of 1969, American Motors was giving out free television sets with the purchase of their problem-plagued Concorde and Hornet (actually, the television sets lasted longer than their cars), Volkswagen had a monopoly on hazardous, poorly heated, and sluggish Beetles and minivans, Ford was churning out biodegradable rustbuckets (and denying they had a secret "J-67" warranty to cover repairs), and the Chrysler Valiant and Dart were the epitome of reliability (except for chronic stalling when it rained). Imagine, when the first guides were written a few years after the APA founding, a good, three-year-old used vehicle could be had for less than $1,500; the average price now tops $7,000.

Chrysler owners' protest in January 1998 against paint peeling led to the creation of Chrysler Lemon Owners' Group (CLOG) in British Columbia and New Brunswick. Although Chrysler cars and minivans are reasonably priced, spacious, and loaded with convenience features, chronic automatic transmission, brake, AC, and paint defects can easily put a $5,000–$7,000 dent in your wallet. As a result of Lemon-Aid *and CLOG efforts, Chrysler no longer routinely blames its workers, bird droppings, or the sun when faced with paint peeling or powertrain claims, and often settles out of court after a review by its Warranty Review Committee (see Part Two).*

I began writing *Lemon-Aid Used Cars* almost 28 years ago as a guide for consumers and as an admonition to automakers and dealers to treat consumers with respect and integrity. Each year, *Lemon-Aid* targets abusive auto industry practices and lobbies for changes. For example, two years ago, we highlighted Chrysler's automatic transmission, paint, AC, and brake problems, and called for Chrysler to pay for repairs on components that failed just after the warranty expired but long before what most experts would consider was reasonable durability (see the chart on p. 0). Following *Lemon-Aid*'s urging, Chrysler set up a special review committee and telephone hot line to compensate first and second owners long after their warranty had expired or they had been refused a refund in the past. That committee is still in operation.

Reacting in a similar fashion to *Lemon-Aid*'s criticism last year, Ford has restructured its customer assistance department and extended the warranty on automatic transmission and 3.8L V6 engine failures affecting primarily its Taurus, Sable, and Windstar. It's a little early to pass judgment on Ford's efforts, except to say that the company has bought back some vehicles and compensated over 40 claimants who had been refused any help previously.

And let's not forget General Motors and its paint delamination problems, shared by Ford and Chrysler. After a five-year campaign by *Lemon-Aid*, GM, Ford, and Chrysler are now providing partial and full refunds to vehicle owners who have had their paint turn chalky white and peel away within the first six years of ownership. GM, the long holdout of the trio, has been sued successfully before small claims courts by plaintiffs armed with copies of internal service bulletins and GM memos produced by *Lemon-Aid*. In fact, the latest case of *Bentley v. GM*, rendered in December 1998, goes further than the other judgments in forcing the automaker to pay for paint repairs for a 1991 pickup (see Part Two, "Jurisprudence").

Dealing with the federal government over the years has been less successful. Transport Canada bureaucrats couldn't even muster the expertise of a 12-year-old in setting up a website to give Canadians important information on recalls, airbag dangers, and addresses of garages that will install airbag shut-off switches. After closing its recall site for two years, and a month after getting blasted in my *www.lemonaidcars.com* website for its incompetence, Transport Canada's Road Safety Branch set up a more user-friendly recall database in March 1999. Cybersurfers can now access the clunky, though functional, recall database for 1970–2000 model vehicles, but, unlike the fast and comprehensive website of the U.S. National Highway Traffic and Safety Administration (NHTSA), owner complaints aren't listed, defect investigations aren't disclosed, and service bulletin summaries aren't provided.

The kicker to this story is that in its two-year site makeover, Transport Canada has re-released on its site a December 1996 joint press release with the Rubber Association of Canada to "advise motorists to think about safe driving in winter"—one month away from spring. And it has the cojones to label the info "new." Our Canadian dollars at work in Ottawa. Shameful!

The junk decades: 1970–1990

Japanese and European cars imported into Canada during the early '70s were junk—as thousands of owners of rust-cankered Fiats, Hondas, Toyotas, and Nissans will confirm.

Nissan, called Datsun at the time, proved to be both incompetent and dishonest in manufacturing and marketing its product line during its first decade in Canada. Not only was its attractively styled 240Z sporty coupe a junk car with multiple brake failures and a rust-weakened chassis, but Nissan had the audacity to add insult to injury by scamming Canadians with two-year-old Datsun 510 sedans marketed as the latest model (a fraudulent practice stopped by the Supreme Court in *Pelletier v. Nissan*). Other importers saw the pickings were easy and followed Nissan's example. And matters went downhill from there. For example, when Hyundai and Lada came along in the early '80s, they set a new benchmark for lousy quality control and a "go to hell" approach to customer relations.

This begs the question as to how foreign automakers peddling unreliable and rust-prone junk got a toehold in the North American car market. The answer is quite simple: foreign automakers smartened up within a remarkably short period of time. They quickly built reliability and durability into their cars and trucks and offered fully loaded and reasonably priced entry-level vehicles.

In the meantime, American automakers pumped out junk—small cars like the Chevy Vega, Firenza, and Fiero; Chrysler's Omni and Horizon; and the Ford Pinto and Bobcat. The Pinto was Ford's rolling Molotov cocktail, catching fire and trapping occupants when hit from the rear. Ford knew of the danger but bowed to its accountants who figured it would cost less to stonewall each future death and burn injury than to recall and redesign the fuel tank.

Savvy Canadian car buyers, turned away by poor-quality American vehicles, bought the cheaper, more reliable, and feature-laden Japanese compacts or re-badged imports like the GM Firefly and Metro, Dodge Colt, and Ford Probe.

Ford re-badged Mazda's MX-6 sporty coupe and sold it as the Ford Probe; a 1996 Probe GT is now worth about $16,000, retaining after three years an impressive 70 percent of its original $22,995 Manufacturer's Suggested Retail Price (MSRP).

Consequently, throughout the '80s, the quality of imports improved dramatically (Hyundai, Audi, and VW excepted), with Toyota and Honda leading the pack. American quality, on the other hand, improved at a snail's pace over the same period. Today, American cars' quality control is about where Japanese cars' was in the mid-'80s. Where the gap hasn't narrowed (and a case can be made that it's actually gotten wider) is in engine, automatic transmission, airbag, and anti-lock brake reliability and fit and finish.

For example, take a look at the following three confidential internal service bulletins depicting serious automatic transmission problems with Chrysler, Ford, and GM cars and trucks. Note how our domestic automakers have allowed the same defect to be carried over year after year.

Chrysler's "Limping" Transmissions

NO: 18-24-95
GROUP: Veh. Performance
DATE: Jun. 23, 1995
SUBJECT:
Improved Transmission Shift Quality

MODELS:

1989-1995	(AA) Acclaim/Sprit/LeBaron Sedan
1989-1993	(AC) Dynasty/New Yorker/New Yorker Salon
1990-1993	(AG) Daytona
1990-1994	(AJ) LeBaron Coupe/LeBaron Convertible
1993-1994	(AP) Sundance/Shadow/Shadow Convertible
1990-1991	(AQ) Chrysler TC
1989-1995	(AS) Caravan/Voyager/Town & Country
1990-1992	(AY) Imperial/New Yorker Fifth Avenue
1993-1995	(ES) Chrysler Voyager (European Market)
**1995	(FJ) Sebring/Avenger/Talon**
1995	(JA) Cirrus/Stratus
1993-1995	(LH) Concorde/Intrepid/Vision/LHS/New Yorker

NOTE: THIS BULLETIN APPLIES TO VEHICLES EQUIPPED WITH THE 41TE OR 42LE TRANSAXLE.

SYMPTOM/CONDITION:

1992 AC. & AY VEHICLES BUILT AFTER FEB. 15, 1992 (MDH 02-15-XX). **1995 FJ VEHICLES** AND ALL OTHER 1993-1995 SUBJECT VEHICLES BUILT BEFORE OCT. 24, 1994, (MDH 10-24-XX) ARE VEHICLES EQUIPPED WITH AN ELECTRONICALLY MODULATED CONVERTOR CLUTCH (EMCC).

Vehicles that operate at speeds where EMCC usage is engaged (vehicle speeds 34 – 41 MPH) may experience early deterioration of the transmission fluid (15,000 – 30,000 miles), exhibit a pronounced shudder during EMCC operation, harsh upshifts/downshifts, and/or harsh torque converter engagements. Performing REPAIR PROCEDURE # 2, which includes updates to the Transmission Control Module (TCM) calibration and eliminates EMCC, will resolve these symptoms/conditions. However, if an overheat condition is identified by the PCM or TCM, EMCC operation will be temporarily enabled.

ALL 1995 FJ VEHICLES AND ALL OTHER 1989-1995 SUBJECT VEHICLES BUILT BEFORE OCT. 24, 1994 (MDH 10-24-XX).

The TCM calibration used in the 1995 model year 41TE and 42LE TCM is being made available for all vehicles dating back to the 1989 model year. The shift quality improvements and default issues that will be corrected by the new TCM calibration are:

1. COASTDOWN TIP-IN BUMP: Vehicle is decelerated almost to a stop (less than 8 MPH), then the driver tips back into the throttle to accelerate, a noticeable bump may be felt.
2. COASTDOWN SHIFT HARSHNESS: Harsh coastdown shifts on some 4-3, 3-2, 2-1 downshifts.
3. 1995 LH WITH 42LE TRANSAXLE – SLUGGISHNESS/LACK OF RESPONSE: On some early 1995 LH vehicles built prior to Oct. 24, 1994, a perceived lack of power or transmission responsiveness may be encountered under normal operating conditions. The transmission may not release the converter clutch as desired with increased throttle. This occurs in 4th gear 35–50 MPH.
4. 1989-1994 WITH 41TE & 42LE TRANSAXLES: Harsh shifts and/or vehicle shudder during 3-2 or 2-1 kickdowns at speeds less than 25 MPH.
5. 1993 WITH 41TE TRANSAXLE: Harsh 3-4 upshifts may occur, especially at highway speeds, while using the speed control.
6. 1989-1994 WITH 41TE TRANSAXLE – HARSH/DELAYED GARAGE SHIFTS: Delay is less than 2 seconds and the shift is harsh after the brief delay. NOTE: Delays greater than 2 seconds are caused by transmission hardware malfunction, i.e., valve body, pump, failed lip seals, or malfunctioning PRNDL or neutral start switch.
7. 1989-1994 WITH 41TE & 42LE TRANSAXLES – POOR SHIFT QUALITY AFTER A BATTERY DISCONNECT: All transmission learned values are reset to the factory default values if battery power is lost to the TCM. The new 1995 calibration will now retain all learned values in memory after battery disconnect. However, if a transmission is rebuilt or a new transmission or TCM is installed, the Quick Learn procedure must be performed to calibrate Clutch Volume Indexes (CVI) on 1993 and later vehicles (1992 and prior vehicle cannot be Quick Learned).

NOTE: BEFORE PERFORMING THE QUICK LEARN PROCEDURE, THE TRANSMISSION MUST BE SHIFTED INTO OVERDRIVE (OD) WITH THE ENGINE RUNNING AND THE TRANSMISSION FLUID SET TO THE CORRECT LEVEL. THIS PROCEDURE WILL PURGE THE AIR IN THE CLUTCH CIRCUITS TO PREVENT ERRONEOUS CLUTCH VOLUME VALUES WHICH COULD CAUSE POOR INITIAL SHIFT QUALITY.

8. EARLY 1993 WITH 41TE & 42LE TRANSAXLE – INTERMITTENT SPEED CONTROL DROP OUT: The new service calibration change corrects this condition (this condition was also covered in Technical Service Bulletin 08-09-93 dated Mar. 12, 1993).
9. 1989-1993 WITH 41TE & 42LE TRANSAXLES: New fault code 35 (failure to achieve pump prime) has been added for improved diagnostic capability, and fault codes 21, 22 and 24 are desensitized to reduce erroneous limp-in conditions.

Chrysler's problems are caused by both hardware and computer software glitches that affect almost its entire product line. Owners of 1991–96 vehicles should claim a "goodwill" repair refund from Chrysler. Hundreds of Chrysler claimants have received compensation up to 75 percent of the repair amount on vehicles not exceeding the above mileage when the transmission failed (see Part Two).

Ford's "Biodegradable" Aluminum Transmissions

SUBJECT: Forward Piston Change
APPLICATION: Ford
DATE: 1995
Forward Piston Change

- No forward or reverse engagement.
- Delayed forward and/or reverse engagement.
- Shifts out of gear when coming to a stop.

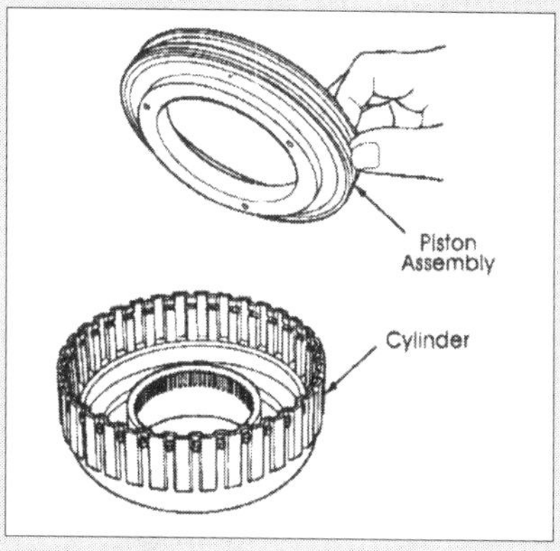

All of these complaints can be attributed to a cracked or broken forward piston clutch. There have been three different versions of the aluminum piston in this location (the original plus two updates). Problems with cracking still persist.
A steel version of this part has been released that should prevent this from happening.
The Ford part number is F4DZ-7A262-A. The aluminum piston should always be replaced with the steel piston.

Note
The part number for the SHO forward piston is F4DZ-7A262-B.

Failures with Ford's aluminum forward piston clutch affect most of the company's products made since 1986, though complaints on the most recent models target mainly the Taurus, Sable, and Windstar. In February 1999, Ford executives promised me that the company would give "goodwill" refunds on a case-by-case basis to new- or used-vehicle owners who had been previously denied compensation for this defect. I am monitoring that commitment.

Finding the best buys

This is an excellent year to buy a used car or minivan. Most vehicles are now equipped with many of the essential safety features that were ignored by automakers two decades ago; overall reliability and durability have improved and are backed by longer warranties that are still in force; and millions of reasonably priced leased vehicles are now entering the used-car market. For example, right now, Ford's popular Crown Victoria, Mercury Grand Marquis, and Windstar minivans are coming off two- and three-year leases and are in abundant supply from both dealers and private parties.

You can reduce your risk of buying a lemon by getting a recommended vehicle that has some of the original warranty in effect. Not only will this protect you from some of the costly defects that are bound to crop up shortly after your purchase, but the warranty allows you to make one final inspection before it expires and requires both the dealer and the automaker to compensate you for all warrantable defects found at that time. The ultimate irony is that because Chrysler dropped its seven-year warranty on its 1995 models, your used 1993 Chrysler will be under warranty until the year 2000—surpassing the 1996 models' three-year guarantee.

GM's "Neutral" Transmissions

File In Section: 7 - Transmission
Bulletin No.: 67-71-64
Date: February, 1997
Subject:
Intermittent Neutral or Loss of Drive at Highway Speeds or from Fourth Gear
(Replace Control Valve Body Assembly)
Models:
1995–96 Buick Skylark Regal, Century, Park Avenue, Riviera, LeSabre
1997 Buick Skylark, Regal, Century, LeSabre
1995 Cadillac DeVille
1995–96 Chevrolet Beretta, Corsica, Lumina APV, Lumina, Monte Carlo 1997 Chevrolet Lumina, Monte Carlo, Venture
1995–96 Oldsmobile Cutlass Ciera, Cutlass Cruiser, LSS, Ninety Eight, Ninety Eight Regency, Eighty Eight, Achieva, Silhouette, Cutlass Supreme
1997 Oldsmobile Eighty Eight, Achieva, Cutlass Supreme, LSS, Silhouette
1995–97 Pontiac Bonneville, Grand Am, Trans Sport, Grand Prix with HYDRA-MATIC 4T60-E Transaxle (RPO M13)
Condition
Some owners may comment about an intermittent, neutral condition while driving at highway speeds or intermittent neutral from fourth gear.
Cause
The 3–2 Manual Downshift valve may be sticking intermittently. When the 3–2 downshift valve sticks, this allows the 2–3 shift valve to float in its bore either exhausting 2ND, 3RD and 4TH clutch oil or sending D-4 oil into the AUX input clutch feed. Exhausting the 3RD and 4TH clutch oil could cause an intermittent loss of drive. Sending D-4 oil into the input circuit could cause a potential tie up condition by engaging the input clutch and 4TH clutch.
Correction
Inspect the transmission fluid and pan for signs of distressed clutches or any abnormal amounts of debris. Remove and disassemble the 1–2 and 2–3 accumulator housing assembly. Inspect the accumulator housing and cover for any signs of scoring, sediment, or abnormalities that could generate debris. If any of the above conditions are found, make the necessary repairs to those items. Then inspect the 2–3 (357) and 3–2 (356) valve line up. If the valves are scored, replace the control valve body assembly.

How's this for a scary scenario: you're cruising down the highway or in city traffic and your 1995–97 GM front-wheel drive car or minivan's transmission drops into "neutral." Another traffic fatality blamed on driver's error?

Want to legally avoid paying tax? Buy privately. You'll save at least $70 per $1,000 paid because, in most provinces, the 7 percent federal Goods and Services Tax (GST) applies only to dealer sales. You may have to hurry to get this benefit, though. The federal government is planning to charge GST on private used-car sales once it harmonizes the GST with provincial sales taxes throughout Canada.

Five Reasons Canadians Buy Used

How popular are used cars in Canada? According to industry analysts, almost 3 million used vehicles change hands each year. Of that number, 70 percent are estimated to be private sales. Why are people buying used? To save money, says the Royal Bank of Canada. While the cost of cars in Canada has risen by an average of 4 percent since 1990, Canadians' take-home pay has grown by only 3.8 percent, meaning fewer and fewer people can afford a new car or minivan.

Following are the main reasons Canadians say they prefer buying used vehicles:

1. There is less initial cash outlay, slower vehicle depreciation, and better and cheaper parts availability.
New vehicle prices are going through the roof due to our declining dollar and the automakers' push for larger profits. The Automotive Industries Association of Canada (AIA) pegs the average cost for a new 1998 model vehicle at $26,139. Insurance for young drivers runs to about $4,000 a year—probably more than the new-car monthly payment. And once you add GST, financing costs, maintenance, taxes, and a host of other expenses, the Canadian Automobile Association (CAA) calculates the yearly outlay at over $6,000, or 30.1¢/km for a compact car. Subcompact cars, like the Geo Metro, would cost about $5,476, or 24.3¢/km, a year; a standard-size car would cost $8,252, or 36.7¢/km, per year. Operating expenses for minivans are a bit lower due to their slower rate of depreciation, which is judged to be the biggest expense for most vehicles.

Be practical. In buying a used car or minivan, keep in mind you're simply buying transportation and function—principally, no-surprise handling, a comfortable ride and interior, cargo and passenger capacity, and reliable performance. Because you only need about one-half the cash or credit required for a new vehicle, it's easy to see that you won't have to invest as much money in a depreciating investment and may be fortunate enough to forgo a loan.

Depreciation savings
If someone were to ask you to invest in stocks or bonds guaranteed to be worth about half their initial purchase value after four or five years, you'd probably head for the door. But this is exactly the trap you're falling into when you buy a new vehicle that could depreciate 20 percent in the first year and 15 percent each year thereafter (minivans and other specialty vehicles like trucks and sport-utilities depreciate about 10 percent each subsequent year). Here's how a new car adds up:

New-Car Cost

Purchase price	$20,000
Provincial tax (8%)	$1,600
Federal GST (7%)	$1,400
Total price	$23,000

Now, the motorist buying a new vehicle is certain that the warranty and status far outweigh any inconvenience. He or she happily forks over $23,000 and takes possession of a new car or minivan.

When you buy used, the situation is altogether different. That same vehicle can be purchased three years later, in good condition and with much of the manufacturer's warranty remaining, for about one-half of its original cost (in the red-hot truck and sport-utility market, depreciation may be much less). Look at what happens to the price:

Used-Car Cost

Manufacturer's Suggested Retail Price (MSRP)	($20,000)
Purchase price (three years old, 60,000 km)	$15,000
Provincial tax (8%)	$1,200
No GST (if sold privately)	—
Total price	$16,200

In this conservative example (some savings exceed 50 percent), the used-vehicle buyer saves $1,400 in federal taxes, $400 in provincial taxes, and gets a reliable, guaranteed set of wheels for about two-thirds of its original price. Furthermore, the depreciation "hit" will be much less in the ensuing years.

Cheap parts

A 1999 Ford Taurus SE equipped with typical options would cost about $25,000. Rebuilding it with factory parts would cost over $72,000. No wonder insurance companies prefer to scrap rather than repair.

Generally, a new, gasoline-powered car or truck can be expected to run, relatively trouble-free, at least 200,000 km (125,000 miles) to 240,000 km (150,000 miles) in its lifetime, and a diesel-powered vehicle can easily triple those figures. Some repairs will crop up at regular intervals, and your yearly running costs, along with preventative maintenance, should average about $700–$800. Buttressing the argument that vehicles get cheaper to operate the longer you keep them, the U.S. Department of Transportation points out that the average vehicle requires one or more major repairs after every five years of use. However, once these repairs are done, it can then be run relatively trouble-free for another five years or more.

Time is on your side in other ways, too. Three years after a model's launching, the replacement-parts market catches up to consumer demand, due to dealers stocking larger inventories and parts wholesalers and independent parts manufacturers expanding their output. Used replacement parts are unquestionably easier to come by, through bargaining with local garages or through a careful search of auto wreckers' yards. A reconditioned or used part usually costs one-third to one-half the cost of a new part. There's generally no difference in the quality of reconditioned mechanical components, and they're often guaranteed for as long as new ones. Also, cross-border shopping in the U.S., buying from discount outlets or independent garages, or ordering through mail order houses can save you big bucks (30–35 percent) on the cost of new parts and another 15 percent on labour.

Body parts are a different story, however. Buyers would be wise to buy only original equipment manufacturer (OEM) parts supplied by automakers in order to get body panels that fit well, protect better in collisions, and have maximum rust-resistance, says *Consumer Reports* in its February 1999 study. Although insurance appraisers often substitute cheaper, lower-quality aftermarket body parts in collision repairs, *Consumer Reports* found that 71 percent of those policyholders who requested OEM parts got them with little or no hassle. It suggests consumers should complain to their provincial Superintendent of Insurance if OEM parts aren't provided. If that doesn't produce the desired results, take out a small claims action or file a class action similar to the one recently filed in the state of Illinois against State Farm (see page 135).

There are generally no problems with parts distribution for American vehicles, unless a particular unit has been off the market for some time or a specific component has experienced a high failure rate (such as electronic control units that regulate engine, transmission, and braking performance). In these cases, though, there are many independent suppliers who can find the missing parts and price them far below dealer cost. Components can also be more easily interchanged on American and Japanese vehicles than on European models.

With some European models, you can count on a lot of aggravation and expense caused by the unacceptably slow distribution of parts and the high markup. Because these companies have a quasi-monopoly on replacement parts, there are few independent suppliers you can turn to for help. And auto wreckers, the last-chance repository for inexpensive car parts, are unlikely to carry foreign parts for vehicles more than three years old or manufactured in small numbers.

Finding parts for Japanese and domestic cars is hardly a problem, though, due to the large number of vehicles produced, the presence of hundreds of independent suppliers, the ease with which relatively simple parts can be interchanged from one model to another, and the large reservoir of used parts stocked by junkyards. Incidentally, when a part is hard to find, the *Mitchell Manual* is a useful guide to substituting parts that can be used for many different models. It's available in some libraries, most auto parts stores, and practically all junkyards.

Auto clubs listed on the Internet are often helpful sources for parts that are otherwise unobtainable. Club members trade and sell specialty parts, keep a list of where rare parts can be found, and are usually well informed as to where independent parts suppliers are located. Most car enthusiast magazines will put you in touch with auto clubs and suppliers of hard-to-find parts.

2. There are lower insurance rates.
The difference in annual insurance costs between a new and used car or minivan may be only a few hundred dollars, but by carefully negotiating the deductible, the smart shopper can further reduce insurance premiums by another couple hundred dollars. For example, as an automobile gets older, the amount of the deductible should increase. It may reach a maximum of $500 per collision. As the deductible increases, the annual premium for collision coverage decreases.

By agreeing to a higher deductible, the motorist agrees to repair the vehicle for all damages where costs do not exceed the deductible. Generally, purchasing used parts (remember, you bought a used vehicle) from a local auto wrecker and having the work done by small, specialized garages keeps repair costs down. What may have been an estimated loss of $500 may be significantly reduced by sharp repair bargaining.

3. There are fewer "hidden" defects.
Have your choice checked out by an independent mechanic (for $75–$100) before paying for a used vehicle. This examination before purchase protects you against any hidden defects the vehicle may have. It's also a tremendous negotiating tool, since it allows you to use the cost of any needed repairs to bargain down the purchase price.

It's easier to get permission to have the vehicle inspected if you promise to give the seller a copy of the inspection report should you decide not to buy it. If you still can't get permission to have the vehicle inspected elsewhere, walk away from the deal, no matter how tempting the selling price. The seller is obviously trying to put something over on you. Ignore the standard excuses that the vehicle isn't insured, the licence plates have expired, or the vehicle has a dead battery.

4. You know the vehicle's history.
Smart customers will want to get answers to the following questions before paying a penny for any used vehicle: What did it first sell for and what is its present value? How much of the original warranty is left? How many times has the vehicle been recalled for safety-related defects? Are parts easily available? Does the vehicle have a history of costly performance-related defects that can be corrected under a secret warranty, through a safety recall campaign, or with an upgraded part? (See Part Three, "Secret Warranties/Service Tips.")

5. Litigation is quick, easy, and relatively inexpensive.
A multitude of federal and provincial consumer protection laws go far beyond whatever protection may be offered by the standard new-vehicle warranty. Furthermore, buyers of used vehicles don't usually have to conform to any arbitrary rules or service guidelines to get this protection.

Let's say you do get stuck with a lemon. Most small claims courts have a jurisdiction limit of $3,000–$10,000, which should cover the cost of most used cars or minivans. Therefore, any dispute between buyer and seller can be settled within a few months, without lawyers or excessive court costs. Furthermore, you're not likely to face a battery of lawyers standing in for the automaker and dealer. Actually, you may not have to face a judge since many cases are settled through court-imposed mediators.

Safety Features

Before buying any used "bargain," determine what degree of active and passive safety the vehicle provides. Passive safety assumes that you will be involved in life-threatening situations and should be warned in time to avoid injury. Daytime running lights and third, centre-mounted brake lights are two passive safety features that do this job extremely well. In fact, NHTSA, an arm of the U.S. Department of Transportation, estimates that centre brake lights—a required safety feature in Canada and the U.S. since 1986—reduce the number of rear-impact crashes by 4.3 percent, thereby preventing 92,000 crashes, 58,000 injuries, and $655 million (U.S.) in property damage annually. The annual cost to the consumer of the brake lamps is about $13.60 (U.S.) a car and about $20 (U.S.) for vans, pickups, and sport-utilities.

Passive safety features also assume some accidents aren't avoidable and that when that accident occurs, the vehicle should provide as much protection as possible to the driver, the vehicle's occupants, and other vehicles that may be struck—without depending on the driver's reactions. Passive safety components that have consistently proven to reduce vehicular deaths and injuries are safety belts and vehicle structures that absorb or deflect crash forces away from the vehicle's occupants.

Advocates of active safety stress that accidents are caused by the proverbial "nut behind the wheel" and believe that safe driving can best bc taught through schools or by private driving courses. Active safety components are generally those mechanical systems that may help avoid accidents, such as high-performance tires and traction control, if the driver is skillful and mature.

The theory of active safety has several drawbacks. First, there's no independent proof that safe driving can be taught successfully. Even if a young driver learns how to master defensive-driving techniques, there's still no assurance that this training will be of any use in an emergency. Second, 40–50 percent of all fatal accidents are caused by drivers who are under the influence of alcohol or drugs. Surely all the high-performance options and specialized driving courses in the world will not provide much protection from impaired drivers who take a bead on your vehicle. Finally, because active safety components get a lot of use—you're likely to need anti-lock brakes 99 times more often than an airbag—they have to be well maintained to remain effective.

Safety features that kill

In the late '60s, Washington forced automakers to include essential safety features like collapsing steering columns and safety windshields in their cars. As the years have passed, the number of mandatory safety features has increased to include seatbelts, airbags, and crashworthy construction. These improvements met with public approval—until three years ago, when reports of deaths and injuries induced by anti-lock brakes (ABS) and airbags showed that defective components and poor engineering negated the potential life-saving benefits associated with these devices.

For example, one out of every five ongoing NHTSA investigations into possible defects in cars and light trucks concerns inadvertent airbag deployment, failure of the airbag to deploy, or injuries suffered when the bag did go off. In fact, airbags are the agency's single largest cause of current investigations, exceeding even the full range of brake problems, which runs second.

Anti-lock brake system (ABS)

ABS pumps the brakes automatically, many times every second, to prevent wheel lockup and enable the driver to maintain steering control. This can mean substantially shorter stopping distances on wet and slippery roads, but not on dry roads. If you anticipate driving a lot on slick roads, anti-locks might be worthwhile. Remember, they're more about steering control than about stopping on a dime. Finally, if you do decide to buy a vehicle with ABS, make sure you can adapt to the excessive pulsation and noise that often occurs when the brakes are applied.

Despite impressive test track performance, the on-the-road safety benefits of car anti-locks are disappointing—they haven't cut down the frequency of crashes or the cost of insurance claims for vehicle damage. Studies by the government, the Insurance Institute for Highway Safety (IIHS), and automakers find that vehicles with anti-locks are in more single-vehicle crashes than those with conventional brakes. It's not clear why this is so, but it is apparent that many drivers don't know how to use anti-locks.

And then there's the hardware and software shortcomings. While anti-locks are sold as an $800–$900 option on most new vehicles, ABS-equipped used models are risky buys because their brakes were rushed to production in the mid-'80s without sufficient testing. This has resulted in owners reporting thousands of incidents of brake failure attributable to poor design and defective components (see "Safety summary/recalls" for the Chrysler Caravan, for example). In fact, *AutoWeek* magazine reports that General Motors has secretly bought back hundreds of vehicles equipped with failure-prone Kelsey-Hayes anti-lock brakes.

General buyback. While publicly blaming drivers for problems with the antilock braking systems on millions of pickup trucks and sport/utility vehicles, General Motors was quietly buying some trucks back from consumers whose dealers could not repair the systems, according to documents filed with the National Highway Traffic Safety Administration.

NHTSA has been investigating ABS defects on 1991-94 Chevy Blazer, GMC Jimmy and Typhoon, and Olds Bravada sport/utes, as well as on Chevy S-10 and GMC Sonoma and Syclone pickups for three years. The agency has received more than 15,000 consumer complaints. In late 1995, NHTSA added 1992-94 Chevy/GMC Suburbans to the probe. In dozens of cases, when owners threatened to contact attorneys or file lemon law claims, GM either bought vehicles back or offered vehicle trade-in assistance to get customers into new vehicles.

Since NHTSA began investigating GM ABS failures in 1994, over 15,000 complaints involving 2,401 crashes have been lodged against GM's 1991–94 pickups, which include the S-10, Sonoma, Blazer, Bravada, and Jimmy. Recently, the 1992–95 Suburban was added to the list.

Furthermore, the replacement of aging ABS components, usually around the fifth year of use, can easily cost $1,000 or more, because mechanics can't troubleshoot them very well and many replacement parts are no longer manufactured. Not surprisingly, mechanics will often junk the old design and put in a new assembly to make sure you don't come back with other failures. If you were trained to brake gently on slippery roads or pump your brakes to avoid a skid, you now have to "unlearn" these old habits and use hard, continuous brake pressure to activate the anti-lock feature.

Airbags

If you're a female, a senior, a person of slight stature, or someone unable to sit at least 25 cm (about 10 inches) away from the airbag housing, don't buy an airbag-equipped used vehicle—unless it has a cut-off switch or can be retrofitted for one, or you know where to go to deactivate the system. Another option is to buy only a 1998–99 vehicle that is equipped with depowered, "Second Generation" airbags, and keep well back from the dashboard. Granted, airbag-equipped used cars and minivans are lifesavers in high-speed collisions, but they also maim and take lives in low-speed, "fender-bender" collisions of 15–20 km/h, according to the following Transport Canada research study, "Airbag Deployment Crashes in Canada" (Paper number 96SI-0-05, presented by Dainius J. Dalmotas of Transport Canada and Kennerly Digges of George Washington University to the Fifteenth International Technical Conference on the Enhanced Safety of Vehicles, at Melbourne, Australia, on May 13–16, 1996).

Air Bag Deployment Crashes in Canada

ABSTRACT

In the fall of 1993 Transport Canada initiated a major field accident study to examine the injury experience of occupants protected by supplementary air bag systems. While the initial findings of this study confirm that belted drivers are afforded added protection against head and facial fracture injury in moderated to severe frontal collisions, **the findings also suggest that these benefits are being negated by a high incidence of bag-induced injury.** Most bag-related injuries consist of AIS 1 facial injuries and AIS 1 to 3 upper extremity injuries. However, they can include AIS ≥ 3 injuries to other body regions if the occupant is close to the deploying air bag. **The incidence of bag-induced injury was greatest among female drivers.**

To further quantify the benefits and drawbacks afforded by air bag systems particularly as a function of collision severity, additional analyses were carried out using U.S. field accident data. Both Canadian and U.S. data examined suggest the protection afforded belted drivers by air bag systems would be greatly enhanced if the deployment thresholds were increased. **Far greater attention to the protection requirements of female drivers needs to be given in federal regulations addressing restraint system performance.**

I can't think of a more alarming phrase than: "...benefits are being negated by a high incidence of bag-induced injury....greatest among female drivers...." Furthermore, General Motors claims that competitors' side airbags are hazardous to children and has asked the U.S. government to investigate. It is believed that GM was targeting BMW's side airbags. Presently, the federal government lacks jurisdiction to regulate side airbags. www.nhtsa.dot.gov/ *is a particularly informative U.S. Department of Transportation website that answers frequently asked questions concerning airbags and keeps a current listing of repair facilities willing to install airbag disabling devices. Forget about Transport Canada's website—it's an embarrassment.*

There's no current federal legislation that prevents one from disabling the airbag; automakers, however, won't sell a shut-off switch without Transport Canada approval. You can get that approval by filling out a "self-certifying" form supplied by the government, auto dealers, or CAA.

Adding insult to potential injury, cut-off switches are estimated to cost $300–$500 and aren't available for all models. (An Alabama class action winding its way through the courts says automakers should pay for the switch installation.) For example, Denyse O'Leary, a Toronto-based freelance auto writer, has discovered that Ford doesn't make on-off switches for her 1992 Taurus. Compounding the problem, some importers, such as BMW, Mercedes-Benz, and Volvo, won't provide the switches, and other importers don't expect to have them for sale before late-1999. In the meantime, don't be surprised to find Canadian dealers reluctant to install the devices until they get clear instructions from automakers. In fact, the American Automobile Association (AAA) estimates that barely 16 percent of the repair facilities it has surveyed will install off-switches.

If you are fed up with the "airbag-disconnect, shut-off switch runaround" and want your airbag disconnected or a shut-off switch installed, here are the steps you can take:

- Have your airbag disconnected by one of the NHTSA-recommended facilities in the U.S. the next time you go on vacation.
- Access the *www.autostream.com/airbagdisconnect.html* website and purchase for less than $100 instructions on how the airbag can be disabled. For about $37 the Alldata website (*www.alldata.com/consumer/tsb/yr.html*) will send you a CD-ROM disc containing automakers' service bulletins with instructions for how a mechanic can disconnect the airbag.

Crashworthiness

The most important safety features are those that reduce the risk of death or serious injury when a crash occurs. This aspect of vehicle design is referred to as crashworthiness. Purchasing a used vehicle with the idea that you'll be involved in an accident is not unreasonable—according to IIHS, the average car will likely have two accidents before ending up as scrap.

The vehicle's structural design is the starting point for protecting you during a serious crash. A good structural design should have a strong occupant compartment, or safety cage, and "crush zones": front and rear ends designed to buckle and bend to absorb crash forces during serious crashes. It's important for these crush zones to keep damage away from the safety cage because the likelihood of injury increases rapidly once this cage begins to collapse. If effectively designed, a longer crush zone lowers both the likelihood of damage to the occupant compartment and the crash forces inside it.

Not all vehicles are equally well designed. Some have crush zones that are too stiff and/or too short and safety cages that aren't strong enough. Such poorly designed crush zones can contribute to the collapse of the occupant compartment in serious crashes. Crash tests demonstrate differences in structural design among vehicles in the same weight class.

Occupants of large vehicles have fewer severe injury claims than do occupants of subcompact and compact vehicles. This was proven conclusively in a 1996 NHTSA study that showed collisions between small cars and light trucks or vans resulted in an 81 percent fatality rate for the occupants of the small cars.

In principal, vehicle weight protects you in two-vehicle crashes. In a head-on crash, for example, the heavier vehicle drives the lighter one backwards, which decreases forces inside the heavy vehicle and increases forces in the lighter one. All heavy vehicles, even poorly designed ones, offer this advantage in two-vehicle collisions but may not offer good protection in single-vehicle crashes.

Interestingly, a vehicle's size or past crashworthiness rating doesn't always guarantee that you won't be injured in an accident. U.S. government crash tests show, for example, that a 1991 Ford Escort small compact gives better full-front collision protection than a 1994 Mercury Villager or Nissan Quest, two large minivans that could easily have been engineered to crash safely at moderate speeds of 57 km/h. Also, different years of the same model can produce startlingly different scores. For example, Chrysler's 1995 Caravan and Voyager minivans earned better crash protection scores than did the 1996–97 Grand Caravan and Grand Voyager.

Two Washington-based agencies monitor how vehicle design affects crash safety. They are NHTSA and IIHS, an insurance research group that collects and analyzes insurance claims data. Crash information from these two groups don't always agree, because while the insurance research group's results incorporate all kinds of accidents, NHTSA figures relate only to 57 km/h (35 mph) frontal and some side collisions. Bear in mind that a car providing good injury protection often produces the highest damage claims because its structure, not the occupants, absorbs most of the force of the collision.

Unsafe designs

Although it sounds hard to believe, automakers will deliberately manufacture a vehicle that will kill or maim simply because in the long run it costs less to pay off victims than to make a safer vehicle. I learned this lesson after reading the court transcripts of *Grimshaw v. Ford* and listening to court testimony of GM engineers who deliberately placed fire-prone "side-saddle" gas tanks in millions of pickups to save $3 per vehicle. In the Grimshaw case, tried in the late '70s, a teenager riding in a Ford Pinto was rear-ended by another car and suffered burns to over 90 percent of his body. Subpoenaed Ford internal documents showed that the automaker had known about the defect for years. In fact, Ford accountants made a cost comparison of the price of recalling and redesigning the defective fuel tanks—$128 million (U.S.)—with what they'd likely have to spend in lawsuits to compensate those injured or killed if they continued to stonewall victims and accident investigators—$69 million (U.S.). Ford chose to stonewall and litigate.

More recent examples of corporate greed triumphing over public safety: airbag designs that maim or kill women, children, and seniors; anti-lock brake systems that don't brake (a major problem with GM minivans, trucks, and sport-utilities; and Chrysler sport-utilities and minivans); flimsy front seats; the absence of rear head restraints; and fire-prone GM pickup fuel tanks and Ford ignition switches. All are examples of hazardous engineering designs that put profit ahead of safety. Incidentally, State Farm Insurance has filed a lawsuit against Ford to recoup the millions of dollars in claims it paid out for fire damage caused by Ford's fire-prone ignition switches.

Recalls

Vehicles are recalled for one of two reasons: they may be unsafe, or they may not conform to federal pollution control regulations. Whatever the reason, though, recalls are a great way to get free repairs—if you know which ones apply to you.

On average, about 800,000 motor vehicles are recalled each year because of safety problems. More than 200 million unsafe vehicles have been recalled by automakers for the free correction of safety-related defects since American recall legislation was passed in 1966. During that time, about one-third of the recalled vehicles never made it back to the dealership for repairs. Auto Service Monitor, an American firm that tracks recalls, estimates that this means there are close to 20 million vehicles still on North American roads that could suddenly careen out of control, catch fire, or fail to brake. Surprisingly, motorists aren't generally motivated to bring in their recalled vehicles—even when the defect is as life-threatening as a fire. For example, NHTSA figures show that Ford's recall of 1.4 million Pintos and Bobcats in the late '70s to correct the much-publicized exploding gas tank defect received a response rate of just 52 percent. NHTSA says the average completion rate is 68 percent. This lack of public concern can be blamed partly on apathy, but car-makers must also bear much of the blame because they downplay the defect's safety implications (see Part Three, the "Safety Summary" for the Chrysler Caravan) and send out only one recall letter to the first recorded owner of the vehicle. If the owner has moved or leased the vehicle, he or she likely won't be notified.

Auto theft and automaker thievery

Automakers are making bundles of money on stolen vehicles. They manufacture products like the Caravan (a kid using a pair of scissors can easily break into the vehicle and then override the ignition), then make money repairing them when they're recovered or selling the victims new ones if they're not found. Insurance companies get to raise their rates and it's the average consumer that gets screwed.

The Caravan heads the list of the most stolen vehicles not because of its popularity but simply because it's so easy to steal. I have received dozens of letters from *Lemon-Aid* readers imploring me to warn 1984–96 Chrysler minivan owners to invest in an ignition kill switch or risk the very real probability that their minivans will be heisted. Chrysler is aware of the problem but prefers to blame the victims rather than make their vehicles more theft-proof.

General Motors is worse. Throughout the '80s and early '90s, its trucks and vans were ranked among those easiest to steal because GM's Saginaw steering column had a cheap, plastic protective covering for the ignition switch instead of a metal plate (used by Ford, for example). This cost-cutting measure was an open invitation to amateur thieves using nothing more than household tools to "rip and ride." The company began embedding a computer chip in the ignition key in the late '80s and thefts declined dramatically.

Less than a decade ago, Honda and Toyota had similar reputations for building easy-to-steal vehicles. Their door lock tumblers and switches used bearings made of low-grade metal that wore down over time. Once worn, any thief using a blank key bought from the local hardware store could easily bypass the lock and ignition switch and rip off any 1982–89 Honda or Toyota truck in 30 seconds or less. Unlike GM, the two Asian automakers quickly changed their design and thefts became less of a problem.

Break-ins and thefts cost Canadians more than $400 million annually; there is a 1 in 130 chance your vehicle will be stolen and only a 60 percent chance you'll ever get it back. No wonder over one-third of new vehicles are equipped with standard or optional security systems. These include fuel cut-off devices, electronic alarms, active and passive units that are armed by the driver or are self-arming, ignition disablers, parts identification, and security keys.

The most effective theft-deterrent systems are not always the most expensive ones. Don't waste your money on expensive anti-theft devices that depend mainly on your vehicle's horn or lights to scare thieves away. They often go off at the wrong time and can be deactivated easily. Furthermore, they frighten the thief away only after the vehicle has been damaged. When it comes to protecting a vehicle's contents, no alarm system can resist a brick through a side window. It takes just 12 minutes for thieves to strip a vehicle's seat, radio, and body parts. As well, few citizens are brave enough to personally stop a theft or testify in court. Your best protection is discretion: take your radio with you or lock it in the trunk, particularly if you own a sport-utility with an upgraded sound system.

Since most vehicles are stolen by amateurs, the best theft deterrent is a visible device that complicates the job while immobilizing the vehicle and sounding an alarm. For less than $150 you can install both a steering-wheel lock and a hidden ignition disabler. The clublike devices cost about $75 and deter thieves by forcing them to carry a hacksaw and adding about a half-minute to the time it takes to steal the average vehicle (they have to cut through the steering wheel or bust the lock). The Winner's Club is an inexpensive steering-wheel lock distributed by Superex (416-497-2850). It's a huge bar that locks the steering wheel and doesn't require any rewiring or other modifications. If your vehicle is stolen, Superex will pay the deductible up to $500. The Nightstick, another steering lock, is carried by Canadian Tire. Ignition disablers are also inexpensive and very effective. They sell for $30–$90, depending on their sophistication.

Choosing the Right Seller

When to buy

Used car sales are seasonal, depending on where you live. Fewer used vehicles are sold in the winter months of December through March than in the spring or summer. Dealers see few customers during this winter period, and used vehicles generally show their worst characteristics: they won't start, have defective heating and defrosting systems, reveal poor suspensions, etc. On the other hand, it's during the fall and spring that dealer stocks of good-quality trade-ins and off-lease returns are at their highest level and private sellers are more active. It's precisely for these reasons that smart consumers shop during these months for real bargains.

Should you use a broker?

My answer is a qualified yes. Even with my 30 years' experience as a consumer advocate and automotive expert, I have used brokers and found most to be well worth the money spent for their services when I consider the runaround time and price haggling they saved me. Interestingly, Toronto's Metro Credit Union (*www.metrocu.com*) has come to the same conclusion and uses its own in-house broker, Dave Lawrence, as part of its Car Facts program to assist members (1-800-777-8506). The result: members have saved an estimated $600,000 since Car Facts began a year ago.

Brokers are independent agents who try to find the new or used vehicle you want at a price below what you'd pay at a dealership, including the extra cost of the broker's services. Broker services appeal to buyers who want to save time and money while avoiding most of the stress and hassle associated with the dealership experience, which for many people is like a swim in shark-infested waters.

Private sellers

Private sellers are your best source for a cheap and reliable used vehicle, because you're on an equal bargaining level with a vendor who isn't trying to profit from your inexperience. This translates into a golden opportunity to negotiate a fair price, which isn't common in many dealer transactions.

Apart from newspaper classified ads, you can track down good private deals and get a good idea of prices through the following:

- word of mouth
- grocery store bulletin boards
- specialty publications (e.g., *Auto Trader, Auto Mart*, or *Buy and Sell Bargain Hunter*)
- the *Canadian Red Book* or the Provincial Automobile Dealers Association *Black Book*

CHEVROLET 1997

Fact S.R.P.	Avg. Whsl	Mod. No.	Description		Avg. Retail
CORVETTE					
48895	**34425**	YY2	Coupe H/B	5.7L-E V8	**38625**
LUMINA					
22340	**12875**	WL5	Sedan	3.1L-E V6	**14975**
23810	**13725**	WL5	Sedan LS	3.1L-E V6	**15825**
25305	**14225**			3.4L-E V6	**16325**
25145	**14275**	WN5	Sedan LTZ	3.1L-E V6	**16375**
26640	**14775**			3.4L-E V6	**16875**
MALIBU					
19995	**12550**	ND5	Sedan		**14550**
23095	**14450**	NE5	Sedan LS		**16450**
MONTE CARLO					
24275	**14800**	WW1	Coupe LS	3.1L-E V6	**17000**
26500	**15950**	WX1	Coupe Z34	3.4L-E V6	**18150**

CHEVROLET 1996

NOTE: Model Numbers shown are the 4th, 5th & 6th Digits of the VIN No.

Fact S.R.P.	Avg. Whsl	Mod. No.	Description		Avg. Retail
BERETTA					
17275	**7950**	LV1	2dr Coupe	2.2L-E 4	**9650**
27939	**8350**			3.1L-E V6	**10050**
BERETTA Z26					
21140	**10000**	LW1	2dr Coupe	3.1L-E V6	**11700**
CAMARO					
20195	**12350**	FP2	2dr Coupe	3.8L-E V6	**14150**
28365	**16350**	FP3	Convt	3.8L-E V6	**18450**
CAMARO Z28					
25530	**16050**	FP2	Coupe Z28	5.7L-E V8	**18050**
32330	**20050**	FP3	Convt Z28	5.7L-E V8	**22450**
CAPRICE CLASSIC					
28345	**14750**	BN5	Sed Class	4.3L-E V8	**16950**
29720	**15050**			5.7L-E V8	**17250**
30980	**16050**	BL8	Wgn Class	5.7L-E V8	**18350**
CAVALIER					
13030	**8250**	JC1	2dr Coupe	2.2L-E 4	**9850**
13530	**8550**	JC5	4dr Sedan	2.2L-E 4	**10150**
16840	**10425**	JF5	4dr Sedan LS	2.2L-E 4	**12025**
17643	**11400**	JF1	2dr Cpe Z24	2.4L-E 4	**13100**
22925	**13200**	JF3	Convt LS	2.4L-E 4	**15250**
CORSICA					
17685	**9900**	LD5	4dr Sedan	2.2L-E 4	**11600**
18620	**10400**			3.1L-E V6	**12100**
CORVETTE					
48080	**29325**	YY2	Coupe H/B	5.7L-E V8	**33425**
56335	**32925**	YY3	Convertible	5.7L-E V8	**37025**
IMPALA SS					
30675	**19375**	BL5	Sedan SS	5.7L-E V8	**21775**
LUMINA					
21455	**11075**	WL5	Sedan	3.1L-E V6	**13075**
22705	**11925**	WN5	Sedan LS	3.1L-E V6	**13925**
LUMINA (Continued)					
24905	**12425**			3.4L-E V6	**14425**
MONTE CARLO					
23625	**12525**	WW1	Coupe LS	3.1L-E V6	**14625**
25826	**13225**			3.4L-E V6	**15325**
25850	**13675**	WX1	Coupe Z34	3.4L-E V6	**15775**

CHEVROLET 1995

NOTE: Model Numbers shown are the 4th, 5th & 6th Digits of the VIN No.

Fact S.R.P.	Avg. Whsl	Mod. No.	Description		Avg. Retail
BERETTA					
16950	**6650**	LV1	2dr Coupe	2.2L-E 4	**8250**
17621	**7150**			3.1L-E V6	**8750**
BERETTA Z26					
20815	**8600**	LW1	2dr Coupe	3.1L-E V6	**10200**
CAMARO					
18995	**10250**	FP2	2dr Coupe	3.4L-E V6	**11950**
26045	**13975**	FP3	Convt	3.4L-E V6	**15975**
CAMARO Z28					
23650	**13950**	FP2	Coupe Z28	5.7L-E V8	**15850**
30450	**17475**	FP3	Convt Z28	5.7L-E V8	**19775**
CAPRICE CLASSIC					
25455	**12350**	BN5	Sed Class	4.3L-E V8	**14550**
26145	**12650**			5.7L-E V8	**14850**
29120	**13625**	BL8	Wgn Class	5.7L-E V8	**16025**
CAVALIER					
12245	**6650**	JC1	2dr Coupe	2.2L-E 4	**8150**
12746	**6950**	JC5	4dr Sedan	2.2L-E 4	**8450**
15645	**8825**	JF5	4dr Sedan LS	2.2L-E 4	**10325**
16895	**9350**	JF1	2dr Cpe Z24	2.3L-E 4	**10950**
22185	**11075**	JF3	Convt LS	2.3L-E 4	**13025**
CORSICA					
17360	**7975**	LD5	4dr Sedan	2.2L-E 4	**9575**
18015	**8475**			3.1L-E V6	**10075**
CORVETTE					
47580	**25150**	YY2	Coupe H/B	5.7L-E V8	**29150**
55020	**28550**	YY3	Convertible	5.7L-E V8	**32550**
86765	**40425**	YZ2	Coupe ZR1	5.7L-E V8	**45425**
IMPALA SS					
29005	**15800**	BL5	Sedan SS	5.7L-E V8	**18100**
LUMINA					
20730	**9375**	WL5	Sedan	3.1L-E V6	**11275**
21950	**10225**	WN5	Sedan LS	3.1L-E V6	**12125**
23695	**10725**			3.4L-E V6	**12625**
MONTE CARLO					
22578	**10400**	WW1	Coupe LS	3.1L-E V6	**12400**
22579	**10900**			3.4L-E V6	**12900**
24678	**11575**	WX1	Coupe Z34	3.4L-E V6	**13575**

For Provincial Valuations - See Inside Back Cover
* Valuation does not include air conditioning
12

Both the Red Book *and the* Black Book *list three prices, from left to right: the MSRP, showing what the vehicle sold for new; the wholesale price negotiated between dealers; and the retail price charged by dealers to walk-in customers. Used vehicles purchased from dealers over the Internet or from private sellers will sell for somewhere between the wholesale and retail prices. When you compare dealer ads with the price guides, you'll see the dealers have added substantial markups that ensnare unsuspecting buyers. Take* Lemon-Aid *along and bargain down the markup as much as you can.*

The best way to determine the price range for a particular model is to read the publications listed above before you buy the vehicle. This will give you a reasonably good idea of the top asking price. Remember, nobody expects to get his or her asking price, be it a dealer or private party. As with price reductions on home listings, a 10–20 percent reduction on the advertised price is common with private sellers while dealers usually won't cut more than 10 percent off their advertised price.

Find out which published used-car guide is the accepted standard in your area of the country. The *Red Book* and *Black Book* serve Canadian dealers, while the *Kelley Blue Book* and *Edmund's Price Guide* are two of the most popular American guides offering comprehensive ratings and prices, gratis through the Internet. You can visit the *Kelley Blue Book* at *www.kbb.com* or *Edmund's* at *www.edmunds.com.*

How can this Vancouver dealer sell a 1993 Nissan Sentra for $7,888 when the Red Book *says its top resale value is only $5,425? There are regional price disparities. Most Canadian price guides have a Montreal-Toronto bias that often results in unrealistically low price quotes for buyers in eastern and western Canada or residents in rural areas where good used cars are often passed down through the family.*

Remember, since April 1996, if you're turning in your car as a trade-in, the dealer must deduct the negotiated value of the trade from the price of the vehicle he's selling you, thereby reducing your GST and PST tax exposure. In private sales, however, the selling price may be arbitrarily fixed by government officials in assessing sales tax. For example, each year Ontario provincial regulations require that used-car buyers prove the price they paid, or have their vehicle independently appraised. There are some common-sense exceptions, though. For example, buyers of vehicles that have an estimated wholesale value of less than $1,000 can pay the retail sales tax on the purchase price they declare without an appraisal. Furthermore, vehicles that are worth more than $1,000 but are damaged or run down can be appraised before payment of sales tax, and the tax will be calculated on either the selling price or the appraised price, whichever is higher. Where there is a dispute, an independent appraiser may be used. Traditional sales tax exemptions remain. Transactions and gifts between family members will have the sales tax rebated. Sales involving status Indians and diplomats, vehicles registered out-of-province, and bequests for estate settlements are not subject to Ontario sales tax.

Private sellers in Ontario need to buy a Used Vehicle Information Package (UVIP) from the Ministry of Consumer and Commercial Relations (cost is about $20). This kit contains lien information, details as to what's covered in a safety inspection, sales tax information, forms for an affidavit, and a bill of sale. These measures have been taken to drive out "curbsiders" (professional sellers posing as private individuals) from the used-car business and to stop buyers from paying less sales tax than they should.

Protect yourself

As a buyer, you should get a printed sales agreement, even if it's just in handwriting, that includes a clause stating there are no outstanding traffic violations or liens against the vehicle. It doesn't make a great deal of difference whether the car will be purchased "as is" or as certified under provincial regulation. A vehicle sold as safety "certified" can still be dangerous to drive or turn into a lemon. The certification process can be sabotaged by incompetent mechanics or poorly calibrated instruments, or if a minimal number of components were checked. "Certified" is not the same as having a warranty to protect you from engine seizure or transmission failure. Certified means only that the vehicle has met minimum safety standards on the day tested.

Make sure the vehicle is lien free and hasn't been written off after an accident. Canada has become a haven for rebuilt U.S. wrecks, and write-offs shipped from provinces where there are stringent disclosure regulations to provinces where there are lax rules, if rules exist at all. Ron Giblin of the Canadian Insurance Crime Prevention Bureau (ICPB) recently told the *Toronto Sun* that ICPB members voluntarily reported 26,709 salvaged vehicles in 1996 for Ontario alone. He estimates the true figure is about double that number.

In most provinces, you can do a lien and registration search yourself. If a lien does exist, you should contact the creditor(s) listed to find out whether any debts have been paid. If a debt is outstanding, you should arrange with the vendor to pay the creditor the outstanding balance. If the debt is larger than the purchase price of the car, it's up to you to decide whether or not you wish to complete the deal. If the seller agrees to clear the title personally, make sure that you receive a written relinquishment of title from the creditor before paying any money to the vendor. Make sure the title doesn't show an "R" for "restored."

Even if all documents are in order, ask the seller to show you the vehicle's original sales contract and a few repair bills in order to ascertain how well the car was maintained. The bills will show you if the odometer was turned back, and will also indicate which repairs are still guaranteed. If none of these can be found—run, don't walk, away. If the contract shows that the car was financed, verify that the loan was paid. If you're still not sure that the vehicle is free of liens, ask your bank or credit union manager to make a check for you. If no clear answer is forthcoming, look for another car.

Repossessed vehicles

Repossessed vehicles are usually found at auctions, but they're sometimes sold by finance companies or banks—institutions that have been held by the courts to be as legally responsible as dealers for the defects found in the vehicles they sell. The biggest problem with repossessed vans, sport-utilities, and pickups, in particular, is that they're likely to have been abused by their financially troubled owners. The full extent of the abuse can't always be ascertained in one brief examination by a mechanic. Before buying a repossessed vehicle, try to find out something about the previous owner and the care the vehicle received. A visit to the local dealer for that particular make, with the vehicle's identification number and a small gratuity in hand, will get you a vehicle history printout from the dealer's PC link with the manufacturer.

Cross-border shopping

There's no ironclad guarantee that you'll save money by purchasing a used vehicle in the U.S. and bringing it into Canada. But, free trade in the other direction is booming. In fact, Dave Lawrence, who I mentioned earlier is an auto consultant with Toronto's Metro Credit Union, tells me that American dealers are drying up the Toronto-area supply of good used vehicles by overbidding for Canadian cars and selling them at a profit in the U.S.:

> Ford credit cars (vehicles coming in off Ford leases and sold by Ford auction open to Ford dealers only) are becoming rarer and rarer. It would seem that U.S. buyers (Ford) are coming up here and buying product out from under Canadian dealers and putting them on a truck and heading south. Due to this wonderful NAFTA, there are no duties to be paid, and with the U.S. buck being worth more, they are scooping all the choice vehicles and driving the market price way up in outbidding each other for these vehicles. Four of my Ford/Mercury dealers are complaining, but that does not help me get good used product for my clients. All that is left are no-name used lots and private sales, and instead of the vehicle being available at "reasonable" rates, now they can write their own price (for wagons) and get it. A 1997 Taurus wagon in average condition with average kilometres should sell for between $12,500 and $14,800. What we are seeing is asking prices around the $19,000 mark. And the ownership or past history is rather dubious...

Safety and pollution control regulations differ considerably between the two countries and it may be impossible to upgrade your used vehicle to Canadian standards. Transport Canada estimates that 15–20 percent of American vehicles can't meet Canadian standards, no matter how they're modified. Find out which ones pass muster by downloading Transport Canada's list of vehicles approved for import from its website. Buyers wishing to know if a vehicle can be imported into Canada from Europe or the U.S. can get a list of admissible vehicles (those that conform to Canadian federal safety and pollution regulations) at *www.tc.gc.ca/roadsafety/rsimp_e.htm#US* or by calling the Registrar of Imported Vehicles at 1-800-511-7755.

Buyers of used vehicles from the U.S. should also be wary of the fact that U.S. vehicles' odometers register in miles instead of kilometres and are likely to have the mileage rolled back. Buyers usually have little effective legal recourse to get their money back.

Furthermore, you won't benefit that much from the lower prices on American-sourced cars because of the higher cost of the American dollar and the substantial Canadian sales taxes (GST, provincial sales taxes, and a gas-guzzling surcharge) that must be paid to bring your vehicle home. Servicing may be a problem as well, since some dealers are reluctant to service cars they didn't sell. They won't refuse service outright, but they'll certainly put their own customers first and give them "goodwill" warranty coverage before you.

If you do decide to import a used vehicle from the U.S., check for liens, make sure the vehicle isn't stolen, and ask an independent dealer selling that make of vehicle to pull the automaker history on the car through its VIN. Unfortunately, there's little cooperation among provincial registrars and no link-up between U.S. and Canadian registrars, so your chances of getting the required information from government sources are pretty slim.

Rental and leased vehicles

The second-best choice for getting a good used vehicle is a rental company. Budget, Hertz, Avis, and National sell cars, minivans, vans, and sport-utilities that have one to two years of service and approximately 80,000 km–100,000 km. These rental companies will gladly provide a vehicle's complete history and allow an independent inspection by a qualified mechanic of the buyer's choice, as well as arrange competitive financing without "boosting" the price.

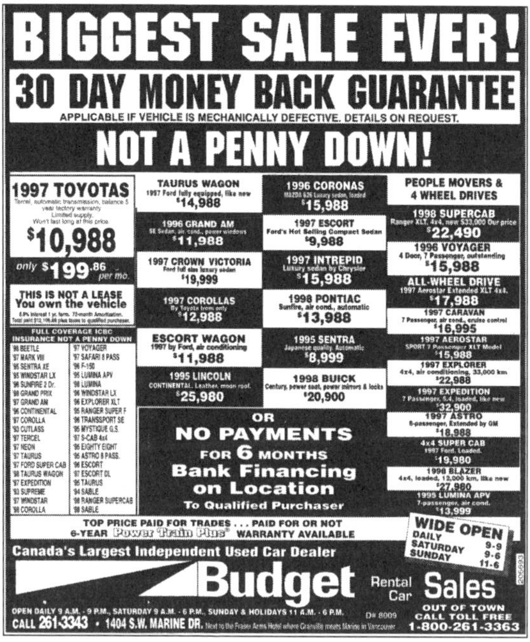

Hmm, not bad. A 30-day money-back guarantee, no payments for six months, and the availability of a six-year powertrain warranty. Plus, that 1997 Crown Victoria listed for $19,999 has a retail value of $19,950 (it sold for $29,895 new). Where do I sign?

Rental vehicles are generally well maintained, sell for a few thousand dollars more than a privately sold vehicle, and come with a strong guarantee, like Budget's 30-day money-back guarantee found in the ad above. Rental car companies also usually settle customer complaints without much hassle so as not to tarnish their image with rental customers.

Vehicles that have just come off a three- or five-year lease are much more competitively priced, generally have less mileage, and are usually as well maintained as rental vehicles. You're also likely to get a better price if you buy directly from the lessee rather than going through the dealership, but remember, you won't have the dealer's leverage to extract post-warranty "goodwill" repairs from the automaker.

Toronto airport limo bargains

Look for limousine bargains at Toronto's Pearson International Airport, where luxury cars have to be sold after five years of use. They are all high-mileage vehicles, but since this is highway driving, it doesn't compromise the vehicles' overall reliability or durability. It does, however, drive down the resale value. For example, a dual-powered (gasoline and natural gas) high mileage (600,000 km–800,000 km) 1993 Lincoln Town Car, Fleetwood (Continentals aren't considered sufficiently reliable or durable), or Cadillac Deville can be bought for $5,000–$6,000. Other airports may offer similar limo bargains, but the choice is more limited.

The limos are cheap because Ford and GM sell them directly to fleets for $10,000–$15,000 less than what dealers pay, and their high mileage depresses their trade-in value. Another advantage is that these airport luxury limos have a complete refurbishing at the three-year mark, so they can run an additional two years without any problems. This revision makes them practically new when it comes time for them to be turned in. Finally, the natural gas conversion cuts fuel costs almost in half.

New-car dealers

Most used-car buyers prefer to deal with private sellers. In fact, the Federation of Automobile Dealer Associations of Canada states that 20 years ago, 86 percent of used cars were sold by new-car dealers—today that number is less than 25 percent, due mainly to the GST, which has driven buyers into the arms of private sellers.

Although going to a new-car dealer is the most expensive way to get a used vehicle (prices are 20–30 percent higher than those for vehicles sold privately), they aren't a bad place to pick up a good used car or minivan if you don't mind paying top dollar. For example, new-car dealers are insured against selling stolen vehicles or vehicles with finance owing or other liens. New-car dealers offer financing, occasionally allow a prospective buyer to have a used vehicle inspected by an independent garage, and offer a much wider choice of models. Repair facilities are also available for warranty work, and if there's a possibility of getting post-warranty "goodwill" compensation from the manufacturer, your dealer can provide additional leverage. Plus, if things do go terribly wrong, dealers have "deeper pockets" than private sellers, so there's a better chance of getting paid should a court judgment be won against the firm.

Automaker "certified" vehicles

Try to get a vehicle that the dealer has had certified by an auto association, and that has had the required repairs carried out. In Alberta, the Alberta Motor Association (AMA) will perform, at a dealer's request, a vehicle inspection. On each occasion, the AMA provides a written report to the dealer, which identifies potential and actual problems, required repairs, or serious defects. Nevertheless, consumers shouldn't let down their guard, says John C. Orr, the Edmonton general manager for the AMA:

> It is up to each dealer to effect the repairs that the dealer deems appropriate. Accordingly, a dealer may
>
> 1) remedy all problems and effect all necessary repairs prior to selling a vehicle;
> 2) remedy some problems or effect some repairs prior to selling a vehicle;
> 3) remedy none of the problems and effect no repairs prior to selling a vehicle; or

4) dispose of a vehicle by "wholesaling" it.

It is important for the public to understand that the AMA does not certify vehicles for dealers and that the AMA does not effect repairs. The buying public should ask the dealer to provide a copy of the AMA report, where the dealer has represented that the vehicle has been AMA inspected, and the buying public should inquire what, if any, repairs have been effected...

The Big Three American automakers and some importers have begun to refurbish and certify used vehicles sold by their dealers. They guarantee the vehicle's mechanical fitness and provide a warranty where length depends on the age of the vehicle. But these vehicles don't come cheap, mainly because automakers force their dealers to bring them up to better-than-average condition before certifying them.

Automaker and dealer leasing

Automakers like Mercedes-Benz, Ford, Chrysler, GM, and Nissan/Infiniti, as well as some dealers, are now leasing three-year-old cars, trucks, sport-utilities, vans, and minivans for additional two- to three-year periods. For a relatively low monthly payment (relative to leasing new), you can get a better-equipped vehicle with some of the automaker's original warranty in effect and a used warranty that kicks in afterwards.

Leasing has been touted as a method of making the high cost of vehicle ownership more affordable, but for most people the pitfalls (whether you lease new or used) far outweigh any advantages. Keep in mind that leasing is costlier than an outright purchase. If you must lease, do it for the shortest time possible and make sure the lease is closed-ended (meaning that you walk away from the vehicle when the lease period ends). Also, make sure there's maximum mileage allowance of at least 25,000 km a year and that the charge per excess kilometre is no higher than 8¢.

Used-car dealers

Used-car dealers usually sell their vehicles for a bit less than what new-car dealers charge, but their vehicles may be worth a lot less, too. They're usually marginal operations that can't invest much money in reconditioning their vehicles, which are often collected from auctions and new-car dealers reluctant to sell the vehicles to their own customers. And they don't always have repair facilities to honour what warranties they do provide. Often, they may offer easier (but more expensive) credit terms than franchised new-car dealers.

Used-car dealers operating in small towns are an entirely different breed. These small, often family-run businesses recondition and resell cars and trucks that usually come from within their community. Routine servicing is often done in-house and more complicated repairs are subcontracted out to specialized garages nearby. These small outlets survive by word-of-mouth advertising and would never last long if they didn't deal fairly with local townsfolk. On the other hand, their prices will likely be higher than elsewhere due to the better quality of used vehicles they offer and the cost of reconditioning and repairing under warranty what they sell.

Superstores

Used-car "superstores," like CarMax, are nationally franchised operations backed by Circuit City and other investors. Although for the past three years superstores have been expanding slowly, they are expected to pick up steam once differences with automakers have been ironed out.

These aren't the places to find "super" bargains when compared with what's offered by traditional new- or used-car dealers. The December 1995 issue of *Smart Money* magazine, for example, found that CarMax prices and warranties came up short 25 out of 26 times when compared to used vehicles sold by local dealers. Furthermore, *Automotive News* reports that studies show used-car superstores tend to drive up prices in the regions where they are established. On the plus side, however, their emergence makes existing dealers more consumer sensitive when it comes to warranties and disclosure. In fact, CarMax's 5-day money-back guarantee and 99-day warranty puts most dealers to shame.

But there are a few clouds on the superstore horizon. CarMax, a subsidiary of Circuit City stores, has yet to earn a profit after six years of sales. Time is also running out for the AutoNation and AutoMax used-car superstore chains to prove they can make a profit. Republic Industries' AutoNation, however, hopes to make its first profit by expanding its outlets and concentrating on a better choice of used vehicles.

Auctions

You need patience and smarts to pick up anything worthwhile at an auction. Government auctions—places where the mythical $50 Jeep is sold—are fun to attend but are risky ventures, because you can't determine the condition of the vehicles put up for bid and government employees often pick over the stock before you.

You swim with the piranha, however, at commercial auctions. They are frequented by "ringers" who bid up the price and professional dealers who pick up cheap, worn-out vehicles unloaded by new-car dealers and independents who are afraid to sell them to their own customers. There are no guarantees, cash is required, and quality is apt to be as low as the price. Remember, too, that auction purchases are subject to provincial and federal sales taxes and the auction's sales commission (3–5 percent), and an administrative fee of $25–$50 may also be charged.

Say what? This ad promises there's no need to purchase anything in order to win one of three $1,000 prizes—to be used ONLY with the purchase of an auctioned vehicle. Watch out! Like phony carpet "Customs Auction" scams, there are many false auto auctions run by used car dealers and independent brokers. Make sure you're dealing with a legitimate auctioneer.

If you are interested in shopping at an auto auction, remember that certain days are reserved for dealers only, so call ahead. You'll find the cars locked in a compound, but you should have ample opportunity to inspect them and, in some cases, take a short drive around the property before the auction begins.

Here are some auctioneers in the Toronto area, around Canada, and just over the border:

Toronto Region
905-790-7653

AB Toronto Auctions
14 Canso Rd, Rexdale
416-245-2277

GTA Auctions and Liquidators
54 Murray Rd, Downsview
416-630-7253

Metro Auto Auction
23 Metropolitan Rd,
Scarborough
416-292-0909

M. Wilson & Co
555 Denison St, Markham
905-479-2886

Canadian Auction Group
3365 Highway 7 East,
Brampton
905-791-9800

Edmonton
780-465-4900

Halifax
902-468-3373

Hamilton
905-560-4851

Moncton
506-857-0198

Rochester, NY
315-539-5006

St. John's
709-364-3250

Winnipeg
204-694-4944

Vancouver
604-580-0011

Paying the Right Price

Get ready for "sticker shock" when pricing popular used cars, pickups, sport-utilities, and vans—most of these vehicles don't depreciate very much and it's easy to get stuck with a cheap one that's been abused through hard off-roading or lack of care. Furthermore, it's such a seller's market out there that even those vehicles with worse-than-average reliability ratings, like early GM Cavaliers, Ford Tauruses, and Chrysler minivans, still command ridiculously high resale prices for the simple reason that they're popular.

If you don't want to pay too much when buying used, you've got the following four alternatives:

- Buy an older vehicle. Choose one that is five years old or more and has a good reliability and durability record, and buy extra protection with an extended warranty. The money you save from the extra years of depreciation and lower insurance premiums will more than make up for the extra warranty cost.
- Look for off-lease vehicles sold privately by owners who want more than what their dealer is offering. If you can't find what you're looking for from the local classified ads, put in your own ad asking for lessees to contact you if they're not satisfied with their dealer's offer.
- Buy a vehicle that's depreciated more than average simply because of its bland styling, lack of high-performance features, or its discontinuation by one of its distributors. For example, many of the Japanese pickups cost less to own than their American-made counterparts, yet are more reliable and more than adequate for most driving chores. And, now that the Dodge Stealth has been dropped by Chrysler, used versions can be picked up for a song and servicing is easily found at Mitsubishi dealerships (in the U.S.) who sell the Stealth's twin, the 3000GT.
- Buy a twin or re-badged model. Following are some of the better-known cars, minivans, vans, sport-utilities, and trucks that are nearly identical but sold under other nameplates. They usually share the same basic design, appearance, dimensions, and mechanical components, but their resale values may differ considerably.

Chrysler

Chrysler Cirrus, Dodge Stratus, Plymouth Breeze

Chrysler Concorde, LHS, New Yorker, Dodge Intrepid, Eagle Vision

Dodge Colt, Eagle Summit

Dodge Colt Wagon, Eagle Summit Wagon

Dodge Dart, Duster, Plymouth Valiant, Duster

Chrysler Sebring, Dodge Plymouth Laser
Chrysler Town & Country, Dodge Grand Caravan, Plymouth Grand Voyager
Dodge Caravan, Plymouth Voyager
Dodge Omni, Plymouth Horizon
Dodge Spirit, Plymouth Acclaim
Dodge Stealth, Mitsubishi 3000GT
Dodge Volaré, Plymouth Aspen
Eagle Talon, Mitsubishi Eclipse, Avenger, Mitsubishi Galant

Ford

Ford Contour, Mercury Mystique
Ford Crown Victoria, Lincoln Town Car, Mercury Grand Marquis
Ford Escort, ZX2, Mercury Tracer, Lynx, Mazda 323
Ford Explorer, Mazda Navajo, Mercury Mountaineer
Ford Mustang, Mercury Capri
Ford Pinto, Mercury Bobcat
Ford Probe, Mazda MX-6
Ford Ranger, Mazda B-Series
Ford Taurus, Mercury Sable
Ford Tempo, Mercury Topaz
Ford Thunderbird, Mercury Cougar
Mercury Villager, Nissan Quest

General Motors

Buick Century, Regal, Olds Intrigue, Pontiac Grand Prix
Buick LeSabre, Olds 88, Pontiac Bonneville
Buick Park Avenue, Riviera, Cadillac Seville, Olds Aurora
Chevrolet Express, GMC Savana
Chevrolet Lumina APV minivan, Olds Silhouette, Pontiac Trans Sport
Chevrolet Lumina, Monte Carlo

The Cavalier (shown above), Sunbird, and Sunfire: identical cars, different prices.

Chevrolet Astro, GMC Safari
Chevrolet Blazer, GMC Jimmy, Olds Bravada
Chevrolet Blazer, Tahoe GMC Yukon
Chevrolet C/K pickup, GMC Sierra pickup
Chevrolet Malibu, Olds Cutlass
Chevrolet Metro, Pontiac Firefly, Suzuki Swift, Sprint
Chevrolet Prizm, Toyota Corolla
Chevrolet S-10 pickup, GMC Sonoma pickup

Chevrolet Camaro, Pontiac Firebird
Chevrolet Cavalier, Pontiac Sunbird, Sunfire
Chevrolet Corsica, Beretta, Pontiac Tempest
Chevrolet Venture, Olds Silhouette, Pontiac Trans Sport
Geo Tracker, Pontiac Sunrunner, Suzuki Sidekick
Olds Achieva, Pontiac Grand Am
Saturn SC, Sl

Imports

Acura Integra, Honda Civic
Acura CL, Honda Accord
Acura SLX, Isuzu Trooper
Honda Odyssey, Isuzu Oasis
Honda Passport, Isuzu Rodeo
Hyundai Elantra, Tiburon
Infiniti I30, Nissan Maxima
Infiniti QX4, Nissan Pathfinder
Lexus ES 300, Toyota Camry
Lexus LX 450, Toyota Land Cruiser
Toyota Paseo, Tercel
VW Golf, Jetta

Prices will vary

In spite of what many guidebooks say (including *Lemon-Aid*), cars and minivans have no "official" selling price. Most dealers will charge as much as the market will bear and use any pretext they can to boost prices—if you let them. Although they pretend to have abandoned negotiated prices and high-pressure sales tactics in favour of "no-haggle prices," dealers will negotiate and it's in your best interest to make sure they do.

Consumer Reports, for example, found that its expert shopper saved hundreds of dollars buying from a conventional dealership as opposed to one which displayed the dealer's invoice cost and a "no-haggle" selling price.

There are several price guidelines, however, and dealers use the one that will make the most profit on each transaction. The most common price quoted is the *Red Book*, which shows the price the vehicle is worth when sold at full retail (similar to the Manufacturer's Suggested Retail Price, or MSRP). The wholesale price, thousands of dollars less, is more likely what the dealer paid. Never mind that few people ever pay that price. The fact that it's there is sufficient reason for most dealers to charge the higher rate. Both price indicators leave considerable room for the dealer's profit margin and some extra padding—inflated preparation charges and administration fees that shouldn't exist for a used vehicle.

TRADE IN DESCRIPTION & LIEN DISCLOSURE		TERMS OF SETTLEMENT	
☐ G.S.T. REGISTRANT	G.S.T. REGISTRANT NO.	TOTAL CASH SALE PRICE	10800 00
YEAR	MI / KM ODOMETER READING		
MAKE NO	MODEL		
SERIAL TRADE		SUBTOTAL	10800 0·
YEAR	MI / KM ODOMETER READING	TRADE-IN ALLOWANCE	
MAKE	MODEL	SUBTOTAL	
SERIAL		G.S.T.	756 00
I HEREWITH TRANSFER TO DEALER ALL MY RIGHTS, TITLE AND OWNERSHIP IN THE ABOVE MOTOR VEHICLE, AND I DECLARE I AM THE SOLE OWNER AND POSSESSOR OF SAME AND THAT THERE IS NO MORTGAGE, LIEN, NOTE OR CLAIM OF ANY KIND OR NATURE ADVERSE TO MY RIGHTS OF, UPON, OR AGAINST SAID VEHICLE OTHER THAN AS STATED BELOW		SUBTOTAL	11556 00
I HEREBY STATE THAT TO THE BEST OF MY KNOWLEDGE THE ODOMETER READING AS STATED ABOVE INDICATES THE TOTAL DISTANCE ACTUALLY TRAVELLED BY THE VEHICLE.			
CUSTOMER SIGNATURE X		ADMINISTRATION FEE	99 00
LIEN PAYABLE TO		PAYOUT LIEN ON TRADE-IN	—
ADDRESS	$	BALANCE DUE	11655 00
		DEPOSIT CASH ☐ CHEQUE ☐ VISA	1000 00
		PAYABLE ON DELIVERY	10655 00
THIS AGREEMENT SUBJECT TO FOLLOWING ADDITIONAL PROVISIONS. ONLY THESE PROVISIONS WILL BE RECOGNIZED.			

Dealer greed knows no bounds. This buyer was charged a $99 "administration fee" for a used car! Just as they do with new cars, dealers know that last-minute, add-on charges are their final chance to stick it to you before the contract is signed. Therefore, they try to extort extra profits through so-called preparation and documentation fees and by charging extra handling costs that give you nothing in return. These charges should have no place in a used-vehicle transaction.

Financing Choices

No one should spend more than 30 percent of his or her annual gross income on the purchase of a used vehicle. By keeping the initial cost of a used vehicle low, the purchaser may be able to pay in cash—a key bargaining tool to use with private individuals selling cars directly. Used-car dealers are not all that impressed by cash sales because they lose their kickback from the finance companies, which is based on the volume and amount of finance business they write up.

In spite of the lower costs involved (as compared to a new-car purchase), not everyone can pay cash for a used car. If you need a loan, consider the advantages and disadvantages of going through the following lending institutions.

Credit Unions. A credit union is the preferred place to borrow money at low interest rates and with easy repayment terms. You'll have to join the union or have an account with it before the loan is approved. You'll also probably have to come up with a larger down payment relative to what other lending institutions require.

In addition to giving you reasonable loan rates, credit unions help car buyers out in a number of other ways. Toronto's Metro Credit Union (*www.metrocu.com*), for example, has offered for the past seven years a Car Facts Centre, which provides free, objective advice on car shopping, purchasing, financing, and leasing. Car Facts advisors provide free consultations in person or by phone. This includes:

- fact sheets on all aspects of buying new or used cars
- a circulating library of car buying and car maintenance books and magazines
- computerized analysis of different buying, leasing, and borrowing options
- *Red Book* used-car price quotations
- dealer invoice price quotations on any make or model of new vehicle, at $20 each

Additionally, Metro has added an AutoBuy program where members can hire an expert "car shopper" who will do the legwork, including the tedious and frustrating dickering with sales staff, and save the member time and hassle. Not to mention get that new or used vehicle at a reduced (fleet) rate, arrange top-dollar prices for trade-ins, provide independent advice on options like rustproofing and extended warranties, carry out lien searches, and even negotiate the best settlement with insurance agents.

Rosemary Edwards, a Car Facts advisor, says Metro's Car Facts program has saved members over $600,000 since its inception. For information on how you can join Metro or start up a similar program with your credit union, contact Rosemary Edwards or David Lawrence at 1-800-777-8506 or 416-252-5621. Metro also holds regular car buying seminars throughout the year in the Greater Toronto Area.

Banks. Banks want to make small loans to consumers who have average incomes and appear to be financially responsible. Their interest rates are very attractive and can be negotiated downwards if you have a good credit history or agree to give them other business. Auto club members can benefit from lower interest rates at banking institutions recommended by the association. Bank loans are seldom made for more than 36–48 months.

In your quest for a bank loan, keep in mind that the Internet offers help for people who need an auto loan and want quick approval but don't like to face a banker. The Bank of Montreal (*www.bmo.com*) was the first Canadian bank to allow car buyers to post loan applications on its website, and it promises to send a loan response within 20 seconds. The service is available to any web surfer, including those who aren't current Bank of Montreal customers.

Dealers. Dealers can finance the cost of a used vehicle at rates that compete with those of banks and finance companies because they get substantial rebates from lenders and agree to take back the vehicle if the creditor defaults on the loan. Some dealers mislead their customers into thinking they can get financing at rates 3–4 percentage points below the prime rate. Actually, the dealer jacks up the base price of the vehicle to compensate for the lower interest charges.

Finance Companies. With their excessive interest rates, finance companies should be the last place to go for a small, short-term loan, but the fact remains that these lenders fill a consumer need created by the restrictive policies of other institutions. The advantages of relaxed credit restrictions and quick loans appeal to many people who can't get financing elsewhere, or who may wish to extend their payments for up to 60 months.

Dealer Scams

Used vehicles are subject to the same deceptive sales practices deployed by dealers who sell new vehicles. One of the more common tricks is to not identify the previous owner, because the vehicle either was used commercially or had been written off as a total loss from an accident. It's also not uncommon to discover that the mileage has been turned back, particularly if the vehicle was part of a company's fleet. These scams can be thwarted if you demand the name of the vehicle's previous owner as a prerequisite to purchasing the vehicle.

It would be impossible to list all the dishonest tricks employed in used-vehicle sales. As soon as the public is alerted to one scheme, crooked sellers use other, more elaborate frauds. Nevertheless, under industry-financed compensation funds, Ontario and B.C. buyers can get substantial refunds if defrauded by a dealer.

Here are some of the more common fraudulent practices you're likely to encounter.

Failing to declare full purchase price. This tactic, used almost exclusively by small, independent dealers, involves the buyer being told by the salesperson that he or she can save on sales tax by listing a lower selling price on the contract. But what if the vehicle turns out to be a lemon or the sales agent has falsified the model year or mileage? The hapless buyer will usually be offered a refund on the fictitious purchase price indicated on the contract only. If the buyer wanted to take the dealer to court, it's quite unlikely that the buyer would get any more than the contract price. Moreover, both the buyer and dealer could be prosecuted for making a false declaration to avoid paying sales tax.

Sales agents posing as private parties ("curbsiders"). Individuals sell about three times as many used vehicles as dealers. Some crooked dealers, though, are using agents to pose as private sellers in order to get a better price for their cars and to avoid paying GST and giving a warranty. Once again, this scam is easy to detect if the seller can't produce the original sales contract or show some repair bills made out in his or her own name. You can also identify a car dealer in the want ads section of the newspaper by checking to see if its telephone number is repeated in many different ads.

Most new-car dealers get very angry when one of these scamming teams hits town. Unfortunately they don't get angry enough, because they continue to sell used cars at wholesale prices to curbsiders who they know are stealing their business and cheating consumers and tax authorities.

If you get taken by one of these scam artists, don't hesitate to sue through small claims court both the newspaper carrying the ad and the provincial department responsible for dealer registration. These parties could be considered negligent for allowing this rip-off artist to operate.

Curbsiders are moving west, buying cars at wholesale prices from dealers, auto auctions, and junkyards (some have been written off as total losses). They then place private classified ads in B.C. and Alberta papers, sell their stock, and leave town. *Vancouver Sun* consumer columnist Gillian Shaw tracked one particular group that simply stole the cars it sold:

> There's a ring operating in the Lower Mainland stealing cars then stealing registration papers and VIN numbers from like cars, putting the two together, and selling them to unsuspecting buyers. Of course the sale is followed a short time later by the cops coming to tow the stolen car and the buyer is left with little more than the pager number of the scam artist. If they buy a stolen car from a dealer, they're covered up to $20,000 by the dealer's compensation fund, but in a private sale, no such luck. And ICBC is saying it's not its problem.

"Free-exchange" privilege. Dealers get a lot of sales mileage out of this deceptive offer. The dealer offers to exchange any defective vehicle for any other vehicle in stock. What really happens, though, is that the dealer won't have any other vehicles selling for the same price and will thus demand a cash bonus for the exchange, or he or she may have nothing but lemons in stock.

"Money-back" guarantee. Once again, the purchaser feels safe in buying a used car with this kind of guarantee, because what could be more honest than a money-back guarantee? Dealers using this technique often charge exorbitant handling charges, rental fees, or mechanical repair costs to the customer who's bought one of these vehicles and then returned it.

"50/50" guarantee. This means that the dealer will pay half the repair costs over a limited period of time. It's a fair offer if an independent garage may do the repairs. If not, the dealer can always inflate the repair costs to double their actual worth and write up a bill for that amount. The buyer winds up paying the full price of repairs, which would probably have been much cheaper at an independent garage. The best kind of used vehicle warranty is 100 percent, with full coverage for a fixed term.

"As is" cars. Buying a vehicle "as is" means that you're aware of mechanical defects and are prepared to accept the responsibility for any damage or injuries caused by the vehicle, and that all costs to fix it shall be paid by you. The courts have held that the "as is" clause is not a blank cheque to cheat buyers and therefore must be interpreted in light of the seller's true intent. That is, was there an attempt to deceive the buyer by including this clause? Did the buyer really know what the "as is" clause could do to his or her future legal rights? It's also been held that the courts may consider verbal representations ("parole evidence") as to the fine quality of the used vehicle made by the seller but never written into the formal contract. Courts generally ignore "as is" clauses when the vehicle has been misrepresented, when the dealer is the seller, or when the defects are so serious that the seller is presumed to have known of their existence.

Odometer tampering. It's often too dangerous for the dealer to turn back the mileage, so independent outfits are hired to pick up the car or visit the dealership and "fix" the odometer, a practice allowed under Canadian federal law. For more recent models a computer chip that turns back the odometer reading can be inserted for less than $50.

Until federal and provincial statutes are toughened and convictions result in severe penalties to anyone engaging in this practice (American laws allow citizens to sue for triple damages plus lawyer and court costs), the best protection from this fraud is to demand that the dealer put the mileage figure on the contract and give you the name and address of the previous owner as well as all repair receipts. It would be smart to demand the same things of a private seller.

Misrepresentation. Used vehicles can be misrepresented in a variety of ways. A used airport commuter minivan may be represented as having been used by a Sunday school class. A mechanically defective pickup that's been rebuilt after several major accidents may have sawdust in its transmission to muffle the "clunks," heavy oil in the motor to stifle the "clanks," and cheap retread tires to eliminate the "thumps." These fraudulent practices may lead to the seller being charged with civil or criminal fraud. The best protection against these dirty tricks is to have the vehicle's quality completely verified by an independent mechanic before completing the sale.

Private Scams

A lot of space in this guide has been used to describe how used car dealers and scam artists cheat uninformed buyers. Of course, private individuals can be dishonest too, so in either case protect yourself at the outset by keeping your deposit small and getting as much information as possible about the vehicle you're considering buying. Then, after a test drive, you may sign a written agreement to purchase the vehicle and give a deposit of sufficient value to cover the seller's advertising costs, subject to cancellation if the automobile fails its inspection.

After you've taken these precautions, watch out for these private sellers' tricks.

Used vehicles that are stolen or have finance owing. Bad economic times mean that a lot of used cars are sold without free title because the original auto loan was never repaid. In Ontario, the Used Vehicle Information Package (mandatory for private sellers) will alert buyers to any problems with the title, but in other provinces, buyers don't have access to this information. Generally, you have to contact the provincial office that registers property and pay a small fee for a computer printout that may or may not be accurate. You'll be asked for the current owner's name and the car's VIN, which is usually found on the driver's side of the dashboard.

Both amateur and professional rip-off artists sell cars that haven't been paid for because it's so easy to do and the profits are enormous. They skip town after sticking you with the finance payments, and by the time the lender tracks you down, most legal recourse is ineffective. Your best bet may be to buy the car again from the lender at a discount and take a tax loss.

This racket can be avoided by asking for proof of purchase and payment from any individual who offers to sell a used car for an incredibly low price. Check the sales contract to determine who granted the original loan and call the lender to see if it's been repaid. Place a call to the provincial Ministry of Transport to ascertain whether the car is registered in the seller's name. Find out if a finance company is named as beneficiary on the auto insurance policy. Finally, call up the original dealer to determine whether there are any outstanding claims.

Stolen cars are almost always sold through private individuals. Car dealers usually can't be fooled, and they have insurance to compensate buyers if they inadvertently sell a stolen vehicle.

Misrepresentation (wrong model year). That bargain-priced car or truck you just bought may be a year older than you think. That's why you should check its true age by looking at the date of manufacture plate found on the driver's side door pillar. If the date of manufacture reads 7/96, it was one of the last 1996 models made before the

September changeover to the 1997 models. Exceptions to this rule are those vehicles that are redesigned or relatively new to the market, which arrive at dealerships in early spring or mid-summer. They're considered to be next year's models but depreciate more quickly due to their earlier launching (a difference that narrows over time). If you have any doubts, look up the 10th character in the VIN, which identifies the model year.

MODEL YEAR INDICATOR (10th VIN POSITION)

Code	Model Year	Code	Model Year
A	1980	S	1995
B	1981	T	1996
C	1982	V	1997
D	1983	W	1998
E	1984	X	1999
F	1985	Y	2000
G	1986	1	2001
H	1987	2	2002
J	1988	3	2003
K	1989	4	2004
L	1990	5	2005
M	1991	6	2006
N	1992	7	2007
P	1993	8	2008
R	1994	9	2009

Now, one last word on protecting yourself from legal liability if you are selling a vehicle to a private individual. Normally, once you take the tags and give the buyer a bill of sale, you're no longer the registered owner. Unfortunately, though, if you don't personally go down to the registry office and make sure the title has changed, you could be sued by an insurance company for damages arising from an accident if you're still the owner of record. *Toronto Sun* columnist Maryanna Lewyckyj tells of one case where a driver sold his clunker for $250 and six months later was sued by Liberty Mutual Insurance for $30,000, representing the damages caused by the subsequent owner. Apparently, the new owner never changed the registration.

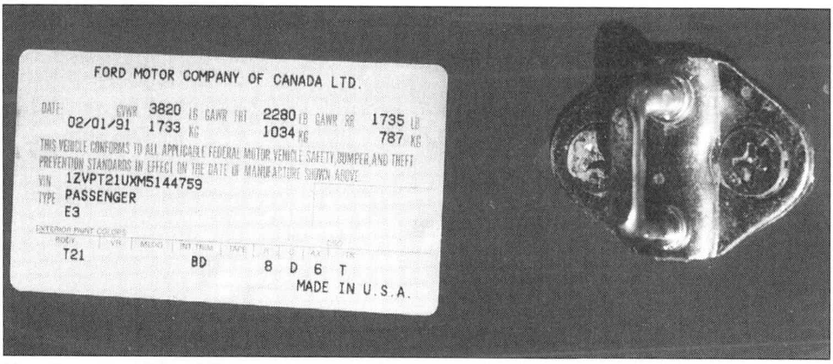

During the '70s, Nissan and Ford conspired with their dealers to sell previous years' cars and trucks as the latest model. Private sellers may continue the fraud unwittingly because the actual model year was never disclosed to them.

Choosing the Right Kind of Vehicle

Import or domestic model?

Whether you buy domestic or imported, overall vehicle quality has improved a great deal during the past decade. Premature rusting is less of a problem and factory-related defects aren't as numerous as they once were. But problems remain. Owners of American-made vehicles still report serious deficiencies, often during the first three years in service. These include electrical system failures caused by faulty computer modules, malfunctioning ABS, failure-prone air conditioning and automatic transmissions, as well as defective powertrains, fuel systems, suspensions, steering, and paint. Further evidence that American cars are more failure-prone than their Japanese-made counterparts can be seen by the service bulletins sent out each month by the automakers. For this guide, I've compared the bulletins of a 1994 Toyota Camry to those of a 1994 Ford Taurus and Sable.

1994 Ford Taurus V6
DSB SUMMARY

1. SEP-89 20 STEPS TO SUCCESSFUL AUTO TRANSMISSION REPAIR
2. JAN-94 A/C - BLOWER MOTOR INOPERATIVE IN AUTOMATIC MODE
3. JAN-94 A/C - MANIFOLD AND TUBE ASSEMBLY REPLACEMENT
4. JUL-95 A/C EVAPORATOR CORE-ON VEHICLE-LEAK TEST
5. OCT-94 A/C SERVICE TIP - OIL VISIBLE AT SPRING LOCK COUPLERS
6. JUL-96 A/C SYSTEM SERVICE TSB LIST
7. MAR-94 A/T (AX4N) - SERVICE PROCEDURE FOR ORANGE TAGGED UNITS
8. MAR-94 A/T (AXOD/AXODE/AX4S) - LEAK AT TORQUE CONVERTER
9. JAN-96 ACCELERATOR CABLE - DIAGNOSTIC PROCEDURE
10. JAN-95 ACCUMULATOR SPRINGS - IDENTIFICATION
11. MAR-96 ADDING REFRIGERANT OIL PROCEDURE
12. MAR-96 AIR BAG - DIAGNOSTIC FAULT CODE 51
13. OCT-95 AIR BAG MODULES - DISCOLORED/MARRED COVERS
14. AUG-97 AIR COMPRESSOR MOAN
15. SEP-95 AIR CONDITIONING - FORD APPROVED FLUSHING
16. FEB-98 AIR CONDITIONING MUSTY AND MILDEW TYPE ODORS
17. SEP-96 AIR CONDITIONING O-RING AVAILABILITY AND APPLICATIONS
18. JAN-95 AIR CONDITIONING O-RING REMOVAL - SERVICE TIP
19. SEP-96 AIR CONDITIONING THICK O-RING APPLICATIONS
20. DEC-93 AIR IN POWER STEERING SYSTEM
21. SEP-87 ALL AUTOMATICS - FRONT BUSHING WEAR
22. MAY-95 ANTENNA - BOWED CONDITION ONLY - WARRANTY REVISION
23. MAR-94 ANTI-LOCK BRAKES - PEDAL FEELS LIKE IT HAS EXTRA TRAVEL
24. OCT-95 ANTI-LOCK BRAKES CYCLING ON ROUGH ROADS
25. JAN-94 AUTOMATIC TRANSMISSION - AX4S - DO NOT SERVICE TAG
26. FEB-90 AUTOMATIC TRANSMISSION FLUID
27. AUG-89 AUTOMATIC TRANSMISSION MATH FORMULAS
28. OCT-89 AUTOMATIC TRANSMISSION MATH PART 2
29. AUG-96 AX4S TRANSAXLE - NEW MAIN CONTROL COVER AND PAN GASKET
30. NOV-94 AX4S TRANSAXLE EXCHANGE PROGRAM
31. FEB-98 AX4S/AX4N NEW TRANSAXLE FLUID

Look at this service bulletin summary for a 1994 Ford Taurus and Mercury Sable. Out of a total of 162 bulletins, 53 bulletins concerned a factory-related defect.

32. NOV-97 AX4S/AX4N, TROUBLE CODES STORED AFTER TRANSAXLE REPLACED
33. JAN-96 AXLE NUT/LUG NUT TIGHTENING SPECIFICATIONS
34. SEP-92 BATTERY REPLACEMENT 72AH TO 84AH
35. MAR-95 BLOWER MOTOR - CHIRP/SQUEAK FROM BLOWER MOTOR
36. NOV-93 BLOWER SPEED CONTROL - AUTOMATIC TEMPERATURE CONTROL
37. SEP-97 BRAKE AND ROTOR SERVICE TIPS
38. DEC-94 BRAKE ROTOR MACHINING EQUIPMENT AND WARRANTY INFORMATION
39. OCT-95 BRAKES - PETROLEUM BASED LUBRICANT CAUSES SWELLING
40. MAR-96 BRAKES - REAR BRAKES NOISE/SQUEAL
41. MAR-95 BRAKES - ROUGHNESS DURING BRAKE APPLICATION
42. NOV-93 BUMPER COVER REPAIR MATERIALS
43. APR-94 BUMPERS - ISOLATOR AND BRACKET ASSEMBLY REPLACEMENT
44. MAY-96 CALIFORNIA REFORMULATED GASOLINE (CARFG) SERVICE TIP
45. JAN-96 CASE BREAKAGE AT REAR PLANET SUPPORT
46. DEC-97 CHIRPING/SQUEAKING FROM BLOWER MOTOR AT LOW SPEEDS
47. JAN-98 CODES 327, 332, MIL LAMP ON, 32 DEGREES F OR BELOW, 3.0L
48. JUL-94 COOLING SYSTEM BY-PASS HOSE LEAK
49. FEB-98 CUSTOMER ASSISTANCE-TELEPHONE NUMBERS FOR TIRE COMPANIES
50. JAN-95 DELAYED ENGAGEMENT, SHIFT ERRORS, POSSIBLE MLPS CODES
51. NOV-93 DRIVEABILITY CONCERNS - DAMAGED FUEL PUMP
52. MAY-94 DRIVER'S SEAT SENSOR CAUSES DOORS TO LOCK WHEN EXITING
53. NOV-93 EGO SENSORS - SILICONE CONTAMINATION
54. NOV-96 ELECTRICAL DIODE IDENTIFICATION AND REPLACEMENT
55. JUN-93 ELECTRICAL INFORMATION - AXOD/E
56. AUG-95 ELECTRONIC SHIFT SOLENOID APPLICATIONS
57. OCT-96 ELIMINATION OF SANDING BRAKE ROTORS AND DRUMS
58. MAY-97 ENGINE COOLING FAN MODULE SERVICE INFORMATION
59. OCT-90 ENGINE TESTING WITH A VACUUM GAUGE - AUTO TRANS.
60. JAN-95 ERRATIC/HIGH LINE PRESSURE
61. OCT-93 FAILSAFE MODES - COMPUTER SHIFTED TRANSMISSIONS
62. JUL-96 FILTERING REFRIGERANT AFTER REPLACING A/C COMPRESSOR
63. OCT-96 FOG/FILM ON WINDSHIELD/INTERIOR GLASS
64. JUL-96 FORD'S POSITION ON PROPYLENE GLYCOL COOLANT
65. JAN-95 FORWARD PISTON CHANGE
66. JAN-94 FORWARD SPRAG UPGRADE
67. NOV-94 FORWARD/REVERSE ENGAGEMENT CONCERN - AXOD, AXOD-E, AX4S
68. DEC-97 FRONT END ACCESSORY DRIVE BELT SLIP, 3.0L 2-VALVE
69. FEB-96 FUEL INFORMATION OF GASOLINE - SERVICE TIP
70. MAR-97 FUEL ECONOMY - CUSTOMER EXPECTATIONS VS. VEHICLE USAGE
71. JAN-96 FUEL PUMP BUZZ/WHINE THROUGH RADIO SPEAKER
72. OCT-97 FUEL SYSTEM - FUEL ODOR INSIDE PASSENGER COMPARTMENT
73. OCT-94 FUEL TANK - FUEL SLOSH NOISE
74. OCT-93 GLASS - IRIDESCENCE OR MOTTLING IN TEMPERED GLASS
75. OCT-95 GROANING NOISE FROM POWER WINDOWS
76. JAN-94 HARSH SHIFTS, NO 3-4 SHIFT, ERRATIC SHIFTS
77. JUL-97 HEADLAMPS OPERATE INTERMITTENTLY
78. AUG-94 HEATED EXHAUST GAS OXYGEN SENSOR APPLICATION CHART
79. JAN-95 HEATER TO ENGINE COOLANT BYPASS HOSE REPLACEMENT
80. JUN-94 HEATER/ A/C - COLD ENGINE LOCKOUT OPERATION
81. MAR-96 HESITATION/NO START/STALLING - INTERMITTENT (HOT)
82. NOV-91 HOW TO USE A PRESSURE GAUGE - AUTOMATIC TRANS.
83. JAN-98 HUB-MOUNT BRAKE ROTOR MACHINING EQUIPMENT AVAILABILITY
84. SEP-97 IDENTIFICATION OF NON-FORD APPROVED REFRIGERANTS
85. APR-94 IGNITION CONTROL MODULE CONNECTOR UPDATE
86. MAR-95 INCORRECT PCM USAGE WILL CAUSE TRANSAXLE DAMAGE
87. FEB-98 INTERMITTENT NEUTRAL CONDITION, NO FORWARD/REVERSE, AX4N
88. AUG-97 LOOSE CATALYST OR MUFFLER HEAT SHIELDS
89. DEC-94 LOW TRANSAXLE FLUID LEVEL IMPROPERLY SETTING DTC
90. AUG-87 METAL SEALING RINGS - AUTOMATIC TRANSMISSIONS
91. JAN-94 MODIFICATION TO IMPROVE GEAR LUBRICATION
92. AUG-95 NEW ROTUNDA FLUID CHANGER - SERVICE TIP

93. JAN-96 NEW TORQUE SPECS FOR INSTALLING NEW DESIGN TIE RODS
94. DEC-94 NO CRANK - POSSIBLE CORROSION AT STARTER SOLENOID
95. FEB-96 NOISE - CLICK FROM TRANSAXLE DURING REVERSE ENGAGEMENT
96. JUL-96 NOISE - FUEL PUMP GURGLING AFTER HOT RESTART
97. DEC-96 NOISE FROM REAR TENSION START - WAGONS ONLY
98. SEP-94 NON-MATING CONDITION - HARNESS CONNECTOR/IGNITION SWITCH
99. SEP-97 ON-VEHICLE HEATER CORE PRESSURE TEST FOR WARRANTY CLAIM
100. NOV-93 PAINT - REGULAR PRODUCTION OPTION PAINT CODES
101. MAR-95 PAINT - IRON PARTICLE REMOVAL
102. NOV-93 PAINT CODES AND SUPPLIER STOCK NUMBERS
103. NOV-94 PAINT PREPARATION PROCEDURE AND MSDS INFORMATION
104. JUL-94 PASSENGER SEAT BACK RATTLES WHEN UNOCCUPIED
105. JUL-94 POWER DOOR LOCK ACTUATORS
106. AUG-97 PREVENTING BRAKE VIBRATION SERVICE TIP
107. JUN-94 PROGRAM 94B48 - FUEL TANK CONTAMINATION
108. JAN-95 RADIATOR - NEW MANUFACTURING PROCESS
109. SEP-97 RADIO AM BAND STATIC WHILE DRIVING
110. MAR-95 REAR SUN GEAR AND DRUM BEARING SERVICE - TRANSAXLE
111. JAN-96 REAR SUN GEAR BEARING - UPDATE
112. DEC-96 REAR TIRE INNER EDGE WEAR
113. AUG-94 REAR VIEW MIRROR - DETACHES FROM WINDSHIELD
114. AUG-97 REAR VIEW MIRROR REATTACHMENT
115. DEC-95 RECALL 95S22 ENGINE COOLING FAN
116. SEP-95 RECYCLED ENGINE COOLANT - SERVICE TIPS
117. DEC-94 RELEASE OF R-134A FLUORESCENT DYE
118. NOV-97 REMOTE KEYLESS ENTRY (RKE) DIAGNOSTIC SERVICE TIPS
119. NOV-93 REMOTE KEYLESS ENTRY SYSTEM
120. FEB-94 REMOTE KEYLESS ENTRY SYSTEM - SERVICE TIPS
121. JAN-95 REMOTE KEYLESS ENTRY SYSTEM TRANSMITTERS
122. OCT-96 REPLACEMENT OF THE MASS AIR FLOW SENSOR AS AN ASSEMBLY
123. NOV-94 REPLACEMENT PROCEDURE FOR FRONT DOOR MOULDING
124. DEC-93 RETURN OF OBSOLETE STARTER SOLENOIDS AND STARTER DRIVES
125. JAN-96 REVISED FORWARD SPRAG RACE
126. JAN-96 REVISED SPLINE DRIVE JOINT - OVERDRIVE DRUM
127. JUL-95 REVISED TORQUE CONVERTER IDENTIFICATION CODES
128. AUG-95 REVISED TRANSAXLE CLUTCH CLEARANCES
129. SEP-94 ROUGH IDLE, HESITATION/STUMBLE, BUCKS/JERKS ON COAST
130. NOV-93 ROUGH IDLE/HESITATION/ POOR HEATER OUTPUT
131. MAR-97 SAFETY REGAL 97S66 ENGINE COOLING FAN MOTOR CIRCUIT
132. JUL-97 SEAT LEATHER CLEANING PROCEDURE
133. JAN-94 SERVICE PARTS RETURN OF OBSOLETE FUEL PRESSURE
134. JUN-94 SERVICE PARTS RETURN OF OBSOLETE FUEL TANKS
135. MAY-97 SERVICE TIP - REFRIGERANT OIL REFILL CAPACITY
136. NOV-96 SERVICE TIP - TORQUE CONVERTER LEAK TEST PROCEDURE
137. APR-97 SERVICE TIPS - ELECTRONIC AUTOMATIC TEMPERATURE CONTROL
138. JUL-97 SERVICE TIPS - WINDNOISE AROUND DOORS
139. AUG-94 SHIFT CONCERNS OR TRANSAXLE DAMAGE - INCORRECT PCM
140. JUN-97 SPEEDOMETER NEEDLE WAVERS AND STICKS
141. NOV-95 SQUEAKS FROM REAR WINDOW AREA
142. SEP-96 SUSPENSION RIDE HEIGHT MEASUREMENT AND ADJUSTMENT
143. SEP-95 TEMPERATURE GAUGE - FLUCTUATION/INACCURATE
144. JUL-95 TEMPERATURE GAUGE READS LOW/ERRATIC
145. DEC-94 THROTTLE POSITION SENSOR - FUNCTION AND DIAGNOSTIC II
146. NOV-94 THUMPING/CLACKING NOISE WHILE BRAKING
147. FEB-94 TORQUE CONVERTER - CHANGE FROM 23 TO 25 TEETH
148. DEC-94 TORQUE CONVERTER CLEANING AND REPLACEMENT GUIDELINES
149. OCT-94 TRANSAXLE - AX4S/AXOD - NEW LUBE TUBES
150. OCT-94 TRANSAXLE (AXOD, AXODE, AX4S) - RETAINING RING ORIENTATION
151. OCT-94 TRANSAXLE LUBE TUBE BRACKET
152. JUL-94 TRANSAXLE - AX4N - FLUID PAN EMBOSSED WITH SHO
153. APR-96 TRANSAXLE - AX4S/AX4N DRIVELINE NOISES

154. DEC-95 TRANSAXLE - BONDED MAIN CONTROL SEPARATOR PLATE GASKETS
155. FEB-96 TRANSAXLE - REVISED GROB SPLINE DRIVE JOINT
156. JUL-95 TRANSAXLE REAR PLANET SUPPORT RETAINING RING I.D
157. JAN-95 TRANSMISSION CONTROLLED CLUTCH PULSING ON/OFF
158. MAR-97 TRANSMISSION FLUID USAGE CHARTS
159. FEB-95 TURBINE SHAFT SPEED AIR GAP MEASUREMENTS CHANGE
160. JUL-95 WHEEL COVER DISTORTED OR MELTED
161. JUN-95 WINDSHIELD, URETHANE MOUNTED, MATERIAL USAGE TIP
162. NOV-97 WIRE HARNESS TERMINAL REPAIR KIT AND WIRE SPLICE REPAIR

Japanese vehicles hold up fairly well, until their fifth year. Then the engine head gasket will probably have to be replaced, the front brakes will need reconditioning, and the rack-and-pinion steering system will likely have to be overhauled. Furthermore, the front-wheel drive constant-velocity joints will probably need replacing—at a cost of about $300 each. Unfortunately, used Japanese-built vehicles are seldom priced reasonably, because their owners paid inflated prices when they were new and want to get some of their money back when they're resold.

1994 Toyota Camry Sedan 4-Door V-6
DSB SUMMARY

1. APR-94 1994 TOYOTA SPECIAL SERVICE TOOLS
2. JAN-95 1995 TSB BINDERS
3. SEP-89 20 STEPS TO SUCCESSFUL AUTO TRANSMISSION REPAIR
4. MAY-96 A/C COMPRESSOR MAINTENANCE FOR STORED VEHICLES
5. OCT-93 A/C COMPRESSOR OIL APPLICATION
6. NOV-94 A540E SECONDARY REGULATOR MODIFICATION
7. MAY-97 AIR CONDITIONING EVAPORATOR ODOR
8. MAR-95 AIRBAG ASSEMBLY REPLACEMENT PROCEDURE
9. SEP-93 ALIGNMENT SPECS - 94 YEAR MODELS
10. SEP-87 ALL AUTOMATICS - FRONT BUSHING WEAR
11. NOV-96 ALTERNATIVE REFRIGERANTS AND RETROFITS
12. JUN-93 APPLYING CAUTION LABEL TO R-12 AIR CONDITIONING SYSTEMS
13. MAY-95 ATM TRANSMISSION FLUID SEEPAGE
14. FEB-90 AUTOMATIC TRANSMISSION FLUID
15. AUG-89 AUTOMATIC TRANSMISSION MATH FORMULAS
16. OCT-89 AUTOMATIC TRANSMISSION MATH PART 2
17. DEC-96 AUTOMATIC TRANSMISSION SERIAL NUMBERS
18. JAN-95 AXLE NUT/LUG NUT TIGHTENING SPECIFICATIONS
19. FEB-95 BATTERY MAINTENANCE FOR IN-STOCK VEHICLES
20. SEP-97 BRAKE BOOSTER PUSH ROD GAUGE
21. JUN-94 BRAKE PAD KITS - NOW AVAILABLE WITHOUT HARDWARE
22. OCT-94 BRAKE REPAIR
23. FEB-94 BRAKES - CAUSE AND REPAIR OF VIBRATION AND PULSATION
24. DEC-96 CARPET CLEANING PROCEDURES
25. JAN-96 CHECKBALL WEAR
26. JUN-95 EMISSION CONTROL LABEL ORDER FORM
27. JUN-94 EMISSION CONTROL LABELS - ORDERING FORM
28. OCT-95 ENGINE SUPPORT BAR IMPROVEMENTS
29. OCT-90 ENGINE TESTING WITH A VACUUM GAUGE - AUTO TRANS

Look at this service bulletin summary for a 1994 Toyota Camry. Out of a total of 79 bulletins, only 13 covered a manufacturing defect.

30. JUN-96 EXHAUST PIPE HEAT SHIELD
31. MAR-96 FRONT BRAKE GROAN NOISE
32. SEP-93 FRONT STABLIZER - NEW LINK NUTS
33. AUG-95 GROOVED B CLUTCH DISC ON AS40E ATM
34. NOV-91 HOW TO USE A PRESSURE GAUGE - AUTOMATIC TRANS.
35. JAN-96 IMPROVED POWER WINDOW REGULATOR ASSEMBLY INSTALLATION
36. AUG-95 INTRODUCTION OF BRAKE SHIM/FITTING KITS
37. JAN-97 MAIN BEARING CHANGES
38. AUG-87 METAL SEALING RINGS - AUTOMATIC TRANSMISSIONS
39. APR-96 MOONROOF PANEL WIND NOISE
40. SEP-93 NEW PDS FOR 1994 MODEL YEAR VEHICLES
41. AUG-94 NEW WIRE HARNESS REPAIR KIT AVAILABLE
42. SEP-94 OVERDRIVE CLUTCH BEARING - IMPROVED DURABILITY
43. DEC-93 PAINT AND REFINISHING FORMULA CODES
44. SEP-95 PAINT REPAIRS ON POLYURETHANE BUMPER
45. OCT-94 POWER WINDOW REGULATOR - NEW PARTS OPTIONS AVAILABLE
46. SEP-93 PRE-DELIVERY SERVICE - SHORT PIN INSTALLATION (NON USA)
47. JUN-94 PRE-DELIVERY SERVICE - SPRING SPACER REMOVAL
48. NOV-94 PROTECTIVE FILM REMOVAL PROCEDURES
49. APR-96 PUBLICATION CORRECTION - GENERAL INFORMATION
50. MAY-95 PUBLICATION CORRECTION INFORMATION
51. JUN-93 R-12 CAUTION LABEL - APPLICATION INSTRUCTIONS
52. MAR-97 REAR BRAKE SQUEAK
53. SEP-93 REAR SUSPENSION - NEW MEMBER NUT
54. MAR-96 REFRIGERANT LEAK DETECTION - SERVICE HINTS
55. JUL-94 REPLACEMENT BATTERY PART NUMBERS
56. JUL-97 REPLACEMENT CERTIFICATION LABELS
57. MAR-94 REPLACEMENT REFRIGERANTS DAMAGE TO SYSTEM
58. JUL-97 REPLACEMENT VIN PLATES
59. OCT-96 SEAT BELT EXTENDER
60. APR-94 SEAT BELT EXTENDER - AVAILABILITY
61. NOV-95 SEAT BELT EXTENDERS
62. OCT-97 SEAT BELT EXTENDERS
63. NOV-94 SPOILER LAMP SENSOR REPLACEMENT
64. AUG-96 STAIN ON PAINT
65. JUL-95 STANDARD BOLT TORQUE SPECIFICATIONS
66. SEP-94 STEERING COLUMN CLICKING NOISE - CORRECTION PROCEDURE
67. MAY-96 STEERING GEAR/WHEEL REMOVAL WITH SRS
68. APR-95 SUSPENSION SQUEAK/GROAN NOISES
69. JAN-94 TIRES CHANGE IN ORIGINAL EQUIPMENT
70. AUG-94 TOYOTA CHECKER APPLICATIONS (REFERENCE GUIDE)
71. DEC-94 TRUNK FINISH PLATE LOOSE
72. OCT-95 TRUNK LID TORSION ROD ADJUSTMENT
73. NOV-93 WATER LEAKAGE - NEW SEALS
74. MAY-95 WATER PUMP SEAL IMPROVEMENTS
75. MAR-96 WHEEL BALANCE ADAPTER KIT
76. SEP-96 WHEEL BALANCE ADAPTER KIT UPDATE
77. JUL-94 WIND NOISE FROM WINDOW - NEW PARTS AVAILABLE
78. NOV-95 WIPER MOTOR CIRCUIT BREAKER INSPECTION
79. JUL-94 WOOD DASH - REPAIR PROCEDURES

Japanese vehicles are redesigned every three to four years, while American automakers often wait a decade or longer. The Big Three American automobile manufacturers know that the Japanese and their dealers build, market, and service vehicles to a higher standard than the Americans do. That's why the Big Three gave up much of the small-car market to Asian producers in the late '80s; it was easier and more profitable for them to buy high-quality Asian products and market them as homegrown. Today, the small, compact, and luxury car markets have been literally conquered by Japanese imports.

Don't buy the myth that parts for imports aren't easily found. It's actually easier to locate parts for used Japanese or domestic vehicles than for new vehicles, due to the large number of units produced, the presence of hundreds of independent suppliers, the ease with which relatively simple parts can be interchanged from one model to another, and the large reservoir of used parts stocked by junkyards. Incidentally, when a part is hard to find, the Mitchell Manual is a useful guide to substituting parts that can be used for many different models. It's available in some libraries, most auto parts stores, and practically all junkyards.

In my *Lemon-Aid New Cars*, the vehicles singled out as the most reliable and durable are Japanese makes, some European imports, and American co-ventures (such as the Chevrolet Metro, which is actually a re-badged Suzuki). But where used vehicles are concerned, the following American-made rear-drive cars are better choices than many imports: Chrysler's V8 New Yorker; Ford's Mustang, T-Bird, Cougar, Crown Victoria, and Mercury Grand Marquis; Lincoln's Town Car; and GM's Camaro, Caprice, and Cadillac Fleetwood. Their depreciated prices are more competitive than the inflated values of Asian and European imports, they're cheap and easy to service almost anywhere, they support lots of power-hungry accessories, and they sustain high mileage with fewer major breakdowns. The only problem is that American rear-drives have become an endangered species, with Ford being the only automaker still churning them out in large numbers.

South Korean cars are poor imitations of Japanese vehicles. They start to fall apart after their third year due to poor-quality body construction, cheap and unreliable electrical components, and parts suppliers who put low prices ahead of reliability and durability. This is particularly evident when one looks at the failure-prone Hyundai Excel.

Hyundai's Sonata. You know Hyundai has serious quality control problems when late-night talk show host Dave Letterman likens Hyundai reliability to the Russian space station Mir.

European cars and minivans deserve much of the bad-mouthing they've received over high parts costs and limited part availability. With some European models, you can count on lots of aggravation and expense due to the unacceptably slow distribution of parts and their high markup. Because these companies have a quasi-monopoly on replacement parts, there are few independent suppliers you can turn to for help. And auto wreckers, the last-chance repository for inexpensive car parts, are unlikely to carry foreign parts for vehicles older than three years or manufactured in small numbers.

Other Buying Considerations

Front-wheel drive

Front-wheel drives direct engine power to the front wheels, which *pull* the vehicle forward, while the rear wheels simply support the rear. The biggest benefit of front-wheel drive (FWD) is foul-weather traction. With the engine and transmission up front, there's lots of extra weight pressing down on the front-drive wheels, increasing tire grip in snow and on wet pavement. However, when driving up a steep hill or towing a boat or trailer, the weight shifts and there's no longer a traction advantage.

Although I recommend a number of FWD vehicles in this guide, I don't like them. Granted, front-wheel drives give a bit more interior room (no transmission hump), provide more carlike handling, and provide better fuel economy than rear-drives. But damage from potholes and fender-benders is usually more extensive, and maintenance costs (premature front tire and brake wear, in particular) are much higher than with rear-drives.

Servicing front-wheel drives can be a real nightmare. Entire steering, suspension, and drivetrain assemblies must be replaced when just one component is defective. Downtime is considerable, the cost of parts is far too high, and the drivetrain and its components are not designed for the do-it-yourself mechanic. A new FWD transmission assembly, called a transaxle, can cost about $2,000 to repair, compared to $700 for a rear-drive transmission.

Accident repairs are a unique problem. FWD transmissions and steering and suspension components are easily damaged and alignment difficulties abound. Repair shops need expensive, specialized equipment to align all four wheels and square up a badly smashed unibody chassis. Even if you can get all four wheels tracking true, the clutch and transaxle can still be misaligned. No wonder many insurance companies prefer to write off a FWD car than repair it. And when that happens, you wind up eating the difference between what you paid for the vehicle and what the insurance company says your vehicle is worth—minus your deductible, of course.

I also feel front-wheel drives are unsafe because braking requires a whole new set of reflexes than what we've developed over the years with rear-wheel drive vehicles. You remember the routine from driving school: *When in a skid, pump the brakes lightly and turn into the direction of the skid.* Well, if you pump the brakes of a FWD in a skid, you've bought the farm.

Rear-wheel drive

Rear-drives direct engine power to the rear wheels, which *push* the vehicle forward. The front wheels steer and support the front of the vehicle. With the engine up front, the transmission in the middle, and the drive axle in the rear, there's plenty of room for larger and more durable drivetrain components. This makes for less crash damage, lower maintenance costs, and higher towing capacities than front-wheel drives.

On the other hand, rear-drives don't have as much weight over the rear wheels as FWDs do (no, putting cement blocks in the bed or trunk will only void your transmission warranty). As such, they can't provide as much traction on wet and icy roads unless equipped with an expensive traction-control system.

Four-wheel drive

Four-wheel drives direct engine power through a transfer case to all four wheels, which *pull and push* the vehicle forward, giving you twice as much traction. The system is activated with a floor-mounted shift lever or a dashboard button. When 4X4 drive isn't engaged, the vehicle is essentially a rear-drive truck. The large transfer case–housing makes the vehicle sit higher, giving you additional ground clearance. Two-wheel rear-drive vehicles have some of the highest payload and towing capabilities, usually slightly more than four-wheel drives.

Many 4X4 customers have been turned off by the typically rough and noisy driveline; a tendency for the vehicle to tip over when cornering at moderate speeds (a Bronco specialty); vague, trucklike handling; high repair costs; and poor fuel economy. Also, extended driving over dry pavement with 4X4-drive engaged will cause the driveline to bind and result in serious damage. Some buyers are turning instead to rear-drive pickups equipped with winches and large, deep-lugged rear tires.

All-wheel drive

Essentially, this is four-wheel drive *all the time.* Used mostly in sedans and minivans, all-wheel drive (AWD) never needs to be deactivated when running over dry pavement and doesn't require the heavy transfer case (although some sport-utilities and pickups do use a special transfer case) that raises ground clearance and cuts fuel economy. AWD-equipped vehicles aren't recommended for off-roading because of their lower ground clearance and fragile driveline parts, which aren't as rugged as four-wheel drive components.

Diesels

Diesel engines become more efficient as the engine load increases, whereas gasoline engines become less so. This is the main reason diesels are best used where the driving cycle includes a lot of city driving, with slow speeds, frequent stops, and long idling times. At full throttle, both engines are essentially equal from a fuel-efficiency standpoint. The gasoline engine, however, leaves the diesel in the dust when it comes to high-speed performance. Many owners of diesel-equipped vehicles are frustrated by excessive repair costs and poor road performance on vehicles that lack turbo. Bear in mind that before the fuel savings can outweigh the high cost of a diesel purchase, the average owner would have to drive 40,000–50,000 kilometres per year.

Fuel economy fantasies

Fuel economy figures are supplied by the federal government and are based on data supplied by the automakers following U.S. Environmental Protection Administration (EPA) testing guidelines approved by Transport Canada. These figures are often off by 10–20 percent due to the testing method chosen. In fact, a recent Ford bulletin warns dealers: "Very few people will drive in a way that is identical to the EPA tests....These [fuel economy] numbers are the result of test procedures that were originally developed to test emissions, not fuel economy."

Rustproofing

Even though automakers have extended their own rust warranties up to seven, in some cases ten, years, it's a good idea to get your car or minivan undercoated each year and keep it in an unheated garage. A full rustproofing job isn't really needed. In fact, you have a greater chance of seeing your rustproofer go belly-up before he makes good on your warranty than of having your untreated vehicle ravaged by premature rusting. Even if the rustproofer stays in business, you're likely to get a song-and-dance about why the warranty won't cover so-called "internal" rusting or why repairs will be delayed until the sheet metal is actually rusted through. If you live in an area where roads are heavily salted in winter or in a coastal region, or if you own a Toyota truck or Ford sport-utility, make sure you regularly have undercoating sprayed on the rocker panels (door bottoms), the rear hatch's bottom edge, and the wheel wells.

"Recommended Beaters" and collectibles

It's funny—in Ontario and Quebec, you are what you drive, and owning a "beater" is seen by others as an admission that you're too poor to own or lease a relatively new car. On the other hand, Maritimers and western Canadians who drive beaters are considered frugal and resourceful individuals. Consequently, there are fewer and fewer garages and parts suppliers in Ontario and Quebec that specialize in repairing older cars and scurrying around the country hunting for ten-year-old parts. But when you head west or east, you immediately notice the popularity of vehicles 10 to 20 years old and the surprisingly large number of independent garages that service them.

My take on this is that the Maritimes' distressed economy forces motorists to buy the least expensive vehicles and keep them for as long as they can, while western Canadians embrace beaters because of the kinder Prairie weather (from a corrosion standpoint) and family tradition. Families in Manitoba, Saskatchewan, and Alberta view their cars and trucks as appliances—not phallic symbols. (B.C. is altogether different, as is its wont.) Their pragmatic repair or recycle disposition and large rural families mean that an old but functional car or truck will be passed down through generations of the same family. Furthermore, western Canadians prefer cars and trucks with rear-wheel drive for winter driving.

As I said before, British Columbians are an automotive breed apart: they aren't forced to be frugal, but the wide availability of well-preserved used cars is too tempting to resist, even at prices that are outrageously inflated. Consequently, you'll find in B.C. lots of classic cars, convertibles, and European imports, along with plenty of garages to service them. Interestingly, B.C. is perhaps the only place where the sun hasn't set on the British automotive empire. (Quick, why do Brits drink warm beer? Because Lucas makes their refrigerators.)

It's getting pretty hard to find a vehicle 10 years old or more that's safe and reliable. Personally, I'd be reluctant to buy any decade-old vehicle anywhere east of Manitoba. Apart from salt being a real body killer, it's just too easy to fall prey to scam artists who cover up major mechanical or body problems resulting from accidents or environmental damage.

Nevertheless, if an independent mechanic says that the more common and costly defects have been repaired, and if you have the time, knowledge, and parts suppliers to do your own maintenance and repairs, you may find an old, beat-up-looking vehicle that runs well (most of the time) in the $300–$1,000 range. Look for one of the following recommended vehicles, avoid the non-recommended ones, take your time, and insist upon an independent check-up. Be wary of the trouble areas listed in parentheses.

If you have a bit more money to spend and want to take less of a risk, look up the recommended vehicles found at the beginning of each vehicle category in Part Three.

Recommended Beaters (Cars to Get You Through the Next Millennium)

Checker—Last made in 1982, the Checker is the poor man's Bentley. Its rear-drive configuration, sturdy, off-the-shelf mechanicals, and thick body panels all add up to bulletproof reliability. Beware of corrosion along door rocker panels and wheel wells.

Chrysler—Dart, Valiant, Duster, Scamp, Diplomat, Caravelle, Newport, rear-drive New Yorker Fifth Avenue, and Gran Fury (engine, electrical system, suspension, brakes, body and frame rust, plus constant stalling when humidity is high).

The Caravelle, Diplomat, and New Yorker Fifth Avenue are reasonably reliable and simple-to-repair throwbacks to a time when land yachts ruled the highways. Powered with 6- and 8-cylinder engines, they will practically run forever with a minimum of care. The fuel-efficient "slant 6" powerplant was too small for this type of car and was changed to a gas-guzzling, but smooth and reliable, V8 after 1983. Handling is vague and sloppy, though, and emergency braking is often accompanied by rear-wheel lockup. Still, what do you want for $500–$2,500 for a 1984–89 "retro rocket"?

Overall reliability is fairly good, and inexpensive parts are available anywhere. Carburetor, ignition, electrical system, brake, and suspension (premature idler-arm wear) problems predominate. It's a good idea to adjust the torsion bars frequently for better suspension performance. Doors, windshield pillars, the bottoms of both front and rear fenders, and the trunk lid rust through more quickly than average.

Chrysler's 1991–93 2000GTX is an above average buy that may cost $4,000–$6,000, depending upon the model year. It's a reliable Japanese-built sedan that was discontinued in 1994. It has a competitive price, modern styling, and high-performance options that put it on par with such benchmark cars as the Honda Accord and Toyota Camry. The base engine provides uninspiring yet adequate performance, but this car is distinguished by its high-performing optional powerplant—a 16-valve version that's smooth and powerful. The manual transmission and clutch are exceptionally smooth. Handling is acceptable in all situations. Excellent braking with anti-lock brakes, but mediocre without. The ride is firm but fairly comfortable on most roads, especially with the adjustable suspension. Interior room is generous for four people and you can carry a fifth passenger in a pinch (in the literal sense).

There are some reports of automatic transmission failures around the 60,000 km mark. The front brakes need more attention than average. The car hasn't been crash-tested.

If you can get past the two-tone paint job, Chrysler's rear-drive Fifth Avenue (shown above), Caravelle Salon, and Diplomat Salon are excellent buys. They have high performance, reliable powertrains and lots of room. Weak spots: brakes, electrical shorts, and ignition glitches.

Ford—Maverick, Comet, Fairmont, Zephyr, Tracer, Mustang, Capri, Cougar, Thunderbird V6, Torino, Marquis, Grand Marquis, LTD, and LTD Crown Victoria (trunk, wheel well, and rocker panel rusting, brakes, steering, and electrical system). The 1991–93 Festiva can also be a good buy as long as you check out the brakes, exhaust system, and body panels for rust and paint; stay away from the 1989–90 versions.

General Motors—Chevette and Acadian (steering, and brakes you have to stand on to stop); rear-drive Nova, Ventura, Skylark, and Phoenix (suspension); front-wheel drive Nova and Spectrum; Camaro, Firebird, Malibu, LeMans, Century, Regal, Cutlass, Monte Carlo, and Grand Prix (suspension and brakes); Bel Air, Impala, Caprice, Laurentian, Catalina, Parisienne, LeSabre, Bonneville, and Delta 88 (suspension and brakes).

Nissan—The Micra is a subcompact commuter car that was sold 1985–91. It uses generic Nissan parts that are fairly reliable and not difficult to find, except for body panels that are more problematic. (Electrical shorts, premature front brake wear, and body rusting along the door rocker panels and wheel wells are the more common deficiencies).

The 1991–95 Axxess is an excellent small minivan that's fairly reliable and easily serviced (electrical system, brakes, manual transmission, and resonator).

Pulsar and NX models are good small-car buys ($500–$5,000, depending upon the year), as long as you pick the right years and stay away from failure-prone and expensive-to-repair turbo models. The Pulsar was replaced by the 1991 NX, a similar small car that also shares Sentra components. For 1983–86 models, overall reliability is poor to very poor. As with other discontinued Nissans, the 1987–93 models have shown remarkable performance improvement and are all the more attractive due to their depreciated prices. Crash test scores were below average for the 1990 and earlier models, while the 1991 and later versions scored quite well. All Pulsars built before 1991 are prone to premature wear on the front brake pads and discs and faulty air conditioners. From 1991 on, the only problems reported concern minor AC malfunctions, premature wearout of front brakes and suspension components, and exhaust systems that don't last very long (two years, tops). Complaints of premature rusting and paint peeling are legion for the 1983–90 models.

Subaru—Sold 1988–95, the Justy does everything reasonably well for an entry-level Subaru. Pairing smooth and nimble handling with precise and predictable steering, the 4X4 system is a boon for people who often need easy-to-engage extra traction and an automatic transmission. Prices range $500–$4,000.

The Justy's engine is anemic, harsh, and noisy. There are annoying steering wheel vibrations caused by the rough-running engine. The non-power-assisted steering requires maximum effort when parking. The ride is uncomfortable when carrying a full load. You lose some steering stability with hard braking, and cornering at high speeds feels tippy on the 4X4. Transmission problems aren't easily diagnosed or repaired due to the transmission's complicated construction. Front shoulder belts are uncomfortable and rear seatbelts are hard to buckle up. The rear seat is for children only. The heater is insufficient and air distribution is inadequate. Trunk space is less than you would expect.

The Justy's reliability record has been about average from the 1991 model onwards. Some owners complain about poor engine idling and frequent cold weather stalling, manual and automatic transmission malfunctions, premature exhaust system rust-out, catalytic converter failures, and paint peeling. With the exception of the CVT, servicing and repairs are made easy due to a very straightforward design.

Toyota (late '80s and early '90s)—All models, except the LE van (brakes, chassis, and body rusting). Chassis rusting and V6 engine head gasket failures are common problems with the 1988–94 sport-utilities and pickups (Toyota will pay for the engine repairs up to 8 years).

Volvo—The 1989–93 240 series is an average buy costing $3,000–$6,000. Problems with entry-level Volvos like the 240 mirror the quality control problems manifested by Saab, the other Swedish automaker, except for the fact that Volvo styling is more bland than bizarre. The 240 is a solid and spacious car. Unfortunately, it doesn't live up to Volvo's advertising as an automotive longevity wonder.

The V6 is an honest, though imperfect, engine. Avoid the turbocharged 4-cylinder engine and failure-prone air conditioning systems. Diesels suffer from cooling system breakdowns and leaky cylinder head gaskets. The brakes on all model years need frequent and expensive servicing, and exhaust systems are notorious for their short life span. The GL and GLE suffer from occasional electrical bugs.

"Beaten" Beaters (Cars That Won't Get You Through the Next Month)

American Motors—Hornet, Gremlin, Concord, Spirit, Pacer, and Eagle 4X4 (faulty engines, transmissions, and steering).

Audi—Fox, 4000, 5000 (engine, transmission, and fuel system).

British Leyland—Austin Marina, MG, MGB, and Triumph (electrical system, engine, transmission, clutch, and chassis rusting).

Chrysler—all years—Cricket, Omni, Horizon, Volaré/Aspen (engine, brakes, steering, chassis rusting); Charger, Cordoba, and Mirada (brakes, body, and electrical system).

The 1985–89 Lancer and LeBaron GTS may only cost $1,000–$2,000, but they're no bargain. In fact, they suffer from many of the same problems as the Aries and Reliant K cars and their 1989 replacements, the Spirit and Acclaim. Although the interior will seat five comfortably and offers all the advantages of the hatchback design, road performance is merely passable with the standard engines and suspension.

Poor reliability causes maintenance costs to mount quickly. The manual transmission is balky and its clutch has a poor durability record. Turbo models are especially risky buys at all times. Head gaskets are prone to leaks, on all engines. Shock absorbers, MacPherson struts, and brakes wear out quickly. Front brake rotors are prone to rusting and warping. The electrical system is troublesome: the distributor pickup and computer modules malfunction constantly, causing stalling and hard starting. Air conditioning components have a short life span, and body hardware is fragile. Crash test scores are below average.

Although they don't cost much—$500–$2,000, depending on the year—steer clear of 1983–89 Aries or Reliants. Uncomplicated mechanical components and a roomy interior made these cars attractive buys when new, but once in service they quickly deteriorate. Both cars use dirt-cheap, low-tech components that tend to break down frequently. They have also performed poorly in crash tests.

All cars have a problem with idle shake caused by the transverse-mounted 4-cylinder engine, as well as driveability problems, especially in cold or damp weather. The "gutless" 2.2L Chrysler-built 4-cylinder motor is a bit more reliable than the 2.6L Mitsubishi engine, which has a tendency to self-destruct. It also suffers from premature wear of the timing belt guide and piston ring, and leaky camshaft oil plugs. The 2.2L powerplant has multiple problems, the main ones being a weak cylinder head gasket and timing belt failures. The turbocharged version often requires expensive repairs. The MAP sensor fails frequently and is costly to replace.

Carbureted engines are renowned for stalling and poor driveability. Exhaust systems, wheel bearings, and air conditioning components aren't durable. Water infiltration around the windshield and into the trunk is common. Serious corrosion generally starts along the trunk line and the edges of the rear wheel wells. Perforations tend to develop along the bottom door edges, around the windshield, and on the floor.

Datsun/Nissan—210, 310, 510, 810, F-10, and 240Z (electrical system, brakes, and rusting).

Eagle—Medallion, Monaco, and Premier. These bargain-priced French imports—$700–$900 for the Medallion and $1,200–$3,500 for the Premier—had a 1988–1992 model run. Sold through Chrysler's Renault connection, they are two of the most failure-prone imports to ever hit our shores. Powertrain, fuel system, electrical system, AC, suspension, and brakes are the worst offenders. The '91 and '92 Monaco and Premier are covered by a 10-year ABS brake warranty.

Fiat—"Fix it, again, Tony." All Fiat models and years are known for temperamental fuel and electrical systems and biodegradable bodies. Alfa Romeos have similar problems.

Ford—Cortina, Pinto, Festiva, Fiesta, Bobcat, and Mustang II (electrical system, engine, and chassis rusting); fire-prone Pinto/Bobcat are mobile Molotov cocktails.

The German import Fiesta and the Festiva built in South Korea are two small imports that only survived a few years in Canada. Parts are practically unobtainable for both vehicles. The 1978–80 Fiesta was a fairly well-made, inexpensive small car that's now worth $300–$500. Highway driving isn't recommended due to its weak engine and brakes. The slightly larger and more refined 1988–90 Festiva is an $800–$2,000 subcompact that's two feet shorter than the

Escort, and weighs about 1,700 lb. Acceleration is slow, but the manual transmission is precise and easy to use. The post–'90 Festiva is only adequate for city use and with one passenger at best. Floorboards, body panels, and exhaust components quickly rust out, brakes and electrical components frequently need replacing, and the tall body and tiny tires make highway cruising a scream. Crash tests of an '89 Festiva showed the driver and passenger would sustain only minor injuries, however.

Imagine, even rich people suffer lemons (Saabs, Jaguars, and front-drive Lincolns and Cadillacs are the main culprits). A Cadillac "lemon"? Sure, just think of the Cimarron and Allanté (above). Can the Catera be far behind?

General Motors—Vega, Astre, Monza, and Firenza (engine, transmission, body, and brakes); Cadillac Cimarron and Allanté (poor-quality components and overpriced); all front-wheel drives (engine, automatic transmission, electronic modules, steering, brakes, and rust/paint peeling); Citation, Skylark, Omega, and Phoenix (engine, brakes, electronic modules, and severe rust canker).

The Pontiac Fiero, sold 1984–88, snares lots of unsuspecting first-time buyers through its attractive, sports-car styling, high-performance pretensions, and $700–$1,500 price. However, one quickly learns to both fear and hate the Fiero as it shows off its fiery disposition (several safety recalls) and "I'll start when I want to" character.

Buick's Reatta, a luxury two-seater sold only 1988–91 (present value: $12,000–$15,000 for the convertible), has higher-than-average maintenance costs, although repairs aren't that dealer dependent, except for the electronically controlled fuel system. The Reatta also has higher-than-average parts costs and body parts are tough to find. GM improved the quality somewhat with its 1991 models, but then discontinued production.

The Geo Storm was GM's Japanese-made small car that only had two model years in Canada (1992–93). Owners report serious body hardware deficiencies (not paint or rust, however), in addition to brake, exhaust, electrical, and ignition problems. Prices vary between $3,500 and $5,000, depending on the model chosen.

Hyundai—Pony and Stellar (electrical system, brakes, poor-quality body components, and no parts). Hyundai's first imports into Canada during the early '80s, these cars were shooting stars that captured lots of sales during their first few years on the market—and then crashed once word got around that they were unsafe and unreliable.

Jaguar—All models (poor body assembly/paint, electrical system, and service).

Lada—These cheap Russian imports ($500–$4,000 for 1990–97 models) incorporate a 20-year-old rear-drive design and use poor-quality body and mechanical components. They're noisy and truck-like on the road, and the dealer network is sparse, making parts hard to get.

Defective fuel and ignition systems (particularly the ignition control module) and a poorly designed electrical system are the most common problems. Frequent carburetor replacements, at $325 plus tax, are rendered necessary due to the butterfly shaft warping or corroding—the result of an inherently defective design. Front brakes need to be replaced often, and transmissions, engine head gaskets, and differentials aren't durable either. Rear window defogger wires corrode easily. Ladas are poorly rustproofed and paint quality is bad, which translates into rapid surface rusting and sheet metal perforation. They've also never been crash-tested.

Lincoln—Versailles (electrical system, suspension, and rusting) and early front-wheel drive Continentals (transmission, electrical system, brakes, and fuel system).

Passport/Isuzu—The I-Mark, Stylus, and Optima compacts were sold from 1988 until 1992 and can be bought for $500–$2,000, depending upon the year and model. Repair costs are higher than average due to the cars' mediocre reliability and the difficulty in finding parts at a reasonable cost. 1988–89 models garnered below-average crash test scores, while the last three model years (1990–92) did quite well.

Renault—All models (fuel and electrical systems, CV joints and brakes, no parts, and few mechanics).

Saab—All models (electrical fires, short circuits, hard-to-find parts, and few competent mechanics).

Skoda—120/130 Series: the best of a bad lot of Eastern European imports (Dacia, Lada, Yugo, etc.).

Volkswagen—Beetle (unsafe front seats, heater that never worked, hazardous fuel tank placement, wheels, and seat tracks); Camper minivan (engine, transmission, fuel system, and heater); Rabbit/Dasher and 411/412: these cars are full of surprises, parts are rare, and the slow and expensive dealer servicing would give anyone nightmares (engine, electrical system, front brakes, and cooling system).

VW's Camper minivan: all the comforts of home during those long roadside waits for Helmut and Franz to come along with a tow truck.

Sold 1987–93, and now costing $500–$2,000, the Fox, with its old design, can't offer the mechanical or interior efficiency that newer small cars provide. Other failings are the absence of power steering and an automatic transmission. It's true that these cars are fun to drive, but they have a checkered reputation as far as reliability and durability are concerned. Furthermore, as time goes on, these cars become more troublesome and difficult to repair. Crashworthiness, as tested by NHTSA, is way below average.

Owners report common failures with engine timing belts, starter motors, the heating system, and brake cylinders. Other problem areas include electrical, engine cooling, and fuel systems. The manual transmission shift linkage needs frequent adjustment, and the front brakes and shocks, as on most VWs, need more frequent servicing than those of comparable makes.

VW's Scirocco is fun to drive, but risky to own. You'll like its nimble handling, strong, responsive engine, and tight steering. You'll like a lot less the car's chronic breakdowns, parts shortages, and poor crashworthiness. Electrical short circuits, chronic fuel supply problems, premature front brake wear, and fragile body parts are common owner complaints. Sold 1985–89, expect to pay $500–$2,500, depending upon the model year.

Collectibles?

Be wary of investing in recent high-performance American and Japanese sports cars. For example, GM took the Corvette ZR-1 off the market due to poor sales after only four years of production.

Although 3,000 vehicles were to be produced annually, that figure dropped to fewer than 500 units. Owners who paid $85,307 for a new 1990 ZR-1 now own a car worth less than $20,000.

Discontinued Japanese sports cars have fared no better, with the possible exception of the limited-production 1965–70 Toyota 2000 GT. Datsun's 1961–69 1500/1600/2000 sports cars, for example, were better-made, more reliable, and better-equipped than the MGBs and MGAs, yet the English roadsters sell for about $15,000 (U.S.), while Nissan's barely go for a third of that price. The 240Z/260Z/280ZX used values decreased in direct proportion to their increased heft. What's the outlook on Nissan's discontinued 1990–96 300ZX? More of the same.

If you're seriously interested in collecting cars, remember that most vehicles jump in value after their twentieth year on the market. So you probably won't lose money if you buy a car of this vintage, especially if it's a convertible (trucks, sport-utilities, and vans are riskier investments). Shop the Internet and car shows—they're also good places to contact wholesalers and restorers. Car clubs also offer a wealth of information, and are regularly listed in major auto magazines like *AutoWeek*, *Car and Driver*, *Motor Trend*, and *Road and Track*. *Hemmings' Motor News* and *Old Cars Weekly* out of Iola, Wisconsin, are two other excellent sources of collector news.

Saving Money and Keeping Safe

You can get a good used vehicle at a reasonable price—it just takes lots of patience and homework. You can further protect yourself by becoming thoroughly familiar with your legal rights as outlined in Part Two and buying a vehicle recommended in Part Two. Following is a summary of the steps to take to keep your risk to a minimum.

Summary

1. Keep your vehicle for at least 10 years.
2. Trade in your vehicle if the dealer's sales tax reduction is more than the potential profit of selling privately.
3. Sell to a private party (10 percent premium).
4. Buy from a private party (10 percent savings).
5. Use an auto broker to save time and money.
6. Buy a *Lemon-Aid*-recommended vehicle for depreciation, parts, and service savings.
7. Buy a three- or four-year-old vehicle with lots of original warranty that can be transferred (35–50 percent savings over a new vehicle).
8. Choose a vehicle that's crashworthy and cheap to insure.
9. Carefully inspect Japanese-built vehicles that have reached their fifth year (engine head gasket, CV joints, steering box, and front brakes).

10. Don't buy an extended warranty ($500–$1,000 savings) unless it's recommended in this guide.
11. Have non-warranty repairs done by independent garages offering lifetime brake and exhaust repairs (50 percent savings).
12. Install used or reconditioned mechanical parts (30–50 percent savings).
13. Keep all the previous owners' repair bills to facilitate warranty claims and help mechanics know what's already been replaced or repaired.
14. Upon delivery, adjust mirrors to eliminate blind spots and adjust head restraints to prevent your head from snapping back in the event of a collision. On airbag-equipped vehicles, move the seat backwards more than half its travel distance and sit at least 10 inches away from the airbag housing. Make sure the spare tire and tire jack haven't been removed from the trunk. Get shoulder/lapbelts for the rear seats; lapbelts alone can be deadly.
15. Make sure the dealer and automaker have your name in their computers as the new owner of record. Ask for a copy of your vehicle's history, stored in the same computers.

Japanese vehicles that have a reputation for bulletproof reliability, like the '96 Honda Civic (shown above), are usually way overpriced. Look instead for a Japanese product sold by an American automaker.

Part Two
THE ART OF COMPLAINING

Chrysler Battles Paint Blemishes by Banning Deodorant

"Paint-line workers at Chrysler Corp.'s Jeep plant were asked to stop using antiperspirant after the company discovered that falling flakes left costly blemishes on the new Jeep...General Motors officials contacted late yesterday said they didn't believe they had a similar problem....

One woman filed a grievance last year after her supervisor asked to check her armpits...."

Newsday, 1991

Ford Admits Its Paint Mistake

"In June 1990, a field survey of about 1,000 F-series trucks (1985–1990 models) was conducted in three locations by Body and Chassis Engineering to assess paint durability. Results showed that about 13% of the F-series trucks displayed peeling paint, which would represent about 90,000 vehicles annually...."

memo to members of the Finance Committee from
A. J. Trottman, Executive Vice President
Ford North American Automotive Operation

Nick Verburg, a British Columbia hauler, in front of his problem-plagued 1991 Dodge Ram truck. His protests succeeded in getting five Getrag transmission replacements—one lasted almost 40 kilometres.

Invest in Protest

Few people really like to complain and even fewer are willing to make their dissatisfaction public. However, with 10 percent of cars estimated by Runzheimer Consultants to be "lemons," and automakers promising more quality than they deliver and then cutting back on warranty coverage, buyers have discovered that they must either "act up" or get shut out. Is it any wonder that Canadian consumers, such as Nick (previous page), are banding together into protest groups and slapping lemon signs on their vehicles to get a fair shake?

Chrysler responds

The 1998 edition of *Lemon-Aid Used Cars* blasted Chrysler Canada for stonewalling owner complaints concerning powertrain, brake, and paint deficiencies that affect Chrysler's entire 1986–97 production. As Chrysler owner complaints continued unabated throughout the fall of 1997, reporters from three different newspapers (Steve Mckinley of the Saint John, New Brunswick, *Times Globe; Toronto Sun*'s business and consumer columnist, Maryanna Lewyckyj; and *Vancouver Sun*'s business columnist, Gillian Shaw) called me for comments on stories they were writing prompted by Chrysler owner complaints.

After speaking with the reporters and a cross-section of Chrysler claimants, I faxed a protest letter to Chrysler Canada President Yves Landry. Three days after receiving the fax, two Chrysler vice presidents agreed to meet me on January 22, 1998, during the Montreal Auto Show.

At the Montreal meeting, former Chrysler Parts, Service, and Engineering Vice President Robert Renaud and Public Affairs Vice President Othmar Stein agreed that Chrysler did, indeed, have some quality control problems (this was a refreshing change from other automaker negotiations I've endured where the first day is wasted with incessant denials that any problem exists) and that they were prepared to "do whatever it takes" to satisfy any outstanding complaints, whether the vehicle was bought new or used or had an expired warranty.

I told both gentlemen that I would ask the Chrysler owners to band together regionally into CLOGs and channel their complaints to the company through Robert Renaud. Renaud agreed to take the calls, but the idea of the owners getting together didn't meet with a great deal of enthusiasm. I argued, however, that it was the only way to have a coherent set of demands, involve the owners directly in the efficient resolution of their complaints, and monitor Chrysler's commitment to fairness. We also agreed to meet again, three weeks later at the Toronto Auto Show, to review how the complaint resolution process was or was not working.

At the Toronto meeting, I met with the same two vice presidents and told them there were apparently 600 complaints outstanding;

however, the team of customer advisors set up by Renaud (bypassing Chrysler's customer relations "hotline," which many owners felt was ineffectual) was successfully whittling that number down. I then gave Renaud a set of powertrain, paint, and brake compensation benchmarks that I felt the company should use in compensating owners. Judging by the cases settled, Chrysler isn't straying far from the durability parameters I gave them. The durability chart that includes those benchmarks, which can be used with any automaker, is found on pages 78–79.

A year later, on February 17, 1999, I met with Chrysler officials a third time. Those present included Othmar Stein, VP, Chrysler Public Affairs; Larry Latta, VP, Chrysler Sales and Marketing; and Lou Spadotto, Chrysler's National Service Manager of Parts, Service, and Engineering. Stein led off the discussion by introducing Latta and Spadotto as the Chrysler executives mandated to oversee the series of measures Chrysler has introduced to improve owner relations and ensure customer complaints are handled fairly and expeditiously. Those measures include:

- The continuance (overseen by Latta and Spadotto) of the Owner Review Committee to ensure that owners may appeal warranty decisions within the Chrysler Owner Relations organization. This committee, first set up by Renaud, who has since been promoted, will still use 519-973-2300 as its primary contact number for Chrysler vehicle owners residing in Canada who wish to have their claims reviewed.
- The committee will continue to handle owner complaints relating to automatic transmission, paint delamination, and brake malfunctions on a case-by-case basis for goodwill assistance, even if the claim has been previously refused, the base warranty has been exceeded, or the owner bought the car used.

Spadotto confirmed that no claim would be automatically rejected because the vehicle owner had repairs done by an independent repair facility. He promised that many other mitigating factors would be weighed before the committee made a final decision to accept or reject a claim.

Owners seeking assistance in the future should initially work through their dealer service manager, request help from a Chrysler regional team representative, or call the Chrysler customer assistance toll-free number, Latta emphasized. Only if these attempts prove unsatisfactory should owners make use of the Review Committee.

Latta stated that he and Spadotto jointly deal with owner complaints involving component failures. Failures that are allegedly safety related, subrogated insurance claims, and cases that have been filed in small claims and higher courts are routinely referred to Chrysler legal affairs specialists. Spadotto later provided documentation relating to Chrysler's 10-year ABS recall/free brake service program and seven-year air conditioning evaporator extended warranty covering the 1993–97 Grand Cherokee, Concorde, Vision, Intrepid, LHS, and New Yorker.

Since that last meeting, Chrysler paint and automatic transmission complaints have subsided. Nevertheless, the company's customer complaint handling will be closely monitored during 1999, and Chrysler's warranty performance will be reviewed in upcoming *Lemon-Aid* guides.

Although CLOG B.C. is still assessing Chrysler's handling of their members' complaints, it's apparent that Chrysler Canada has been both reasonable and efficient in settling complaints that were brought up for review by individuals and the CLOG groups.

Ford responds

Lemon-Aid Used Cars 1999 targeted Ford of Canada for its denial of claims by owners of 1992–96 Tauruses, Sables, and Windstars who had experienced 3.8L V6 engine head gasket and automatic transmission failures. Subsequently, Ford formally extended the warranty on head gasket claims to up to 5 years/100,000 km (letters were sent out under the company's Owner Notification Program) for 1994–95 Tauruses, Sables, and Windstars, and pledged to review on a case-by-case basis all engine head gasket and automatic transmission claims (principally, forward clutch piston failures) not fitting the above parameters.

When I met with Ford executives a second time, in February 1999, they flatly refused to formally extend the warranty relating to inadequate automatic transmission planetary gear lubrication, clutch slave cylinder, and Villager engine exhaust manifold stud failures, preferring to treat post-warranty claims on an individual basis. At that same meeting, they were very forthcoming in providing me with information on their free fixes available under a number of Owner Notification Programs on late-model cars, trucks, and minivans. These free fixes cover automatic transmission downshifts, front coil springs, engine head gaskets, engine intake manifolds, engine mount bolts, engine knocking, engine block heaters, seatbelt buckles, fuel tank stops, fuel hose replacement, cooling fans, GEM modules, cooling system by-pass kits, rear brake shields, and air cleaners (see following charts).

"M"
EXTENDED WARRANTY PROGRAMS (AS OF 2/23/99)

PROGRAM #	AFFECTED COMPONENT	VEHICLE LINE	PARAMETERS
98M04	Front Coil Springs	All 1995–1997 4cyl. Contour/Mystique	Through December 31, 2004
98M02	Front Coil Springs	Certain 1993 Taurus/Sable	December 31, 2001, regardless of distance
98M01	Headgasket 3.8L Engine	Certain 1994–1995 Taurus/Sable 3.8L engine *1994 Continental 1995 Windstar	5 years or 100,000 km, whichever occurs first *6 years or 120,000 km, whichever occurs first
97M91	Intake Manifold	Certain 1996–1997 Crown Victoria (taxi), Thunderbird, Cougar, Mustang 4.6L Engine	7 years unlimited distance
97M90	Seat Belt Buckles	Certain 1989–1990 Continental Certain 1989 Taurus/Sable	12 years, regardless of distance travelled
96M89	Engine Knock	Certain 1995–1996 Villager	7 years or 160,000 km, whichever occurs first

Owners seeking assistance or compensation for any of the above problems should call the Ford assistance toll-free number at: 1-800-565-3673 (FORD). Since Ford has designated specific members of its staff to deal with engine head gasket and transmission forward clutch piston problems, it would be helpful if owners identify one of the above as their problem, or mention they were referred by Phil Edmonston, at the beginning of their call.

"B"
EXTENDED WARRANTY PROGRAMS (AS OF 2/23/99)

PROGRAM #	AFFECTED COMPONENT	VEHICLE LINE	PARAMETERS
98B37	Transmission Downshift	Certain 1999 Windstar 3.8L	Through July 31, 1999
98B36	Engine Mount Bolts	Certain 1999 Windstar 3.8L	Through July 31, 1999
98B35	Fuel Tank Stop	Certain 1999 F-800 (33KGVW)	Through July 31, 1999
98B32	Rear Brake Shields	Certain 1999 F250-550 SuperDuty 4x4 Trucks	April 30, 1999, regardless of distance
98B31	Fuel Hose Replacement	Certain 1992–1998 B-series chassis 5.9L	November 30, 1999, regardless of distance
98B30	Cooling Fan	Certain 1998 Contour/Mystique	April 30, 1999, regardless of distance
98B29	GEM Module	1999 F250/350 Super Duty 4x4	Through July 31, 1999
98B26	Cooling System By-pass	Certain 1996–1997 3.0L Taurus Methanol Engine	Through February 28, 1999, regardless of distance
98B23	Cooling System By-pass Kit	Certain 1996–1997 3.0L Taurus/Sable Gas/Ethanol	Through February 28, 1999, regardless of distance
98B21	Engine Block Heater	Certain 1995–1998 Contour/Mystique 1998 ZX2	Through February 28, 1999, regardless of distance
97B16	Air Cleaners	1994–1997 H.D. F-series with 7.3 DIT	Through February 28, 1999, or 160,000 km whichever occurs first

Since that last meeting, Ford engine head gasket and automatic transmission complaints have moderated, and almost all of the original 75 complaints I sent to Ford in January have been settled for approximately 75 percent of the amount claimed. If this trend continues, Ford will have gone a long way in regaining its customers' loyalty and respect.

But, as with Chrysler, I intend to continue monitoring Ford's customer complaint handling during 1999 and will revisit the matter in upcoming *Lemon-Aid* guides.

Urgent: provincial lemon laws

Lemon-Aid criticism and CLOG activism aren't all that unusual. There have been other successful Canadian grass-roots associations like the Rusty Ford Owners Association, Friends of Firenza, Automobile Protection Association, and other consumer protection groups that spontaneously appeared in the '70s and '80s in response to a combination of poor-quality vehicles and arrogant and unresponsive customer relations staffers.

Although these groups can be surprisingly effective over the short term, the most efficient response to poor quality control is provincial lemon law legislation (already enacted in most American states), where automakers are forced to take back a vehicle and compensate the owner if the dealership can't fix the problem after three attempts. The law would also contain a provision that all "secret warranties" or "goodwill programs" must be divulged to all owners equally, as is the case in the United States.

Getting Your Money Back

When letter writing and public protests fail to elicit a satisfactory response, your only recourse is to sue, or at least threaten to sue. Yet we all know that you can win in court and still end up losing your shirt. That's why this section of the book is dedicated to helping you get your money back without having to go to court or getting frazzled by the seller's broken promises and "benign neglect." But if going to court is your only recourse, you'll find here the jurisprudence you need to get an out-of-court settlement or win your case without spending a fortune on lawyers or research.

First of all, forget about the seller's claim that "there's no warranty" or "because it was sold 'as is' you can't claim a refund." This is all hogwash. Any sales contract for a used vehicle can be cancelled for one or more of the following reasons:

- The vehicle was misrepresented.
- The vehicle was unfit for the purpose for which it was purchased.
- The vehicle was not reasonably durable, considering its selling price, how well it was maintained, the mileage driven, and the type of driving.
- The vehicle was seriously defective at the time of purchase.

For example, if the seller claims that a minivan can pull a 2,000-pound trailer and you discover that it can barely tow half that weight, you can cancel the contract for misrepresentation. The same principle applies to a seller's exaggerated claims concerning a vehicle's off-road capability, fuel economy, and reliability; and to "demonstrators" that are in fact used cars with false (rolled back) odometer readings.

It's essential that printed evidence and/or witnesses (relatives are not excluded) be available to confirm that the false representation actually occurred. These misrepresentations must also concern an important fact that substantially affects the vehicle's performance, reliability, or value.

Some vehicles are meant to be driven hard, but when they fail to live up to the advertised hype, sellers often blame the owner for having pushed the vehicle beyond its limits. Therefore, when you attempt to set aside the contract, the testimony of an independent mechanic is essential to prove that the vehicle's poor performance isn't caused by negligent maintenance or abusive driving.

Another hurdle to overcome is the "reasonable diligence" rule that requires one to file suit within a reasonable time after the purchase or after the defect is discovered. This delay must not exceed a few months if there have been no negotiations with the seller. The delay for filing the lawsuit can be extended if the seller has been promising to correct the defects for some time or has done some minor repairs while negotiating a final settlement.

Warranties

Unscrupulous dealers and con artists won't tell you this, but it's true: *every vehicle sold has some kind of warranty that can be enforced in court.* In fact, every new or used vehicle sold in Canada is covered by both provincial and federal warranties that protect you from misrepresentation and a host of other evils. Furthermore, Canadian law and jurisprudence presume that car dealers, unlike private sellers, are aware of the defects present in the vehicles they sell. That's why they're paid a commission. The vehicles they sell are expected to be reasonably durable and merchantable. What is reasonably durable depends on the price paid, miles driven, the purchaser's driving habits, and how well the vehicle was maintained by the owner. Judges carefully weigh all these factors in awarding compensation or cancelling a sale.

There is no standard warranty period; however, some provinces require that dealers provide minimal warranty coverage. Safety restraints like airbags and seatbelts often have coverage extended for the lifetime of the vehicle. On the other hand, you can lose warranty coverage through abusive driving or inadequate maintenance, or by purchasing aftermarket products and services, such as gas-saving gadgets, rustproofing, paint protectors, air conditioning, and van conversions.

The manufacturer's or dealer's warranty is a legal promise that the product sold will perform in the normal and customary manner for which it was designed. This promise remains in force regardless of the number of subsequent owners, as long as the warranty's original time/mileage limits haven't expired.

Sometimes dealers will do all sorts of minor repairs that don't correct the problem, and then, after the warranty runs out, they'll tell you that major repairs are needed. You can avoid this nasty surprise by repeatedly bringing in your vehicle to the dealership before the warranty ends. During each visit insist that a written work order include the specific nature of the problem, as *you* see it, and that it carries the notation that this is the second, third, or fourth time the same problem has been brought to the dealer's attention. Write it down yourself, if need be. This allows you to show a pattern of nonperformance by the dealer during the warranty period and establishes that it's a serious and chronic problem. When the warranty expires, you have the legal right to demand that it be extended on those items consistently reappearing on your handful of work orders. *Lowe v. Chrysler* is an excellent judgment that reinforces this important principle.

Extended (supplementary) warranties

Supplementary warranties that provide extended coverage may be sold by the manufacturer, dealer, or an independent third party, and are automatically transferred when the vehicle is sold. They cost between $500 and $1,500, and should be purchased only if the vehicle you're buying is off its original warranty, has a poor repair history (see Part Three), or if you're reluctant to use the small claims courts when trouble arises that is factory related. Don't let the dealer pressure you into deciding right away. Generally, you can purchase an extended warranty anytime during the period the manufacturer's warranty is still in effect. Although dealers have a quasi-monopoly on selling these warranties, it still pays to shop around. One B.C. reader wrote that Global sells extended warranties at a competitive price.

Dealers love to sell extended warranties because about a third to one-half of the warranty's cost represents dealer markup. Out of the remainder comes the sponsor's administration costs and profit margin, calculated at another 25 percent. What's left is a minuscule 25 percent of the original amount paid to the dealer. It's estimated that of the car buyers who purchase an extended service contract, fewer than half actually use it.

It's often difficult to collect on supplementary warranties because independent companies not tied to the automakers frequently go out of business. When this happens, and the company's insurance policy won't cover your claim, take the dealer to small claims court and ask for the repair cost and the refund of the original warranty payment. Your argument for holding the dealer responsible is a simple one: by accepting a commission for acting as an agent of the defunct company, the selling dealer took on the obligations of the company as well.

Emission control warranties

These little-publicized warranties come with all new vehicles and cover the emission control system for up to 8 years/130,000 km. Unfortunately, although owner manuals vaguely mention the emissions warranty, most don't specify which parts are guaranteed. The U.S. Environmental Protection Agency (EPA) has ruled that all major motor and fuel-system components are covered. These include fuel metering, ignition spark advance, restart, evaporative emissions, positive crankcase ventilation, engine electronics (computer modules), and catalytic converters, as well as hoses, clamps, brackets, pipes, gaskets, belts, seals, and connectors. Unlike the United States, Canada has no government definition and lets each manufacturer decide which components are covered. Many of the dealer service bulletins listed in Part Three under "Service Tips" show which parts failures are covered under the emissions warranty.

Secret warranties

Secret warranties have been around since automobiles were first mass-produced. They're set up to provide free repairs to fix performance-related defects caused by substandard materials, faulty design, or assembly-line errors. In 1974, *Lemon-Aid* exposed Ford's J-67 secret seven-year rust warranty, which covered the company's 1970–74 models. After first denying that it had such a warranty, Ford admitted two years later that it was indeed in place and negotiated a $2.8 million settlement with this author to compensate owners of rust-cankered Fords who had formed the Rusty Ford Owners Association. And you know what? Twenty-five years later, hundreds of secret warranties continue to exist among most automakers and *Lemon-Aid* is still the only consumer publication blowing the whistle on hundreds of current programs that secretly allocate funds for the repair of engine, transmission, fuel pump, and paint defects on cars, sport-utilities, trucks, minivans, and vans.

Even mundane little repairs that can still cost you a hundred bucks or more are covered. Take, for example, the elimination of foul, musty, or mildew odours emitted by your air conditioning unit. Despite what the dealer may say, it's covered by the base warranty. In fact, as you can see from the following GM, Ford, and Chrysler service bulletins, automakers have secret warranty policies that pay for AC service adjustments.

File In Section: 1 - HVAC
Bulletin No.: 53-12-12A
Date: December, 1996
Subject:
Air Conditioning Odor at Start Up in Humid Climates
(Disinfect Evaporator Core, Install Delayed Blower Control Package)

Models:
1993-96 Passenger Cars (Except GEO)
1993-96 Light Duty Models (Except Tracker)

Condition
Some owners may comment on odors emitted from the air conditioning system, primarily at start up in hot, humid climates.

Cause
This odor may be the result of microbial growth on the evaporator core. When the blower motor fan is turned on, the microbial growth may release an unpleasant musty odor into the passenger compartment.

Correction
To remove odors of this type, it is necessary to eliminate the microbial growth and prevent its recurrence- To accomplish this, these two procedures must be completed.
- Deodorize the evaporator core using Deodorizing Aerosol Kit, P/N 12377951 (AC Delco 15-102).
- Install the new A/C Delayed Blower Control Package, P/N 12370470, (AC Delco 15-8632).

GM's AC warranty service will take about an hour.

Article No.
98-2-7
02/02/98
AIR CONDITIONING - MUSTY AND MILDEW TYPE ODORS - SERVICE PROCEDURE
FORD:
1992-93 FESTIVA
1992-94 TEMPO
1992-97 CROWN VICTORIA, ESCORT MUSTANG, PROBE, TAURUS, THUNDERBIRD
1994-96 ASPIRE
1995-97 CONTOUR
LINCOLN-MERCURY:
1992-94 TOPAZ
1992-97 CONTINENTAL, COUGAR, GRAND MARQUIS, SABLE, TOWN CAR, TRACER
1993-97 MARK VIII
1995-97 MYSTIQUE
LIGHT TRUCK:
1992-95 BRONCO
1992-97 AEROSTAR, ECONOLINE, EXPLORER, F SUPER DUTY F-150, F-250 HD, F-250 LD, F-350, RANGER
1995-98 WINDSTAR
1997 MOUNTAINEER
1997-98 EXPEDITION
1998 NAVIGATOR
MEDIUM/HEAVY TRUCK:
1992-97 F & B SERIES

This TSB article is being republished in its entirety to include the Expedition, Navigator and 1998 Windstar.

ISSUE: Musty and mildew type odors may come from the air conditioner and heater system. This odor is caused by mildew-type fungi growth in the A/C evaporator. It is most noticeable when the A/C is first turned on.

ACTION: A new Disodorizer(R) which encapsulates the mildew is now available to reduce these odors. This is a seasonal repair. Engineering effort to provide a long term repair for A/C odors will continue. Apply the Ford A/C Disodorizer(R) product (F6AZ-19G210 AA or YN-18) to the A/C system as directed in the following A/C Odor Treatment Procedure.

WARRANTY STATUS: Eligible Under The Provisions Of Bumper To Bumper Warranty Coverage

Ford's AC warranty service is a 20-minute job.

NO: 24-11-97
GROUP: Heater & A/C
DATE: Jul. 11. 1997
SUBJECT:
A/C Evaporator Odor

MODELS:
1992 - 1995 (AA) Spirit/Acclaim/LeBaron Sedan
1994 - 1998 (AB) Ram Van/Waqon
1992 - 1993 (AC) Dynasty/New Yorker/New Yorker Salon
1992 - 1993 (AG) Daytona
1992 - 1995 (AJ) LeBaron Coupe/LeBaron Convertible
1992 - 1998 (AN) Dakota
1992 - 1994 (AP) Shadow/Shadow Convertible/Sundance
1992 - 1995 (AS) Town & Country/Caravan/Voyager
1992 - 1993 (AY) Imperial/New Yorker Fifth Avenue
1994 - 1998 (BR) Ram Pickup
1993 - 1995 (ES) Chrysler Voyager (European Market)
1996 - 1998 (GS) Chrysler Voyager (International Market)
1995 - 1998 (JA) Cirrus/Stratus/Breeze
1996 - 1998 (JX) Sebring Convertible
1993 - 1998 (LH) Concorde/Intrepid/LHS/New Yorker/Vision
1996 - 1998 (NS) Town & Country/Caravan/Voyager
1995 - 1998 (PL) Neon
1997 - 1998 (TJ) Wrangler
1992 - 1998 (XJ) Cherokee
1994 - 1998 (ZG) Grand Cherokee (International Market)
1993 - 1998 (ZJ) Grand Cherokee/Grand Wagoneer

SYMPTOM/CONDITION:
Some vehicle operators may experience a musty odor from the A/C system, primarily at start up in hot and humid climates. This odor may be the result of microbial growth on the evaporator core. During normal A/C system operation, condensation forms in and around the A/C evaporator. When airborne pollutants mix with this condensation, bacteria and fungi growth begins and odor results.

DIAGNOSIS:
Operate the A/C system, if a musty odor is experienced, perform the Repair Procedure.

PARTS REQUIRED:
1 04897625AA Cleaner, Aerosol Evaporator
POLICY: Reimbursable within the provisions of the warranty.

Chrysler's AC warranty service may take 30–90 minutes and is classed: "Failure Code: XX–Service Adjustment."

Most automakers are reluctant to make secret warranty extensions public because they feel it would weaken confidence in their product and increase their legal liability. The closest they come to an admission is sending a "goodwill policy," "product improvement program," or "special policy" service bulletin to dealers or first owners of record. Consequently, the only motorists who get compensated for repairs to defective parts are the original owners who haven't moved or leased their vehicle, or those who yell the loudest and present automaker service bulletins (see Appendix II to order your own bulletins).

If you're refused compensation, keep in mind that secret warranty extensions are an admission of manufacturing negligence. Try to compromise with a pro rata adjustment from the manufacturer. If polite negotiations fail, challenge the refusal in court on the grounds that you should not be penalized for failing to make a reimbursement claim under a secret warranty you never knew existed!

Here are a few examples of the latest and most comprehensive secret warranties that have come across my desk this year. Keep in mind that an up-to-date listing of other secret warranties and service programs can be found in Part Three under the heading "Secret Warranties/Service Tips."

Chrysler/Jeep

- **1989–91 Cherokee and Wagoneer; 1990–93 Dynasty, New Yorker, Salon, Fifth Avenue, and Imperial; 1991–92 Monaco and Premier; and 1991–93 minivans. Problem**: Malfunctioning ABS. **Warranty coverage**: Chrysler has extended the warranty to up to 10 years/160,000 km (except for the brake actuator piston assembly and the pump motor assembly, which have lifetime warranties). Owners will also be reimbursed for previous ABS repairs.

Ford

- **1994 Lincoln Continental; and 1994–95 Taurus, Sable, Windstar. Problem**: Defective 3.8L engine head gaskets may cause loss of engine coolant, engine overheating, or destruction of the engine. **Warranty coverage**: Ford will replace the defective components at no charge up to 5 years/100,000 km. Vehicles that have exceeded the mileage limit will still be covered if repaired before December 31, 1998. Continentals are eligible up to 6 years/120,000 km.

Ford/Nissan

- **1995–96 Villager and Quest minivans equipped with 3.0L engines. Problem**: Excessive engine noise is caused by a poorly designed connecting rod that's insufficiently lubricated. The design on the VG30E was changed in January 1996 to provide a quieter operation; the new design should have worked its way into production by March of that same year (check the month of production on the doorplate). **Warranty coverage**: Ford will replace at no charge any 3.0L V6 engine block if customers complain of engine knock following a cold start. Ford is replacing the entire engine, including the cylinder head, and says it will issue an owner notification (not a recall) or set up a warranty extension program. Nissan, on the other hand, doesn't plan to notify customers of this free replacement program.

Ford Motor Company of Canada, Limited
Ford du Canada Limitée

The Canadian Road
P.0. Box 2000
Oakville, Ontario
L6J 5E4

Mail Date

XXXXXXXXXXXXXXXXX C 96M89

J. Sample
123 Main Street
Anywhere, Prov.
ANA NAN

As part of our ongoing efforts to increase customer satisfaction, Ford Motor Company of Canada, Limited (Ford) is providing a no-charge Service Program, number 96M89, to owners of certain 1995 and 1996 Model Year Villagers. Our records show you own the vehicle with the serial number shown above your name and address.

Reason for Program

Ford has learned through reports from dealers and customers and follow-up studies that engines in some Villagers may develop an abnormal engine noise. The noise is called "piston slap" and sounds like loud knocking when the engine is cold and just started. It is an annoyance sound and may go away as the engine warms.

If your Villager's engine develops the noise condition, it will not cause any problem in the starting or operation of the engine. It is an annoying engine noise condition. We assure you that you can continue to drive your Villager until a service appointment can be scheduled with your dealer at your convenience.

What We Will Do

Ford is providing additional warranty coverage for this condition only. If your Villager's engine should experience this condition, your dealer will verify the condition according to the instructions provided by Ford and, if necessary, replace the engine with a Ford Quality Renewal product at no cost to you.

The no charge coverage for this condition is available for 7 years or for 160,000 kilometres of service, whichever comes first from the vehicle original warranty start date and is automatically transferred to all subsequent owners at no charge. This coverage extends the warranty more than twice the length of the original coverage. All other engine or vehicle concerns have the Basic warranty coverage (3 years/60,000 kilometres, whichever comes first from the original warranty start date).

Both Ford and Nissan also have a serious problem with engine exhaust manifold studs breaking and costing over $3,000 to repair. Both companies will offer "goodwill" compensation, if pressed—particularly if owners show up with service bulletins in hand (see Part Three).

Ford Motor Company of Canada, Limited
Ford du Canada Limitée

The Canadian Road
P0. Box 2000
Oakville, Ontario
L6J 5E4

Mail Date

XXXXXXXXXXXXXXXXX C 98M01

J. Sample
123 Main Street
Anywhere, Prov.
ANA NAN

As part of our ongoing efforts to increase customer satisfaction, Ford Motor Company of Canada, Limited (Ford) is providing extended coverage under service program number 98M01, to owners of 1994 and 1995 Taurus and Sable and 1995 Windstar vehicles equipped with a 3.8L FWD Engine. Our records show you own the vehicle with the serial number shown above your name and address.

Reason For This Program:	Premature failure of the head gaskets may occur and may result in engine overheating, or, in extreme cases could cause engine failure if the overheat condition is ignored. The extended coverage is Ford Motor Company's commitment to customer satisfaction and we are taking this action as part of our ongoing efforts to maintain your confidence in our products.
No Charge Coverage:	Ford is providing extended coverage for this condition only. If your vehicle's engine should experience this condition, your dealer will verify the condition and replace the engine head gaskets, if necessary. The extended coverage for this condition is available on affected Taurus, Sable and Windstar vehicles for 5 years or 100,000 km. from the vehicle's original warranty start date and is automatically transferred to subsequent owners. On vehicles for which the extended coverage has expired, vehicles are eligible for this program through December 31, 1998, regardless of distance travelled.
What You Should Do:	PLEASE KEEP THIS LETTER. If your vehicle's engine head gasket should exhibit the condition described above within the extended coverage period, contact your dealer. The dealer will replace the engine head gaskets, if necessary, after verification of this specific condition. If you should lose this letter the dealer will still honour the provisions of this coverage.

Continental owners get an extra year and 20,000 more kilometres. Hmm, that doesn't seem fair.

General Motors

• **All models, all years**. **Problem**: Faulty paint jobs that cause paint to turn white and peel off of horizontal panels. GM internal dealer memo shows problem is a factory defect. **Warranty coverage**. GM will repaint the vehicle for free up to 6 years with no mileage limitation. Thereafter, the automaker plays out a "Let's Make a Deal" scenario and usually offers 50 percent refunds.

Paint Should Last Six Years, Says GM!

PONTIAC

PONTIAC DIVISION
General Motors Corporation
One Pontiac Plaza
Pontiac, Michigan 48340-2952

October 16, 1992

TO: All Pontiac Dealers

SUBJECT: Partners in Satisfaction (PICS)
Dealer Authorization

Pontiac continually reviews the Warranty Management System to ensure that Warranty Administration achieves its purposes, including high levels of customer satisfaction with after sale treatment.

Following a recent review, Pontiac has decided to provide dealers authorization for cases involving paint repairs for vehicles up to six (6) years from the date of delivery, without regard for mileage. This is a change from the current PICS dealer self-authorization which allows paint repair goodwill adjustments to be made up to 6 years/60,000 miles. Dealers who have a deductible override capabilities may also waive deductibles as they see appropriate on this type of repair.

Paint repairs are only to be authorized beyond the warranty period by the Dealership Service Manager on a case-by-case basis as with any other goodwill policy adjustment.

Assistance should only be considered for cases involving evidence of a defect in materials or workmanship by the manufacturer. Assistance should not be considered for conditions related to wear and tear and/or lack of maintenance (such as fading, stone chips, scratches, environmental damage, etc.).

Please contact your Zone representative if you have specific questions.

Perry S. White

Perry S. White
Director of Service/
Customer Satisfaction

This confidential memo to Pontiac dealers applies to all of GM's vehicles, and can be used as a benchmark for what the automaker considers its obligation when faced with paint claims. Note GM doesn't use any "weasel" words, like "acid rain" or "UV ray deterioration," to avoid its responsibility. Plus, GM says owners don't have to pay any deductible.

- **1992–93 Cavalier/Sunbird with 2.2L 4-cylinder engines. Problem**: Faulty head gaskets may cause loss of engine coolant, engine overheating, or destruction of the engine. **Warranty coverage**: GM will replace the faulty head gasket or repair the engine damage caused by head gasket failure at no charge up to 7 years/160,000 km.

Dear General Motors Customer:

As the owner of a 1992 or 1993 Chevrolet Cavalier equipped with a 2.2L engine, your satisfaction with our product is of the utmost concern to us. Your vehicle was provided with a new vehicle warranty, which covers certain parts of your vehicle for a specified period. These warranties are of considerable value to you if you should experience problems with your vehicle.

This letter is intended to make you aware that some 1992 and 1993 Chevrolet Cavalier models with 2.2L engines may develop a failure of the cylinder head gasket that allows coolant to leak from the cylinder head gasket to engine block joint. Early evidence of this would be a loss of coolant in the coolant reservoir and an odor of coolant from the engine compartment, or a low coolant lamp. There may also be visible coolant deposits at the cylinder head to engine block joint.

General Motors of Canada Limited is, therefore, taking the following action:

We are providing owners with special coverage. If the above-mentioned condition occurs within seven (7) years of the date your vehicle was originally placed in service or 160,000 km, whichever occurs first, your vehicle will be repaired for you at no charge.

This special policy applies only to repairs requiring cylinder head gasket replacement as a result of cylinder head gasket failure that results in an engine coolant leak. It does not cover engine damage from continuing to operate the engine in an overheated condition after loss of coolant.

This is not a recall campaign. Do not take your vehicle to your GM dealer as a result of this letter unless you believe that your vehicle has the condition as described above. Keep this letter with your other important glovebox literature for future reference.

If you have already paid for some or all of the cost to have the cylinder head gasket replaced and in-service time was less than seven (7) years and 160,000 km, you should contact your GM dealer. You may be eligible for partial or complete reimbursement of costs if genuine GM parts were used in the repair. If the work was done by someone other than a GM dealership the amount of reimbursement may be limited to the amount the repair would have cost GM to have it completed by a GM dealership. Please provide your dealer with your original paid receipts or invoices verifying the repair, the amount charged, proof of payment, and the date of payment of those charges by March 1, 1997.

Repairs and adjustments qualifying under this Special Policy coverage must be performed by your GM dealer.

Confidential service bulletins

The above special warranties are confirmed by technical service bulletins (confidential, for the most part) automakers send to dealers to advise them of special warranties and to help them quickly diagnose and correct factory defects. These bulletins also disclose how much of the repair the dealer can charge back to the manufacturer and which parts are available free of charge. Armed with these bulletins, motorists can use less expensive, independent garages to diagnose and repair their vehicles or to negotiate compensation for defects that the bulletins point out are the manufacturer's fault.

The major problem with these bulletins is that they're difficult to get. Dealers and automakers are reluctant to provide this kind of detailed technical information because it allows customers to second-guess a mechanic's work or to buttress the owners' demands for compensation. However, as long as their involvement isn't disclosed, some dealers will discreetly provide copies of service bulletins to help their customers fight for compensation from the auto manufacturer.

How long should parts/repairs last?

Let's say you can't find a service bulletin that says your problem is factory related or covered by a special compensation program. Or a part lasts just a little longer than its guarantee, but not as long as is generally expected. Can you get a refund if the same problem reappears shortly after it has been repaired? The answer is "yes," if you can prove the part failed prematurely.

Automakers, mechanics, and the courts have their own benchmarks as to what's a reasonable period of time or amount of mileage one should expect a part or adjustment to last. The following table shows what most automakers consider is reasonable durability as expressed by their original warranties and secret warranties that are often called "goodwill" or "special policy" programs.

Estimated Part Durability

ACCESSORIES	
Air conditioner	7 years
Cellular phone	5 years
Cruise control	5 years/ 100,000 km
Power antenna	5 years
Power doors, windows	5 years
Radio	5 years
BODY	
Paint peeling	7 years
Rust (perforations)	7 years
Rust (surface)	5 years
Vinyl roof	5 years
Water/wind/air leaks	5 years
BRAKE SYSTEM	
Brake drum	120,000 km
Brake drum, turn	40,000 km
Brake drum linings	35,000 km
Disc brake calipers	30,000 km
Disc brake pads	30,000 km
Master cylinder, rebuild	100,000 km
Wheel cylinder, rebuild	80,000 km
ENGINE AND DRIVETRAIN	
Constant velocity joint	5 years/ 100,000 km
Differential	7 years/ 150,000 km
Engine (gas)	7 years/ 160,000 km
Radiator	4 years/ 80,000 km
Transfer case	7 years/ 150,000 km
Transmission (auto.)	7 years/ 150,000 km
Transmission (man.)	7 years/ 200,000 km
Transmission oil cooler	5 years/ 100,000 km
Universal joint	5 years/ 100,000 km
EXHAUST SYSTEM	
Catalytic converter	5 years/ 100,000 km or more
Muffler	2 years/ 40,000 km
Tailpipe	3 years/ 60,000 km

FUEL SYSTEM

Carburetor	5 years/ 120,000 km
Fuel filter	2 years/ 40,000 km
Fuel pump	5 years/ 80,000 km
Injectors	5 years/ 80,000 km

IGNITION SYSTEM

Cable set	60,000 km
Electronic module	5 years/ 80,000 km
Spark plugs	20,000 km
Tune-up	20,000 km

SAFETY COMPONENTS

Airbags	life of vehicle
ABS brakes	7 years/ 150,000 km
ABS computer	10 years/ 160,000 km
Seatbelts	life of vehicle

STEERING AND SUSPENSION

Alignment	1 year/20,000 km
Ball joints	80,000 km
Power steering	5 years/ 80,000 km
Shock absorber	2 years/ 40,000 km
Struts	5 years/ 80,000 km
Tires (radial)	5 years/ 80,000 km
Wheel bearing	3 years/ 60,000 km

VISIBILITY

Aim headlights	20,000 km
Halogen/fog lights	3 years/ 60,000 km
Sealed beam	2 years/ 40,000 km
Windshield wiper motor	5 years/ 80,000 km

The previous guidelines are extrapolated from Chrysler's payout to dozens of Chrysler Lemon Owner Group (CLOG) members from December 1997 through March 1998, in addition to Chrysler's original seven-year powertrain warranty applicable 1991–1995. Other sources for this chart were the Ford and GM transmission warranties outlined in their secret warranties and GM, Mercury, Nissan, and Toyota engine "special programs" laid out in their internal service bulletins.

Safety features generally have a lifetime warranty, with the exception of ABS brakes, which are a wear item. Nevertheless, the Chrysler 10-year "free service program" portion of its ABS recall announced several years ago can serve as a handy benchmark as to how long one can expect these components to last.

Airbags are a different matter. Those that are deployed in an accident and the personal injury and interior damage their deployment will likely have caused are covered by your accident insurance policy. However, if there is a sudden deployment for no apparent reason, the automaker and dealer should be held jointly responsible for all injuries and damages caused by the airbag. This will likely lead to a more generous settlement from the two parties and prevent your insurance premiums from being jacked up. Inadvertent deployment may occur after passing over a bump in the road, slamming the car door, having wet carpets in your Cadillac (no kidding), or, in some Chrysler minivans, simply putting the key in the ignition. This happens more often than you might imagine, judging by the hundreds of recalls and thousands of complaints recorded on the U.S. National Highway Traffic and Safety Administration's website at *www.nhtsa.dot.gov/cars/problems/complain/compmmy1.cfm*.

Finally, the manufacturer's emission warranty serves as the primary guideline governing how long a vast array of electronic and mechanical components should last. Look first at your owner's manual for an indication of which parts on your vehicle are covered. If you come up with few specifics, use the provincial government's guidelines in provinces where emissions testing is mandatory. Keep in mind that these durability benchmarks, secret warranties, and emissions warranties all apply to subsequent owners.

Recall repairs

Let the automaker know who and where you are. If you've bought a used vehicle or moved, it's a smart idea to pay a visit to your local dealer and get a "report card" on which recalls, free service campaigns and warranties apply to it. Simply give the service advisor your vehicle identification number (VIN)—found on the dash just below the windshield on the driver's side, or on your insurance card—and have the number run through the automaker's computer system ("Function 70" for Chrysler, "OASIS" for Ford, and "CRIS" for GM). Ask for a computer printout of the vehicle's history (have it faxed to you, if you're so equipped) and make sure you're listed in the automaker's computer as the new owner. This ensures that you'll receive notices of warranty extensions and emissions and safety recalls.

Still, don't expect to be welcomed with open arms when your vehicle develops a safety- or emissions-related problem that's not yet part of a recall campaign. Automakers and dealers generally take a restrictive view of what constitutes a safety or emissions defect and frequently charge for repairs which should be free under federal safety or emissions legislation. To counter this tendency, look at the following list of typical defects that are clearly safety related, and if you experience similar problems, tell the dealer you expect your repair to be paid by the manufacturer:

- airbag malfunctions
- corrosion affecting safe operation
- disconnected or stuck accelerators
- electrical shorts
- faulty windshield wipers
- fuel leaks
- problems with original axles, drive shafts, seats, seat recliners, or defrosters
- seatbelt problems
- stalling or sudden acceleration
- sudden steering or brake loss
- suspension failures
- trailer coupling failures

U.S. recall campaigns force automakers to pay the entire cost of fixing a vehicle's safety-related defect. This includes used vehicles, and has no cut-off limitation. Recalls may be voluntary or ordered by the U.S. Department of Transportation. Canadian regulation has an added twist: Transport Canada can only order automakers to notify owners that their vehicles may be unsafe; it can't force them to correct the problem. Fortunately, most U.S.-ordered recalls are carried out in Canada, and when Transport Canada makes a defect determination on its own, automakers generally comply with an owner notification letter.

Voluntary recall campaigns are a real problem, though. They aren't as rigorously monitored as government-ordered recalls, and dealers and automakers routinely deny there's a recall. Also, the company's so-called "fix" often doesn't correct the hazard. Take, as an example, Chrysler's voluntary recall to strengthen the rear latches on as many as 4.5 million 1984–95 minivans. Almost 50 percent of affected owners were still waiting for Chrysler to fix their minivans nearly two years after the company volunteered to correct the defect.

Safety defect information

If you wish to report a safety defect or want recall info, you may access Transport Canada's website on the Internet. You'll get recall information in French or English as well as general information relating to road safety and importing a vehicle into Canada. The web page can be accessed by typing *www.tc.gc.ca/roadsafety/Recalls/search_e.asp.* Cybersurfers can now access the recall database for 1970–2000 model vehicles, but, unlike NHTSA's website, owner complaints aren't listed, defect investigations aren't disclosed, and service bulletin summaries aren't provided. You can also call Transport Canada at 1-800-333-0510 (toll free within Canada) or 613-993-9851 (within the Ottawa region or outside Canada) to get additional information.

Unfortunately, there are some problems with Ottawa's database, and attitude. First, when calling Ottawa through the toll-free line, Transport Canada bureaucrats insist that the dealer must already have refused you the recall info before they will give it to you. You won't be told if others have reported similar safety problems affecting your vehicle. And more often than not, if you suspect your car has a safety defect, you'll be asked to take it to the dealer for a safety exam (where there's a good chance the problem will be covered up or you'll be blamed for the malfunction).

If you're not happy with Ottawa's treatment of your recall inquiry, try the U.S. government's NHTSA website. It's more complete than Transport Canada's site (NHTSA's database was established in 1972 and is updated daily). You can search the database specific to your vehicle and be thoroughly briefed on recalls, crash ratings, safety and performance defects reported by other car owners, and a host of other safety-related items. The web address is *www.nhtsa.dot.gov/cars/problems/recalls/recmmy1.cfm* for recalls and *www.nhtsa.dot.gov/cars/problems/complain/compmmy1.cfm* for the complaint database.

NHTSA's fax-back service provides the same info through a local line that can be accessed from Canada—although long-distance charges will apply (most calls take 5–10 minutes to complete). For calls placed within the U.S., the toll-free hotline is 1-800-424-9393 (1-800-424-9153 for the hearing impaired). The following local numbers get you into the automatic response service just as quickly, and can be reached 24 hours a day: 202-366-0123 (202-366-7800 for the hearing impaired).

Three Steps to a Settlement

Step 1: Informal negotiations

If your vehicle was misrepresented, has major defects, or wasn't properly repaired under warranty, the first thing you should do is give the seller (the dealer and automaker or a private party) a written summary (by registered mail or fax) of the outstanding problems and stipulate a time period in which they will need to be corrected or your money will be refunded. Keep a copy for yourself along with all your repair records. Be sure to check all of the sales and warranty documents you were given to see if they conform to provincial laws. Any errors, omissions, or violations can be used to get a settlement with the dealer in lieu of making a formal complaint.

At the beginning, try to work things out informally and, in your attempt to reach a settlement, keep in mind the cardinal rule: ask only for what is fair and don't try to make anyone look bad.

- Listen. The really tough part of negotiating is listening. Listen to the automaker's rep or the dealership principal and try to understand their problem while thinking of a cooperative solution. This means frequently restating the other side's position so they realize you understand their offer.
- Line up evidence and allies. Be sure to line up your proof and allies (like work orders, service bulletins, and independent garage reports) before making your claim.
- Be reasonable and give as well as take. Consumers are frequently given a "Let's Make a Deal" spiel where the initial offer of 50 percent is often boosted to 75 percent compensation if the customer will agree—at that very moment—to pay 25 percent of the repair.
- Keep your demands reasonable but add a request for consequential damages (frustration, inconvenience, rental cars, missed work/vacation, etc.) and keep it as a throwaway claim to be used at a critical juncture in the talks.
- Know when to shut up.
- Don't set up an unrealistic timetable.

Furthermore, take a close look at the following confidential GM bulletin, which outlines to dealers the strategy they should follow and questions they should ask before offering their clients any goodwill warranty consideration.

FILE IN SECTION: Warranty Administration
BULLETIN NO.: 57-05-01
DATE: April 1995
SUBJECT:
Dealer/Retailer Administration of Case-by-Case individual Goodwill Adjustments
MODELS:
General Motors Passenger Cars and Trucks
All General Motors Marketing Divisions **(Except GM of Canada)**
ATTENTION: Service Manager/Warranty Claims Administrator
Listed below is the definition for goodwill adjustments, general administrative guidelines, identifying the facts and division claim authorization empowerment information.
DEFINITION:
Individual, case-by-case, goodwill adjustments are intended to recognize that circumstances, outside the parameters of the written warranty, may exist where special consideration is in order to enhance customer satisfaction and loyalty. Such goodwill adjustments should not be confused with special policy adjustments provided to all involved owners through a special bulletin and direct mail. Case-by-case individual goodwill adjustments are not legal obligations like the terms of General Motors warranties.
GENERAL ADMINISTRATIVE GUIDELINES:

- Maximum dealer/retailer authorization empowerment, common to all GM divisions, is up to 2 years/24,000 miles for non-paint conditions and up to 2 years/unlimited mileage for paint, beyond the limits of the Bumper to Bumper base new vehicle warranty. Maximum authorization empowerment regarding GMC and Chevrolet medium duty truck authorization empowerment will be 2 years unlimited mileage for both non-paint and paint due to the unlimited mileage base warranty.
- Some dealers/retailers may be empowered to limits less than the maximum while a few select dealers/retailers may not be empowered at all. Individual dealer/retailer authorization empowerment levels are subject to change upon divisional notification.
- Any claim involving "individual goodwill adjustments" submitted after the effective date of March 15, 1995, which exceeds the dealer/retailer goodwill adjustment authorization empowerment limit will require wholesale review and authorization. This includes work in process for which you may have already committed to perform.
- Bumper to Bumper base new vehicle warranty coverage does not include current and past model year warranty coverages on specific components that carry warranty coverage longer than the base new vehicle warranty coverage, i.e., "Federal and California Emissions Warranties," "Sheet Metal Coverage" or Diesel Engine Coverage."
- Paint Labor Operations excluded from authorization empowerment. See below:
  Dealer/retailer authorization empowerment for paint claims covers exterior body panels only.

IDENTIFYING THE FACTS:
Your authorization empowerment should be exercised when it appears that the condition could be the result of a defect in materials or workmanship rather than conditions occurring from aging, physical damage, lack of proper maintenance or owner abuse. Vehicle concerns should be assessed on their own merits within the context of the current owner's reasonable expectations and with appropriate concern for the cost to General Motors.
In exercising your judgment in each case, the following questions are some which should be considered:

- Is the vehicle covered by other service or extended warranty contracts? If "yes," repairs must be performed under that contract.
- Does the vehicle show a lack of proper care and maintenance?
- Is the overall condition of the vehicle such that the cost to repair will approximate the vehicle's value?
- If the vehicle was purchased used by the current owner, is it likely the vehicle exhibited the condition before purchase?
- Did the condition occur while the vehicle was in the possession of a second or third owner and no vehicle history is available?
- Could the customer have done anything to prevent or cause the condition?
- Has the vehicle been used in ways for which it was not intended?
- Did the condition result from an after market conversion or alteration?
- Did the condition result from use of non-GM parts or accessories?
- Is there evidence of vehicle mileage alteration?
- Is the vehicle condition a normal operating characteristic?
- Is the condition largely a result of normal aging, wear and usage?
- Is the condition a result of unusual localized environmental, geographical or climate factors?

Additional Questions When Paint Condition Is the Concern:

- Is the condition largely the result of normal paint aging or environmental damage (fading, gloss loss, chalking, cracking, flaking, yellowing)?
- Is the condition only on a physically damaged panel that the customer declines to repair?

The above list is not intended to be all inclusive. However, if the answer to any of these or similar questions is "yes," then it probably is not a concern you should address at General Motors' expense.
PARTIAL PARTICIPATION:
In situations beyond the warranty period, but within your claim authorization empowerment, customers have received value from use of the vehicle. It would be reasonable to consider partial payment by the customer. The judgment belongs to you.

- Determine what the customer expects before making an offer to assist.
- Evaluate the reasonableness of the customer's expectations.
- Determine what offer would satisfy the customer as a fair and equitable adjustment.
- Make your evaluation of the situation through direct communication with the customer.
- Always strive for a mutually acceptable agreement.

Be prepared to answer the above questions that GM suggests its dealers ask you before doling out any "goodwill" repair refunds. In fact, you can save time and improve your chances for getting compensation by incorporating some of the questions and answers in a fact sheet accompanying your complaint letter.

Finally, when negotiating, speak in a calm, polite manner and try to avoid polarizing the issue. Talk about how "we can work together" on the problem. Let a compromise slowly emerge—don't come in with a hardline set of demands. Don't demand the settlement offer in writing, but make sure that you're accompanied by a friend who can confirm the offer in court if it isn't honoured (relatives may testify in court). Be prepared to act upon the offer without delay so that you won't be blamed for its withdrawal.

Dealer/service manager

If you bought a used vehicle from a dealer who sells the same make new, you stand a good chance of getting free repairs, particularly if the vehicle is still under warranty or is covered by a "goodwill" program, or you intend to plead premature failure of a specific part based upon the parameters listed in the Estimated Part Durability chart found on pages 78–79.

Service managers are directly responsible to the dealers and manufacturers and make the first determination of what work is covered under warranty. They are paid to save the dealers and automakers money and to mollify irate clients—almost an impossible balancing act, wherein my sympathies are more with the dealers than with the automakers. When service managers agree to warranty coverage, it's because you've convinced them that they must do so. This can be done by getting them to access the vehicle's history from the manufacturer's computer and by you presenting the facts of your case in a confident, forthright manner with as many dealer service bulletins and NHTSA owner complaint print-outs as you can find for support.

Don't use your salesperson as a runner, since the sales staff are generally quite distant from the service staff and usually have less pull than you do. If the service manager can't or won't set things right, your next step is to convene a mini-summit with the service manager, the dealership principal, and the automaker's rep. By getting the automaker involved, you run less risk of having the dealer fob you off on the manufacturer and you can often get an agreement where the seller and automaker pay two-thirds of the repair cost.

Dealers selling a brand of used vehicle that they don't sell new and independent dealers give you less latitude. You have to make the case that the vehicle's defects were present at the time of purchase or should have been known to the seller, or that the vehicle doesn't conform to the representations made when it was purchased. Emphasize that you intend to use the courts if necessary to obtain a refund—most independent sellers would rather settle than risk a lawsuit with all the attendant publicity. An independent estimate of the vehicle's defects and cost of repairs is essential if you want to convince the seller that you're serious in your claim and stand a good chance of winning your case in court. The estimated cost of repairs is also useful in challenging a dealer who agrees to pay half the repair costs and then jacks up the costs a hundred percent so that you wind up paying the whole shot.

Step 2: Sending a registered letter or fax

This is the next step to take if your claim is refused. Send the dealer and manufacturer a polite, registered letter or a fax that asks for compensation for repairs that have been done or need to be done, insurance costs while the vehicle is being repaired, towing charges, supplementary transportation costs like taxis and rented cars, and damages for inconvenience.

Specify five days (but allow 10 days) for either party to respond. If no satisfactory offer is made, file suit in small claims court. Make the manufacturer a party to the lawsuit, especially if the emissions warranty, a secret warranty extension, a safety recall campaign, or extensive chassis rusting is involved. The two sample claim letters and fact sheet that follow can be useful in getting compensation for a defective used vehicle or for unsatisfactory repairs. Include a reference in your letter to any court decisions you find in this section of the book that support your claim and add a fact sheet if you're seeking a "goodwill" payout from the automaker and dealer.

Used-Car Complaint Letter/Fax

Without Prejudice

Date: ________________
Name: ________________

Please be advised that I am dissatisfied with my used vehicle, a (state model), for the following reasons:

1. ____
2. ____
3. ____
4. ____
5. ____

In compliance with the provincial consumer protection laws and the "implied warranty" set down by the Supreme Court of Canada in *Donoghue v. Stevenson* and *Longpré v. St-Jacques Automobile*, I hereby request that these defects be repaired without charge.

This vehicle has not been reasonably durable and is, therefore, not as represented to me.

Should you fail to repair these defects in a satisfactory manner and within a reasonable period of time, I shall get an estimate of the repairs from an independent source and claim them in court, without further delay.

I have dealt with your company because of its honesty, competence, and sincere regard for its clients. I am sure that my case is the exception and not the rule.

A response within the next five (5) days would be appreciated.

Sincerely,

____

(signed with telephone or fax number)

Secret Warranty Claim Letter/Fax

Without Prejudice

Date: __________
Name: __________

Please be advised that I am dissatisfied with my vehicle, a _____________, bought from you on _________.

It has had the following recurring problems that I believe are factory-related defects, as confirmed by internal service bulletins sent to dealers, and are covered by your "goodwill" policies:

1. ____
2. ____
3. ____

If your "goodwill" program has ended, I ask that my claim be accepted nevertheless, inasmuch as I was never informed of your policy while it was in effect and should not be penalized for not knowing it existed.

I hereby formally put you on notice under federal and provincial consumer protection statutes that your refusal to apply this extended warranty coverage in my case would be an unfair warranty practice within the purview of the above-cited laws.

I have enclosed several estimates (my bill) showing that this problem is factory related and will (has) cost $___________ to correct. I would appreciate your refunding me the estimated (paid) amount, failing which, I reserve the right to have the repair done elsewhere and claim reimbursement, plus consequential and punitive damages from you in court, without further delay.

A response within the next five (5) days would be appreciated.

Sincerely,

____
(signed with telephone or fax number)

Claim Fact Sheet

Claimant: ________________ Date: ________________

Reference: Failure of _______________ to last a reasonable period of time.

Dealer/Automaker: Request for "Goodwill" repair/refund for the following reasons:

VEHICLE

- Vehicle has NOT performed as represented by seller
- Vehicle is NOT covered by any other warranty/service contract
- Vehicle shows no signs of mileage alteration or accident damage
- Vehicle has been properly maintained and driven
- Vehicle was NOT used for purposes for which it was not intended

PROBLEM

- Problem is recurring and seller has been given ample opportunity to fix it
- Problem is NOT a normal operating condition or part of normal maintenance
- Problem was NOT caused or worsened by owner's actions
- Problem is NOT caused by aftermarket conversion, use of non-OEM parts and accessories, or servicing by an independent
- Problem is NOT the result of normal aging, wear, usage, or environmental factors
- Problem likely existed at the time of purchase

____

(signed with telephone or fax number)

Step 3: Mediation and arbitration

If the formality of a courtroom puts you off or you're not sure that your claim is all that solid and don't want to pay legal costs to find out, consider using mediation or arbitration sponsored by the Better Business Bureau, Automobile Protection Association, Canadian Automobile Association, small claims court (often a prerequisite to going to trial), or provincial and territorial governments.

Involving other agencies

Remember, it's not the mediator or arbitrator's fault if you lose your case due to poor preparation. Ask government and independent consumer protection agencies, like the one listed on the following page, to evaluate how well you're prepared before going to your first hearing.

Automobile Protection Association
292 St-Joseph Blvd West, Montreal, Quebec H2V 2N7
and
2 Carlton St., Suite 1319, Toronto, Ontario M5B 1S3
Tel.: 416-204-1444; Fax: 416-204-1985; Website: *www.apa.ca*

A nonprofit Canadian motorist protection organization, the Automobile Protection Association (APA) mediates thousands of complaints yearly, mostly for its own members. The group also negotiates class settlements on behalf of motorists across Canada after being alerted to common problems identified through the complaints it receives from members and non-members alike.

Government consumer affairs
Investigation, mediation, and some litigation are the primary areas in which consumer affairs offices can be helpful. Despite severe budget restraints to these offices, consumer protection legislation still stands in most of the provinces, and resourceful consumers can use these laws along with media coverage to prod provincial consumer affairs offices into action. Furthermore, provincial bureaucrats aren't as well shielded from criticism as are their federal counterparts. A call to your MPP, your MLA, or the minister's executive assistant can often get things rolling.

Federal consumer protection is a crock. Although the revised Competition Act has some bite when challenging misleading advertising and a number of other nefarious business practices, reorganized Consumer and Corporate Affairs departments have been de-fanged and de-gummed.

Online services/Internet/websites
America Online and CompuServe are two online service providers with active consumer forums that use experts to answer consumer queries and to provide legal as well as technical advice. The Internet offers the same information but uses a worldwide database. If you or someone you know is able to create a website, you might consider using this site to attract attention to your plight and arm yourself for arbitration or court.

Several years ago, Debra and Edward Goldgehn's 1985 Ford Ranger caught fire and burned completely. The couple's suspicion that it was a factory-related defect was later confirmed by a TV show that reported a series of similar Ford fires. The couple created their own website called "Flaming Fords" and began amassing an incredible database containing reports of similar fires, class action lawsuits, expert witnesses, and actions taken outside the U.S. (For example, Ford had already recalled a number of its vehicles in order to fix the problem here in Canada.) Shortly thereafter, Ford USA recalled 8.7 million cars and trucks, representing the largest recall ever announced by a single auto manufacturer. Ford says that the Internet pressure was coincidental and not a factor in its decision to recall the vehicles in the States. Right, and Elvis is building Fords in Oakville.

Michael Hos, a dissatisfied Acura owner from the United States, became fed up with what he felt was Acura's stonewalling of his complaints. Rather than get angry, he got organized and set up a website called "Acura 1997 CL 3.0L: My Lemon" to collect other owners' comments and list some of the most common Acura problem areas. Within six months, Acura settled. Here's what Hos has to say:

> Phil,
>
> Hi, remember me? I'm the guy that had the 1997 Acura CL lemon. Well, as it turns out Acura settled with me. I got my attorney fees taken care of, and I'm walking away with $4000 in my pocket after the experience. I was so glad to get rid of this car you have no idea.
>
> I have pulled down my "anti-Acura" website and washed my hands clean of the entire ordeal. I turned in the CL about 2 weeks ago, and my checks should be here next week....
>
> ...At any rate, I wanted to say thanks for all your input into the case and the encouragement to continue on. The last few months of driving the car were horrible, I'm glad it's done.

In a follow-up posting, Hos writes:

> As far as my website goes, I think it was a major part of them settling early. I had a counter placed on it which showed them how many people had visited the site. Anyone can set up a web page like mine pretty easily. I have web space on my university's computer, so it was free for me to use. Folks without space should expect to spend about $20 a month for space, or if they have their own e-mail account, web space is usually provided for free. If they don't know how to set up their web page, paying someone to do it will be kinda pricey, a few hundred bucks should cover it. The main thing it needs to have is the counter, and it also needs to be slander free. I had only facts on my web page as I didn't want to get involved in a slander suit. They also need to register the site with all the major search engines so it comes up when looking for the manufacturer. Submitit.com offers such services for free. Putting in a <Meta> tag into the page also helps move it up the search engines' list of hits. Posting to newsgroups also is helpful. I also wrote to JD Power, NHTSA, *Consumer Reports,* and any other consumer oriented agency I could think of.
>
> When we settled before going to court, I had to sign the settlement papers saying I would pull down my site. They would not settle with me until I did that. This shows how much power the site can have. I would also put the manufacturer's phone number and address on it so viewers of the site can contact the manufacturer.
>
> For the most part everyone who read my site took my side of the story and agreed that Acura should pay up. I did have a few folks that were mad I was slamming Acura, but I wasn't concerned with them.

I have about 200 e-mail responses that people have mailed to me over the last few months.

As a side note, I think the only real reason they settled with me, in addition to the page being up, was my attorney. I had 7 charges filed against them in Superior Court. Also keep in mind that I paid nothing for my attorney until after we settled. My bill for him is $1500, but that's included in the settlement. Also, I'm only 23, so anyone can do this if they are persistent.

> **PROJECTS AND CAUSES**
>
> A BAD FIRE occured in our 1994 Jeep Cherokee because of driving with a compact spare tire while in 4WD (could not get out of 4WD). Compact spare is standard equipment but appropriate warning not in Owners Manual. No appropriate answers from Chrysler Canada yet. Seeking others with similar experience.

The Globe and Mail.

Classified ads

Use your local paper's "Personals" column to pressure the seller and gather data from others who may have experienced a problem similar to your own. This alerts others to the potential problem, helps build a core base for a class action or group meeting with the automaker, and puts pressure on the dealer or manufacturer to settle.

Going to Court

When to sue

If the seller you've been negotiating with agrees to make things right, give him or her a deadline and then have an independent garage check the repairs. If no offer is made within 10 working days, file suit in court. Make the manufacturer a party to the lawsuit only if the original, unexpired warranty was transferred to you, your claim falls under the emissions warranty, a secret warranty extension or a safety recall campaign exists, or extensive chassis rusting is involved.

Choosing the right court

You must decide what remedy to pursue; that is, whether you want a partial refund or a cancellation of the sale. To determine the refund amount, add the estimated cost of repairing existing mechanical defects to the cost of prior repairs. Don't exaggerate your losses or claim for repairs that are considered routine maintenance.

A suit for cancellation of sale involves practical problems. The court requires that the vehicle be "tendered" back to the seller at the time the lawsuit is filed. This means you are without transportation for as long as the case continues, unless you purchase another car in the interim. If you lose the case, you must then take back the old car and pay storage fees. You could go from having no car to having two, one of which is a clunker.

Generally, if the cost of repairs or the sales contract amount falls within the small claims court limit (discussed later), the case should be filed there to keep costs to a minimum and to obtain a speedy hearing. Small claims court judgments aren't easily appealed, lawyers aren't necessary, filing fees are minimal (about $125), and cases are usually heard within a few months.

If the damages exceed the small claims court limit and there's no way to reduce them, you'll have to go to a higher court—where costs quickly add up and lengthy delays of a few years or more are commonplace.

Small courts, big victories

How effective is just threatening a small claims court lawsuit in getting a refund? Well, the Maritimer who wrote the letter below after getting a $900 out-of-court engine gasket settlement is so encouraged that he's now prepared to file a *second* action for the automatic transmission failure of his 1992 Taurus.

> January, 19, 1999
>
> Good morning Phil,
>
> I have the pleasure of reporting to you this morning about another happy automotive consumer that has been helped through your good advice and support.
>
> I received a call from a Halifax solicitor representing Ford. Their offer was to settle for $900 my claim for $1,148....
>
> ...When I receive the offer and accept it, I am then going to contact the solicitor representing the dealer and indicate my action against them is still outstanding and ask them to pay me the difference ($1,148 less $900) or at least make me an offer.
>
> In addition, I will argue that the dealer made money selling me the car with a defective engine head gasket, yet, I was held out to dry when I requested the dealer's help (I was told it was a manufacturer's, not the dealer's, problem). At the time they told me this, they would have been well aware of the service bulletins and service recall on the '94 and '95 Taurus and Sable, yet they did nothing for me/told me nothing about this....
>
> I am going to have another go at Ford for the transmission problem that cost $1,200 to repair (after I get the above action finalized).
>
> Phil, you have been a terrific help to me in the above claim and without your advice and encouragement, I would not have achieved the above result.
>
> Bob B.
> Halifax, Nova Scotia

Before we get into the details of launching a small claims action, read the following letter from Steve Wheeler and Eleanor Crawley of Woodlawn, Ontario, which describes how they got almost $4,000 refunded after following the *Lemon-Aid* court tips:

> We purchased an '88 Tempo....Within six days $1,700 (Cdn.) had been spent on repairs (on an emergency trip to Cleveland). The head and head gasket, a tension pulley, and an axle all had to be replaced during the first few days. The car never made it to Cleveland and my wife was stranded in motels or stuck sitting beside the interstate or in garages waiting for repairs....The dealer said he would only reimburse us if we signed an agreement which gave up any future legal claims on repair on the Tempo. We wouldn't sign and negotiations ended.
>
> We filed in small claims court for the refund of $2,800, the car's price with tax, the $737 we paid for repairs while the car was under warranty (dealer had already paid $1,000 using his credit card over the phone), and expenses incurred while my wife was stranded, for a total of $3,796 plus $50 for the small claim. On the advice of the lawyer, we named the company, the dealer (who also owned the company), and the mechanic as defendants.
>
> We had written the dealer during the negotiations, using the example of the Used Car Letter in the *Lemon-Aid Used Car Guide 1996*, and had cited the Sale of Goods Act and several sample cases from the book.
>
> In the preliminary hearing, the dealer's defence was that my wife had misunderstood the warranty and that he didn't warranty any work not done at his mechanic's garage.
>
> Yesterday, our day in court came. We got there at 9:30 a.m. and left at 5 p.m. We were awarded $3,800 and the dealer had to take the car back. The judge based his decision on the Sale of Goods Act and found that the Tempo was "not fit for the purpose for which it was sold." [The judgment can be found under: *Eleanor A. Crawley v. Armatage Motors Ltd., Frank Marchington and Harold Lavign,* Ottawa small claims court, July 10, 1996.]
>
> ...The process took from April till July. I found the Companies Branch of the Consumer and Commercial Relations Department of the Ontario government helpful. They have a brochure, which they'll mail out, that contains relevant sections of the Business Practices Act, including recision. The local MOT [Ministry of Transportation] branch mailed me a copy of the safety check procedures for mechanics and I used the free 1/2 hour of legal consultation from a lawyer provided by DIAL-A-LAW (1-800-387-2992) and the Lawyer's Referral Service (1-800-268-8326) to start out, since I had never been to small claims court before....Without the first few chapters of *Lemon-Aid Used Car Guide 1996*, we wouldn't have known where to start or how to proceed.
>
> A few things we noticed which might help someone else stuck with a lemon: get inspections done as quickly as possible after you realize you don't want the car. Park it and don't drive it as soon as

possible. Get a mechanic to accompany you to court to back up his inspection; the judges seem to be more impressed by a personal appearance than a written report. Keep good records of all meetings and communications for the recounting of events in court. We found that the dealer who was trying to confuse the issue had no answers when we itemized clearly the negotiations and attempts at agreement....

I hope that this letter and the case will encourage anyone else who gets stuck with a lemon.

There are small claims courts in most counties of every province and you can make a claim in the county where the problem happened or where the defendant lives and conducts business. The first step is to make sure that your claim doesn't exceed the dollar limit of the court (the limits differ from province to province). Then, you should go to the small claims court office and ask for a claim form. Instructions on how to fill it out accompany the form. Remember, you must identify the defendant correctly. It's a practice of some dishonest firms to change a company's name to escape liability; for example, it would be impossible to sue Joe's Garage (1999) if your contract is with Joe's Garage Inc. (1984).

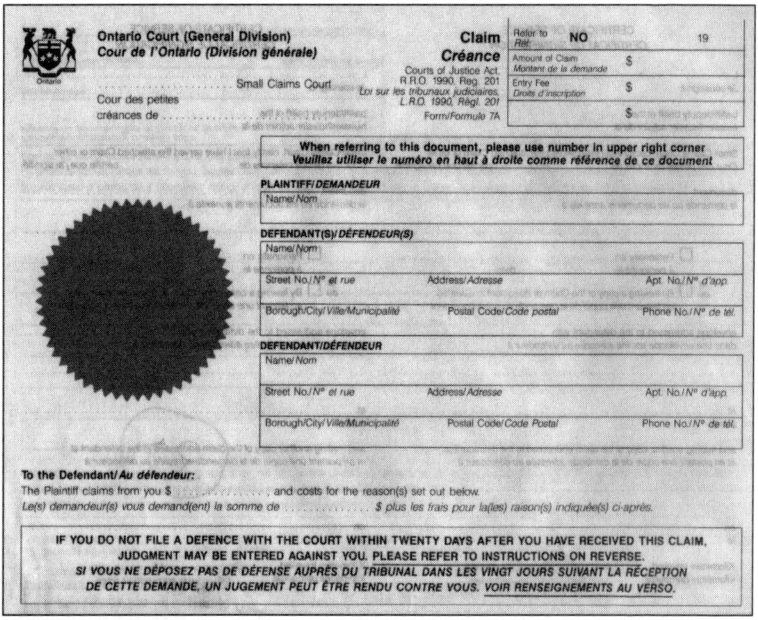

Ontario Court (General Division)
Cour de l'Ontario (Division générale)

.... Small Claims Court
Cour des petites créances de ....

Claim
Créance
Courts of Justice Act, R.R.O. 1990, Reg. 201
Loi sur les tribunaux judiciaires, L.R.O. 1990, Régl. 201
Form/*Formule* 7A

Refer to / *Réf.*	**NO**	19
Amount of Claim / *Montant de la demande*	$	
Entry Fee / *Droits d'inscription*	$	
	$	

When referring to this document, please use number in upper right corner
Veuillez utiliser le numéro en haut à droite comme référence de ce document

PLAINTIFF/*DEMANDEUR*
Name/*Nom*

DEFENDANT(S)/*DÉFENDEUR(S)*
Name/*Nom*
Street No./*N° et rue* — Address/*Adresse* — Apt. No./*N° d'app.*
Borough/City/*Ville/Municipalité* — Postal Code/*Code postal* — Phone No./*N° de tél.*

DEFENDANT/*DÉFENDEUR*
Name/*Nom*
Street No./*N° et rue* — Address/*Adresse* — Apt. No./*N° d'app.*
Borough/City/*Ville/Municipalité* — Postal Code/*Code Postal* — Phone No./*N° de tél.*

To the Defendant/*Au défendeur:*
The Plaintiff claims from you $ ...., and costs for the reason(s) set out below.
Le(s) demandeur(s) vous demand(ent) la somme de .... $ plus les frais pour la(les) raison(s) indiquée(s) ci-après.

IF YOU DO NOT FILE A DEFENCE WITH THE COURT WITHIN TWENTY DAYS AFTER YOU HAVE RECEIVED THIS CLAIM, JUDGMENT MAY BE ENTERED AGAINST YOU. PLEASE REFER TO INSTRUCTIONS ON REVERSE.
SI VOUS NE DÉPOSEZ PAS DE DÉFENSE AUPRÈS DU TRIBUNAL DANS LES VINGT JOURS SUIVANT LA RÉCEPTION DE CETTE DEMANDE, UN JUGEMENT PEUT ÊTRE RENDU CONTRE VOUS. VOIR RENSEIGNEMENTS AU VERSO.

Canadian small claims courts offer simple claim forms, like this one, that cost only about $125 for a $6,000 claim, and actions take only a few months to be heard. The beauty of a small claims action is that most provincial courts force the automaker into mediation with you before proceeding to trial. Many claims are settled through mediation for one-half to two-thirds of the amount demanded in the action.

At this point, it would be a smart idea to hire a lawyer or a paralegal for a brief walk-through of small claims procedures to ensure that you've prepared your case properly and that you know what objections will likely be raised by the other side. If you'd like a lawyer to do all the work for you, there are a number of inexpensive law firms around the country that are experienced in small claims litigation. In Toronto, some law offices charge a flat fee of $750 for the basic small claims lawsuit and trial.

Remember that you're entitled to bring to court any evidence relevant to your case, including written documents, such as a bill of sale or receipt, contract, or letter. If your car has developed severe rust problems, bring a photograph (signed and dated by the photographer) to court. You may also have witnesses testify in court. It's important to discuss a witness's testimony prior to the court date. If a witness can't attend the court date, he or she can write a report and sign it for representation in court. This situation usually applies to an expert witness, such as an independent mechanic who has evaluated your car's problems.

If you lose your case in spite of all your preparation and research, some small claims court statutes allow cases to be retried, at a nominal cost, in exceptional circumstances—if a new witness has come forward, additional evidence has been discovered, or key documents (previously not available) have become accessible. In Ontario, this little-known provision is Rule 18.4 (1).B.

Class actions

Class action suits allow a single individual to sue a company, government, or other entity on behalf of hundreds or even thousands of other individuals with similar claims. Although class action suits have been used for three decades in the United States, they are a recent arrival in Canada. Quebec was the first province to adopt this legal remedy, in the early '80s, while Ontario and British Columbia followed in 1992 and 1995, respectively.

Class actions allow for contingency fees where consumers can enter into no-win-no-pay agreements with lawyers. Usually, if you lose, you pay your expenses and move on. However, Ontario judges can require losing class action plaintiffs to pay the defendant's fees as well. (B.C. legislation requires that both sides pay their own legal costs.)

One of the more recent successful class actions in Canada concerned the recovery of condo owner deposits in Toronto. In *Windisman v. Toronto College Park Ltd.*, 544 condominium residents recovered $2.6 million, representing the interest earned on their deposit payments for apartments, parking spaces, and storage lockers. Other Canadian class actions presently winding their way through the courts involve breast-implant manufacturers, Red Lobster restaurants, appliance manufacturers, the Toronto Transit Commission, and Chrysler Canada.

How to file a class action

- Like any other lawsuit, a lawyer files the plaintiff's statement of claim against a defendant, and the plaintiff applies to the court to certify the lawsuit as a class action.
- The presiding judge will then decide whether there is an "identifiable class of two or more persons," whether a class action is the "preferable procedure," and whether the plaintiff truly represents the class. If you meet all the above criteria, the judge will issue a certification order and designate you as the Class Representative.
- Other class members must be notified of the lawsuit. Small groups can be contacted by mail, but larger groups may require notification through newspaper ads backed up by a toll-free telephone line. Members of the class must then be given an opportunity to opt out of the lawsuit. If they don't, they remain part of the class.

If the class wins, individual cases may then be heard to assess damages, or a notification will be sent to each member to apply for his or her part of the settlement or award.

It may take from three to five years before a final judgment is rendered. However, appeals may double that time period. Lawyers typically charge the class one-third of the amount obtained.

Presenting your case

As plaintiff, you will get the first opportunity to rise and state your case to the judge. You should ask for the exclusion of witnesses from the courtroom (this increases the chance that the other side will give contradictory testimony) and then proceed to lay out your proof, concluding within 5 to 10 minutes. Hang your narrative on the documents you produce (you can call them "P-1," "P-2," for "Proof 1," "Proof 2," and so on) and let them serve as note cards.

The first three documents should be the sales contract, all work orders relating to the problems you've had, and your registered complaint letter or fax. After that, you may wish to produce the dealer's or automaker's response, a report or work orders from an independent garage supporting your position, and copies of your maintenance records and dealer service bulletins in order to show the problems are factory related and not maintenance items. Conclude your presentation by simply restating the claim as it's written on your court complaint.

On the day of the trial, bring in a mechanic to confirm the defects exist and to provide an estimate of repair costs. If the repairs have already been carried out, he or she can explain what caused the defects and justify the bill for repairing them. This should be done by presenting the defective parts, if possible. The mechanic must convince the judge that the defects were present at the time the car was sold and were not caused by poor maintenance or abusive driving habits.

When the dealer gets on the stand, ask for the exclusion of all witnesses and try to ferret out the following facts:

- When and from whom was the used car last purchased?
- For how much was it bought?
- What was done to recondition the car, and at what cost?
- What was the *Red Book* value of the car when first bought from the previous owner and when sold to the plaintiff?

This line of questioning should show the judge the considerable profits the dealer made by buying the car below the market value, not spending much to recondition it, and reselling it far above the *Red Book* value.

Before the dealer leaves the stand, get him to confirm whether he or his salespeople made any representations, either verbally or through a newspaper ad, extolling the vehicle's qualities. With witnesses excluded, it's quite likely that the dealer's witnesses will contradict him when it's their turn to testify.

Other witnesses who can help your case are the car's previous owner (who can testify as to its deficiencies when sold) and any of your co-workers or friends who can testify as to how well you maintained your vehicle, how you drove it, and the seriousness of the defects.

Dos and Don'ts

- Don't sue an auto dealer if he has no money, isn't bonded, is bankrupt, has changed his company, or is willing to negotiate a settlement.
- Don't accept a false sales tax receipt. Crooked car dealers try to get their customers to take a receipt that indicates a selling price far below the price paid.
- Don't threaten or insult the dealer. This will only gain sympathy for him and hurt your own credibility.
- Do complain to the provincial transportation and consumer affairs ministries about possible violations of provincial laws.
- Do bring in the RCMP if you suspect the odometer has been tampered with.
- Do contact local consumer groups and the APA for used car jurisprudence and help in mediating the complaint.
- Do publicize any favourable court judgment as a warning to other dealers and as encouragement for other consumers.

Settlements

You may be asked to sign a document, called a release, which proves that a final settlement has been made. Generally, once you sign the release you can't sue the other person for that particular debt or injury. If you're the debtor, it's very important that you make sure the other person signs the release when you pay him or her. If you're the creditor collecting on the debt, you must sign the release, but don't do so until you've received the money. Release the debtor from that particular debt, but don't release him or her from all future debts.

Sample Settlement Form

I, John Doe, hereby acknowledge the payment of $300 by Jane Smith to compensate me for the defects in the vehicle I bought from her on _________. In accepting this payment, I hereby drop all present and future claims against Ms. Smith arising from the purchase of this vehicle.

________________ ____________________________

Date **John Doe**

____

Jane Smith

Deadbeat defendants

If you're dealing with a professional crook, the real work begins once you win your case. You may have to garnish (seize) part of the defendant's bank account or wages or ask the sheriff to serve a writ of execution. This writ allows the sheriff to demand full settlement plus court costs and, failing that, to seize the defendant's goods to cover the amount of the judgment. But here's the catch: property that's needed to earn a living (car, tools, machinery, etc.), household goods, and anything encumbered by a lien are exempt from seizure.

Professional deadbeats can tell the sheriff that practically everything they own is exempt—and it will take another action, at the plaintiff's expense, before the regular courts have the defendant questioned under oath. If he's found to be lying, he can then be sent to jail for perjury or contempt of court, but the small claims court judgment will remain unpaid.

Key Court Decisions

The following Canadian lawsuits and judgments cover typical problems that are likely to arise. Use them as leverage when negotiating a settlement or as a reference should your claim go to trial. Legal principles applying to Canadian and American law are similar; however, Quebec court decisions may be based on legal principles that don't apply outside that province. Therefore, do what most lawyers do: present all the court judgments that may be helpful and let the presiding judge or the defendant's lawyer sort out those they feel may not apply.

Additional court judgments can be found in the legal reference section of your city's main public library or at a nearby university law library. Ask the librarian for help in choosing the legal phrases that best describe your claim.

Prior-ownership disclosure

Lemon-Aid readers have told me that some government agencies won't give out prior-ownership information, claiming that it's confidential. Fortunately, the courts have ruled otherwise.

General Motors Acceptance Corp. of Canada v. Saskatchewan Government Insurance. GMAC asked the Saskatchewan Government Insurance Corporation (SGI) for the names and last-known addresses of registered owners of its vehicles. SGI refused to provide this information on the basis that it was personal information exempt from disclosure by virtue of s. 24 of the Freedom of Information and Protection of Privacy Act. On appeal, the court noted that the dominant objective of the act was disclosure, though there were limited exceptions. The court ruled that a name was not in itself personal information and neither were the details of a licence or permit. Furthermore, s. 24 stated that the legislative purpose was not to restrict access to information that had been available in the past, which included vehicle registration information of the kind sought in this case.

Using an expert (whistleblower)

GM v. Elwell, U.S. Supreme Court, January 1998. Auto industry whistleblowers have gained an ally: the U.S. Supreme Court ruled that they are free to testify against their former employers in other states or in Canada, despite having signed an employee agreement not to help plaintiffs in litigation against their employer. Justice Ruth Bader Ginsburg wrote the majority opinion that the employee-employer agreements were valid only in the state where they were signed. Elwell, a former GM engineer active in assessing product liability lawsuits for the automaker over a 15-year period, has been particularly effective, since he retired, in helping plaintiffs win against GM in cases involving fire-prone fuel tanks in pickups.

Delay (in bringing lawsuit)

Bouchard v. Vaillancourt (1961), C.S., 171. If the seller has defrauded the consumer, delays for initiating a lawsuit are longer than in cases where fraud is not involved.

See also:

- *Ginn v. Canbec Auto* (1976), C.S., 1416
- *Lemire v. Pelchat* (1957), R.C.S., 823
- *Ennis v. Klassen,* 70 D.L.R. (4th), 325

Damages (Punitive)

Punitive damages allow the plaintiff to get compensation that exceeds his or her losses. Judges sometimes award punitive damages in Canada as a deterrent to those who carry out dishonest or negligent practices; however, these kinds of judgments are more common in the U.S.

Bolduc v. Racicot, April 2, 1982, Quebec Provincial Court (Hauterive), No. 655-02-000364-817, Judge Cloutier. The plaintiff was given a verbal estimate of $500 for repairs, but the garage presented a bill for $1,255. The Court reduced the bill to $900, with the consent of the plaintiff, and then awarded another $300 as punitive damages against the garage for failing to conform with the requirements of the Consumer Protection Act by not providing a written estimate, by doing more work than was requested, by not guaranteeing the work, and by not presenting a bill in conformity.

Travis Eby v. J. S. Saville Holdings and Guelph City Imports Subaru/Lada, and Cory Tanguay, Ontario Court (General Division), No. 9358/96, September 5, 1997, Justice Clarke. The purchaser of a used 1989 Pontiac Sprint discovered the vehicle had 83,000 km and 12 previous owners, instead of 16,000 km and only one previous owner, as had been represented by the dealer. Judge Clarke awarded the plaintiff $4,000 in damages as the difference in what the buyer would have paid had the true facts been known. Interestingly, under the Ontario Business Practices Act, the plaintiff was awarded another $1,000 in exemplary damages (rarely given in Canadian courts) because the judge found the dealer's staff made

> false, misleading or deceptive consumer representations as to the quality of the vehicle...failed to state material facts, which they knew or ought to have known and/or employed exaggeration, innuendo, and ambiguity with respect to material facts. I also find that s.4 of the Act applies. I find such business practices were unfair and warrant the court awarding exemplary or punitive damages....

Vichek v. Koshel (1988), 44 C.C.L.T. 314 (B.C.S.C., Van Reg. No. B842974). The plaintiff was seriously injured when she was thrown from an all-terrain cycle she had been riding as a passenger. The court allowed for punitive damages because the manufacturer was well aware of the injuries likely to be caused by the cycle.

Defects (Safety Related)

Airbags

Although safety experts agree that you are likely to need anti-lock brakes 99 times more often than an airbag, the bag's advantage is that it doesn't depend upon driver skill or reaction time and it deploys when it's too late for braking. Nevertheless, there have been thousands of reports of airbags that have failed to deploy or have accidentally gone off and caused massive injuries. In fact, General Motors recalled almost a million Cadillacs, Cavaliers, and Sunfires for just that problem of accidental deployment—caused by wet carpeting in Cadillacs and passing over a bump in the road for the other vehicles.

Safety experts at the NHTSA once estimated that 25,000 people were injured by airbags between 1988 and 1991. Additionally, recent NHTSA-run crash tests indicate that all of Chrysler's minivan airbags produced in 1997 and earlier deploy with such excessive force they may cause disabling or fatal injuries. In February 1999 crash tests, the deploying passenger-side airbag in a 1997 Caravan caused neck injuries to a small, belted, female dummy that would have disabled or killed a person, the agency said. The suspect airbags may be in as many as 1.9 million minivans. Coincidentally, a U.S. national auto safety group, the Insurance Institute for Highway Safety (IIHS), has launched an exhaustive investigation into reports that inflating airbags have seriously injured motorists.

Hundreds of lawsuits have been filed claiming airbags don't function as designed (not deploying when they should, or deploying when they shouldn't) and over 60 suits have been filed claiming the device caused or aggravated injuries after actually deploying as designed. Chrysler, the first automaker to install airbags as standard equipment, is the target of most of these lawsuits. So far, the successful suits against Chrysler relate to poor design rather than malfunctions and fall into two categories: severe burns and premature deployment at low speeds.

Severe burns

Claimants have won substantial jury awards for first- or second-degree burns caused by the Thiokol-designed airbag directing hot gases at the driver's hands and wrists. Used on Chrysler's 1988–91 models, these airbags have vent holes that direct hot gases at the three o'clock and nine o'clock hand positions. In late 1990, the vent holes were relocated to the twelve o'clock position. A class action lawsuit asking for damages arising from the earlier Thiokol design was filed in Philadelphia County and has recently resulted in a verdict of $60 million in compensatory damages and another $3.75 million in punitive damages for 80,000 Pennsylvania Chrysler owners who purchased their vehicles between 1988 and 1990. The jury ruled that vehicles sold during that time came with airbags that when deployed could severely burn the hands and wrists of drivers. Owners were awarded $730 each to replace the defective airbags, although Pennsylvania's Consumer Protection Law could triple the damages. Martin D'Urso from the law firm of Kohn, Swift & Graf and Isaac Green from Moody & Anderson pleaded the case for the plaintiffs.

Premature airbag deployment

All automakers are worried they may face a slew of huge damage awards in the future following a $750,000 jury award for damages in the death of a five-year-old from a deploying airbag. In *Crespo v. Chrysler*, a New York jury concluded that the 1995 minivan's airbag design contributed to the child's death because it deployed too early (at a speed of between 9 and 12 mph). Safety experts contend that the airbag should deploy within a range of 15 to 20 mph. Chrysler submitted, however, that there are no standards as to what speeds should trigger the airbag's deployment and claimed most automakers program their airbags to deploy at crash speeds of 8 to 14 mph. The jury rejected Chrysler's argument and awarded half the damages sought by the child's family, despite the fact that the child was unbelted and not seated in a safety seat.

Axle

Fuller v. Ford of Canada Ltd. (1978), 2 O.R. (2d), 764. A new 1974 Econoline truck had an axle failure that caused an accident. The court held both the manufacturer and the seller responsible.

Battery

Marin v. Varta Batteries Ltd. (1983), 28 Sask. R. 173; 5 D.L.R. (4th) 427 (Q.B.). An exploding battery caused serious injuries to the plaintiff's face. The battery maker admitted liability and the court awarded the plaintiff $21,000 in damages.

Brakes

Marton Properties v. Northbridge Chrysler Plymouth Ltd., March 2, 1979, British Columbia Supreme Court. The plaintiff's used Chrysler had serious brake defects that warranted cancellation of the sale contract. The court ordered the vehicle returned and the purchase price refunded.

Morschler v. Masser's Garage (1956), 2 D.L.R. (2d), 484. The plaintiff was awarded damages to a third party from a repairer who had failed to fix a car's brakes properly.

Phillips v. Ford of Canada and Elgin Motors (1970), 2 O.R., 714. Ford was held responsible for the injuries caused by power-brake defects that wouldn't be apparent to the plaintiff. The dealer and Ford had a duty to warn the plaintiff that if the power-brake unit failed, the back-up brakes would be inadequate.

Brakes (ABS)

Chrysler has come under fire over the past several years for installing defective ABS brakes in its minivans. The American judgment below is the first of a series of lawsuits alleging that the company's base braking system is also faulty.

Santos v. Chrysler Corporation, Suffolk County, New York, Superior Court, February 1996. The plaintiff's wife and three children were killed when his 1988 Caravan's rear brakes locked as he applied them to avoid rear-ending another vehicle. The jury award for $19.2 million following Paul Santos's pleadings that Chrysler "knowingly built the vehicle with a deadly defect that caused the rear brakes to lock before the front brakes." Chrysler's defence was that Santos drove the Caravan 100 miles with a broken windshield wiper and steered directly into oncoming traffic.

Fires

Ford

Brown v. Ford of Canada Ltd. (1979), 27 N.B.R. (2d), 550. While under warranty, the plaintiff's 1977 van suddenly caught fire. The court held the manufacturer only partially negligent because the plaintiff's son, who was a mechanic, should have spotted the defect.

Chabot v. Ford Motor Company of Canada Ltd. (1983), 39 O.R. (2d) 162. This case contains an excellent and exhaustive review of a manufacturer's liability for defectively manufactured vehicles. It establishes liability for losses arising from a defect while the vehicle is being maintained by the dealer. Although the plaintiff could not prove that a defective part caused a fire, Judge Eberle's presumption that the 1979 F-250 truck ignited from some sort of manufacturing defect implied that the manufacturer was negligent (*re ipsa loquitur*).

The court also ruled that Ford breached a fundamental term of the sales contract by providing an unfit vehicle and that Ford also failed to meet its obligations as set by the Manitoba Sale of Goods Act (Ontario Sale of Goods Act R.S.O. 1980 c 462, 5-15).

In his decision, the judge made a number of interesting and critical observations about the motives and strategies of a large manufacturer like Ford when it prepares for a case and presents expert evidence.

In cases where companies try to avoid liability through their own limited warranties or by invoking the exclusions expressed in the warranty, this decision confirms the *Kravitz v. General Motors* decision from a common law standpoint. That case struck down these exclusions under Quebec law.

See also:

- *CPR v. Kerr*, 49 S.C.R., 33 at 36
- *Gougeon v. Peugeot Canada*, July 20, 1973, Quebec Superior Court, No. 12736, Judge Kaufman
- *Lazanik v. Ford*, June 15, 1965, C.S.M. 623-664, Judge Challies
- *Parent v. Lapointe* (1952), I S.C.R., 381
- *Rioux v. General Motors*, March 9, 1970, C.S.M. 739-005, 6
- *Touchette v. Pizzagalli* (1938), S.C.R. 433
- *Zelezen v. Model Auto Sales Ltd.*, November 26, 1971, C.S.M. 722-487, Judge Nichols

Chrysler
La Paix v. Chrysler Canada, April 5, 1982, Quebec Provincial Court, No. 500-02-040677-796, Judge Prenouveau. A new 1976 Volaré suddenly caught fire when the owner tried to start the car. It had been in use for only 10 months. Chrysler Canada refused all liability, claiming the fire was probably due to poor starting technique. The court held Chrysler responsible for $4,039 in damages after it was proven that the fire was caused by unburned gasoline catching fire in the muffler.

General Motors
Racine v. Durand Pontiac-Buick and General Motors, December 15, 1977, Quebec Provincial Court, No. 02-015218-774, Judge Lacoste. The plaintiff claimed $700 for a catalytic converter that exploded. General Motors and the dealer claimed that the explosion was due to poor maintenance and that the 5-year/50,000 mile emissions warranty on the converter did not cover such maintenance-related defects. The judge disagreed, stating that the catalytic converter exploded due to a defective PCV valve for which General Motors and the dealer had to take responsibility. The fact that the converter could overheat and catch fire whenever the engine was badly tuned caused the judge to suggest that a warning be placed on all General Motors vehicles with the converter. The plaintiff was awarded $700.

Saab
Delage vs. Saab, San Francisco Superior Court, November 1997. A jury awarded Delage $1.4 million for damages caused by an electrical fire in his 1988 Saab 9000, even though the car was never examined. After hearing testimony from eight other Saab owners whose cars had caught on fire, the jury concluded that a defective fuse box caused the fire. Related to this, the *San Francisco Chronicle* published a Saab internal memo that indicated a main connection in the 9000 box (located behind the glove compartment) could loosen, overheat, and ignite the insulation.

Parking gear slippage
General Motors v. Colton, October 17, 1980, Quebec Court of Appeal, No. 500-09-000692-772. A 1970 Oldsmobile parked on the plaintiff's inclined driveway rolled back and injured the owner. Judgment was against General Motors for $29,000.

Rusting undercarriage
McGouey v. Lawson Motors Ltd. (1982), 42 N.B.R. (2d) 225. Seven weeks after McGouey bought a used Fiat, the frame collapsed from excessive corrosion. The purchaser was awarded a complete refund of the purchase price and damages.

R. v. Ford Motor Company of Canada. In this case, the federal Ministry of Transportation successfully pleaded that rusting of the undercarriage of Ford's 1965–1974 midsize cars was a safety defect. Ford appealed the case for over a decade and agreed to notify owners of the defect (but not fix the vehicles) only after losing all its appeals.

Seatbelt design/defects

Door-mounted seatbelts, motorized seatbelts, and lapbelts have come under fire during the past decade from safety advocates and the courts, after insurance studies and trauma specialists showed that they increase the severity of accident injuries.

Door-mounted seatbelts

Federal safety agencies are currently looking at the reliability and safety of door-mounted seatbelts in light of accident reports showing a high failure rate and increased severity of injuries. GM, for example, is being sued for $33 million over the inadequate performance of the driver's door-mounted seatbelt on a 1990 Pontiac Sunbird, says the Washington-based Center for Auto Safety. The driver was killed when the car door opened, rendering the seatbelt inoperative. The suit claims that GM was negligent in the design that caused the belt to be undone, leaving the driver unprotected and unrestrained.

Motorized seatbelts

These front seat restraints run along a channel and cross the shoulder when the ignition is turned on. A lapbelt has to be fastened separately. These automatic belts are set so high above the door frame that they're uncomfortable to wear and a nightmare to adjust. Their most serious shortcoming, however, is that they give the driver a false sense of security and the separate lapbelt often goes unfastened (the U.S. government says that only 29 percent of people with motorized seatbelts use the lap portion). This has resulted in a number of accidents where the driver or front passenger has been decapitated or paralyzed by the shoulder belt.

Three-point seatbelts, failure to provide

In *Garrett v. Ford*, a Baltimore, Maryland, federal court jury rendered a $3.2-million verdict against Ford for installing a dangerously designed lapbelt and for not including back seat shoulder harnesses in a 1985 Escort. The lapbelts were designed to cross at the waist instead of the pelvis, an error that contributed to the passengers' paralysis. If shoulder harnesses had been installed, the lapbelts would not have aggravated the injuries, says the Center for Auto Safety.

Ford is involved in another ongoing trial, under appeal, where the company is being sued for compensatory and punitive damages arising from the death of a 72-year-old nun, Sister Mary Margaret

LeGlise. She was riding in the rear seat of a 1987 Ford Tempo with her lapbelt fastened when the accident occurred. At trial, California Superior Court judge Jeffrey Miller flew into Ford for keeping its settlements in other cases secret (referring to a $6-million payout to settle a San Diego case) in order to avoid embarrassment and other lawsuits. Judge Miller further noted that Ford could not claim that the technology to install three-point belts had not been perfected when the company had, for years, put the safer three-point belts in its European models. Judge Miller concluded,

> Ford knew that the rear seat lapbelt not only did not provide adequate protection but actually caused injury to its users. Ford also knew through its own internal engineering analysis and experimentation that the three-point restraint system was vastly superior and would save many lives.

Steering

Holley v. Ford of Canada, June 18, 1980, Nova Scotia Supreme Court, Judge Cowan. The plaintiff purchased a used 1976 Ford Custom 200 truck that was subsequently involved in an accident due to the failure of its steering assembly. The plaintiff was awarded $4,100 in damages.

White v. International Harvester (1969), W.W.R. 235 (Alta. T.D.). This new truck's steering assembly had problems and was adjusted by the dealer. However, 4,200 miles later, the truck suddenly veered out of control due to a defective steering component. The plaintiff was awarded both costs and damages.

Sudden acceleration

Willar v. Ford (1991), 118 N.B.R. (2d) 323. The owner sued Ford after his vehicle's throttle jammed open twice, causing several accidents. The court held that Ford, not the dealer, was responsible for the plaintiff's damages.

Tire iron placement adding to collision injuries

Gallant v. Beitz (1983), 42 O.R. (2d) 86. This judgment states clearly that automakers are presumed to know their vehicles will be involved in accidents and must take reasonable steps to minimize injuries caused by flaws inherent in the design of their vehicles.

Tires

Chase v. Goodyear Tire and Rubber Co. and Goodyear Canada Inc., New Brunswick Court of Queen's Bench, Trial Division, No. S/C/513/89, April 12, 1991. The plaintiff was injured when the tire he was inflating exploded. There was no evidence of a manufacturing defect on the original tire or on the retread. Goodyear was found negligent because it failed to warn its customers of the danger inherent in inflating a retreaded tire that had worn-out and damaged radial cords, thereby breaching its duty to warn. The plaintiff was awarded $65,000 for his injuries.

Gagnon v. Canadian Tire (1979), C.P. 251, Quebec Provincial Court. The plaintiff was awarded $1,310 as compensation for a radial tire that exploded when he was rocking his car out of a snowbank.

Murphy v. D. & B. Holdings (1979), 31 N.S.R. (2d) 380; 52 A.P.R. 380 (S.C). The court held that the manufacturer was liable for damages for permitting a customer to use its tires in a dangerous manner. (The tires were under-inflated and spun the wheels.)

Transmissions jump from "Park" to "Reverse"
McEvoy v. Ford Motor Company et al., Supreme Court of B.C., No. 3841989, September 6, 1989, Judge Hinds. McEvoy was killed when his new Ford pickup backed over him after the transmission jumped from "Park" to "Reverse." He had left the engine running and was unloading cargo when the right front wheel crushed his chest. Justice Hinds found Ford to be 65 percent negligent for the following reasons:

- The RCMP mechanic testified that the gearshift lever could easily be mispositioned into an "unstable Park" position.
- It was determined that Ford was aware of this defect as early as 1971, according to internal company documents presented during the trial.
- Ford had a duty to warn its customers of the hazard caused by the C-6 automatic transmission; the warning in the owner's manual was deemed insufficient to satisfy this duty.
- Ford and its Canadian distributor had breached their duty to the consumer.

Defects (Body/Performance Related)

When a used vehicle no longer falls within the limits of the warranty expressed by the manufacturer or dealer, it doesn't necessarily mean that the manufacturer can't be held liable for damages caused by defective design. As mentioned before, the manufacturer is always liable for the replacement or repair of defective parts if independent testimony can show that the part was incorrectly manufactured or designed. The existence of a secret warranty extension or service bulletins will usually help to prove that the part has a high failure rate. For example, in *Lowe v. Chrysler*, internal service bulletins were instrumental in showing an Ontario small claims court judge that Chrysler had a history of automatic transmission failures since 1989!

In addition to replacing or repairing the part that failed, an automaker can also be held responsible for any damages arising from the part's failure. This means that loss of wages, supplementary transportation costs, and damages for personal inconvenience can be awarded.

Paint delamination/peeling

A defect that first appeared on Ford sport-utilities, vans, and pickups in the '80s, paint delamination occurs when the top coat of paint separates from the primer coat or turns a chalky colour, mostly along horizontal surfaces and often as a result of intense sunlight. When the paint peels, the entire vehicle must be repainted after a new primer resurfacer has been added. For some vehicles, the labour alone can run about 20 hours at a cost of $75 an hour.

The same paint problem affects mostly 1986–97 Chrysler, Ford, and GM vehicles equally; however, each company has responded differently to owners' requests for compensation. To help you prepare the best arguments for negotiations or court, each automaker is profiled separately, beginning with an analysis of the problem and a website reference, followed by copies of lawsuits, judgments, or dealer service bulletins that will help your claim. Pay particular attention to GM's "Goodwill Administration" service bulletin (see page 83). It gives you a peek into how all the automakers play mind games with their customers in order to keep their warranty payout as low as possible.

Chrysler

Chrysler's paint deficiencies include paint delamination, cracking, and fading between the third and fifth year of ownership. One Chrysler service bulletin that can serve as a guide for paint claims is Bulletin 23-11-90, "Base Clearcoat Paint Damage Resulting from Alkaline Spotting, Industrial Fallout, or Chemical Etching," issued September 10, 1990. It contains a number of illustrations relating to the kinds of paint problems that Chrysler will repair under its base warranty on all 1990 model year vehicles and puts Chrysler on record as accepting so-called environmental paint damage.

That bulletin and other helpful documents can be downloaded from the Internet at *www.wam.umd.edu/~gluckman/Chrysler/* (Chrysler) and *www.goofball.com/keane/badpaint/* (Jeep).

A class action lawsuit has also been filed in the state of Washington, seeking damages for all Chrysler owners who have owned or leased paint-delaminated 1986–97 models. The suit, *Schurk, Chanes, Jansen, and Ricker v. Chrysler*, No. 97-2-04113-9-SEA, was filed in the Superior Court of King County, Washington on October 2, 1997 (contact Steve Berman or Clyde Platt with the Seattle, Washington, law firm of Hagens and Berman at 206-623-7292).

The 29-page Statement of Claim uses many photos and internal bulletins and memos to show that Chrysler engaged in

> ...unlawful, unfair, and fraudulent business practices and unfair competition by treating different members of the class differently

with respect to repairs it agrees to perform as "goodwill gestures," and by effecting partial repairs that do not address the true nature and extent of the delamination defect.

In explaining the delamination defect, the plaintiffs' lawyers maintain that Chrysler knew the defect also subjects the paint to softening, chipping, and other damage well before the delamination is detected.

Chrysler did not, however, disclose the defect to potential or actual purchasers, and has until recently denied the existence of any defect, blaming the problem on "acid rain," "the environment," or other factors that are beyond Chrysler's control.

Instead, Chrysler has pursued the practice of only selectively repainting vehicles for a few customers. In doing so, the company has turned away many more customers with the same problem with the explanation that Chrysler bears no responsibility for the condition. It has not notified purchasers or lessors of the defect, or offered them a refund for the difference between what these damaged cars were worth and the actual purchase or leased price. Rather, it has engaged in a uniform course of conduct designed to deceive class members who have asked about the cause of the delamination, by falsely stating that it was caused by environmental factors, road conditions, the age of the car, normal wear and tear, and other similar false and misleading responses.

Only a few customers who have repeatedly complained about their defective paint have received offers from Chrysler to pay for repainting their vehicles. Chrysler also sometimes offers "patch" paint jobs; these are only cosmetic cures and progressive damage is certain to follow such temporary measures. Even then, Chrysler offers to pay for only a portion of the expense of such partial repairs and pays consumers with the same extent of damage different amounts.

In researching the root cause of Chrysler's delamination defect, the lawsuit quotes Chrysler's Northwest Division Customer Sales Manager, Michael Mackey.

Mackey identified the cause of the delamination as an "incompatibility between the primer and the paint" that prevented adhesion between the two. He also said that Chrysler had known about the problem for "quite some time." Mackey further stated that Chrysler encountered the problem when it was forced to begin using water-based paints due to a ban on the use of leaded paints. In order to rectify the problem, Mackey said it would be necessary to strip the existing paint and primer and to put on a new primer coat using a process known as "etching."

This 1992 Chrysler minivan's paint peeling typifies Chrysler's peeling problem, while the 1988 Dodge Ram (below) shows the effects of paint delamination. Both problems are common to Ford and GM vehicles.

Beginning in the early '80s, Chrysler changed the constituents of its exterior car paints in order to enhance the corrosion resistance of its vehicles, to reduce the expense of the painting process by using the new, corrosion-resistant paint, and to comply with environmental regulations. In doing so, Chrysler streamlined the painting process for its vehicles. Up until that time, Chrysler painted its vehicles using a three-coat process: a bottom-layer electrocoat applied directly to the sheet metal, a spray primer, and a colour coat. Chrysler then began to move to a two-coat process. It substituted Uniprime

("HBEC" or "Ecoat")—a "high-build electrocoat" manufactured by PPG Industries and other paint manufacturers—for both the bottom coat and the spray primer. In eliminating the intermediate layer of paint, Chrysler partly offset the increased cost of using Uniprime: the time and expense required for the painting processes were reduced, as were chemical emissions. One of its competitors, Ford Motor Company, which had also begun at about the same time to use Uniprime made by PPG, estimated a savings of $6–$16 per vehicle by eliminating the middle layer of paint. Chrysler made this change without thoroughly testing the process or the integrity of the vehicle finishes it produced.

Chrysler began to use the new two-stage "enamel electrocoat" process at its plants in Newark, Delaware, Belvedere, Illinois, and Windsor, Ontario, in or around 1983–85; in or around January 1985, the company announced that it intended to coat all of its products with Uniprime. However, the new enamel electrocoat process was flawed.

Uniprime is typically applied electrostatically, with the result that the topmost or colour coat does not adhere properly. Uniprime is further defective because its use renders the bond with the finish coat subject to progressive deterioration when exposed to ultraviolet radiation produced even in climates with less-than-average sunlight conditions. Ultraviolet light causes Uniprime to oxidize, which weakens its bond with the colour coat of paint. In addition, the painting process is subject to a number of variables, including timing of application and thickness of the topcoat, temperature fluctuations during the painting process, and humidity inside the paint booth—variables that are critical to producing a durable finish but that were not properly controlled. These conditions added to the progressive bonding failure inherent in the use of Uniprime as the undercoat and primer.

Chrysler knew that its new painting process was flawed almost as soon as it was used in production. The experience of auto manufacturers like Ford, which had begun to use Uniprime in its painting process, illustrates the knowledge of the existence and cause of the paint-peeling problem by the automobile industry.

Contrary to Chrysler's claim that the delamination problem is not widespread, a memo to members of the Finance Committee from A. J. Trottman, executive vice president of Ford North American Automotive Operations, indicates what others in the automotive industry experienced on one model vehicle painted using the same two-step system as used on Chrysler vehicles:

> In June 1990, a field survey of about 1,000 F-series trucks (1985–1990 models) was conducted in three locations by Body and Chassis Engineering to assess paint durability. Results showed that about 13% of the F-series trucks displayed peeling paint, which would represent about 90,000 vehicles annually.

Chrysler's 1992 Warranty Manual provided that it covered "the cost of all parts and labor needed to repair or adjust any Chrysler supplied item...that proves defective in material, workmanship or factory preparation." Chrysler's actions and knowledge, as alleged above, constituted a breach of these warranty conditions.

Ford

Louisiana attorney Danny Becknel, along with other lawyers, has filed three separate class action lawsuits against Chrysler, Ford, and GM. He's also filed suit against PPG Industries, a company out of Pittsburg that he says sold defective car paint to the three companies.

Becknel claims in his suits that in the late '80s and early '90s, the Big Three bought Uniprime from PPG Industries without thoroughly testing it. He says that when the companies bought Uniprime, they switched from a three-coat car painting process (bottom coat, spray primer, and colour coat) to a two-coat process that eliminated the middle coat (spray primer). Becknel claims that eliminating that middle layer of spray primer saved $6–$16 a car: "This paint seems to be a minor cost but when you multiply it by 10 to 15 million times a year, it's a big number."

Faced with an estimated 13 percent failure rate of its painting process, Ford repainted its delaminated 1983–93 cars, minivans, vans, F-series trucks, Explorers, Rangers, and Broncos free of charge for five years under a secret "Owner Dialogue Program." Ford whistleblowers say the company discontinued the program in January 1995 because it was proving to be too costly. Nevertheless, owners who cry foul and threaten small claims action are still routinely given initial offers of 50 percent compensation, and eventually complete refunds if they press further.

In your negotiations with Ford, be sure to refer to Ford's admission of the delamination problem found in the above Ford service bulletin and in the Washington-state Chrysler class action.

Apart from the Becknel class action, Ford hasn't been a party to many lawsuits, preferring to settle before the cases come to trial. Nevertheless, you can get the latest information and internal documents relating to Ford's secret warranty by logging onto the following site set up by dissatisfied Ford owners, called "Ford Paint Peel": *www.ihs2000.com/~peel*.

General Motors

Confidential U.S. dealer service bulletins and memos confirm the 6-year/unlimited mileage benchmark that GM uses to accept or reject secret warranty paint claims. Of course, GM wants its customers to jump through hoops to benefit from its secret paint warranty, hence the line of questioning contained in its service bulletin reproduced on page 83.

Following is the dealer service bulletin that GM put out several years ago to guide dealers in determining whether paint delamination is a factory defect or is due to other external causes like acid rain, stone chips, etc. Pay particular attention to GM's explanation of the cause of the delamination problem. In effect, the automaker admits that it didn't apply sufficient primer to protect the clearcoat from ultraviolet light. Also, look at the masking-tape test GM recommends in its "Problem Identification" section to diagnose clearcoat delamination; it's the same test used by Ford and Chrysler.

Preliminary Information
FILE IN SECTION: 10-Body
BULLETIN NO.: 23-10-54
DATE: June 1995
SUBJECT:
Service Procedures for the Repair of Paint Colorcoat Delamination from ELPO Primer (Repair Surfaces Above the Body Side Moldings)
MODELS:
1988-91 Buick Century (Plant Code 6)
1988-92 Buick LeSabre (Plant Code H), Skylark
1988-90 Chevrolet Celebrity
1988-92 Chevrolet Beretta, Corsica, Skylark
1988-90 Oldsmobile Cutlass Supreme
1988-91 Oldsmobile Calais, Ciera (Plant Code 6)
1988-92 Oldsmobile Achieva, Eighty Eight (Plant Code H)
1988-90 Pontiac Grand Prix
1988-91 Pontiac 6000 (Plant Code 6)
1988-92 Pontiac Firebird, Grand Am, Tempest
1988-92 Chevrolet and GMC Truck C/K, R/V, S/T M/L, G Models
1991-92 Oldsmobile Bravada
This bulletin is being revised to provide a process change for the repair of paint delamination from ELPO primer. Please discard previous publications:

Buick	92-10-57
Chevrolet	92-300A-10
GM of Canada	93-10-100
GMC Truck	92-10-134A
Oldsmobile	93-T-05
Pontiac	93-10-59

EFFECTIVE DATE:
THE REVISED REPAIR PROCEDURES, NEW LABOR OPERATION NUMBERS AND REVISED TIME AND MATERIAL ALLOWANCES CONTAINED IN THIS BULLETIN ARE EFFECTIVE WITH REPAIR ORDERS WRITTEN ON OR AFTER JUNE 20, 1995.
CONDITION:
This bulletin is being issued to assure that the correct procedure is followed to repair a condition known as delamination. Some of the listed passenger cars, light duty trucks, and vans may have DELAMINATION (peeling) of the paint colorcoat from the ELPO primer depending upon variable factors including prolonged exposure to sunlight and humidity.
Blues, Grays, Silvers and Black Metallics are the colors that have the highest potential for this condition. On rare occasions, other colors may be involved.
Important:
Delamination is different than other paint conditions and/or damage. A proper problem identification is necessary, and the service procedure that follows is specific to the proper repair of delamination and must be followed. The information in this bulletin covers paint delamination of the colorcoat from the ELPO primer ONLY. It does not address any other paint conditions. Procedures for the repair of other paint conditions (stone chips, scratches, environmental damage, clearcoat peeling, runs, dirt, fading, etc.) will not effectively repair delamination and customer dissatisfaction will result.
CAUSE:
***This condition may occur on vehicles produced in plants where the paint process did not call for application of a primer surfacer. Under certain conditions, ultraviolet light can penetrate the colorcoat, sometimes causing a reaction and separation of portions of the colorcoat from the ELPO (electrocoat) primer.**
PROBLEM IDENTIFICATION:
On a clean surface, at or above room temperature, firmly apply a 2" wide piece of masking tape and pull upward quickly. DO NOT USE duct tape, cloth backed tape or other aggressive tapes. If the colorcoat flakes or peels away from the ELPO (leaving the ELPO intact) the colorcoat is delaminating and the vehicle should be repaired.

Before settling your paint claim with GM or any other automaker, download the latest information from dissatisfied customers who've banded together and set up their own self-help websites. Follow the links at *www.lemonaidcars.com.*

Getting a Settlement

The following settlement advice applies mainly to paint defects, but you can use these tips for any other vehicle defect that you believe is the automaker's/dealer's responsibility. If you're not sure that the problem is a factory-related deficiency or a maintenance item, have it checked out by an independent garage or get a dealer service bulletin summary for your vehicle. The summary may include specific bulletins relating to the diagnosis, correction, and ordering of upgraded parts needed to fix your problem.

1. If you know the problem is factory related, take your vehicle to the dealer and ask for a written, signed estimate. When you're handed the estimate, ask that the paint job be done for free under the manufacturer's "goodwill" program. (Ford's euphemism for this secret warranty is "Owner Dialogue Program," GM's term is "Special Policy," and Chrysler just calls it "goodwill." Don't use the term "secret warranty" yet, you'll just make the dealer and automaker angry and evasive.)
2. Your request will probably be met with a refusal, an offer to repaint the vehicle for half the cost, or, if you're lucky, an agreement to repaint the vehicle free of charge. If you accept half-cost, make sure that it's based on the original estimate you have in hand, since some dealers jack up their estimates so that your 50 percent is really 100 percent of the true cost.
3. If the dealer/automaker has already refused your claim and the repair hasn't been done yet, get an additional estimate from an independent garage that shows the problem is factory related.
4. Again, if the repair has yet to be done, send a registered claim letter or fax to the automaker (send a copy to the dealer), claiming the average of both estimates. If the repair has been done at your expense, send a registered claim letter or fax with a copy of your bill. A sample letter/fax can be found on page 87.
5. If you don't receive a satisfactory response within a week, deposit a copy of the estimate or paid bill and letter/fax claim before the small claims court and await a trial date. This means that the automaker/dealer will have to appear, no lawyer is required, costs should be minimal (under $100), and a mediation hearing or trial will be scheduled in a few months followed by a judgment a few weeks later (the time varies among different regions).

Things that you can collect to help your case: photographs, maintenance work orders, previous work orders dealing with your problem, dealer service bulletins, and an independent expert (the garage or body shop that did the estimate or repair is best, but you can also use a local teacher who teaches automotive repair).

Other situations

- If the vehicle has just been repainted but the dealer says that "goodwill" coverage was denied by the automaker, pay for the repair with a certified cheque and write "under protest" on the cheque. Remember, though, if the dealer does the repair, you won't have an independent expert who can affirm that the problem was factory related or that it was a result of premature wearout. Plus, the dealer can say that you or the environment caused the paint problem. In these cases, internal service bulletins can make or break your case.
- If the dealer/automaker offers a partial repair or refund, take it. Then sue for the rest, including any "deductible" you were forced to pay; GM's own memo says the deductible can be waived (see page 76). Remember, if a partial repair has been done under warranty, it counts as an admission of responsibility, no matter what "goodwill" euphemism is used. Also, the repaired component/body panel should be just as durable as if it were new. Hence, the clock starts ticking again until you reach the original warranty parameter—again, no matter what the dealer's repair warranty limit says.
- It's a lot easier to get the automaker to pay to replace a defective part than it is to be compensated for a missed day of work or a ruined vacation. Manufacturers hate to pay for consequential expenses apart from towing bills because they can't control the amount of the refund. Fortunately, the courts have taken the position that all expenses (damages) flowing from a problem covered by a warranty or service bulletin are the manufacturer's/dealer's responsibility under both common law and Quebec civil law. Nevertheless, don't risk a fair settlement for some outlandish claim of "emotional distress," "pain and suffering," etc. If you have invoices to prove actual consequential damages, then use them. If not, don't be greedy.

Very seldom do automakers contest these paint claims before small claims court, opting instead to settle once the court claim is bounced from their customer relations people to their legal affairs department. At that time, you'll probably be offered an out-of-court settlement for 50–75 percent of your claim.

Stand fast and make reference to the service bulletins you intend to subpoena in order to publicly contest in court the unfair nature of this "secret warranty" program (automaker lawyers cringe at the idea of trying to explain why consumers aren't made aware of these bulletins). One hundred percent restitution will probably result. A good example is the *Shields v. General Motors of Canada* judgment rendered January 6, 1998. (See following section.)

Bentley v. Dave Wheaton Pontiac Buick GMC Ltd and General Motors of Canada, Victoria Registry No. 24779, British Columbia Small Claims Court, December 1, 1998, Judge Higinbotham. This is the third, and most recent, small claims judgment against GM. It builds upon the Ontario *Shields v. General Motors of Canada* decision and cites other jurisprudence as to how long paint should last on a house. If you're wondering why Ford and Chrysler haven't been hit by similar judgments, remember that they usually settle.

Reasons for judgment

In this case the claimant purchased a vehicle, a pickup truck, from a dealership, Dave Wheaton Pontiac Buick G.M.C. Ltd., a new vehicle, in 1991. There was an admitted defect in the paint which did not become apparent until later. General Motors is also a defendant in this action and discovered in a general sense this problem of delamination in the paint on some vehicles in 1992, about one year after the claimant purchased the vehicle in question.

The specific problem with this vehicle was observed early in 1994. It was brought to the attention of Wheaton when the vehicle was brought in for other maintenance two weeks after the warranty expired. At that time the problem was relatively minor. I say relatively in the sense that compared to what later occurred it was minor.

Mr. Palfry, who is the manager of the paint and body shop for the retailer Wheaton, was made aware of the problem. There is no dispute about that, and he sold the claimant a tub of touch-up paint.

The paint continued to deteriorate and in late 1996 was severely peeled.

In January of 1997, the claimant became aware that this problem was general to certain GM vehicles, vehicles of certain colours produced at a certain time by the defendant company.

The claimant took the vehicle back in but was told that it was too late. The warranty had expired and even the discretionary goodwill warranty was over. I do not think the claimant was told about the discretionary goodwill warranty, but in fact it was a policy of GM to extend the warranty for this sort of claim in certain circumstances, but it was discretionary. In any case, the claimant was told that it was too late for it to be fixed under warranty. Neither the dealership nor the manufacturer would accept responsibility at that time. As a result, this action was commenced.

I make the following findings: There was a latent defect in the vehicle relating to the paint, which revealed itself over time and in far less time than a good paint job would be expected to last.

The dealer or the manufacturer, had the manufacturer been informed by the dealer, ought to have advised the claimant in May of 1994 that they were aware of this general delamination problem and ought to have advised as to what warranty extension might be available. I will say more about this in a moment. In any event, the dealer and the manufacturer, if the dealer had notified the manufacturer of the problem, would have known that the delamination commenced within the three year period and ought to have honoured the warranty.

Despite my findings, as I said earlier that the dealer and manufacturer ought to have given guidance to the claimant in May of 1994 as to a possible warranty claim, I do not find any cause of action arising directly from this finding. It would simply have been good business practice for them to have advised the claimant.

If liability is to be found against General Motors, it must be for a breach of the warranty. On the other hand, if liability is to be found against Dave Wheaton, it must be because as a seller the dealership breached an implied condition of the Sale of Goods Act.

On the issue of the warranty supplied by General Motors, I note that it covers "repairs, or adjustments to correct any vehicle defect related to material or workmanship occurring during the warranty period." It appears from the evidence that General Motors' major concern with this particular case was that the problem with this specific vehicle was not brought to their attention during the warranty period, not that the problem with the paint did not actually arise within the warranty period.

I find the claimant has established that the paint problem was brought to the attention of the dealer in mid-May of 1994. The warranty had expired two weeks earlier. I accept that the defect occurred during the warranty period on two bases. First, as submitted by the claimant, Ms. Bentley, the defect occurred at the time of manufacture and continues to this day.

Similar finding was made in *Shields v. General Motors of Canada*, a decision of the Ontario Court, General Division, number 1398 of 1996.

Secondly, in any event, based on the condition of the paint in mid-May 1994, the defect likely became apparent during the three year warranty period. I therefore find that General Motors has breached the warranty and the claimant is entitled to damages.

As for the dealership, defendant's counsel argues that there has been no breach of the Sale of Goods Act in that the vehicle when sold was of merchantable quality and reasonably fit for the purpose for which it was intended.

I agree as to merchantable quality. And as to the other implied condition, I also agree if what is meant by "reasonably fit for the purpose" is that it was a truck that operated and was capable of hauling cargo and passengers, but I am of the view that every seller of new vehicles knows that the purchaser expects the vehicle to be reasonably fit for the purpose of resale at some future time, depending upon the age and quality of the vehicle. This vehicle was not and is not reasonably fit for resale given those factors. It is not reasonably fit due to the latent defect in its quality, a defect which existed in incipient form at the time of sale.

I note that in *McCready Products v. Sherwin-Williams*, (1984) 53 A.R. 304, a decision of the Alberta Queen's Bench, referred to in the article by Fridman submitted by counsel, in that case paint that weathered and faded in less than three years was found unfit for its purpose. That was house paint. The same is true here. Even though the paint in question here was only a component of the item purchased, it was a very important component having a great deal to do with the value of the vehicle.

I therefore find the defendant Dave Wheaton Pontiac Buick G.M.C. Ltd. also liable for damages. The liability of both defendants is joint and several.

I turn now to the question of damages. The claimants have averaged three estimates they have placed before the court and claim the amount of twenty-three hundred seventy-three dollars and sixty-one cents.

The defendant says damages are lower as a different allowance to dealers are made under the warranty. The defendants cannot now get the benefit of this, in my opinion, as responsibility under the warranty was denied by them.

I prefer to assess damages by taking the defendant's estimate or figure of fifteen hundred eighty-eight dollars twenty-six cents, a sum which the dealer could charge under the warranty to GM, and multiply that figure by a factor admitted to by the defendant as to what another body shop would—the number of hours another body shop would employ in order to obtain a realistic assessment.

It was stated in evidence that twenty-two to twenty-five hours is required to repair this damage, of which two hours are the actual painting.

The defendant's estimate is based on sixteen point one hours, because that's all they can claim under the warranty. There is therefore a difference of approximately nine hours in the estimates based on the upper level of twenty-five hours required by another body shop. I accept the proportion of paint to labour as stated by the witness and therefore accept that the defendant's estimate is based on fourteen point one hours of labour.

I also accept that the acceptable labour rate is fifty dollars and fifty-five cents and the painting rate is twenty-three dollars and seventy-one cents. It is the nine hours of labour that is in issue here in the assessment of damages.

The costs of the paint and materials I accept is two hundred and sixty-two dollars and forty-two cents.

I am therefore going to base damages on the defendant's estimate of one thousand eight hundred and ten dollars and sixty-two cents, which includes taxes, plus an additional nine hours labour at fifty dollars and fifty-five cents per hour, plus taxes, or an additional five hundred eighteen dollars eighty-four cents, bringing the total to two thousand three hundred and twenty-nine dollars and forty-six cents, very close to the estimate given by the claimant.

I am making no adjustment for betterment as it is known because, in my opinion, this is offset by the fact that for seven and a half years, or at least most of those seven and a half years, the vehicle was essentially unmarketable, unsaleable without substantial loss.

The claimant will therefore have judgment against both defendants, joint and several, in the amount of two thousand three hundred twenty-nine dollars forty-six cents, plus costs. No interest is awarded as it is inapplicable to this type of claim.

Shields v. General Motors of Canada, No. 1398/96, Ontario Court (General Division), Oshawa Small Claims Court, 33 King Street West, Oshawa, Ontario L1H 1A1, July 24, 1997, Robert Zochodne, Deputy Judge. The owner of a 1991 Pontiac Grand Prix purchased the vehicle used with over 100,000 km on its odometer. Commencing in 1995, the paint began to bubble and then flake and eventually peel off. Deputy Judge Robert Zochodne awarded the plaintiff $1,205.72 and struck down every one of GM's environmental/acid rain/UV-rays arguments. Other important aspects of this 12-page judgment that GM did not appeal:

1. The judge admitted many of the service bulletins referred to in *Lemon-Aid* as proof of GM's negligence.
2. Although the vehicle had 156,000 km when the case went to court, GM still offered to pay 50 percent of the paint repairs if the plaintiff dropped his suit.
3. Deputy Judge Zochodne ruled that the failure to protect the paint from the damaging effects of UV rays is akin to engineering a car that won't start in cold weather. In essence, vehicles must be built to withstand the rigours of the environment.
4. Here's an interesting twist: the original warranty covered defects that were present at the time it was in effect. The judge, taking statements found in the GM bulletins, ruled the UV problem was factory related, and therefore, *it existed during the warranty period and thereby represented a latent defect* that appeared once the warranty expired.
5. The subsequent purchaser was not prevented from making the warranty claim, even though the warranty had long since expired from a time and mileage standpoint and he was the second owner.

Williams v. General Motors of Canada, British Columbia Provincial Small Claims Court, No. 15691, Nanaimo Registry, October 27, 1997, Presiding Judge J. D. Cowling. The claimants sought damages for the paint delamination of a blue 1988 Grand Am that they purchased new on October 1, 1988. They first noticed the paint chipping from their vehicle in September 1994. They called GM's 1-800 customer relations hotline and were told there were no known paint issues regarding their model of vehicle.

By 1995, the chipping problem had expanded into extensive delamination of paint on the upper horizontal surfaces of the car. The plaintiff's local dealer then admitted it was a common problem, but added that the "goodwill" program was discontinued as of August 1995. The plaintiffs were quoted $1,939.60 to have their car repainted.

Subsequent inquiry by the claimants revealed that as of October 1992, the defendant, GM, had issued a product service bulletin that described the problem as follows:

> Blues, Grays, Silvers, and Black Metallics are the colours that have the highest potential for this condition.
>
> ...may have delamination (peeling) of the paint colourcoat from the ELPO primer, depending upon variable factors including prolonged exposure to sunlight and humidity.
>
> This condition may occur on vehicles produced in plants where the paint process does not call for application of a primer surfacer. Under certain conditions, ultraviolet light can penetrate the colourcoat, sometimes causing a reaction and separation of portions of the colourcoat from the ELPO (electrocoat) primer.

Other product service bulletins from the defendant's U.S. branch indicate the paint delamination problem was not unique to cars in Canada. In the U.S., the Center for Auto Safety negotiated an agreement with GM to implement a post-warranty adjustment program for paint problems.

The nature of the delamination problem is such that, once detected in a particular vehicle, spot repainting is not a viable remedy, but rather, the removal of the topcoat and primer surfaces is required.

> The purchase of this car by the claimants is a sale by description within the meaning of s. 17 of the Sale of Goods Act. Pursuant to s. 18 (b) of the Act there is an "...implied condition that the goods are of merchantable quality..." and further by virtue of s. 18 (c) an "...implied condition that goods will be durable for a reasonable period of time."
>
> I find that, having regard to the type of car purchased by the claimants, the condition in which they have maintained it, the extent of the paint problem it suffers from, and the reason for the problem, the claimants have established a breach of the section 18 (b) and (c) warranties. I consider the reasoning of O'Donnell J. in *Thauberger v. Simon Fraser Sales Ltd. et al* (1977) 3 B.C.L.R. 193 with respect to what was then s. 20 (b) of the Act to be directly on point in this case. I acknowledge that the paint surface in Thauberger failed after 21 months and not six years as here, however, I consider that the buyer of a new car in 1988 may reasonably expect its paint surface not to delaminate under normal use within six years.
>
> The claimants are entitled to judgment against the defendant in the sum of $1,229.70. They are entitled to Court Order Interest on this amount from September 1, 1995.

Other defects

Beauvais v. Banque Continentale du Canada, August 5, 1982, Quebec Provincial Court (Abitibi), No. 625-32-000172-814, Judge St-Pierre. The plaintiff recovered $500 from the small claims court as compensation for a defective, repossessed car sold by the bank. The purchase price was $800, but the plaintiff reduced the amount to $500 in order to meet the small claims court limit.

Fissel v. Ideal Auto Sales Ltd. (1991), 91 Sask. R. 266. Shortly after the vehicle was purchased, the car's motor seized and the dealer refused to replace it, even though the car was returned on several occasions. The court ruled that the dealer had breached the statutory warranties in s. 11 (4) and (7) of the Consumer Products Warranties Act. The purchasers were entitled to cancel the sale and recover the full purchase price.

Friskin v. Chevrolet Oldsmobile, 72 D.L.R. (3d), 289. A Manitoba used-car buyer asked that his contract be cancelled because of a chronic stalling problem. The garage owner did his best to correct it. Despite the seller's good intentions, the Manitoba Consumer Protection Act allowed for cancellation.

Graves v. C&R Motors Ltd., April 8, 1980, British Columbia County Court, Judge Skipp. The plaintiff bought a used car on the condition that certain deficiencies be remedied. They never were. He was promised a refund, but it never arrived. The plaintiff brought suit claiming that the dealer's deceptive activities violated the provincial Trade Practices Act. The court agreed, concluding that a deceptive act that occurs before, during, or after the transaction can lead to the cancellation of the contract.

Green v. Holiday Motors (1975), 4 W.W.R., 445. The plaintiff was sold a used car that had been used for drag racing. The seller didn't make this fact known. The motor blew up. The judge ruled that the Manitoba Consumer Protection Act allowed damages of $1,000.

Hachey v. Galbraith Equipment Company (1991), 33 M.V.R. (2d) 242. The plaintiff bought a used truck from the dealer to use in hauling gravel. Shortly afterwards, the steering failed. The plaintiff's suit was successful because expert testimony showed that the truck wasn't roadworthy. The dealer was found liable for damages for being in breach of the implied condition of fitness for the purpose the truck was purchased, as set out in s. 15 (a) of the New Brunswick Sale of Goods Act.

Henzel v. Brussels Motors (1973), 1 O.R., 339 (C.C.). The dealer sold this used car while brandishing a copy of the mechanical fitness certificate as proof that the car was in good shape. The plaintiff was awarded his money back because the court held the certificate to be a warranty that was breached by the car's subsequent defects.

Johnston v. Bodasing Corporation Limited, Ontario County Court (Bruce), No. 15/11/83, February 23, 1983, Judge McKay. The plaintiff bought a used 1979 Buick Riviera, for $8,500, that was represented as being "reliable." Two weeks after the purchase, the motor self-destructed. Judge McKay awarded the plaintiff $2,318 as compensation to fix the Riviera's defects.

One feature of this particular decision is the trial judge found that the Sale of Goods Act applied, notwithstanding the fact that the vendor used a standard contract that said there were no warranties or representations. The judge also accepted the decision in *Kendal v. Lillico* (1969), 2 Appeal Cases, 31, which indicates that the Sale of Goods Act covers not only defects that the seller ought to have detected, but also latent defects that even his utmost skill and judgment could not have detected. This places a very heavy onus on the vendor and it should prove useful in actions of this type in other common-law provinces where laws similar to Ontario's Sale of Goods Act are in force.

Kelly v. Mack Canada, 53 D.L.R. (4th), 476. Kelly bought two trucks from Mack Sales. The first, a used White Freightliner tractor and trailer, was purchased for $29,742. It cost him over $12,000 in repairs during the first five months, and another $9,000 was estimated for future engine repairs. Mack Sales convinced Kelly to trade in the old truck for a new Mack truck. Kelly did this, but shortly thereafter, the new truck had similar problems. Kelly sued for the return of all his money, arguing that the two transactions were really one.

The Ontario Court of Appeal agreed and awarded Kelly a complete refund. It stated: "There was such a congeries of defects that there had been a breach of the implied conditions set out in the Sale of Goods Act."

Although Mack Sales argued that the contract contained a clause excluding any implied warranties, the court determined that the breach was of such magnitude that the dealer could not rely upon that clause. The dealer then argued that since the client used the trucks, the depreciation of both should be taken into account in reducing the award. This was refused on the grounds that the plaintiff never had the product he bargained for and in no way did he profit from the transaction. The court also awarded Kelly compensation for loss of income while the trucks were being repaired, as well as the interest on all of the money tied up in both transactions from the time of purchase until final judgment.

Mathieu v. Autos M.L., February 15, 1982, Quebec Small Claims Court (Levis), No. 230-32-000906-817, Judge Bosse. This consumer bought a seven-year-old car with 80,000 kilometres on it. Two months later, he spent $441 on motor repairs. Judge Bosse awarded him $441 because the vehicle "did not withstand normal usage over a reasonable period of time," as prescribed by Articles 37, 38, and 53 of Quebec's Consumer Protection Act.

Morrison v. Hillside Motors (1973) Ltd. (1981), 35 Nfld. & P.E.I.R. 361. A used car advertised to be in A-1 condition and carrying a 50/50 warranty developed a number of problems. The court decided that the purchaser should be partially compensated because of the ad's claim. In deciding how much compensation to award, the presiding judge considered the warranty's wording, the amount paid for the vehicle, the year of the vehicle, its average life, the type of defect that occurred, and how long the purchaser had use of the vehicle before its defects became evident. Although this judgment was rendered in Newfoundland, judges throughout Canada have used a similar approach for more than a decade.

Neilson v. Maclin Motors, 71 D.L.R. (3d), 744. The plaintiff bought a used truck on the strength of the seller's allegations that the motor had been rebuilt and that it had 210 hp. The engine failed. The judge awarded damages and cancelled the contract because the motor had not been rebuilt, it did not have 210 hp, and the transmission was defective.

Parent v. Le Grand Trianon and Ford Credit (1982), C.P., 194, Judge Bertrand Gagnon. Nineteen months after paying $3,300 for a used 1974 LTD, the plaintiff sued the Ford dealer for his money back because the car was prematurely rusted out. The dealer replied that rust was normal, there was no warranty, and the claim was too late. The court held that the garage was still responsible, for the following reasons:

- When purchased, the car had been repainted by the dealer to camouflage rust and perforations.
- During the 19 months, the plaintiff and the dealer continued to explore ways the rusting could be stopped.
- It wasn't until just before the lawsuit that the plaintiff found out how bad the rust was.
- Ford and its dealers admitted they knew that many of their 1970–74 cars had serious premature corrosion problems.

The plaintiff was awarded $1,500, representing the cost of rust repairs.

"As is" clauses

Since 1907, Canadian courts have ruled that a seller can't exclude the implied warranty as to fitness by including such phrases as "there are no other warranties or guarantees, promises or agreements, than those contained herein." *Sawyer-Massey Co. v. Thibault* (1907), 5 W.L.R. 241.

Adams v. J&D's Used Cars Ltd. (1983), 26 Sask. R. 40 (Q.B.). Shortly after purchase, the engine and transmission failed. The court ruled that the inclusion of "as is" in the sales contract had no legal effect. The dealer had breached the implied warranty set out in Saskatchewan's Consumer Products Warranties Act. The sale was cancelled and all monies refunded.

Bedard v. Zelnick Associates Inc., April 15, 1977, Quebec Provincial Court, No. 500-02-026998-760, Judge Decary. This dealer was aware of serious motor problems and chassis rusting on the vehicle when it was sold to the plaintiff. Judgment was in favour of the plaintiff for $2,300, including towing costs and repairs, despite the fact that the dealer had sold the vehicle "as is—no warranty."

Gillespie v. MacDonald Pontiac Buick GMC Ltd. (1986), 71 N.B.R. (2d) 1. Prior to purchasing the used car, the buyer noticed that the engine was leaking oil. The seller promised to fix the leak after the vehicle was sold. When the customer brought it back, the seller refused to fix the car, claiming that it had been sold "as is." The court ruled that the seller's promise to repair overruled the "as is" limitation, and ordered a refund of the purchase price.

McCann v. B.J.'s Auto Sales, Ontario Provincial Court (Guelph), 1987. Plaintiff McCann bought a used 1977 Thunderbird "as is," without a warranty and without a roadworthiness certificate. A Ford dealer told McCann that it would cost her $2,200 to put the vehicle in shape to drive. The court found the car to be in "deplorable condition" and awarded McCann $1,000 plus $125 in legal fees.

Thériault v. Roy (1990), 109 N.B.R. (2d) 75. The dealer assured the purchaser that the used vehicle sold "as is" was in good condition. Three weeks later, the motor failed. Inasmuch as the buyer had complained about excessive oil consumption one day after purchase, the court ruled that the seller had breached New Brunswick's Consumer Product Warranty and Liability Act. The act protects buyers with an implied warranty as to durability for a reasonable period of time.

Tricco v. Haynes (1971), 2 Nfld. and P.E.I.R. 53. A used car sold "as is" and inspected before its purchase later turned out to be a lemon. The court ruled that the selling party would have to take the vehicle back because there was a fundamental breach of warranty by the seller, a private individual.

Misrepresentation

Mechanical fitness certificate

Belanger v. Fournier Chrysler Dodge Ltée. (1975), 25 N.B.R. (2d), 673. A used car sold in New Brunswick carried an inspection sticker that stated the body and mechanical components were in good condition.

Actually, the chassis was rusted through and the police ordered the vehicle off the road. The plaintiff was awarded the purchase price.

Beliveau v. Duval, November 26, 1973, Quebec Provincial Court, No. 02-046408-72, Judge Casgrain. The plaintiff bought a used Jeep with a spare motor that the defendant said could be used if the original engine failed. It did and the spare motor was the wrong size. Judgment was awarded against the defendant for misrepresenting the condition of the Jeep and giving false assurances that the spare engine was adequate. Both parties were private individuals.

Bellerose v. Bouvier (1955), B.R., 175. In this judgment, the court recognized the right of a consumer to ask for a reduction or reimbursement of part of the purchase price if there's a fraudulent representation of part of the transaction. The plaintiff didn't have to ask for the contract to be cancelled.

See also:

- *Nissan Automobile v. Pelletier* (1974) 503 (reaffirmed by the Supreme Court of Canada)
- *Bérubé v. Enright*, January 25, 1982, Quebec Provincial Court, No. 500-02-050980-791, Judge Filion

Ennis v. Klassen, 70 D.L.R. (4th), 325. Ennis bought a $9,000 used BMW that was represented as a 1980 733i. He learned two days later that it was really a 1979 BMW 728—a model without fuel injection that wasn't supposed to be allowed into Canada because of federal safety standards. It was also impossible to find parts for this vehicle. He kept the car in his garage for two years thereafter.

Klassen, a private seller, maintained that the lawsuit was tardy, that Ennis had inspected the vehicle before purchase, and that the whole case revolved around innocent misrepresentation.

Manitoba's Court of Appeal disagreed on the grounds that "[t]he misrepresentation does go to the root of the contract. The plaintiff was led to believe that he was buying a BMW model 733 at a fair retail price for a second-hand vehicle of the advertised model, type and vintage. He received nothing of the kind, and so he may rescind [cancel] the sale."

As for the action being tardy, the Appeal Court also disagreed, stating: "The plaintiff no longer used the vehicle, and took immediate steps on the path to rescision." Ennis got a complete refund.

See also:

- *F&B Transport Ltd. v. White Truck Sales Manitoba Ltd.* (1975), 49 D.L.R. (2d) 670
- *Adams v. Canadian Co-Operative Implements Ltd.* (1980), 20 A.R. 533

Ferraro v. Marsilio, May 15, 1981, Quebec Provincial Court, No. 500-02-026887-807, Judge Thinel. The plaintiff was awarded $3,900, representing a refund of the purchase price of a used car that had been misrepresented by the salesman as being in good condition. Actually, the vehicle had been heavily damaged in an accident. The court held that the plaintiff's inexperience had been exploited.

Gagnon v. Leveille, May 2, 1975, Quebec Provincial Court No. 103-879, Judge Dussault. The plaintiff was told that the old motor had been replaced with a new one. Actually, another old motor had been used. The court ordered the plaintiff to pay $1 for the old motor and the defendant to pay $850 for a new one.

Keilty v. East-Court Sales Limited, Ontario Court of Justice, 1996 C. J. Lexis 2247, June 14, 1996. The plaintiff purchased a used Lincoln Town Car that the dealer claimed was an "executive-driven car" from the Ford Motor Company for $20,000. When the car broke down, the owner ran a title check and found the vehicle had been a Budget rental prior to being sold to the defendant by Ford at an auction. The judge ruled that the seller was guilty of fraudulent misrepresentation and could not rely upon the sales contract's exclusionary clause to protect himself from the result of the fraud. The purchase price was refunded.

L.G. Wilson Motors v. Woods (1970), 2 N.B.R. (2d), 581 (S.C.). A safety certification carried out by a provincial inspector doesn't guarantee that a used vehicle is safe or that the inspector was competent. The buyer should have made his own inspection. The plaintiff's claim was rejected.

Presley v. MacDonald (1963), 1 O.R. (2d), 619. In this case, an Ontario judge ruled that, under s. 49 (1) of the Highway Traffic Act, the mechanical fitness certificate made the seller half liable for damages caused in a subsequent accident.

Miscellaneous misrepresentation

Beale v. Taylor (1967), 3 All E.R., 253 (C.A.). The seller innocently misrepresented a used car as being in good condition when it really was two cars welded together. The court ordered the sale set aside.

Bottcher v. Selig, 26 N.S.R. (2d) 347 (C.A.). A used car was represented as needing only a quart of oil every 1,000 miles. The plaintiff was awarded damages for serious oil leaks. In addition to misrepresenting the condition of the vehicle, this Nova Scotia court held that the seller's promise was an implied warranty.

Hanson and Campbell v. Cornell Chevrolet Oldsmobile Ltd. and Canadian Imperial Bank of Commerce and Weaver (1983), 4 W.W.R. 285. The purchaser of a used car in B.C. had it seized when he moved to Alberta. He then brought suit against the B.C. dealer that sold him the vehicle, which had a valid chattel mortgage registered against it. The court held that the British Columbia Sale of Goods Act, s. 16, protects consumers whether or not the goods remain in the province. The important element was that the contract was agreed to in B.C. As a result, the dealer had to compensate the client for his losses.

Henzel v. Brussels Motors Ltd. (1973), 1 O.R., 339 (C.C.). A used car sold with a mechanical fitness certificate had defective brakes. The court ordered the purchase price refunded.

Sandilands v. Guelph Datsun 1980 Ltd. (1982), 35 O.R. (2d) 25. The dealer represented a used car as having "good" brakes. Four months after purchase, the vehicle was taken off the road by the provincial Ministry of Transportation because of defective brakes and rust perforations. The court held that the contract should be cancelled because the dealer's misrepresentation was an "unfair practice" according to Section 4(1)(a) of the Ontario Business Practices Act. It was pointed out that the dealer had given false or misleading information when he did the following:

1. He supplied a car equipped with brakes that wouldn't work in an emergency and that hadn't had an adequate mechanical inspection.
2. He supplied a car that hadn't had its rust defects repaired, although he had contracted to do so. The stipulation in the contract to provide a full body repair job was supplementary to the safety check.
3. He concerned himself with outfitting a car that he knew wasn't worth making roadworthy.

The dealer was forced to take back the car and pay storage and transport costs.

Truck misrepresentation

MacDonald v. Equilease Co. Ltd., January 18, 1979, Ontario Supreme Court, Judge O'Driscoll. The plaintiff leased a truck that was misrepresented as having an axle stronger than it really was. The court awarded the plaintiff damages for repairs and set aside the lease.

Seich v. Festival Ford Sales Ltd. (1978), 6 Alta. L.R. (2d), 262. The plaintiff bought a used truck from the defendant after being assured that it had a new motor and transmission. It didn't and the court awarded the plaintiff $6,400.

Vehicle not as ordered
Whether you're buying a new or used vehicle, the seller can't misrepresent it. Anything that varies from what one would commonly expect or expect from the seller's representation must be disclosed prior to signing the contract. Typical scenarios are odometer turnbacks, accident damage, used or leased cars sold as new, and new vehicles that are the wrong colour and model year or that lack options or standard features that were promised.

Chenel v. Bel Automobile Inc. (1981), August 27, 1976, Quebec Superior Court (Quebec), Judge Desmeules. The plaintiff did not receive his new Ford truck with the Jacob brakes, which were essential for transporting sand in hilly regions. The court awarded the plaintiff $27,000, representing the purchase price of the vehicle less the money he earned while using the truck.

Lasky v. Royal City Chrysler Plymouth, Ontario High Court of Justice, February 18, 1987, 59 O.R. (2d) 323. The plaintiff bought a 4-cylinder 1983 Dodge 600 that was represented by the salesman as a 6-cylinder model. After putting 40,000 km on the vehicle over a 22-month period, the buyer was given her money back, without interest, under the provincial Business Practices Act.

White v. Munn Motors (1960), 45 M.P.R. 253 (Nfld. T.D.). The dealer misrepresented a half-ton truck as having a three-quarter-ton capacity. The owner was given his money back because he didn't get what he paid for.

Wrong model year
Adelaide Motors v. Alexander (1962), 48 M.P.R. 258 (Nfld. T.D.). The customer bought a new 1960 car that was misrepresented as being a 1961 model. The court ruled that the plaintiff was entitled to damages.

Ginn v. Canbec Auto (1974), Quebec Superior Court (Montreal), No. 500-05-014597-74-4. Relying on expert testimony provided by the APA, the court awarded $10,000 to the purchaser of a new 1972 BMW that was fraudulently sold as a 1973 model. This case demonstrates beyond a doubt that auto retailers are culpable for the misrepresentation of their products.

Woods v. Borstel (1962), 34 D.L.R. (2d) 68 (Alta. C.A.). The customer bought a new 1958 car that was misrepresented as being a 1959 model. The court ruled that the seller's representation of model year was a warranty as defined under s. 2(1)(n) of the Alberta Sale of Goods Act. The plaintiff was entitled to all damages, but the contract remained in force.

In Ontario, two separate, successful lawsuits claimed that Ford systematically cheated consumers by selling 1970 Cortinas as new 1971 models. Each purchaser received $400 in compensation in small claims court. In *Brosseau v. Lewis Motors,* Ottawa-Carleton, and *King v. Paddy Shanahan Ford,* Toronto, the court pointed to deceptive and misleading sales tactics used to sell the Cortinas.

See also:

- *Ennis v. Klassen,* 70 D.L.R. (4th), 325

False Advertising

Dealer posing as a private seller

A recent small claims judgment slams a practice called "curbsiding," whereby a professional sales agent posing as a private party sells cars from a residence. Frank Longstaff, the lawyer for the plaintiff, told me that a number of other New Brunswickers who fell for this scam have contacted his office after reading about the case in their local paper.

It was held that the defendant's practice of buying used cars wholesale from a dealer and then quickly advertising them for sale to consumers was sufficient to label him a distributor as defined in New Brunswick's Consumer Product Warranty and Liability Act. This made him subject to the act and, hence, required him to stand behind any warranty, either oral or in writing, he may have given the purchaser in his advertisement. It also meant that, pursuant to sections 10 (1) and 12 (1) of the act, he gave an implied warranty that the product was "fit for the purpose" and would be "durable for a reasonable period of time." Prior to this decision, private sellers (curbsiders) had not been subject to the act.

Van Wart v. MacDonald and MacDonald, the Court of Queen's Bench, Trial Division, Judicial District of Saint John, New Brunswick, July 25, 1994, Judge H. H. McLellan. Nineteen-year-old Van Wart bought a 1987 Firefly for $2,000, with her parents' assistance, after responding to a newspaper advertisement. MacDonald claimed that the car belonged to his wife and sold it "as is, where is." At trial, it was learned that MacDonald had bought the vehicle five days earlier from Brett Chev-Olds for $1,200. Brett's sales manager said MacDonald had bought eight cars from him the previous year and that the Firefly purchase was accompanied by a bill of sale stating, "This vehicle is junk. There is no expectation of driving this vehicle."

Judge McLellan ordered the car be taken back and the $2,000 purchase price and sales tax refunded, and made MacDonald responsible for Van Wart's legal bills.

False sales prices

R. v. Lanthier and Lalonde, Sessions Court (Montreal). A Montreal Ford dealer was fined $300 for placing ads showing tremendous reductions on new cars sold at prices below the manufacturer's suggested retail price. After criminal charges were laid, it was shown that the advertised savings were lies (judgment on file with Consumer and Corporate Affairs). The federal government has fined several dozen suburban Toronto Chrysler dealers for similar deceptions.

Odometer tampering

Odometer tampering is a criminal offence under the federal Weights and Measures Act. The Department of Consumer and Corporate Affairs uses the RCMP to investigate all such cases. Many violators have been caught and successfully prosecuted. Nevertheless, the federal law is weak because fines are so low that they practically represent a licence to operate illegally, and an individual can escape prosecution by pleading that the odometer was broken and had to be changed.

Bouchard v. South Park Mercury Sales (1978), 3 W.W.R., 78. The odometer figure written on the contract was incorrect. The dealer pleaded ignorance but the judge ruled that the car's owner should receive damages to compensate for the extra mileage.

Delmaire v. Westcoast Honda, British Columbia Small Claims Court (Victoria), August 13, 1986. The plaintiff bought a used 1979 Rabbit "as is" and found that the odometer had been turned back. He argued that although he had no proof that the dealer was responsible, the dealer was an expert and should have suspected the low mileage. The judge agreed and awarded $1,299 for a rebuilt engine.

Zelezen v. Model Auto Sales, November 26, 1971, Quebec Superior Court, No. 722-487, Judge Nicols. Not only did the judge order the car returned and the purchase price refunded because of odometer tampering, he also ordered the dealer to pay compensation for the plaintiff's injuries from an accident caused by the car's defects.

Repairs

Faulty diagnosis

Imagine that before taking a holiday you have your van checked out and are assured that it's in good condition. But while en route to your vacation spot, it dies on the highway and ruins your vacation. When the checkup is incorrect and leads to financial losses (damages), the garage owner is responsible for those losses as well as for the refund of the diagnostic costs. Of course, you'd better show the defect that caused your troubles was present and detectable at the time the vehicle was checked.

Ali v. 301077 Ontario Ltd. c.o.b. Mocar Motors, Ontario Provincial Court (Civil Division), Justice Sigurdson, February 10, 1984. The plaintiff bought from a private party a used automobile that had been inspected by Mocar. The car later broke down. Judge Sigurdson ruled that Mocar was guilty of negligent misrepresentation, and ordered the company to pay the repair bills for the plaintiff's car.

Babcock v. Servacar (1970) 1 O.R., 125. A motorist took a car he was planning to buy to an Ottawa Esso Diagnostic Clinic to determine its condition. He did the recommended repairs and then had serious problems with the car while on vacation. The judge ruled that the clinic would have to pay for the repairs and reimburse the diagnostic cost as well. The court held that the garage's advertising claims gave rise to a contractual warranty that promised to root out the car's defects before it was purchased.

Chan v. W. Gordon Inc., October 15, 1981, Quebec Provincial Court, No. 02-000002-811, Judge Page. The plaintiff was awarded $300 to cover repairs to a 1973 Volvo that was taken to Gordon's Service Station prior to purchase. The garage mistakenly diagnosed the car as being sound and roadworthy.

Davies v. Alberta Motor Association, Alberta Provincial Court, Civil Division, August 13, 1991, No. P9090106097, Judge Moore. The plaintiff had a used 1985 Nissan Pulsar NX checked out by the Alberta Motor Association's (AMA) Vehicle Inspection Service prior to buying it. The car passed with flying colours. A month later, the clutch was replaced and numerous electrical problems ensued. At that time, another garage discovered that the car had been involved in a major accident, had a bent frame and a leaking radiator, and was unsafe to drive. The court awarded the plaintiff $1,578.40 plus three years of interest. The judge held that the AMA set itself out as an expert and should have spotted the car's defects. The AMA's defence that it was not responsible for errors was thrown out. The court held that a disclaimer clause could not protect the association from being accountable for a fundamental breach of contract.

Lowe v. Fairview Chrysler-Dodge Limited and Chrysler Canada Limited, Ontario Court (General Division), Burlington Small Claims Court, Claim N0, 1224/95, May 14, 1996. The court ruled the following:

- Internal dealer service bulletins are admissible in court in order to prove that a problem exists and that certain parts should be checked out.
- If a problem is reported prior to a warranty's expiration, it is automatically carried over after the warranty ends.
- It is not up to the car owner to tell the dealer/Chrysler Canada what the specific problem is.
- Repairs carried out by an independent garage can be refunded if the dealer/Chrysler Canada unfairly refuses to apply the warranty.

- The dealer/Chrysler Canada cannot dispute the cost of the independent repair if it fails to cross-examine the independent repairer.
- Auto owners can ask for and win compensation for their inconvenience (which in this judgment amounted to $150).

Court awards quickly add up. Although the plaintiff was given $1,985.94, with court costs and prejudgment interest, plus costs of inconvenience fixed at $150, the final amount came to $2,266.04.

McCormick v. Servacar and Esso, February 9, 1981, Quebec Small Claims Court, No. 500-32-008700-801, Judge Prevost. Before purchasing a used 1978 Ford Fairmont, the plaintiff had it checked out by an Esso Diagnostic Clinic. The clinic said the car was in good condition, except for dirty transmission fluid. Ten days after purchasing the car, the transmission was replaced. The judge ordered Servacar and Esso to pay the $300 for repairs, reasoning that if it were not for the clinic's incompetence, negligence, or lack of experience, the defect could have been detected earlier.

Fuel-system fire

Lenz v. J.B.'s Automotive Service Center Ltd. (1987), 51 Alta. L.R. (2d) 16 (Q.B.). The plaintiff's vehicle caught fire three weeks after the fuel system was worked on. The court held that the garage was liable because the fire occurred in the same area that had been repaired.

Oil filter improperly installed

Proulx v. Salois Automobile, December 6, 1974, Quebec Provincial Court, No. 02-004322-74, Judge Lacourciere. The plaintiff was awarded damages to pay for motor repairs caused by an improperly installed oil filter. The dealer's mechanic was ordered to pay $1,000 in damages.

Repair misrepresentation

Smith v. Marc Patrick Motors and Canadian Tire Corporation, September 20, 1984, Ontario Small Claims Court (Niagara North), Judge Kingstone. The plaintiff bought a used 1978 Honda Civic that was sold as having a "new" engine. It failed shortly thereafter. Canadian Tire had not installed a new engine. Judge Kingstone ordered both defendants to split the cost of the repairs ($1,000 plus 13 percent interest) for misrepresenting the engine.

Repair charges excessive (flat-rate manuals)

R. v. Birchcliff Lincoln Mercury Ltd., Ontario Court of Appeal, July 7, 1987, 60 O. R. (2d) 610, 220; 220 A. C. 274. This new-car dealer advertised a $38 hourly labour charge and then calculated the number of hours using a flat-rate guide that allowed more hours than those actually worked. This was held by the Appeal Court to be false advertising under the federal Combines Act.

Repairs not done properly

Berta v. Gold Seal Engine Rebuilders, Ontario Small Claims Court (Toronto), No. C217/83, August 22, 1985, Justice Sigurdson. The plaintiff's engine was improperly repaired. Judge Sigurdson awarded $2,047.03 to the plaintiff after hearing the garage's employees testify in a "less than truthful" manner.

Canbec Auto v. De Levo, January 18, 1973, Quebec Provincial Court, No. 03-058009-72, Judge Page. The dealer sued the defendant for money owing on a bill for a general inspection and correction of defects. The car owner refused to pay the bill because the vehicle was still running poorly. Judgment was rendered in favour of the defendant. The court held that the garage had an obligation of results, where the repairs would be expected to correct the problems complained of by the car's owner.

Collier v. McMaster's Auto Sales, No. W/C 157/90, New Brunswick Court of Queen's Bench, April 26, 1991, Judge McLellan. The judge ruled that the dealer's failure to repair the car properly under warranty justified the owner getting the engine repaired elsewhere at the dealer's expense. The failure to repair the plaintiff's car properly amounted to a breach of the manufacturer's expressed warranty.

Humble v. Blais International Inc., January 24, 1996, Saskatchewan Court of Queen's Bench, Judicial District of Battleford, S. J. No. 33, DRS 96-05260, Q. B. No. 655 of 1991 J. C. B., Judge Laing. The tractor dealer installed a new engine camshaft using old parts, contrary to the express advice of the manufacturer. The improper repair caused the engine to seize three years later. Justice Laing awarded $7,000 to the Humble family for the tractor engine's reduced life expectancy and an additional $10,000 in damages because the garage didn't have a lawful lien when it repossessed the tractor for nonpayment. That unlawful seizure violated parts of the Farm Debt Review Act, the Garage Keepers Act, and the Saskatchewan Farm Security Act. The $10,000 was compensation for deterioration to the tractor's components, the cost of renting a replacement tractor, and lost opportunity to engage in snow-clearing work while the garage held the tractor.

Mudge v. Corner Brook Garage, 8 Nfld. & P.E.I.R., 374. The truck owner sued the garage successfully for negligent repairs to the rear frame assembly.

Pelleray v. Heritage Ford Sales Ltd., Ontario Small Claims Court (Scarborough), SC7688/91, March 22, 1993. The owner of a 1986 Ford Aerostar was awarded $1,691.29 as compensation for repairs to his automatic transmission, which had failed three times after being repaired by the dealer.

Raiches Steel Works v. J. Clark & Son, 16 N.B.R. (2d), 535. The brakes failed three weeks after being repaired. The judge found the mechanic negligent and awarded damages of $6,423.

Sigurdson v. Hillcrest Service & Acklands (1977), 1 W.W.R., 740. An accident was caused by a defective brake hose. The garage that installed the hose claimed that the parts distributor was responsible. The judge ruled that Saskatchewan's Sale of Goods Act made both garage and distributor responsible for damages.

Sylvain v. Carrosseries d'Automobiles Guy Inc. (1981), C.P. 333, Judge Page. The garage had an obligation to ensure that the paint job on the plaintiff's car would be satisfactory. It was not. The court held the garage liable for the cost of repainting the car elsewhere.

Tremblay v. Au Grand Salon, September 19, 1973, Quebec Provincial Court, No. 02-054410-73, Judge Bousquet. The plaintiff asked that his brakes be checked and paid $30 for the inspection. After the inspection, the brakes failed and an accident ensued. The court held the dealer responsible, although no repairs had been done.

See also:
- *Acme v. Coziol* (1962) B.R.
- *Chomedy Ford v. Mash,* March 11, 1976, Quebec Provincial Court, No. 500-02-007671-767
- *Gagnon v. Ford Motor Co.* (1974) C.S., 422
- *Lazanick v. Ford Motor Co.,* June 15, 1965, Quebec Superior Court No. 623504

Water in fuel

Claridge v. Three Way Motors, Manitoba, July 20, 1987, No. 221/86 (Thompson). The owner of a 1980 Audi 5000 had starting problems and rusty fuel-injection nozzles caused by what he believed to be gasoline contaminated by water. He didn't have any direct proof that the gas station's pumps were contaminated but the judge ruled that the balance of probabilities indicated that this was likely because the station owner had had problems with the storage pumps in the past.

Illegal/Unfair Insurance Company Practices

Inadequate compensation

The following two lawsuits may be helpful in any disputes you may have with an insurance company over its failure to pay a claim or use quality, original equipment manufacturer (OEM) replacement parts in repairs. Don't get the impression that this is just a State Farm problem. These abuses are widespread.

Campbell v. State Farm Mutual Insurance Automobile Insurance Co. This Utah case saw a $147.6 million jury award against State Farm cut to $26 million by the presiding judge who didn't want the case overturned on appeal because the original award may have been considered excessive.

In his verdict of December 19, 1997, Judge William B. Bohling called State Farm "greedy, callous, clandestine, fraudulent, and dishonest," after evidence showed the company refused to pay off a claim until the plaintiff sued it for bad-faith dealings. The Campbell attorneys successfully asserted that State Farm had a national plan to cheat policyholders, which included using inferior car parts in repairs, misleading consumers about policy benefits, and lowballing settlements. This evidence led the judge to conclude that State Farm "appeared to have preyed on the weakest of the herd" in cheating "the most vulnerable" policyholders with its "calculated and callous attitude towards settling valid claims." He concluded:

> It became a matter of plain evidence that State Farm has sold as its product, peace of mind, and has used as its advertising slogan, 'like a good neighbor.' State Farm's action amounts to betraying the trust that it invites its policyholders to place in it.

Poor-quality replacement parts

In a $2-billion Illinois class action, State Farm has been accused of breaching its promise to restore policyholders' vehicles to their "pre-loss condition," after it was found the company used aftermarket bumpers, door panels, and other parts that failed to meet automakers' specifications for fit, finish, corrosion protection, and safety. If the suit is successful, over 20 million car owners whose vehicles were repaired since 1980 will be eligible for refunds or repairs.

State Farm has settled similar lawsuits in Cook County, Illinois, and San Diego.

Part Three
CAR AND MINIVAN RATINGS

"Friends don't let their friends drive junk!"

Sign on a used-car
dealer's lot on York Road
in London, Ontario

A used car or minivan must first live up to the promises made by the manufacturer and dealer. Ideally, it should be crashworthy, reasonably durable (lasting at least 10 years), cost no more than $800 per year to maintain, and provide you with a fair resale value a few years down the road. Parts should be reasonably priced and easily available, and competent servicing shouldn't be hard to find. We also factor into the rating the relative availability of a particular vehicle, those models and years that are the best buys, the estimated annual maintenance and repair costs averaged over five years, and alternative vehicles that will give you as much or more for less money.

A 1995 or later rear-drive Ford Aerostar (shown above) represents a cheaper and more reliable buy than the front-drive Ford Windstar.

Models are rated on a scale from Recommended to Not Recommended. Recommended vehicles are those that are an excellent choice in their class and promise their owners relatively trouble-free service (like post-1991 Ford Escorts). Vehicles that are given an Above Average or Average rating are good second choices if a Recommended vehicle isn't your first choice or is too expensive. Vehicles given a Not Recommended rating are best avoided no matter how low the price, even though they may be attractively styled and loaded with convenience features (Chrysler minivans, for example); they're likely to suffer from a variety of durability and performance problems that will make them expensive and frustrating to own. Sometimes, however, a Not Recommended model will improve over several model years and garner a better rating (as recent Ford Aerostar and GM Astro and Safari minivans have done). Keep in mind, too, that the reliability and quality averages are rising, so that a newer vehicle rated Average is a far better buy than an older model with an Average rating.

Some enterprising readers of *Lemon-Aid* use the ratings as a buying opportunity. Dave Ingram, a friend from North Vancouver and founder of Cen-Ta, uses my Not Recommended list as a shopping guide for vehicles: he buys them up at depressed prices and refurbishes them through garages offering lifetime warranties on major components that I rate as weak. He's done that with several used Cadillacs and seems happy with the system. I doubt many buyers possess Dave's mechanical acumen and intestinal fortitude.

Reliability data is compiled from a number of sources: confidential service bulletins; owner complaints sent to the author by the 2,000-plus readers who filled out the readers' survey in Appendix II, "Survey and Bulletin Search"; vehicle-owner comments posted on the Internet; and survey reports and tests done by auto associations, consumer groups, and government organizations. Some auto columnists feel this isn't a scientific sampling, and they're quite right. Nevertheless, it seems to have been right on the mark over the past 28 years. Not all vehicles sold during the last decade are profiled; those that are newer to the market or relatively rare may only receive an abbreviated mention until sufficient owner or service bulletin information becomes available. Best and worst buys for each model category (e.g., "Small" or "Medium") are listed in a summary at the beginning of each rating section. Vehicles may also be profiled in the "Recommended Beaters" and "Beaten Beaters" sections on pages 52–59.

Strengths and weaknesses

Unlike other auto guides, *Lemon-Aid* pinpoints potential parts failures and explains why those parts fail. We also give parts numbers for upgraded parts (why replace poor-quality brake pads with the same ones, for example?) and offer troubleshooting tips direct from the automakers' bulletins, so your mechanic won't replace parts unrelated to your troubles before coming upon the defective component that is actually responsible. To complement the "Secret Warranties/Service Tips" and vehicle "Profile" tables, we look at a vehicle's overall road performance and reliability, providing details as to which specific mechanical, electrical, or body part fails repeatedly. This helps an independent mechanic check out the likely trouble spots before you make your purchase.

Dealer service bulletins

Imagine spearing your service manager with a confidential bulletin, detailing that pesky defect your car has that he always claimed was your fault. Here we summarize problem areas addressed by confidential dealer service bulletins for more recent models. Important bulletins offering substantial repair refunds are reproduced in the "Secret Warranties/Service Tips" section.

Safety summary/Recalls

Data from independent crash tests, insurance claims statistics, ongoing safety investigations, owner safety complaints, and safety recalls make up this section. Vehicles are rated according to how well they performed in U.S. government 57 km/h frontal crash tests (the impact is the same as if two identical vehicles, each travelling at 57 km/h, collided head-on). Information recorded during the crash tests measures the likelihood of serious injury, and vehicles are classified by the estimated chance of injury for the driver or passenger. For the past several years, vehicles have been given a one- to five-star rating by the NHTSA, with five stars indicating the best protection. Cars and minivans that are identical but carry different nameplates from the same manufacturer can be expected to perform similarly in these crash tests. On the other hand, sometimes the same vehicle tested from one year to the next will post dramatically different results even though the model has remained relatively unchanged. Safety experts admit that this happens occasionally and that consumers should look at the trend established over several model years.

Lemon-Aid is unique in that it includes estimated head, chest, and, for some model years, leg trauma as life-threatening factors in determining a model's safety rating. This explains why the crash rating scores shown below are often lower in *Lemon-Aid*. Also, this guide's safety rating applies only to the driver.

❶	❷	③	④	⑤
Multiple injuries	One injury	Average protection	Above average protection	Excellent protection

Both safety and emissions recalls are listed in chronological order. If your vehicle is listed and hasn't been fixed, the dealer and manufacturer must pay for the inspection and correction of the defect regardless of the vehicle's mileage, model year, or number of previous owners. It's not the law, but it's general practice.

Keep in mind that recalls affecting only a few hundred vehicles haven't been listed, and be watchful for the cut-off years. Even if your model year isn't listed, it may be currently under investigation or may have been recalled since this year's guide was published. Also, safety probes may be upgraded or dropped—download Transport Canada's or NHTSA's data files from the Internet for an update: *www.tc.gc.ca/roadsafety/Recalls/search_e.asp* (Transport Canada), *www.nhtsa.dot.gov/cars/problems/recalls/recmmy1.cfm* (NHTSA).

Secret warranties/Service tips

It's not enough to know which parts on your vehicle are likely to fail. You should also know which repairs will be done for free by the dealer and automaker even though you aren't the original owner and the manufacturer's warranty has long since expired.

Welcome to the hidden world of secret warranties found in confidential dealer service bulletins (DSBs) or gleaned from owner feedback. A summary of all the important DSBs for each model year is listed along with selected diagrams. These bulletins target defects—related to safety, emissions, and performance—that service managers would have you believe don't exist or are your responsibility. They also list the upgraded parts that will best repair the vehicle you plan to buy or have just bought.

Service bulletins cover repairs that may be eligible for warranty coverage in one or more of the following five categories:

- emissions warranty (5–8 years)
- safety component warranty (covers seatbelts and airbags and usually lasts the lifetime of the vehicle)
- body warranty (paint: six years; rust perforations: seven years)
- secret warranty (coverage varies)
- factory defects (depends on mileage, use, and repair cost)

Use these bulletins to get free repairs—even if the vehicle has changed hands several times—and to alert an independent mechanic about which defects to look for. They're also great tools for getting compensation from automakers and dealer service managers after the warranty has expired, since they prove that a failure is factory related and, therefore, not part of routine maintenance or the effect of an environmental anomaly (like bird droppings and acid rain).

Their diagnostic shortcuts and lists of upgraded parts make these bulletins invaluable in helping mechanics and do-it-yourselfers troubleshoot problems inexpensively and replace the right part the first time. Auto owners can also use the DSBs listed here to verify that a repair was diagnosed correctly, the correct upgraded replacement part was used, and the labour costs were fair.

Getting your own bulletins

If you want a personalized DSB summary for your vehicle, fill out the "*Lemon Aid* Survey/DSB Summary Request" order form in Appendix II. For a $15 fee (this includes my computer time and fax or mailing costs), you'll receive an exhaustive summary of all DSBs that concern your vehicle. From the summary, you can then order every DSB listed that addresses your concerns, for $5 per bulletin (no fax charge).

Vehicle profiles

These tables cover the various aspects of vehicle ownership at a glance. Included for each model year are details on crashworthiness, repair histories for major mechanical and body components (specific defective parts are listed in the "Strengths and weaknesses" section for each vehicle rated), and which model years have secret warranties or should be bought with an extended warranty.

Prices

Dealer profit margins on used cars vary considerably—giving lots of room to negotiate a fair price if you take the time to find out what the vehicle is really worth. Three prices are given for each model year: the vehicle's selling price when new as suggested by the manufacturer, its maximum price used (↑), and its lowest price used (↓).

Used prices are based on private sales as of May 1999. Prices are for the lowest-priced standard model that is in good condition with a maximum of 20,000 km for each calendar year. Be watchful for price differences reflecting each model's equipment upgrades designated by a numerical or alphabetical abbreviation. For example, L, LX, and LXT usually mean more standard features are included. Numerical progression, like the 2300 series of Mazda trucks or the Mercedes 300, usually relates to engine size.

The original selling price (manufacturer's suggested retail price) is also given as a helpful reference point. Sellers overprice some vehicles (mostly Japanese imports, minivans, and sport-utilities) in order to get back some of the money *they* overpaid in the first place. This is particularly true in the prairie provinces and B.C.

Prices reflect the auto markets in Quebec and Ontario, where the majority of used-vehicle transactions take place. Residents in eastern Canada should add 10 percent, and western Canadians should add 15–20 percent to the listed price. Why the higher cost? Less competition and inflated new-vehicle prices in these regions. Don't be too disheartened, though; some of what was overpaid will be recouped when you resell the vehicle down the road.

Why are *Lemon-Aid*'s prices lower than the prices found in dealer guides (*Red Book*, etc.)? The answer is simple: dealer guides inflate their prices so that you can bargain the price down and wind up convinced that you made a great deal.

I use newspaper classified ads from Quebec, Ontario, and B.C., as well as auction reports, for my used values. I then check these figures with the *Red Book* and *Black Book*; I don't start with the *Red Book*'s retail or wholesale figures (like real estate listings, vehicle prices are inflated about 10 percent for wholesale-private sales and almost 20 percent for retail-dealer sales—compare the two and you'll see what I mean). I then project what the value will be by year's end, and that lowers my prices further. I almost always fall way under the *Red Book*'s value, but not far under the *Black Book*'s price. In the *Lemon-Aid* guide, I print a top and bottom price to give the buyer some margin for negotiation, as well as to account for regional differences in prices, the sudden popularity of certain models or vehicle classes (sport-utilities, trucks, minivans, etc.), and the appreciated value of used cars generally.

Depreciation is the biggest (and often ignored) expense new-car owners encounter when trading in a vehicle, but smart car shoppers who know which cars and minivans depreciate the least can save a bundle.

Most new cars depreciate 30–40 percent during the first two years of ownership despite the fact that good-quality used cars are in high demand. On the other hand, some minivans and most vans, pickups, and sport-utilities lose little of their value even after four years of ownership.

No evaluation method is foolproof, so check dealer prices with local, private classified ads and add the option values listed below to come up with a fairly representative offer. Interestingly, the value of anti-lock brakes in trade-ins has plummeted in the last few years, undoubtedly as a result of insurance studies showing that ABS has failed to reduce collision fatalities and injuries.

	Model Year					
Option	**1998**	**1997**	**1996**	**1995**	**1994**	**1993**
Air conditioning	$900	$600	$500	$400	$300	$300
AM-FM-CD	300	200	175	150	100	100
Anti-lock brakes	300	175	150	125	100	50
Automatic transmission	500	400	300	275	250	200
Cruise control	225	125	100	75	50	50
Electric six-way seat	300	175	150	125	100	50
Leather upholstery	700	400	325	225	200	100
Level control (suspension)	250	150	125	100	75	50
Paint protector	0	0	0	0	0	0
Power antenna	125	75	50	50	25	0
Power door locks	300	175	150	125	100	50
Power windows	325	175	150	125	100	50
Rustproofing	50	25	25	0	0	0
Sunroof	300	150	125	75	50	50
T-top roof	1,000	700	500	400	300	200
Tilt steering	175	100	75	75	50	50
Tinted windows	50	50	25	0	0	0
Traction control	400	275	175	150	125	100
Wire wheels/locks	275	175	150	125	100	75

It will be easier for you to match the lower used prices if you buy privately. Dealers rarely sell much below the maximum prices. They inflate their prices to cover the costs of reconditioning and paying future warranty claims and to make you feel better. If you can come within 5–10 percent of this guide's price, you'll have done well.

Extended warranties and secret warranties

Usually, but not always, an extended warranty is advised for those model years that aren't rated Recommended. In shopping for an extended warranty, don't be surprised to discover that dealers have the market practically sewn up. You can bargain the price down by getting competing dealers to bid against each other, contacting them by fax or through their Internet sites. Be wary of extended-warranty companies that aren't backed by the major automakers.

Model years that are eligible for free repairs under a secret warranty are listed in the "Profile" section and further detailed in the "Secret Warranties/Service Tips" section. A **Y** signifies that one or more secret warranties exist or that an extended warranty is needed. An **N** means that no secret warranty applies or that there's no need to buy an extended warranty.

Reliability

The older a vehicle, the greater the chance that a major component like the engine or transmission will fail as a result of high mileage and environmental wear and tear. Surprisingly, there's a host of other expensive-to-repair failures that are just as likely to occur in a new vehicle as in an older one. Air conditioners, electronic computer modules, electrical systems, and brakes are the most troublesome components, manifesting problems early in a vehicle's life. Other deficiencies that will appear early, due to sloppy manufacturing and a harsh environment, include failure-prone body hardware (trim, finish, locks, doors, and windows), water leaks, wind noise, and paint peeling/discoloration.

The following legend shows a vehicle's relative degree of overall reliability and which mechanical and body parts are subject to premature failure. Note that the numbers lighten as the rating becomes more positive.

❶	❷	③	④	⑤
Unacceptable	Below Average	Average	Above Average	Excellent

SMALL CARS

The proverbial "econobox," this size of car is for city dwellers who want economy at any price. Small cars offer excellent gas economy, easy manoeuvrability in urban areas, and a low retail price.

One of the more alarming characteristics of a small car's highway performance is its extreme vulnerability to strong lateral winds, which may make the car difficult to keep on course. Most of these cars can carry only two passengers in comfort—rear seating is limited—and there is insufficient luggage capacity. As well, engine and road noise are fairly excessive.

Crash safety may be compromised by the small size and light weight of these vehicles. Nevertheless, engineering measures that direct crash forces away from occupants and the addition of airbags have made many small cars safer in collisions than some larger cars.

Recommended

Chrysler Colt, Summit, Vista (1993–95)
Ford Escort, Lynx, Tracer (1994–98)
Honda Civic, CRX, del Sol (1993–98)
Mazda 323, Protegé (1996–98)
Nissan Sentra (1995–98)
Subaru Chaser, DL, GL, Impreza, XT (1996–98)
Suzuki Esteem (1996–98)
Toyota Corolla (1995–98)
Toyota Paseo (1994–98)
Toyota Tercel (1993–98)

Above Average

GM/Suzuki Firefly, Metro, Sprint, Swift (1995–98)
Honda Civic, CRX, del Sol (1992)
Hyundai Accent (1995–98)
Hyundai Elantra (1996–98)
Mazda 323, Protegé (1991–95)
Nissan Sentra (1991–94)
Toyota Corolla (1991–94)
Toyota Paseo (1992–93)
Toyota Tercel (1991–92)
Volkswagen Cabrio, Golf, Jetta (1997–98)

Average

Chrysler Colt, Summit, Vista (1989–92)
Ford Escort, Lynx, Tracer (1992–93)
GM Cavalier, Sunbird, Sunfire (1996–98)
GM/Saturn Saturn (1997–98)
GM/Suzuki Firefly, Metro, Sprint, Swift (1990–94)
Honda Civic, CRX, del Sol (1989–91)
Hyundai Elantra (1991–95)
Mazda 323, Protegé (1985–90)
Nissan Sentra (1988–90)
Subaru Chaser, DL, GL, Loyale, Impreza, XT (1993–95)
Toyota Corolla (1985–90)
Toyota Tercel (1987–90)
Volkswagen Cabrio, Golf, Jetta (1994–96)

Below Average

Chrysler Colt, Summit, Vista (1985–88)
Chrysler Duster, Shadow, Sundance (1993–94)
Ford Aspire (1994–97)
GM/Suzuki Firefly, Metro, Sprint, Swift (1987–89)
Honda Civic, CRX, del Sol (1984–88)
Hyundai Excel (1992–94)
Subaru Chaser, DL, GL, Loyale, XT (1984–92)
Volkswagen Cabrio, Golf, Jetta (1993)

Not Recommended

Chrysler Charger, Horizon, Omni (1983–90)
Chrysler Duster, Shadow, Sundance (1987–92)
Chrysler Neon (1995–98)
Ford Escort, Lynx, Tracer (1981–91)
GM Cavalier, Sunbird, Sunfire (1984–95)
GM/Saturn Saturn (1992–96)
Hyundai Excel (1986–91)
Nissan Sentra (1983–87)
Volkswagen Cabrio, Golf, Jetta (1985–92)

Chrysler's 1996 Neon isn't recommended due to its head-gasket-a-day engine, transmission, and body component failures.

CHRYSLER/EAGLE

Charger, Horizon, Omni

Rating: Not Recommended (1983–90). All of these cars are virtually identical and have been off the market for almost a decade. **Maintenance/Repair costs**: Inexpensive. Repairs can be done by independents. **Parts**: Except for body parts, easily found and relatively cheap.

Strengths and weaknesses: Gone and best forgotten, these cars are relatively roomy for subcompacts (except for rear seating and an awkward driving position); they give good, but not spectacular, fuel economy; they perform well on the road with the 2.2L powerplant; and they can be repaired almost anywhere—which is good, because they tend to break down almost everywhere. Expect serious reliability problems with the 2.2L engine (see "Secret Warranties/Service Tips"), in addition to fuel, electrical, and ignition system problems that are difficult to diagnose and even harder to repair due to the poor quality of components. The power-steering rack develops leaks and the exhaust system rusts quickly. Other problems include rapid brake wear and unreliable electronic components. Air conditioners and turbochargers often malfunction and are expensive to troubleshoot and repair.

All 1978–90 models suffer from extensive surface rust, with perforations found around rear wheels and the rear hatch. Many owners also complain of severe underbody rusting that affects safety. All years have such poor body assembly that doors are constantly sticking shut (some owners have had to climb out the windows), locks fall off, handles detach, and water and air leaks are legion.

Safety summary/Recalls: **Recalls**: **1985–87 Turbo**—The fuel hose connection may leak. **1986**—The rear suspension may partially separate from the vehicle, causing a sudden loss of control. **1987**—A faulty pressure regulator may leak fuel. **1989–90**—2.2L and 2.5L engines leak oil at the valve cover; this can be corrected with a new cylinder head cover kit.

Secret Warranties/Service Tips

1988–90—2.2L and 2.5L EFI engines that run roughly at idle may require a new EGR valve. **1989–90**—2.2L EFI engines with a spark knock during hot engine idle in drive may need a new engine controller. **All models/years**—2.2L, 2.5L, and 3.0L V6 engines that surge and buck at 35–55 mph with A413 and A670 transmissions may require driveability kit #4419447. • All front-drives with automatic transmissions that have delayed engagement and no Drive or Reverse after start-up should have the front transmission pump replaced.

Charger, Horizon, Omni Profile

	1983	1984	1985	1986	1987	1988	1989	1990
Cost Price ($)								
Omni/Horizon	6,600	7,000	7,397	7,397	7,295	7,795	8,550	9,500
Charger	7,487	7,705	7,999	8,280	8,412	—	—	—
Used Values ($)								
Omni/Horizon ↑	500	500	500	600	700	1,000	1,200	1,400
Omni/Horizon ↓	500	500	500	500	600	900	1,100	1,300
Charger ↑	600	700	800	900	1,100	—	—	—
Charger ↓	600	700	700	800	1,000	—	—	—
Extended Warranty	Y	Y	Y	Y	Y	Y	Y	Y
Secret Warranty	N	N	N	N	N	N	N	N
Reliability	❶	❶	❶	❶	❶	❶	❶	❶
Crash Safety	❷	❷	❷	❷	❷	❷	❷	❷

Colt, Summit, Vista

Rating: Recommended (1993–95); Average (1989–92); Below Average (1985–88). The 1997–98 Mirage is the standout among these small cars and "mini" minivans. The 1996–98 models were sold only in the States. **Maintenance/Repair costs**: Repairs aren't difficult to perform and can be done by cheaper independent garages or Mitsubishi dealers. **Parts**: Easily found and relatively inexpensive, although Summit and Colt wagon body panels may be in short supply.

Strengths and weaknesses: Some of the best small cars and wagons that Chrysler doesn't make (they're all Mitsubishi imports). Although the E and DL models aren't sparkling performers, they remain competitive in the subcompact arena and can be purchased for a lot less than their Japanese cousins. The 1985–88 Colts don't handle as well or offer as much rear leg room and cargo space as the more rounded, aerodynamically styled, contemporary 1993–98 Colts. The 1989–92 versions got a small horsepower boost that gives them a more spirited performance; however, passenger and cargo room are still at a premium.

This well-proven design has accumulated few problems over the years, due primarily to good quality control. Body construction and assembly are also quite good, although inadequate soundproofing allows the intrusion of lots of engine and road noise. Early models tend to have some front brake, electrical system, and engine problems, while the later 1989–96 versions are mostly beset with fuel supply, electrical system, and brake malfunctions.

A troublesome turbocharged 1.6L motor was first offered in 1984. The engines on high-mileage cars often burn oil because of worn piston rings. Models equipped with automatic transmissions vibrate badly when idling in gear. Air conditioners are unreliable and expensive to repair. Carburetors can be finicky, too. Front brakes wear rapidly. There are many reports of ignition troubles for the 1984–85 models. Problems carried over to the 1986–92 models are premature engine and exhaust system wear. Since then, Colts have shown few serious defects. In fact, the redesigned 1997 Mirage (the Colt was dropped in 1994) carries an upgraded automatic transmission, uses a stiffer body, and is much quieter and better appointed than its predecessor. Post-1986 LX, Vista, and 4X4 models are plagued by similar defects, with the addition of transmission and fuel system malfunctions.

The Vista, and Summit Wagon are small minivans with five- to seven-passenger seating. They are more reasonably priced, practical, and fuel efficient than many other small wagons. The five-passenger Colt wagon is similar to the Summit and Nissan Axxess in that it offers the extra versatility of a third seat in back and a tall body. As such, it makes a great car for a small family, while being able to haul small loads to the cottage or wherever. The wagon series is available in four-wheel drive and uses practically the same mechanical components as the other Colt models. If the wagon's price is too steep, check out the Nissan Axxess, which offers similar advantages. The Summit wagon is essentially a wagon version of the Colt 200 and sells at a premium. The 1.8L or 2.4L engine goes best with the 5-speed manual transmission. With an automatic, there is a significant fuel penalty and the 2.4L takes a while to change gears, particularly when going uphill.

On post-1990 Colts and Summits, the 2.4L head gasket may fail prematurely. Shocks aren't very durable, braking isn't impressive, and the front brakes have a short life span. Emission components, like the oxygen sensor, often fail after two years of use. A few owners have reported automatic transmission failures. Be especially wary of the troublesome 4X4 powertrain on 1989–91 wagons and the 16-valve turbo dropped in 1990. Owners of the 1993 wagon complain of poor heating and defrosting. Surface rust is common, as are rust perforations on door bottoms, the front edge of the hood, and the rear hatch.

Owners of 1992–97 models report premature piston ring wear and excessive engine noise caused by carbon buildup on the top of the piston. Front brakes and shock absorbers continue to have a short life span.

Safety summary/Recalls: **1989–93 Colts and Summits**—Automatic shoulder belts malfunction. • The wagon's sliding doors have child-proof door locks. Unfortunately, some Summit wagons may have defective sliding door latches. • 1994 Mirage and Summit inadvertent airbag deployment killed driver. **Recalls: 1986–91 Colts and 1989–91 Summits**—Takata seatbelts need replacing. **1992–93 Colts and Summits**—Automatic seatbelts may not move into place.

1992–96—Transfer case oil leaks could cause bearing seizure or wheel lockup. **1993**—Shoulder belt–guide rail cable may jam.

Secret Warranties/Service Tips

1989–90—Cold-start driveability problems can be fixed by cleaning the engine of carbon deposits and putting in an upgraded valve cover (#MD 118-125). • Squeaky front suspension: replace both stabilizer bar bushings. **1991**—Hard shifting or gear clashing can be corrected by installing an improved 1–2 synchronizer hub and sleeve, a 1–2 synchronizer spring, and a 3–4 synchronizer spring on cars equipped with manual transaxles. **1992**—Wagon fuel tanks may be slow to fill because the in-tank baffle impedes fuel flow. • Wagons may have excessive wind noise around the upper front door frame. **1992–93**—A front-end popping noise means the stabilizer ball joint grease has deteriorated. Add new grease and install an upgraded dust cover. **1992–94**—Carbon buildup on the piston top can be reduced by adding a bottle of Mopar Fuel Injector Cleaner to a full tank of gas. • The sliding door may not open from inside due to a faulty connecting door latch rod clip. **1993**—Wagons with poor heating/defrosting require upgraded distribution ducts. **All models/years**—A common problem with the 1.6L engine is that the exhaust manifold can come loose, causing an exhaust leak. When you first spot the trouble, simply tighten the exhaust manifold bolts. Chrysler may cover costs related to premature engine head gasket, ring, and valve wear if the five-year emissions warranty is applicable.

Colt, Summit, Vista Profile

	1988	1989	1990	1991	1992	1993	1994	1995
Cost Price ($)								
Colt/Summit	8,526	9,098	8,672	7,723	7,820	8,745	10,995	12,350
Colt 200	—	9,874	10,181	9,195	9,380	—	—	—
Colt Wagon	—	—	—	—	12,919	13,146	15,205	—
Vista Wagon	—	13,662	14,850	13,875	—	—	—	—
Summit Wagon	—	—	—	—	12,919	13,600	15,205	18,555
Used Values ($)								
Colt/Summit ↑	1,200	1,600	2,000	3,000	3,500	4,000	5,000	6,000
Colt/Summit ↓	1,000	1,500	1,800	2,500	3,000	3,500	4,500	5,500
Colt 200 ↑	—	2,200	2,500	3,000	4,000	—	—	—
Colt 200 ↓	—	2,000	2,300	2,600	3,100	—	—	—
Colt Wagon ↑	—	—	—	—	4,500	5,500	7,500	—
Colt Wagon ↓	—	—	—	—	4,000	5,000	6,000	—
Vista Wagon ↑	—	2,500	3,000	4,000	—	—	—	—
Vista Wagon ↓	—	2,000	2,500	3,500	—	—	—	—
Summit Wagon ↑	—	—	—	—	4,500	5,500	7,500	9,000
Summit Wagon ↓	—	—	—	—	3,500	5,000	6,500	8,000
Extended Warranty	N	N	N	N	N	N	N	N
Secret Warranty	N	N	N	N	N	N	Y	Y
Reliability	❷	❷	❷	❷	③	③	③	④
Air conditioning	❷	❷	❷	❷	③	③	③	④
Body integrity	❷	❷	❷	❷	❷	❷	❷	③
Braking system	❷	❷	❷	❷	❷	❷	❷	❷

Electrical system	❷	❷	③	③	④	④	⑤	⑤
Engines	❷	❷	③	③	③	③	③	③
Exhaust/Converter	③	③	③	④	④	⑤	⑤	⑤
Fuel system	③	③	④	④	④	⑤	⑤	⑤
Ignition system	③	③	⑤	⑤	⑤	⑤	⑤	⑤
Manual transmission	④	④	⑤	⑤	⑤	⑤	⑤	⑤
- automatic	③	③	③	③	❷	❷	③	③
Rust/Paint	❷	❷	❷	❷	③	④	④	⑤
Steering	④	④	④	④	⑤	⑤	⑤	⑤
Suspension	❷	❷	③	③	③	④	③	③
Crash Safety								
Colt Sedan	❷	❶	❶	❶	❶	❶	❶	—
Colt Wagon	❷	❷	❷	❷	❷	❷	❷	—
Vista Wagon	❷	❷	❷	❷	⑤	—	—	—

Duster, Shadow, Sundance

Rating: Below Average (1993–94); Not Recommended (1987–92). 1994 was the last model year, then these cars were replaced by the Neon (somewhat akin to "jumping from the frying pan into the fire"). **Maintenance/Repair costs**: Average. Repairs aren't dealer dependent. **Parts**: Average cost, but can be bought for less through independent suppliers.

Strengths and weaknesses: Fairly rapid depreciation makes these cars relatively inexpensive to buy. They're better built than other Chryslers in their class, but they still have many of the same mechanical and body weaknesses that Chrysler owners have suffered with throughout the years.

On early models, the turbocharged engine isn't very reliable and can be quite expensive to repair. The fuel-injection system can be temperamental and the MacPherson struts leak or wear out prematurely. Front brakes are particularly prone to rapid wear, and parking brake cables need frequent service. Convertible tops on the 1991 Shadow tend to leak profusely, and non-metallic paint chips easily.

Owners of 1989–94 models report cylinder head, oil pan gasket, and rear crankshaft seal leakage; the air conditioning compressor rarely lasts more than two years; windshield wiper fluid often freezes in hoses; and the power-steering assembly seldom lasts longer than five years. The upgraded 1992 41TE automatic transmission is still problematic: engine head gaskets often need replacing; oil leaks and oil pump failures are common; rear brakes are noisy; the heating, air conditioning, and ventilation systems often malfunction; electrical components (notably electronic modules) aren't very durable; and excessive suspension vibrations are common. Body assembly is sloppy, paint discolours or peels prematurely, door moldings fall off, and doors freeze shut.

Safety summary/Recalls: According to the Center for Auto Safety (CFAS), hundreds of owners have complained about engine fires caused by fuel and oil leaks in 1978–90 front-drive Chryslers. • Brake failures are also common, according to the CFAS. **Recalls**: **1987 Turbo**—Fuel can leak at supply hose/pressure regulator/fuel rail connections. **1987 Shelby**—The parking brake may be faulty. **1988**—The front passenger seatbelt retractor may be faulty, compromising child safety seat protection. **1988–89**—The automatic shoulder restraint system could malfunction in a collision. **1989–91**—The possibility of engine oil leakage means that owners can get a new engine valve cover gasket. **1991**—Guide pin bolts on front disc brake calipers may be too loose. **1991–92**—Steering wheel cracks may cause the wheel to loosen. **1991 Shadow**—Both airbag front-impact sensors may be improperly mounted. • Outboard front seatbelts may not latch properly. **1991–92**—The driver's seatback attaching bolt may fail. **1992**—Coupling bolts are faulty in the steering column shaft.

Secret Warranties/Service Tips

1990—Clearcoat paint delamination cause and cure are outlined in a special bulletin. **1990–91**—Surging or bucking at 35–55 mph with A413 or A670 automatic transmission can be fixed with driveability kit (#4419447). • 2.2L and 2.5L engines with oil leaks at the valve cover need a new cover. **1990–92**—Erratic idle speeds occurring after deceleration from a steady cruising speed can be corrected by replacing the idle air control motor. **1991–93**—The serpentine belt may come off the pulley after driving through snow; install an upgraded shield, screw, and retainers. **1991–94**—Engines that stall following a cold start may need an upgraded Park/Neutral/start switch. • Poor AC performance while the AC blower continues to operate is likely due to the evaporator freezing. **1992**—If the heater and ventilation systems change to the defrost mode during acceleration, trailer towing, or hill climbing, the installation of a revised vacuum check valve should cure the problem. • Long crank times, a rough idle, and hesitation could be corrected by replacing the intake manifold assembly. **1992–93**—A buzzing heard when the 41TE automatic transmission shifts into Reverse can be fixed by replacing the valve body assembly or the valve body separator plate. • A deceleration shudder can be eliminated by replacing the powertrain control module with an upgraded version. • 3.0L engines that burn oil or produce a smoky exhaust at idle can be fixed by installing snap rings on the exhaust valve guides and replacing all of the valve guide stems or the cylinder head. **1992–94**—Rough idling after a cold start with 2.2L and 2.5L engines can be corrected by installing an upgraded powertrain control module (PCM). • Harsh automatic shifts can be tamed by installing the following revised parts: kickdown, accumulator, reverse servo cushion springs, and accumulator piston. **1993**—Failure of the fuel pump check valve can cause start-up die-out, reduced power, or erratic shifting. **1993–94**—Acceleration shudder may be caused by automatic transmission front pump leakage. • Improved automatic shifting can be achieved by installing an upgraded transmission control module. **1994**—Poor AC performance is likely due to defective compressor suction or discharge lines. **All models/years**—A rotten-egg odour coming from the exhaust is likely the result of a malfunctioning catalytic converter, possibly covered under the emissions warranty.

Duster, Shadow, Sundance Profile

	1987	1988	1989	1990	1991	1992	1993	1994
Cost Price ($)								
Shadow/Sundance	9,402	9,795	10,150	11,650	8,995	9,295	10,750	11,170
Duster V6	—	—	—	—	—	11,400	12,825	13,985
Used Values ($)								
Shadow/Sundance ↑	1,500	2,000	2,500	3,000	4,000	4,500	5,000	5,500
Shadow/Sundance ↓	1,000	1,500	2,000	2,500	3,500	4,000	4,500	5,000
Duster V6 ↑	—	—	—	—	—	5,000	5,500	6,000
Duster V6 ↓	—	—	—	—	—	4,500	5,000	5,500
Extended Warranty	Y	Y	Y	Y	Y	Y	Y	Y
Secret Warranty	N	N	N	N	Y	Y	Y	Y
Reliability	❶	❷	❷	❷	❷	❷	❷	③
Air conditioning	❶	❷	❷	❷	❷	❷	❷	❷
Body integrity	❶	❶	❶	❶	❶	❶	❷	③
Braking system	❶	❶	❷	❷	❷	❷	③	④
Electrical system	❶	❶	❷	❷	❷	❷	③	③
Engines	❶	❷	③	③	③	③	③	④
Exhaust/Converter	❷	❷	❷	❷	❷	③	③	③
Fuel system	❶	❷	❷	❷	❷	③	③	③
Ignition system	❶	❷	③	③	❷	③	④	⑤
Manual transmission	③	③	③	④	④	⑤	⑤	⑤
- automatic	❷	❷	❷	❷	❷	❷	❷	③
Rust/Paint	❶	❶	❶	❶	❶	❶	❷	③
Steering	❷	❷	❷	③	③	③	④	④
Suspension	❷	❷	③	③	③	③	④	⑤
Crash Safety	❶	—	—	—	⑤	⑤	❷	—

Neon

Rating: Not Recommended (1995–98). A low-quality econobox that eats engine head gaskets for breakfast. Choose a model with touring suspension for the best ride. The Highline and Sport versions are more feature laden. **Maintenance/Repair costs**: Higher than average. Repairs must be done by dealers. **Parts**: Easily found and relatively inexpensive.

Strengths and weaknesses: A small, noisy car with big quality problems, the Neon does offer a spacious interior and responsive steering and handling. Nevertheless, it uses an antiquated 3-speed automatic gearbox, a DOHC engine that has to be pushed hard to do as well as the SOHC engine, and a mushy base suspension. Furthermore, a perusal of service bulletins and owner comments clearly shows that these cars have a plethora of serious factory-related defects, including a biodegradable engine head gasket (covering by a 7-year/160,000 km secret warranty), an abrupt-shifting and unreliable automatic transmission, an air conditioning system that often requires expensive servicing, a multitude of electrical glitches, lots of interior noise and leaks, uneven fit and finish, and poor-quality trim items that break or fall off easily. The finish is not as good as on most other subcompacts; Chrysler uses only two coats of paint versus the Japanese practice of applying three coats. Furthermore, the thickness of the coat varies considerably and can chip easily.

Except for the addition of a cast aluminum oil pan to reduce engine vibrations, redesigned wheel covers, improved sound system, and a centre console with armrest for 1997 models, the Neon remains basically unchanged since it was first launched.

Dealer service bulletins: **1995**—Engine sags, hesitates, shudders, and surges. • Rough idle especially bad during cold start-up. • Oil leaks at the cam position sensor. • Buzzy manual and automatic gearshift lever, and harsh-shifting, erratic automatic transmission. • Noisy clutch pedal and wheel cover. • Rear suspension bottoms out. • Steering wheel shakes and accelerator pedal vibrates. • Premature front brake pad wear. • AC freeze-up, poor performance, and evaporator odours. • Inaccurate fuel gauge readings, instrument panel glare, and water leaks. • Water leaks into the passenger compartment on left-hand turns. • Flickering headlights. (Neons are so cheap that the electrical system can't take the AC load and often the headlights will flicker annoyingly. DSB #08-25-95, issued June 3, 1994, says too bad; it's operating as designed.) • Discharged battery. • Intermittent wiper operation. • Excessive engine noise in passenger compartment, exhaust noise/hiss, B-pillar wind noise, power-steering rattles, steering column and right engine-mount click, and chattering steering column tilt lever. • More noise: front seat rattle and front seatback squeak, hood prop-rod rattle, idle air control motor whistle, front brake moan, poorly fitted deck-lid rattle, and poor AM reception (static). • Glue oozes out at backlight/windshield molding, and backlight molding lifts off. **1996**—Cold-start hesitation, engine misfiring, and erratic idling. • Excessive engine vibration and exhaust noise. • Transmission slippage from second to third gear during light acceleration. • Speed control overshoots or undershoots. • Rear brake chirps or howls. • AC evaporator produces a high-pitched whistle. • Fuel tank won't fill or is slow to fill. • Delayed windshield washer fluid output. • Interior-window film buildup. • Water leaks at cowl cover seam. **1997**—Rear brake howl. • Front foot well creak/rattle. • Scratched door glass. • Improper AC compressor engagement. • Loss of power steering in heavy rain or when passing through puddles. • Poor radio reception. • Warning that premium fuel may cause stalling, long cold-start times, hesitation, and warm-up sags. • Front suspension popping/creaking noise. **1998**—AC compressor lockup at low mileage. • Sag, hesitation, harsh AC operation, and headlight flickers. • Steering wheel/column rattles and clunks. • Cold-start power-steering noise. • Emissions recall due to radiator fan malfunction. • Front brake squeal, creep, or groan. • Paint fogging. • Warning that premium fuel may cause stalling, long cold-start times, hesitation, and warm-up sags. • Popping noise when passing over bumps or making turns. • Remote keyless transmitter battery failure. • Sunroof shade rattles in open position. • Vehicle overheats or radiator fan runs continuously.

Safety summary/Recalls: 1995–98 models performed poorly in IIHS 40 mph crash tests. NHTSA officials have opened a formal investigation into reports of oversensitive airbag deployments on 1995 Neons. **1995**—Steering column fires. • Ninety-eight reports of "inappropriate" airbag deployment, causing 13 crashes and injuring 28 people. • Sudden steering loss. • Small horn buttons are hard to find in an emergency. • Headlight switch is a "hide and go seek" affair. • Axle shafts may suddenly fail, as the owner of a 1995 Neon discovered:

> I was driving in traffic and the car just stopped. I couldn't move forward or backwards but the engine was still going. It turned out that the left front axle of the car had collapsed. I was just lucky that I wasn't driving on the highway when this occurred, because I would have been severely injured or killed. I had the car towed to the dealer, and they repaired the broken axle, but I just don't feel safe in the vehicle....

1995–96—Engine compartment fires. • Electrical short in dashboard. • Sudden acceleration. • ABS brake failures. • Inadvertent airbag deployment. • Airbag failed to deploy. • Seatbelt failed to restrain driver. • Driver's seatbelt tightens uncomfortably. • Driver's side seatbelt pulled out from buckle during collision. • Premature front brake pad/rotor wearout. • Excessive wearout of brake rotors apparently caused by premature pitting of the rotors due to bubbles created during the casting process. • Front brake caliper sticking. • Chronic engine head gasket failures. • Engine camshaft seal oil leaks • Engine motor mount and exhaust donut gasket failures. • Stuck in idle. • Chronic stalling. • Throttle system failures. • Faulty cruise control. • Chronic transmission failures. • Transmission suddenly downshifts to first gear when accelerating at 90 km/h. • Sudden steering loss. • Steering locks up every time it rains. • Trunk springs won't hold lid up. • Door hinge failures. • Premature wheel bearing and steering knuckle wear. • Chronic light and gauge failures. • Faulty Goodyear tires. • Fuel gauge failures. • Defective brake master cylinder. • Window rattling, goes off track, or shatters when door is closed. • AC condenser and compressor failures. • Water leaks in from dash on driver's side. • Poor door fit allows water to enter cabin. **1997**—Circuit board behind dash caught fire. • Sudden acceleration. • Cruise control won't disengage when braking. • Chronic stalling. • Airbag failed to deploy. • Gas tank leaks fuel. • O-ring in the fuel rail leaks fuel. • Chronic sudden brake failures. • Noisy brakes. • Brake rotor failures. • Steering belt failures. • Sudden steering lockup after passing over speed bump. • Engine head gasket failures. • Window seal failures. • Fuel pump failures. • Seatbelt failed to restrain occupant. • Catalytic converter failure. • Hood fell, injuring driver's arm. **1998**—Sudden acceleration. • Airbag failed to deploy. • Driver-side window exploded in warm weather. • Front right wheel bolt fell out causing wheel to bend. • Trunk lid may fall. • Engine surging and stalling. • Engine loses speed rapidly when going uphill. • Excessive engine carbon buildup. • Timing belt broke causing extensive engine damage. • Engine mounts broke and head gaskets leaked. • Erratic automatic transmission performance. • There is a partial steering hangup when making a right turn. **Recalls**: **1995**—Corroded fuel and rear brake tubes may fail. • Steering columns could snap loose from the car's frame. • Defective master cylinder piston. • Faulty PCM may cause chronic stalling. • Rear brake lines may be out of position. • Faulty rear ABS brake master cylinder seal. **1996**—Engine wiring harness may short, causing stalling. **1997**—Airbag module should be replaced to prevent airbags from deploying while vehicle is parked. **1997–98**—Faulty fan relay could cause engine overheating. **1998**—Rear suspension cross-member may be missing spot welds.

Secret Warranties/Service Tips

1995—Tips on how to fix paint fogging and stained white bumpers are offered. • Faulty radios will be replaced for free under a secret warranty. **1995–97**—If water drips into the vehicle from the roof-rail weatherstrip channel, Chrysler will provide, under warranty, a free anti-drip roof-rail retainer channel.

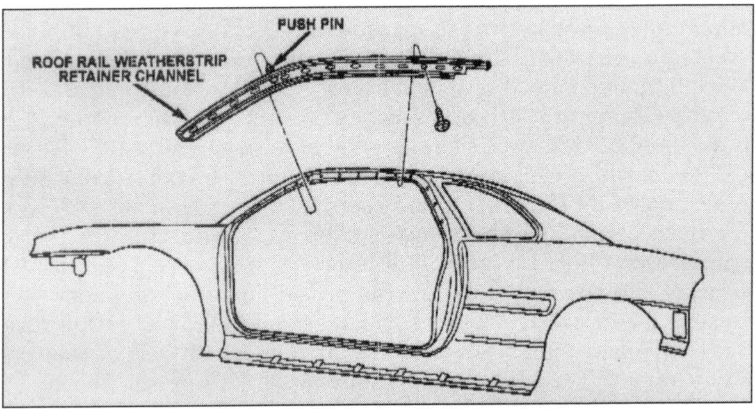

This correction should take one hour and is reimbursable under the warranty.

1996—Water could enter the air cleaner housing, be ingested into the engine, and cause serious engine damage. To prevent this from occurring, the dealer will drill a hole in the housing and seal the hood to cowl weatherstrip. This 30-minute correction is free of charge under Chrysler Customer Satisfaction Notice #660. It's not a safety recall, so you may have a hard time getting Chrysler to acknowledge the problem.

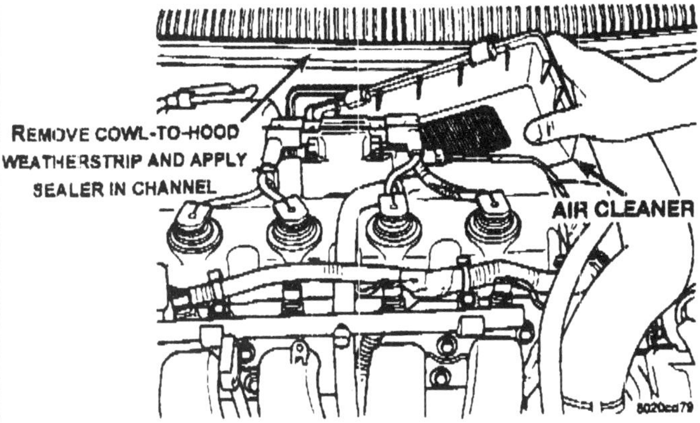

Figure 2

Get this fixed, or you'll face huge engine expenses.

Neon Profile

	1995	1996	1997	1998
Cost Price ($)				
Base	11,866	12,835	14,750	15,350
Sport 4d	16,126	15,515	16,900	17,500
Used Values ($)				
Base ↑	6,000	8,000	11,000	13,000
Base ↓	5,000	7,000	9,500	11,000
Sport 4d ↑	8,000	10,000	12,000	14,500
Sport 4d ↓	6,500	9,000	11,000	13,000
Extended Warranty	Y	Y	Y	Y
Secret Warranty	Y	Y	Y	Y
Reliability	❶	❶	❷	❷
Air conditioning	❶	❶	❷	③
Body integrity	❷	❷	❷	❷
Braking system	❶	❶	❷	③
Electrical system	❶	❶	❷	③
Engines	❶	❶	❷	❷
Exhaust/Converter	③	④	⑤	⑤
Fuel system	❷	❷	③	③
Ignition system	③	④	⑤	⑤
Manual transmission	⑤	⑤	⑤	⑤

- automatic	❷	❷	③	③
Rust/Paint	❷	❷	③	③
Steering	❶	❷	③	③
Suspension	⑤	⑤	⑤	⑤
Crash Safety	③	④	③	③
Side Impact	—	—	—	❷

FORD

Aspire

Rating: Below Average (1994–97). The Aspire's size, engine, and drivetrain limitations restrict it to an urban environment, and its low quality control restricts it to the driveway. **Maintenance/Repair costs**: Higher than average. Repairs are dealer dependent. **Parts**: As with other discontinued models, body panels are rare and expensive.

Strengths and weaknesses: Discontinued in 1997, the Korean-built Aspire comes in three- and five-door versions and is powered by a fuel-injected, 4-cylinder engine hooked up to either a standard 5-speed manual transmission or an optional 3-speed automatic. With its smooth-shifting 5-speed manual transmission and fuel-injection, the Aspire's base powerplant is adequate for short, city errands. This car has plenty of front seat room and excellent front and rear visibility. Corrosion protection is enhanced with galvanized steel panels, sealers, and coatings. Except for a minor facelift and upgraded side-impact protection, the 1997 Aspire is much like its predecessors.

The Aspire's engine lacks high-end torque, and the widely spaced gear ratios on the manual shifter rob it of much-needed mid-range power. It takes 13 seconds to reach 100 km/h—and an additional 4 seconds with the automatic gearbox, which is jerky when pushed. Air conditioning slows down the car even more. Steering is heavy and vague at high speeds. Unfortunately, the three-door version doesn't offer power steering. Expect excessive low-speed engine and road/tire noise at higher speeds. Its fuel economy is not all that impressive when compared to entry-level Hondas and Nissans, but the higher cost of these vehicles wipes out any fuel savings.

Dealer service bulletins: **1994**—Rough idle, hesitation, excessive fuel consumption, and poor heater output likely caused by the thermostat stuck in open position or opening before it should. **1995**—Automatic transmission fails to upshift in cold weather. • Brake roughness upon application. • Malfunctioning air conditioning and excessive compressor noise. • No-start due to faulty ignition switch. **1996**—Fog/film on windshield/interior glass. • MTX transmission clicks, clunks, and rattles when in Reverse. • Musty and mildewy odours. **1997**—Excessive manual transmission noise. • Delayed automatic transmission engagement. • Door glass won't roll down.

Safety summary/Recalls: Owners report this car is unstable in a crosswind, and the tiny brakes don't inspire confidence. NHTSA has opened an official investigation of fuel vapour valve leaks on the 1996–97 Aspires. **1994**—Wiring harness fire. • Driver's airbag deployed and caused fractured eye socket, detached retina, and several dislocated bones. • Airbag deployment caused severe burns. • Airbag failed to deploy. • Stalling caused by malfunctioning throttle position sensor. • Input shaft and reverse gear failure. • Steering column locks up, causing ignition switch problems. • Left front window shattered. • Tie rod broke and caused accident. • Front brake rotor warpage every 12,000 km. • Ball joint rusted away. • Frequent headlight switch failures. • Window shatters when door is closed. • Inadequate ventilation; poor defrosting. • Headlights are too dim. • Passenger-side seat unbolted from floor board. **1995**—Under-hood fire. • Frequent complaints that the airbag failed to deploy. • Severe burns and broken wrist when airbag deployed. • Stalling while cruising on the highway with AC operating. • Steering wheel lockup. • Many reports of fuel odour in the interior. • Premature brake pad wearout. • Wheel bearing failure caused vehicle to careen out of control. • Several reports the transmission shift lever broke off. • Transmission failures. • Cracked bracket causes clutch cable to break. • Rotten-egg smell permeates the interior. • Water enters interior when windows are down. • Chronic distributor short circuits. • Lights dim when using defroster. • Headlight failure when switching from low to high beam. • Inside windshield ices up due to faulty defroster. • Rusted-out hatchback hinge. • Rolled-down window shatters when door is closed. **1996**—Inadvertent airbag deployment. • Steering wheel lockup. • Brakes require extended stopping distances. • Excessive and premature brake pad and rotor warpage. • Front suspension geometry causes premature tire wear. • Gasoline smell in the interior. **1997**—Brake failure. • Wiring harness short caused an under-hood fire. • Frequent distributor cap failures cause chronic engine stalling. • Hood flew up and hit windshield. • Chronic fuel vapour odour in car's interior. **Recalls**: **1994**—Fuel-supply, return, and vapour hoses or lines may leak. Steel fuel lines will be repositioned; get additional corrosion protection, if needed.

Secret Warranties/Service Tips

1994—The upper steering column cover causes the warning flasher to stick. • Fuel leaks from fuel and vapour hoses/lines. **1996–97**—Fuel odour in the interior may be caused by a cracked seam weld on the vapour vent valve located in the left quarter panel area.

Aspire Profile

	1994	1995	1996	1997
Cost Price ($)				
Base	9,995	10,895	11,295	11,995
Used Values ($)				
Base ↑	4,500	6,000	7,000	8,000
Base ↓	4,000	5,500	6,500	7,500
Extended Warranty	Y	Y	Y	Y
Secret Warranty	Y	Y	Y	Y
Reliability	❷	❷	❷	❷
Air conditioning	❷	❷	③	③
Body integrity	❷	❷	❷	❷
Braking system	❷	❷	❷	❷
Electrical system	❷	❷	❷	③
Engines	④	④	④	④
Exhaust/Converter	③	④	⑤	⑤
Fuel system	❷	❷	❷	❷
Ignition system	④	⑤	⑤	⑤
Manual transmission	❷	❷	⑤	⑤
- automatic	❷	❷	❷	❷
Rust/Paint	❷	❷	④	⑤
Steering	❷	❷	❷	❷
Suspension	③	④	⑤	⑤
Crash Safety	④	④	④	④

Escort, Lynx, Tracer

Rating: Recommended (1994–98); Average (1992–93); Not Recommended (1981–91). The LX became the base Escort during the 1994 model year. Lots of upgraded 1997 Escorts are coming off lease now; they represent an exceptionally good buy. For the 2000 model year, the Escort will be replaced by the Focus. **Maintenance/Repair costs**: Higher than average on pre-1991s; below average for later models. Repairs can be done by independents or Ford or Mazda dealers. **Parts**: Expensive, but easily found.

Strengths and weaknesses: These front-drive small cars are usually reasonably priced and economical to operate, and provide a comfortable though busy ride and adequate front seating for two adults. However, they have a "Dr. Jekyll and Mr. Hyde" disposition, depending on which model year you buy. From 1982 through 1991, these subcompacts were dull performers with uninspiring interiors. Worse, they had a nasty reputation for being unreliable and expensive to repair.

Be wary of early Escorts (1984–87) that use Mazda's 2.0L diesel 4-cylinder; it's weak, noisy, difficult to service, and prone to expensive cylinder head and gasket repairs. From 1987 until 1991, model year quality continued to go downhill. The 1.9L engine used from 1985 to 1990 gives respectable highway performance but its failure rate is still much higher than average. The radiator and other cooling components, including the fan switch and motor, are failure prone. Carburetors and fuel-injection systems are temperamental. Ignition modules are often defective. Power-steering racks fail prematurely. Front and rear wheel alignment is difficult. Exhaust systems rust rapidly.

The 1991 model's changeover to mostly Mazda 323 components gave it a longer wheelbase, making for a more comfortable ride and a bit roomier interior. The 1.9L engine runs more smoothly, as well. The GT and Tracer LTS are equipped with Mazda's powerful 127-hp, 1.8L, 4-cylinder engine, and their overall highway performance and fuel economy are far superior to what previous models offered. Wagon versions are particularly versatile and spacious. The front seats on the Tracer GS and LS are very comfortable and the cargo area is especially spacious in the wagon. Rear seat room is a bit cramped on all other models.

Unfortunately, quality control and reliability didn't improve but continued to dramatically decline over the following years. Owner complaints relating to the 1991–96 model years concern primarily the automatic transmission and engine, the cooling system, brakes, electrical short circuits, corroded fuel pump wiring, air conditioning, the fuel pump, and ignition system failures.

The 1997 models were significantly improved quality-wise and are more attractively designed and ride and handle better and featured improved comfort, a quieter interior, and more standard equipment. The car is four inches longer, mostly taken up by a larger trunk. New standard features included power steering; rear heat ducts; intermittent wipers; a battery-saver system; a 24-watt, four-speaker stereo unit that can be upgraded to an 80-watt system hooked to a trunk-mounted CD player; and solar glass. Three-door, five-door, and GT models were dropped and the ZX2, a revived coupe, made its debut in late 1997 as a 1998 model.

Safety summary/Recalls: **Escort recalls**: **1983–93**—Ford will replace fire-prone ignition switches on designated Escorts/Tracers. **1985**—Rocker arm oil leakage could cause a fire. • Cars with 1.9L engines may fail to return to idle. **1985–86**—Manual gearshift lever may accidentally slip into Neutral in cold weather. **1985–87**—Driver's seat could come loose in an accident. **1986–88**—Fuel line leakage could cause a fire. **1987**—Stainless-steel lug nuts may fracture, causing the wheel to fall off. **1990**—Windshield could come out in a frontal collision. **1991**—Accelerator pedal could stick wide open. • Steering column could lock up. **1991–92**—Fuel vapour could escape from a full fuel tank, causing a fire. **1991–94**—Fuel pump wiring corrosion could cause chronic stalling. **1992**—Faulty brake light switch. **1993**—Faulty driver's seat could fail in a collision. **1995**—Airbag mounting bolts may be missing or improperly torqued. **Tracer recalls**: **1991—** Fuel vapour could escape from fuel tank, causing a fire. **1992—** Steering column may lock up.

Secret Warranties/Service Tips

Escort: **1983–94**—Hesitation, a rough idle, or poor heater output may all be caused by a faulty thermostat. **1986–89**—Hard starting/stalling may be caused by a defective fuel pump or sender assembly. **1988**—Owner notification M59 says that any fuel pump malfunction that drains the battery, makes for hard starts, or cuts engine performance can be corrected by installing an upgraded fuel pump diode (#E8FZ-14A411-A). **1990**—Owner notification B91 once allowed for the free modification of faulty dealer AC units with condenser fans that rotate backwards. **1991**—Eliminate knocking heard after a cold start or when returning to idle speed after freeway driving by installing new engine camshafts (after ruling out the possibility of lifter noise). **1991–93**—Metallic ticking heard after initial start-up or when returning to idle speed may be caused by a faulty lifter, low amount of oil in the crankcase, incorrect oil filter, or oil deterioration. **1991–94**—Under a recall and Service Program #94B55, Ford will install at no charge a fused jumper harness in the fuel pump electrical circuit. This will prevent short-circuit problems such as stalling, erratic instrument gauge readings, and extensive wiring damage, which are caused by water intrusion. • Fix a timing belt that's noisy during cold weather by installing an upgraded, more rigid belt tensioner. • A missing or loose front valance panel will be replaced or secured with longer bolts free of charge. **1992–93**—A high idle rpm after heavy use may be corrected by installing a new idle air control valve. **1994–98**—Tips on eliminating wind noise around doors are given in DSB #97-15-1. **1998**—An erratic transaxle shift may simply be caused by a pinched wire.

Article No.
98-21-13
10/26/98
TRANSAXLE - ERRATIC SHIFT WITH DIAGNOSTIC
TROUBLE CODE (DTC) P0712
FORD:
1998 ESCORT
LINCOLN-MERCURY:
1998 TRACER
ISSUE
An erratic transaxle shift with a Diagnostic Trouble Code (DTC) of P0712 may occur on some vehicles. This may be caused by the Transmission Fluid Temperature (TFT) wire being pinched on the spring clip attached to the valve body.
ACTION
Repair the TFT wire with a plastic wrap. Refer to the following Service Procedure for details.
SERVICE PROCEDURE
1. Verify that DTC P0712 is stored in memory.
2. Check the TFT circuit for shorts to ground or opens. If either exist, proceed to Step 3. If neither exist, follow the appropriate Pinpoint Test in the Powertrain Control/Emissions Diagnosis (PC/ED) Service Manual.
3. Remove the transaxle pan by referring to the 1998 Escort/Tracer Workshop Manual.

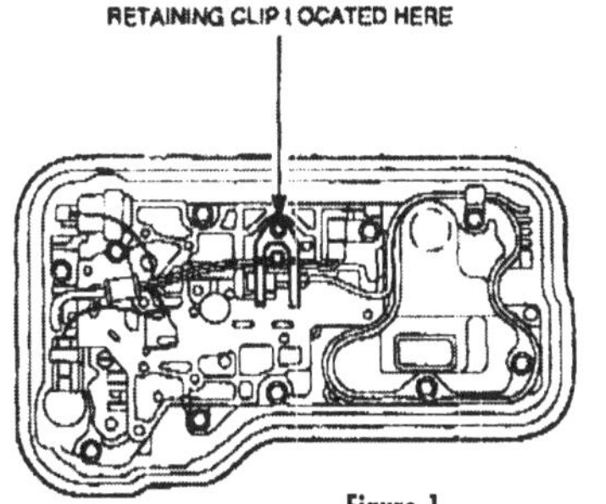

Figure 1

4. Remove the clip that retains the TFT sensor wire (Figure 1).

Escort, Lynx, Tracer Profile

	1991	1992	1993	1994	1995	1996	1997	1998
Cost Price ($)								
Base/LX	10,095	11,195	10,795	12,195	12,995	13,595	14,595	14,895
GT	13,895	15,095	15,472	13,995	14,295	15,295	—	—
Used Values ($)								
Base/LX ↑	2,500	3,500	4,500	6,000	8,000	9,000	10,500	12,000
Base/LX ↓	2,000	3,000	4,000	5,000	7,000	8,000	9,500	11,000
GT ↑	4,000	5,500	7,500	9,000	10,000	11,000	—	—
GT ↓	3,000	4,700	6,500	8,000	9,000	10,000	—	—
Extended Warranty	Y	Y	Y	N	N	N	N	N
Secret Warranty	Y	Y	Y	Y	Y	Y	Y	Y
Reliability	❷	③	③	④	④	⑤	⑤	⑤
Air conditioning	❶	❷	❷	③	③	③	③	④
Body integrity	❷	❷	❷	❷	❷	③	③	③
Braking system	❶	❶	❶	❷	③	③	④	④
Electrical system	❶	❷	❷	❷	❷	❷	③	④

Engines	❷	③	③	③	⑤	⑤	⑤	⑤
Exhaust/Converter	③	④	⑤	⑤	⑤	⑤	⑤	⑤
Fuel system	③	③	③	③	④	⑤	⑤	⑤
Ignition system	❷	③	③	③	④	⑤	⑤	⑤
Manual transmission	⑤	⑤	⑤	⑤	⑤	⑤	⑤	⑤
- automatic	❷	③	③	④	⑤	⑤	⑤	⑤
Rust/Paint	❶	❷	❷	③	③	③	③	④
Steering	③	⑤	⑤	⑤	⑤	⑤	⑤	⑤
Suspension	❷	③	③	④	⑤	⑤	⑤	⑤
Crash Safety								
Escort 4d	⑤	⑤	❶	❶	④	④	③	③
Tracer 4d	⑤	⑤	❶	❶	—	—	④	③
Side Impact								
Escort 4d	—	—	—	—	—	—	③	③
Tracer 4d	—	—	—	—	—	—	③	③
Escort ZX2 2d	—	—	—	—	—	—	—	③

Note: The low crash scores represent severe leg trauma.

GENERAL MOTORS

Cavalier, Sunbird, Sunfire

Rating: Average (1996–98); Not Recommended (1984–95). Repair bills will run you bankrupt if you get a pre-1995 or maintenance schedules aren't followed to the letter. An incredibly slow depreciation rate means that recent models are no bargains. Still, I'd buy a Cavalier, Sunbird, or Sunfire over a Saturn any day. The base Sunbird became the LE in 1989, and then changed its name to the Sunfire in 1995. The Z24 convertible was replaced by the LS in 1995. **Maintenance/Repair costs**: Average; repairs aren't dealer dependent; however, ABS troubleshooting is a real head-scratcher. **Parts**: Reasonably priced; often bought for much less from independent suppliers.

Strengths and weaknesses: Snappy road performance (with the right engine and transmission hookup) has been marred by abysmally poor reliability. The basic versions are lackluster performers. The 2.0L 4-cylinder engine gets overwhelmed by the demands of passing and merging. On top of that, major reliability weaknesses afflict all mechanical and body components. Engine, transmission, electronic module, and brake failures are common for all model years. 2.0L engine blocks crack, cylinder heads leak, and the turbocharged version frequently needs expensive repairs. Oil leakage from the rear crankshaft seal is common and oil filters on all model years tend to wear out quickly.

1986–89 model year improvements simply changed the nature, not the frequency, of engine and automatic transmission breakdowns.

Power-steering rack and front suspension components aren't durable. Consider replacing the original components with front gas struts and rear cargo coil springs to improve handling and durability. Owners also report that a change to high-octane fuel can help improve engine performance and reduce knocking and engine run-on. The cooling, exhaust, ignition, and fuel systems have had more than the average number of problems. The manifold heat shield tends to be noisy. Front brakes wear out quickly, and rear brakes tend to lock the rear wheels in emergency stops (one cause being the seizure of the rear brake adjusters). The Cavalier's optional 2.8L V6 with 3-speed automatic transmission is the best highway performer, but intake manifold gasket failures, premature head gasket wear, and transmission malfunctions compromise driving pleasure.

For 1990–94 versions, the Cavalier's base 2.2L 4-cylinder and optional 3.1L replaced the failure-prone 2.0L and 2.8L powerplants. Unfortunately, the newer engines already have a checkered reputation, highlighted by reports of chronic head gasket failures causing coolant leakage, overheating, engine seizure, hard starting, stalling, and surging—problems covered by a secret warranty. Air conditioning and hood latch failures, seatbelt defects, and a plethora of body deficiencies are also commonplace. Door bottoms and wheel housings are particularly vulnerable to rust perforation. Premature paint peeling and cracking, discoloration, and surface rust have been regular problems since these cars were introduced.

The 1995–98 versions benefited from some engine tweaks, standard dual airbags and ABS, a longer wheelbase (but shorter body), a stiffer structure, and an improved suspension. Unfortunately, though, the engine head gasket failures have apparently been carried over to the later models as well, plus quality control is still woefully deficient.

Dealer service bulletins: **1993**—3.1L engines hesitate or stall. • No Reverse or slipping while in Reverse with 3T40 automatic transmissions. • Rear brake squawk. **1994**—Body fit deficiencies. • Premature front brake wear. • Faulty ignition switches. • Electrical short circuits. **1995**—Engine failing to crank and no-start. • Brake vibration and/or pedal pulsation. • Longer-life front brake linings. • Low voltage reading and dim lights at idle. • Grinding/growling when in Park on an incline. • Knocking noise when traversing rough roads. • Rear window panel squeaking. • Rear quarter-panel road noise. • Loose door trim panel. • Poor paint application and rust spots. **1996**—Excessive engine roll. • Low speed knocking noise when passing over rough roads. • Air conditioning odours. • Radio frequency interference diagnosis tips. • Water leak diagnostic guide. • Loose door trim panel. • Tips on silencing rear quarter-panel road noise. • Condensation within exterior light housing. • Rear edge of hood rubs windshield when opening. • Twisted seatbelt webbing.

1997—Axle seal leakage. • Scuffed interior quarter trim panels on convertibles. • Coolant odour or leakage. • Deck lid hard to open or close. • Automatic transmission failure. • Low-engine-coolant light flashes. • Popping noise originating from the engine compartment. • Rear seatback rattles/squeaks. • Rear shock noise coming from trunk area. • Engine cranks, but won't run.

Safety summary/Recalls: **1995**—Steering locks up. • Faulty cruise control. • Windshield wipers fail in cold weather. • **1995–96**—ABS brake failure. • Airbags fail to deploy or deploy accidentally. • Inoperative horn. • Transmission slips out of Park. • Sudden acceleration, stalling. • Passenger side seatbacks won't stay upright. **1996**—The NHTSA has received 96 complaints of inadvertent airbag deployment causing 10 crashes and injuries to 53 people. • Engine gaskets leak oil. • Mufflers fail frequently. **1997**—Engine fires. • Inadvertent airbag deployment. • Airbag failed to deploy. • During a collision, driver sustained serious leg injuries when the seat pushed her lower body under the instrument panel. • Sudden acceleration. • Chronic stalling and hesitation, particularly when it rains. • Premature engine cylinder failure; leaks oil. • Premature oil pump failure. • Faulty fuel pump relay provokes stalling. • Failure of the engine mounts and ignition switch. • Cruise control failure. • Restart after stalling causes engine to race. • Sudden loss of power. • Steering wheel locked up. • Transmission failures. • Noisy rear end due to faulty suspension struts. • Faulty master cylinder and modulator assembly led to brake failure. • Frequent front brake rotor warpage, premature pad wear, and excessive vibrations when braking. • ABS brake failure. • Horn failed. • Cracked water pump. • Dashboard cracking, rattling, and popping. • Defective instrument panel control module. • Low beam switch failure. • Inoperative lighting due to rotted out wiring harness. • Door hinges don't hold door open. • Door came ajar while driving. • Windshield wiper failure. • When turned off, windshield wiper stops in the field of vision. • Lug nuts are easily broken when changing tire. **Recalls**: **1983–84**—The floor pan anchor on cars with a manually adjusted driver's seat could break, causing the driver's seat to tip backwards suddenly and causing a potential loss of vehicle control. • Models with 2.0L engines may suddenly accelerate due to a kinked accelerator control cable. **1985**—Air cleaner plastic trim could catch fire. **1986**—A defective headlight switch can cause headlights to flicker or fail. **1987**—It is possible that the fuel feed/return lines will crack, leaking fuel. • The parking brake lever may fail, allowing the vehicle to roll away unexpectedly. • Cars with 2.0L engines may have a frozen accelerator cable that could cause sudden acceleration. **1989**—The fuel tank, which could leak fuel, will be inspected and replaced, if necessary, for free. **1991**—Cracked front-door shoulder belt guide loops could pull loose in an accident. • Front door frames that anchor the seatbelt housing could collapse in an accident, resulting in seatbelt failure. **1992**—The hood could open suddenly, blocking the driver's view. **1993**—Faulty rear brake hoses could cause brake failure. **1993–94**—Horn pad may fall apart. GM has a "safety improvement campaign" to correct the defect for free. **1994**—Loose drive-axle spindle nuts may cause steering knuckle tire-wheel assembly to separate from the axle. **1995**—Front suspension lower control-arm assemblies may be defective. **1995–96**—Front or rear hazard warning lights could be faulty. **1996**—Accelerator cable may be kinked, requiring cxcessive pedal effort. **1996–97**—Airbag sensor is too sensitive. • Rear suspension trailing arm bolts can fatigue and break. **1997**—Driver's wiper blades may be five inches too short. **1988 Sunbird**—Backup lights may be inoperative. **1992–93**—Vehicles equipped with 2.0L engines may have defective throttle-cable assemblies that could stick open in cold weather.

Secret Warranties/Service Tips

1985–88—Premature brake lining wear may be caused by a maladjusted cruise control cutoff or brake light switch. **1985–89**—No first gear and/or slips in first may mean that you need new forward clutch piston seals. **1985–90**—Exhaust boom or moan can be corrected by installing a mass dampener (#10137382). • Noise from rear springs requires the installation of upgraded rear spring insulators (#22555689). **1986–87**—Erratic idle and 2.0L engine surging require the replacement of the PROM or TCC solenoid; another possible cause is a defective mass air flow sensor. **1987–88**—Difficult cold starts with 2.0L engines may require a new drop-in manifold deflector plate (#10112342). • A sagging headliner needs service package #22541347. **1987–90**—A rattle or buzz from the instrument panel may require a new upgraded brake-booster check valve (#18012017). **1988–89**—2.0L engine valve train noise can be reduced by adjusting or replacing the rocker arms. **1988–94**—Water leaks at the front upper door frame are treated in depth in a DSB issued in October 1994. **1989**—Product Campaign 89C16 provides for the free replacement of the 5-speed manual transmission. **1990**—Heater and AC blower noise can be reduced by replacing the blower assembly. **1990–91**—Poor starting may be caused by the spring in early starter drives compressing too easily; install an upgraded starter motor drive assembly (#10473700). **1991–94**—A crunching noise coming from the right side of the dash when turning the steering wheel can be silenced by installing a new steering grommet on the right side. **1992–93**—Engine head gasket failures may cause coolant leakage, overheating, and serious engine failure. GM will repair the defect at no charge under the following secret warranty.

GM

1908 Colonel Sam Drive
Oshawa, Ontario L1H 8P7

Dear General Motors Customer:

As the owner of a 1992 or 1993 Chevrolet Cavalier equipped with a 2.2L engine, your satisfaction with our product is of the utmost concern to us. Your vehicle was provided with a new vehicle warranty, which covers certain parts of your vehicle for a specified period. These warranties are of considerable value to you if you should experience problems with your vehicle.

This letter is intended to make you aware that some 1992 and 1993 Chevrolet Cavalier models with 2.2L engines may develop a failure of the cylinder head gasket that allows coolant to leak from the cylinder head gasket to engine block joint. Early evidence of this would be a loss of coolant in the coolant reservoir and an odor of coolant from the engine compartment, or a low coolant light. There may also be visible coolant deposits at the cylinder head to engine block joint.

General Motors of Canada Limited is therefore taking the following action:

We are providing owners with special coverage. If the above-mentioned condition occurs within seven (7) years of the date your vehicle was originally placed in service or 160,000 km, whichever occurs first, your vehicle will be repaired for you at no charge.

This special policy applies only to repairs requiring cylinder head gasket replacement as a result of cylinder head gasket failure that results in an engine coolant leak. It does not cover engine damage from continuing to operate the engine in an overheated condition after loss of coolant.

This is not a recall campaign. Do not take your vehicle to your GM dealer as a result of this letter unless you believe that your vehicle has the condition as described above. Keep this letter with your other important glove box literature for future reference.

If you have already paid for some or all of the cost to have the cylinder head gasket replaced and in-service time was less than seven (7) years and 160,000 km, you should contact your GM dealer. You may be eligible for partial or complete reimbursement of costs if genuine GM parts were used in the repair. If the work was done by someone other than a GM dealership the amount of reimbursement may be limited to the amount the repair would have cost GM to have it completed by a GM dealership. Please provide your dealer with your original paid receipts or invoices verifying the repair, the amount charged, proof of payment, and the date of payment of those charges by March 1, 1997.

Repairs and adjustments qualifying under this Special Policy coverage must be performed by your GM dealer.

The same head gasket problem has been reported by owners of 1994–95 models; GM hasn't formally included them in this program yet, probably because most vehicles may still be covered by the base warranty. Whatever the reason, remember this: GM IS 100 PERCENT RESPONSIBLE FOR REPAIRING THIS DEFECT.

1992–97—Paint delamination, peeling, or fading: GM lets dealers repaint the entire car at no charge to the owner (up to $650), regardless of whether it was bought new or used. Head office permission isn't needed before work commences (see pages 76, 113). **1993–94**—Excessive engine vibrations at idle or a clunk upon acceleration is most likely due to a defective engine mount. **1994**—A squeaking noise heard when going over bumps, accelerating, or shifting can be stopped by replacing the exhaust manifold pipe seal. • Water leaks into the front footwell are discussed in depth in a December 1993 DSB. **1986–92 Cavalier**—Vehicles equipped with the 3T40 automatic transmission may experience slippage in manual Low or Reverse; install service package #8628222, which includes a Low/Reverse clutch release spring (#8664961) and clutch retainer and snap ring (#656/657). **1987–88**—Engine stalling in Reverse or Drive with the THM 125C automatic transmission may mean that you need an upgraded auxiliary valve body filter (#8664921). • Poor AC cooling may be caused by a faulty pressure switch O-ring seal. **1992**—Engine hesitation or roughness, particularly at idle, may be corrected by installing a new lower intake manifold gasket (#10103647). **1995–97**—Axle seal leakage may be caused by a pinched transaxle vent hose. • Delayed automatic transmission engagement after a cold soak signals the need to install a revised forward clutch housing assembly. • Repair tips are offered for scuffed interior quarter trim panels on convertibles. • A sticking deck lid may need an upgraded lid release cable. • A dome light that won't shut off probably has a corroded door jamb switch. • The left-hand mirror may not adjust if the lever has become disengaged. • A popping noise originating from the engine compartment may mean that the torque strut-mount attaching bolts are loose. • Troubleshooting tips on silencing rear shock noise and rear seatback rattles and squeaks are available. **1996–97**—Coolant odour or leakage may occur at the joint where the radiator outlet pipe is connected to the coolant pump cover or at the joint between the cooling system air bleed pipe and the coolant outlet. **1997**—A low-engine-coolant light may come on to signal that the cooling system surge tank is defective. **All models/years**—Squeaking front brakes can be silenced by replacing the semi-metallic front brake linings with quieter linings (#12321424). The new linings will be 20 percent less durable (DSB #86-5-20). • A rotten-egg odour coming from the exhaust is probably caused by a malfunctioning catalytic converter; this repair is covered by GM's emissions warranty.

Cavalier, Sunbird, Sunfire Profile

	1991	1992	1993	1994	1995	1996	1997	1998
Cost Price ($)								
Cavalier	10,998	10,598	10,498	10,998	12,245	13,030	14,390	14,765
Z24	14,998	15,798	14,798	17,298	16,895	17,643	19,000	19,295
Z24/Conv.	—	18,400	18,098	20,298	22,185	22,925	24,285	25,880
Sunbird/LE	11,398	11,398	10,498	11,498	—	—	—	—
Sunbird GT	14,981	15,398	14,998	—	—	—	—	—
Sunfire	—	—	—	—	12,945	13,380	15,340	15,960
Used Values ($)								
Cavalier ↑	3,000	3,500	4,500	5,000	7,500	9,500	11,000	13,500
Cavalier ↓	2,500	3,000	4,000	4,500	7,000	9,000	10,500	12,500
Z24 ↑	5,000	6,000	7,500	8,500	10,500	13,000	14,500	16,500
Z24 ↓	4,500	5,500	7,000	8,000	10,000	12,000	13,500	16,000
Z24/Conv. ↑	—	8,000	10,500	12,000	12,500	15,000	17,000	20,000
Z24/Conv. ↓	—	7,500	9,500	11,000	12,000	14,000	16,000	19,000
Sunbird/LE ↑	3,000	3,500	4,500	5,500	—	—	—	—
Sunbird/LE ↓	2,500	3,000	4,000	4,500	—	—	—	—

Sunbird GT ↑	5,000	6,000	7,500	—	—	—	—	—
Sunbird GT ↓	4,000	5,000	6,500	—	—	—	—	—
Sunfire ↑	—	—	—	—	8,000	9,500	11,500	14,500
Sunfire ↓	—	—	—	—	6,500	8,500	10,500	13,500
Extended Warranty	Y	Y	Y	Y	Y	Y	Y	Y
Secret Warranty	Y	Y	Y	Y	Y	Y	Y	Y
Reliability	❶	❶	❷	❷	❷	③	③	③
Air conditioning	③	③	③	③	③	③	④	④
Body integrity	❶	❶	❶	❷	❷	❷	❷	❷
Braking system	❶	❶	❶	❷	❷	❷	❷	❷
Electrical system	❶	❶	❶	❷	❷	❷	❷	❷
Engines	❷	❷	❷	❷	❷	❷	③	④
Exhaust/Converter	❷	❷	③	③	③	④	④	⑤
Fuel system	❶	❶	❶	③	③	③	④	④
Ignition system	❷	❷	❷	❷	❷	③	③	③
Manual transmission	③	③	③	③	③	④	④	④
- automatic	❶	❶	❶	❷	❷	❷	③	③
Rust/Paint	❶	❶	❶	❷	❷	❷	❷	❷
Steering	❷	❷	❷	❷	❷	③	③	④
Suspension	❶	❶	❶	❷	❷	③	③	④
Crash Safety								
Cavalier 4d	⑤	⑤	❷	❷	—	③	④	④
Side Impact								
Cavalier 4d	—	—	—	—	—	—	❶	❶

Note: NHTSA says the Sunbird, Sunfire safety ratings should be identical to the Cavalier's score.

GENERAL MOTORS/SATURN

Saturn

Rating: Average (1997–98); Not Recommended (1992–96). Don't go anywhere near a used Saturn unless you're armed to the teeth with a comprehensive extended warranty. As bizarre as it may appear, the Saturn division has a better reputation than the car it sells. The car's poor quality control is masked by generous amounts of "goodwill" refunds that quickly dry up when the car changes hands. The Geo Storm, Honda Civic LX, and Toyota Corolla perform well and offer better quality for about $1,500–$2,500 more. **Maintenance/Repair costs**: Average. Repairs aren't dealer dependent, unless you're seeking some Saturn "goodwill" refunds. **Parts**: Higher-than-average cost, but not hard to find through independent suppliers.

Strengths and weaknesses: Conceived as an all-American effort to beat the Japanese in the small car market, the Saturn compact isn't any better built than the other GM home-grown compacts we've learned to dread over the past 30 years. The car is far from high-tech; it's remained virtually unchanged, except for a minor face-lift and a bit more leg room, since it was launched in 1991.

Saturns have exhibited a plethora of serious body and mechanical problems which GM has masked by generously applying its base warranty. Second owners aren't treated as well, however, and complaints are frequent where owners of used Saturns report they had to pay dearly for GM's powertrain and body mistakes. Servicing quality is spotty. And if the rumours of GM merging Saturn with one of its other divisions are true, warranty claim servicing will become even more problematic, and routine dealer servicing will likely deteriorate as well.

What's so great about Saturns? Its advertising agency.

Granted, these are competitively priced, roomy, and comfortable small cars. They handle nimbly in good weather where acceleration isn't a prime consideration. On the other hand, these lightweight vehicles are tricky to handle in snow or ice.

Powered by a 4-cylinder aluminum engine, and a multi-valve variation of the same powerplant in the coupe, these cars are remarkably fuel efficient with the base engine hooked to a manual transmission. The rack-and-pinion steering is fairly precise and predictable, while the suspension gives a firm, but not harsh, ride.

The loud, coarse, standard single-cam engine gives barely adequate acceleration times (12.2 seconds to reach 100 km/h) with the manual transmission. This time is increased to a glacial 13.7 seconds with the 4-speed automatic gearbox that robs the engine of what little power it produces. Other generic problems affecting all model years are stalling and hard starting. One Ontario owner of a 1994 Saturn plagued with chronic stalling problems had this to say:

> After stalling in rush hour traffic, car was towed as hazard lights failed after ten minutes; unable to restart. Car was then towed. Cause indicated: bad fuel tank, pump, battery, ECM and ignition module. Replaced probe, module, pump, tank, valve, module, and battery.
>
> Work order indicated that they performed various diagnostic checks, including: updated computer program, checked fuel pressure, checked ignition module circuit for proper operation, checked circuit from ignition switch to PCM for poor connection, and tried good ignition switch. Monitored voltage at back of ignition switch circuit for Fuse #1 PCM, replaced fuel tank and fuel pump as per tac, and replaced ECM. Car returned Feb. 27, 1998, but that evening car stalled again.
>
> Car stalled Feb. 28, March 1, 2, and 3 and finally, after stalling a dozen or so times on my way to work and almost getting into an accident with a bus, I told the Customer Service in Toronto that they could come get the car from my work as I was not driving it home....

The 5-speed manual transmission sometimes has trouble going into Reverse. With the automatic, there's lots of gearbox shudder when the kickdown is engaged while passing.

Dealer service bulletins indicate that four major quality problems are likely to crop up: self-destructing engines; chronically malfunctioning automatic transmissions; myriad electrical short circuits; and a host of body defects, led by paint delamination, rattles, and wind and water leaks. On early models, the doors were poorly fitted, rear seats had to be lowered half an inch to give much-needed head room, engine mounts were changed to reduce vibrations, the shifter mechanism on the manual transmission was unreliable, and the reclining front seats were recalled because they could suddenly slip backwards. Rear head room is still a bit tight, and owners report that the headliner in the rear tends to sag. Despite the dent-resistant plastic body panels, Saturn owners complain that the fascia chips and discolours.

GM's minor refinements to the 1997 Saturns don't appear to justify the higher price. The sedans and wagons got upgraded engine mounts to lessen noise and vibration and the coupe was given the sedan's 102-inch wheelbase, providing additional leg room, particularly in the rear. More effective silencing materials were added forward of the dash, and the air inlet was redirected away from the passenger compartment. Upgraded seatbelts on all 1997 models are less likely to trip passengers trying to access the rear seat.

Dealer service bulletins: **1995**—Engine flares, loss of power, harsh shifts into Reverse. • Engine knock and/or rattle at normal temperature. • Engine stalls when cold, difficult to restart. • Excessive engine vibration at idle. • ATF leak at automatic transaxle oil pressure filter. • Possible causes of brake pulsation. • More recent info on brake vibration and/or pedal pulsation. • Steering pull, torque steer and wander. • Squealing sound from front of engine during cold starts. • Buzzing noise at rear of vehicle. • Possible other causes of buzzing, rattling, and fluttering noises. • Hoot noise upon light deceleration. • Clicking, ticking noise from instrument cluster odometer. • Ignition noise heard through radio speakers. • Popping noise from front seatback frame. • Rattle or chatter coming from passenger side. • Rattle, pop, or click from front end whenever passing over rough roads. Squeak/squawk noise from rear of vehicle. • Whistling produced by antenna mast. • Troubleshooting tips to silence wind noise at highway speeds. • Low voltage reading or dim lights at idle. • Instrument panel centre air outlet closes. • Hard-to-operate sunroof sunshade. • Water drips onto seat/carpet with door open. • Water leaks into right front footwell from front of dash. • Water leak at front upper door frame. **1996**—Excessive engine vibration at idle. • Radio frequency interference diagnosis. **1997**—Automatic transmission malfunctions. • Failure-prone cylinder head. • Cruise control drops out intermittently or won't reset after drop-out. • Faulty switch for front-door jamb causes the dome light to stay on and prevents the alarm from arming. • Ignition key can be removed when vehicle is in gear. • Defective ignition lock cylinders create havoc with accessories (radio, power windows, AC, etc.). • Ignition key may bind in the ignition's Run position. • Excessive front brake noise or pulsation. • Knocking/rattling from front of floor to front of dash area. • Loose rear exterior door panels. • Runs out of fuel while fuel gauge reads one-quarter full. • Automatic transmission whine.

Safety summary/Recalls: Owners say horn isn't well positioned for emergency use and isn't sensitive enough. NHTSA officials have opened an investigation into 1994 Saturn seatback failures. **1992**—Faulty alternators cause vehicle to suddenly stall. • Motorized-seatbelt failure. • Shifter inoperative due to sudden cable breakage. • **1993**—Rear axle broke and wheel flew off. • Rear wheel fell off. • Sudden steering loss. • Airbag failed to deploy. • Driver and passenger seatbacks collapsed in a rear-end collision. • ABS brake failure. • Intermittent cruise control failures. • Driver-side power door lock would not unlock. • Chronic premature transmission failures. **1994**—Airbag failed to deploy. • Severe burns from airbag deployment. • Driver's seatback collapsed several times. • Passenger seatback collapsed in a collision. • Chronic stalling. • Manual transmission jumps out of third and fifth gear when driving. • Windshield wiper failure due to plastic fitting in the linkage pivot drive assembly wearing down. • Automatic seatbelt chokes occupant. • Premature front brake rotor warpage. **1995**—Reports of stuck accelerators. • Plastic fuel line leaks fuel. • Airbag failed to deploy. • Seatbelt failed to restrain driver in collision. • Gear lever slips out of gears and is hard to put into Reverse. • Manual transmission jumps out of third and fifth gear. • Brake rotor warpage and failures. • Steering wheel shakes uncontrollably at 65 km/h. • Sudden head gasket failure caused other engine components to self-destruct. • Prematurely worn engine timing chain. • AC fan knob broke in half. • Windshield wiper nut fell out, causing the wiper to fail. • **1996**—Airbag failed to deploy. • Rack-and-pinion steering gear failure. • Frequent brake failures. • Firestone tire blowouts. • Inadequate defrosting. • Engine suddenly accelerates to 3,500 rpm at idle or when driving. • Engine suddenly stalls and can't be restarted. • Car slips out of fifth gear. **1997**—Steering wheel came apart while car was driven. • Several complaints of total loss of steering control. • Engine suddenly loses power while cruising on the highway. • In a rear-end collision, driver's seatback broke, causing serious injuries. **Recalls**: **1991**—Seatback recliners will be strengthened at no charge. **1991–92**—The seatbelt retractor may not fully retract the belt. **1991–93**—Faulty trunk lock assemblies will be replaced. • GM dealers will install at no charge a fusible link wiring harness to prevent an engine fire. **1992**—Automatic shift lever may show the wrong gear. **1993**—Sudden brake loss may be caused by faulty brake-booster housing. • The windshield wiper and brake-booster assemblies will be inspected and replaced, if necessary, at no charge under two separate safety Recalls. • Battery cable terminal may be defective. **1995 automatic**—Possible to shift from Park with key removed or to remove key when lever is in gear. **1996–97**—Horn may not operate or can cause a fire; dealer will replace the horn assembly. **1996 wagon**—Some welds between roof and reinforcement panels may not meet specifications. **1997**—Belted front passenger's seat not secure during a frontal impact.

Secret Warranties/Service Tips

1991—Engine misses, surges, or backfires may be caused by a poor ground at the electronic distributorless ignition (DIS). Install another DIS module. • Headlights that stay on when the switch is turned off can be fixed by installing an upgraded switch. • Wind noise from the front and rear doors may be caused by insufficient sealing under the mirror patch gasket, missing sealer at certain locations (for example, the door frame to door assembly at the beltline), insufficient contact of the secondary seals to door openings, or the glass run channels not sealing to the glass at the upper corners. **1991–92**—Owners have reported excessive noise and vibration levels coming from the steering wheel, seat, and floor pan. Saturn officials say that the noise and vibration could be caused by the following: pre-loaded powertrain mounts (1991); a pre-loaded engine strut cradle bracket (1992); improperly positioned or worn exhaust system isolators and muffler band clamp/block (1991–92); lower cooling module grommets and improper positioning of wiring harnesses and upper cooling system module grommets (1991–92); improper routing of AC hoses or hood release cable and air inlet snorkel (1991–92); debris in accessory drive belt pulleys (1991); improper adjustment of hood stop(s) (1991–92); PCV or brake-booster check valve noise (1991); DOHC (LLO) automatic transaxle mount assembly replacement (1991); or a malfunctioning engine, electrical, or fuel system (1991–92). **1991–94**—Rough running or surging after a cold start may signal the need to clean carbon or fuel deposits from the engine's intake valves. • The many causes of hard-to-crank windows are covered in DSB #94-T-19. • Inoperative electric door locks may have been shorted by water contamination. Since the design and positioning of the relay for the power door lock allows this to develop, Saturn is accepting claims on a case-by-case basis. • Water leaks into the front footwell and at the front upper door frame are treated in depth in two DSBs published in August and November 1994. • Whistling noises are also treated in two different bulletins published in June and October 1994. **1991–95**—Erratic cruise control operation can be corrected by replacing the cruise control module assembly. • If the engine stalls within five minutes of starting, or when coming to a stop, or is difficult to restart, the oil viscosity or engine's hydraulic lifters may be at fault. • Engine squealing after a cold start can be corrected by installing an upgraded belt idler pulley assembly. **1991–97**—Excessive front brake noise or pulsation requires the installation of upgraded brake pads, according to DSB #96-T-40A. • Engines that run hot or have coolant mixed in the engine oil probably have a defective engine cylinder—a factory-related goof, according to GM's service bulletin on the following page.

BULLETIN NO.: 96-T-65A
ISSUE DATE: February 1997
GROUP/SEQ. NO. Engine-15
CORPORATION NO.: 686204R
SUBJECT:
Engine Runs Hot and Engine Oil Mixed with Engine Coolant in Engine Coolant Recovery Reservoir (Replace Cylinder Head Assembly)
This bulletin is revised to replace an incorrect part number for the one gallon container of DEX-COOL(TM) and supersedes bulletin 96-T-65, which should be discarded.
MODELS AFFECTED:
1991–1997 Saturns equipped with SOHC (LKO-1991-1994, L24-1995-1997) engines
CONDITION:
Engine may run hot and/or have engine oil mixed with engine coolant. This condition may be noticeable when checking coolant recovery reservoir level.
CAUSE:
Some 1991–1997 SOHC engines may develop a crack on or near the camshaft journals and surrounding casting areas allowing engine oil to mix with engine coolant. These cracks may be caused by "folds" in the aluminum that occur during the head casting process.
CLAIM INFORMATION

Case Type	Description	Labor Operation Code	Time
VW	Replace Cylinder Head Assembly	T9715	11.2 hrs
Add:	with A/C		0.8 hrs
	with power steering		0.3 hrs

To receive credit for this repair during the warranty coverage period, submit a claim through the Saturn Dealer System as shown.

• If your Saturn runs out of fuel while the fuel gauge reads one-quarter full, it's likely you have a plugged EVAP canister vent which should be repaired free of charge under the emissions warranty. • A whistle or groaning noise heard at highway speeds may be caused by the radio antenna mast. **1992**—An engine that stalls, hesitates, or surges during light acceleration may require new PCM calibrations. • Harsh Reverse engagement can also be corrected with new PCM calibrations. • Engine rattling can be fixed by changing the motor mounts. **1993–94**—A popping noise coming from the base of the left-hand A-pillar, hinge pillar, and engine compartment is caused by a slight flexing in the area where the three are joined together. **1993–97**—Troubleshooting tips are available on diagnosing delayed or harsh automatic transmission shifting into Reverse. **1994**—White Saturns may have yellow stains or spotting along the fenders, fender extension, or quarter panel. If so, the company will change the affected part and repaint the area at no charge to the owner. • A Saturn equipped with manual transmission may have the transaxle stuck in gear due to a defective shift control housing. **1994–95**—Excessive engine knocking can be corrected by changing the clearance between the piston pin and connecting rod bushing. **1995–97**—Electrical accessories may lose power after the car is started or while it's on the road. GM blames the problem on a defective ignition lock cylinder. **1996–97**—If your security alarm won't work properly or your dome light won't go out, GM suggests you change the door jamb switches. **All models/years**—A rotten-egg odour coming from the exhaust is likely the result of a malfunctioning catalytic converter, which you can have replaced free of charge under GM's emissions warranty. • Paint delamination, peeling, or fading (see pages 76, 113).

Saturn Profile

	1992	1993	1994	1995	1996	1997	1998
Cost Price ($)							
SL	9,895	10,995	11,595	12,395	12,998	13,948	14,188
SL1	11,080	12,350	12,825	13,795	14,398	14,948	15,118
SL2	13,865	15,095	14,510	15,495	16,148	16,758	17,128
SC	14,620	15,840	13,895	14,795	15,348	16,028	16,418
Used Values ($)							
SL ↑	3,000	4,000	6,000	7,500	9,000	9,500	11,000
SL ↓	2,500	3,500	5,500	7,000	8,500	8,500	10,000
SL1 ↑	4,000	5,000	7,000	8,500	10,000	11,000	12,000
SL1 ↓	3,000	4,000	6,000	8,000	9,000	10,000	11,000
SL2 ↑	5,000	6,000	8,000	9,500	11,000	12,500	14,000
SL2 ↓	4,000	5,500	7,000	8,500	10,000	11,500	13,000
SC ↑	5,500	6,500	8,500	10,000	11,500	13,000	14,500
SC ↓	5,000	6,000	7,500	9,000	10,500	12,000	13,500
Extended Warranty	Y	Y	Y	Y	Y	Y	Y
Secret Warranty	Y	Y	Y	Y	Y	Y	Y
Reliability	❷	❷	❷	❷	❷	③	④
Air conditioning	❷	❷	❷	③	④	④	④
Automatic transmission	❷	❷	❷	❷	③	③	③
Body integrity	❷	❷	❷	❷	❷	❷	③
Braking system	❷	❷	❷	❷	❷	③	③
Electrical system	❷	❷	❷	❷	❷	③	③
Engines	❷	❷	❷	❷	❷	③	③
Exhaust/Converter	❷	③	③	④	④	④	⑤
Fuel system	❷	❷	③	③	④	④	④
Ignition system	❷	❷	❷	❷	❷	③	③
Rust/Paint	❷	❷	❷	❷	❷	③	③
Steering	③	③	④	④	④	⑤	⑤
Suspension	③	③	④	④	④	④	④
Crash Safety							
SL2	❷	❶	❶	④	④	④	⑤
Side Impact	—	—	—	—	—	③	③

GENERAL MOTORS/SUZUKI

Firefly, Metro, Sprint, Swift

Rating: Above Average (1995–98); Average (1990–94); Below Average (1987–89). Stay away from AC-equipped versions, unless you want to invest in an AC repair facility. Look at the redesigned 1995 version for better quality, a new body style, standard dual airbags, and a peppier 4-cylinder engine. The convertible version packs plenty of fun and performance into a reasonably priced subcompact body. **Maintenance/Repair costs**: Average. **Parts**: Expensive; sometimes drivetrain and body components are back-ordered several weeks.

Strengths and weaknesses: Unlike Ford's Festiva and Aspire imports, GM's Suzuki-based mini-cars are still being churned out in their Ingersoll, Ontario, manufacturing plant. These tiny, economical, 3- and 4-cylinder, front-wheel drive, hatchback econoboxes, equipped with either a manual or an automatic transmission, offer acceptable performance and economy for urban dwellers. In fact, these little squirts should be considered primarily city vehicles due to their small size, small tires, low ground clearance, and high-speed average handling. Interior garnishing is decent but plain, and there's plenty of room for two passengers, with four fitting in without too much discomfort. The turbocharged convertible model is an excellent choice for high-performance thrills in an easy-to-handle ragtop.

Mechanically speaking, the GM-Suzuki partnership hasn't hurt quality control—though, it hasn't helped either. Trouble spots on the 1988–95 models: excessive oil consumption after the fourth year of service; automatic transmission and differential failures around 80,000 km; electrical system shorts; a faulty AC and cooling system (fogging of the side windows and windshield due to inadequate heat distribution is a common complaint); premature brake, clutch, and exhaust system wearout; and minor fuel supply malfunctions. Body construction is the pits on early models.

Owners of 1990–94 models report that hatches vibrate when windows are open and the rear hatch seems to want to open on its own; push buttons for the lights and wipers tend to fly off the dashboard; fuel economy is often exaggerated by dealers; front metallic brake pads are noisy and the front discs warp easily; occasionally the third gear is hard to engage; and the fuel-injection system performs poorly. Owners also point out that the tiny radiator can't stand up to the rigours of northern climates. If it's not checked, the cooling system will eventually fail, causing great damage to the aluminum engine.

The cars were redesigned for the 1995 model year and were built with more care. Additionally, they gained more horsepower, improved road feel, a more comfortable ride, quieter operation, and more cabin space. Remaining problems are premature front brake wear, electrical system and AC malfunctions, and sub-par body assembly highlighted by paint peeling and discoloration, early rusting, and poorly fitted body panels leading to rattles, and air and water leaks.

Safety summary/Recalls: Side window defogging is slow and sometimes inadequate. Driver's window continually pops out of its mount. **Recalls**: **1986**—The push-pull headlight switch will be replaced at no charge. **1989–1991**—Takata seatbelt will be replaced. **1989–93**—Hood may fly up when car is in motion. **1995**—Faulty rear brake drums may lead to sudden wheel separation. **1995 Metro hatchbacks without ABS**—Faulty rear brake drums could cause wheel separation. **1997 Metro with automatic transmissions**—Dealers will replace the shifter assembly because the shift lever can move too easily out of Park.

Secret Warranties/Service Tips

1989–90—If you have hard starts or no-starts in cold weather, install Suzuki's cold-start harness set (#36690-60A00). **1989–92**—Stalling or loss of power shortly after starting may be due to high pressure in the hydraulic lifter assemblies. **1991–92**—Excessive vehicle vibration when the vehicle is in Reverse is most likely due to poor insulation between the engine/transaxle assembly and the vehicle's chassis. Install upgraded engine mounts. **1994**—Vehicles that won't start usually have defective oxygen sensors—an emissions warranty item. • When the battery runs down in wintertime, it can set off the car alarm. Buy a heavy-duty battery. **All models/years**—Paint delamination, fading, peeling, hazing, and "microchecking" (see pages 76, 113).

Firefly, Metro, Sprint, Swift Profile

	1991	1992	1993	1994	1995	1996	1997	1998
Cost Price ($)								
Firefly/Sprint	7,765	7,765	—	9,145	10,395	10,995	11,495	11,680
Metro	—	6,999	7,995	8,995	10,395	10,995	11,495	11,680
Swift	8,835	7,495	8,537	8,995	10,495	10,995	10,995	11,495
Used Values ($)								
Firefly/Sprint ↑	1,500	2,500	4,000	—	5,800	7,400	7,500	9,500
Firefly/Sprint ↓	1,000	2,000	3,500	—	4,700	6,200	7,000	8,500
Metro ↑	—	2,300	3,600	5,000	6,000	7,600	7,700	9,600
Metro ↓	—	2,000	3,000	4,500	5,000	6,500	7,200	8,600
Swift ↑	1,900	2,800	3,600	5,000	6,000	7,600	7,700	9,500
Swift ↓	1,400	2,500	3,000	4,500	5,000	6,500	7,200	8,500
Extended Warranty	Y	Y	Y	N	N	N	N	N
Secret Warranty	Y	Y	Y	Y	Y	Y	N	N
Reliability	❷	❷	❷	❷	③	④	④	⑤
Air conditioning	❷	❷	❷	❷	❷	❷	❷	③
Body integrity	❶	❷	❷	❷	❷	③	④	④
Braking system	❷	❷	❷	❷	❷	③	③	④
Electrical system	❷	❷	❷	❷	③	③	③	④

Engines	❷	❷	❷	❷	❷	④	⑤	⑤
Exhaust/Converter	❷	❷	❷	❷	③	④	⑤	⑤
Fuel system	❷	❷	③	③	④	④	⑤	⑤
Ignition system	③	④	④	③	④	④	④	④
Manual transmission	③	③	③	③	④	④	⑤	⑤
- automatic	❶	❷	❷	❷	③	④	④	④
Rust/Paint	❶	❷	❷	❷	❷	③	④	⑤
Steering	④	④	⑤	⑤	⑤	⑤	⑤	⑤
Suspension	④	④	⑤	⑤	⑤	⑤	⑤	⑤
Crash Safety	⑤	⑤	⑤	⑤	④	④	④	—

HONDA

Civic, CRX, del Sol

Rating: Recommended (1993–98); Above Average (1992); Average (1989–91); Below Average (1984–88). The Si was dropped in 1995. Depreciation is so minimal that, if you want to get a good buy at a fair price, the 1993 is your best bet. **Maintenance/Repair costs**: Average. Repairs can be carried out by independent garages, but the 16-valve engine's complexity means that dealer servicing is a must. To avoid costly engine repairs, owners must check the engine timing belt every 2 years/40,000 kilometres and replace it every 100,000 kilometres ($300). **Parts**: Parts are a bit more expensive than most other cars in this class, but they aren't hard to find.

Strengths and weaknesses: The quintessential econobox, Civics have distinguished themselves by providing sports car acceleration and handling with excellent fuel economy and quality control that's better than what American automakers can deliver. Other advantages: a roomy, practical trunk and simple, inexpensive maintenance.

CRXs are a sportier version of the Civic, equipped with a more refined 1.5L and 1.6L engine and stiffer suspension. They give improved handling at the expense of fuel economy, interior space, and a comfortable ride. They also require valve adjustments every 30,000 km.

Some other Civic disadvantages: power steering isn't offered with the manual 5-speed transmission; there's no ABS available; you won't find much backseat head room; engine noise may seem excessive when under load; and the Civic's high resale value means bargains are rare.

1984–92 Civics suffer from failing camshafts/crankshafts and prematurely worn piston rings. The 12-valve engine is prone to valve problems and is costly to repair. Early fuel-injection units were also problematic until the system was redesigned in 1988. Manual transmission shifter bushings need frequent replacement and the automatic version needs careful attention once the 5-year/150,000 km point has been reached.

The 1988 redesign improved handling and increased interior room. There isn't a great deal of torque with the 1.6L engine below 3,500 rpm, however, and serious generic problems present since the car's debut continue to plague the later models. First, the front brakes continue to wear out quickly and are often noisy when applied, causing excessive steering wheel vibration. Premature constant-velocity-joint and boot wear on all cars is another problem area that needs careful inspection before purchasing. The rack-and-pinion steering assembly often needs replacement around the five-year mark. There have also been reports of premature clutch wear with 4X4 Civics.

A large number of 1988–91 Honda Civics and CRXs have faulty distributor igniters. When the igniter fails, the car stalls, and it may be impossible to restart. The only remedy is to call a tow truck and replace the distributor.

1993–98 Civics continue to be both rugged and reliable. The few problems that are reported concern front brake noise and premature wear, AC malfunctions, and minor body faults.

What are minor body faults with recent models turn into major rust problems with older Civics, where simple surface rust rapidly turns into perforations. The underbody is also prone to corrosion, which leads to the severe structural damage that compromises safety. The fuel tank, front suspension, and steering components, along with body attachment points, should be examined carefully in any Civic more than a decade old. Since 1988, Hondas have been much more resistant to rusting and overall body construction has been vastly improved. All hatchbacks let in too much wind/road noise due to poor sound insulation. Owners complain of water leaking into the trunk area through the rear taillights on the 1990 DX four-door sedan. The two-piece tailgate rattles and is complicated, for no good reason.

Safety summary/Recalls: **Recalls**: **1986–1991 Civic**—Takata seatbelts will be replaced. **1989–1991 CRX**—Takata seatbelts will be replaced. **1990**—Front windshield will be replaced. **1992–94**—The automatic transmission's shift lever position may not match the actual gear that's engaged. **1992–95**—Hood could fly up. **1994**—Passenger-side airbag module may carry a defective inflator. **1996**—Faulty brake booster vacuum hose. **1996–98**—Passenger-side airbag may not deploy properly. **1993 Civic, del Sol**—The automatic transmission's shift lever position may not match the actual gear that's engaged.

Secret Warranties/Service Tips

Civic, CRX, del Sol: **1988–90**—Excessive valve noise will be corrected with a cam holder kit (#04101-PM3-308). • The dashboard cracks at the centre bolt hole. **1988–91**—Distributor igniters are faulty (see "Strengths and weaknesses"). • New front brake pads that minimize front brake squeal are available for all Civic and CRX models. • A clicking noise heard while making a left or right turn may be caused by a worn outboard drive shaft joint. **1988–93**—A creaking noise coming from the window regulator can be corrected by installing an upgraded regulator spiral spring. **1989–91**—An oil leak around the spark plug well will be corrected with a cam holder kit (#04101-PM3-308). **1990**—The horn sounds by itself when the temperature drops. **1992**—A growling noise coming from the wheel area may mean that water has entered the wheel bearing through the hubcap and has damaged the bearing. • A steering wheel shake or body vibration when braking may indicate that the rear brake drum hub is crowned, causing excessive runout when the wheel nuts are torqued. Other factors could be a bent rear wheel, over-torqued wheel nuts, or excessive rust buildup on the brake rotors. **1992–95**—Power steering pump fluid leakage requires a new O-ring, which may be eligible for "goodwill" consideration, says Honda. **1992–97**—Water leaking into the footwell from under the corner of the dash can be stopped by applying sealer to the seam where the side panel joins the bulkhead.

Bulletin No.
92-050
Issue Date
FEB 3, 1997
Mode [NEW]
1992 and
LATER
CIVIC
Applicable To [NEW]
ALL
File Under
BODY
Water Leak From Seam Under Corner of Dash
(Supersedes 92-050, dated May 15, 1995)

SYMPTOM
A water leak into the footwell from under the left or right corner of the dash.

PROBABLE CAUSE
Not enough sealer on the cowl side of the panel-to-bulkhead seam.

CORRECTIVE ACTION
[NEW]
Apply sealer to the seams where the side panel joins the bulkhead. Refer to section 20 of the appropriate Service Manual for removal and reinstallation of body components.
[NEW]
1. Look for a leak from the lower corners of the dashboard while an assistant runs water over the windshield and cowl. Confirm which side is leaking.
[NEW]
2. Remove the windshield wiper arms and the cowl cover.
[NEW]
3. Remove the side sill, the inner fender, and the front fender from the side that is leaking.

4. Insert a scraping tool through the access holes and remove all loose sealer. For better visibility, shine a flashlight through the holes.

Honda says this leak can be repaired within 90 minutes and is eligible for "goodwill" warranty consideration.

• An abnormally long crank time before the car starts may be caused by a leaking check valve inside the fuel pump. **1993**—Poor AM reception or a popping sound coming from the speakers is likely due to a poor ground connection between the antenna collar and car body. **1994–97**—If the AC doesn't blow cold air, Honda will replace both the evaporator and the receiver/dryer free of charge (see Honda Accord). • When operating a manual or power-assisted front window, the rear edge of the glass comes out of the channel (see Honda Accord). • **1996**—A clunking noise in the front suspension can be fixed by installing upgraded upper arm flange bolts.

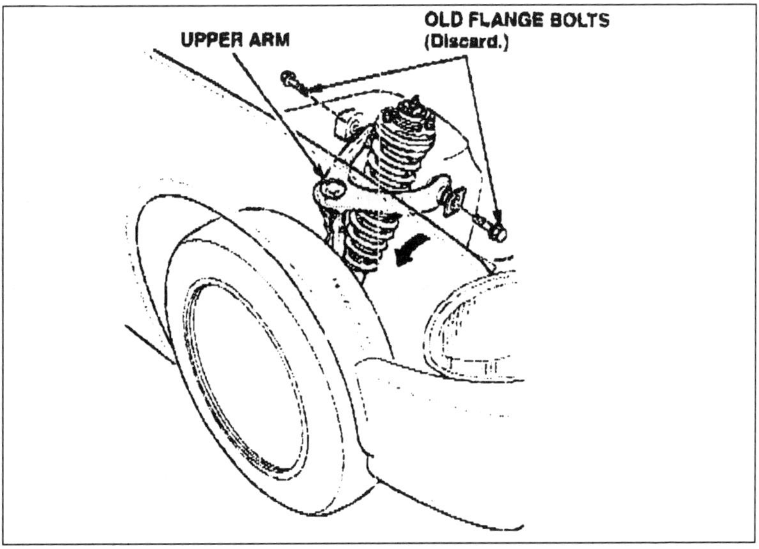

Honda's service bulletin specifically singles out this repair as also eligible for "goodwill" consideration. Refer to "Defect Code 042" and ask Honda to pay the tab.

1996–97—In a settlement with the U. S. Environmental Protection Agency, Honda paid fines totaling $17.1 million and extended its emissions warranty on 1.6 million 1995–97 models to 14 years or 150,000 miles. This means that costly engine components and exhaust system parts like catalytic converters will be replaced free of charge, as long as the 14-year/150,000 mile limit hasn't been exceeded. Additionally, the automaker will provide a full engine check and emissions-related repairs at 50,000 to 75,000 miles and will give free tune-ups at 75,000 to 150,000 miles. It is estimated the free checkups, repairs, and tune-ups will cost Honda over $250 million. The story of the settlement was first reported on page 6 of the June 15, 1998, edition of *Automotive News.* Canadian owners may wish to use this settlement as leverage for free repairs in Canada, or use the full-terms of the settlement when visiting the States. One thing is certain, neither Transport Canada nor Environment Canada are sufficiently enthused to render any assistance. **All models/years**—Most Honda DSBs allow for special warranty consideration on a "goodwill" basis even after the warranty has expired or the car has changed hands. Referring to this euphemism will increase your chances of getting some kind of refund for repairs that are obviously related to a factory defect.

Civic, CRX, del Sol Profile

	1991	1992	1993	1994	1995	1996	1997	1998
Cost Price ($)								
Civic	8,355	9,112	10,395	10,595	11,495	12,995	13,495	14,000
Si	14,705	13,320	14,995	16,595	17,495	17,895	17,995	18,300
CRX/del Sol	12,980	—	16,795	18,595	20,295	20,495	20,995	—
Used Values ($)								
Civic ↑	3,500	4,000	6,000	7,500	8,000	9,500	11,000	12,500
Civic ↓	3,000	3,500	5,500	7,000	7,500	9,000	10,500	12,000
Si ↑	5,500	6,500	8,500	10,000	12,500	13,500	15,000	16,000
Si ↓	5,000	6,000	7,500	9,500	12,000	13,000	14,000	15,000
CRX/del Sol ↑	6,000	—	10,000	12,000	15,000	16,500	18,000	—
CRX/del Sol ↓	5,500	—	9,000	11,000	14,000	16,000	17,000	—
Extended Warranty	Y	N	N	N	N	N	N	N
Secret Warranty	N	N	N	N	N	Y	Y	N
Reliability	③	③	④	⑤	⑤	⑤	⑤	⑤
Air conditioning	❷	③	③	④	④	⑤	⑤	⑤
Body integrity	❷	❷	③	③	③	③	④	④
Braking system	❷	❷	③	③	③	④	⑤	⑤
Electrical system	③	③	③	④	④	④	⑤	⑤
Engines	③	③	③	③	⑤	⑤	⑤	⑤
Exhaust/Converter	❷	❷	③	③	④	⑤	⑤	⑤
Fuel system	③	③	④	④	④	④	⑤	⑤
Ignition system	③	③	④	④	④	④	⑤	⑤
Manual transmission	③	③	③	④	④	⑤	⑤	⑤
automatic	③	③	③	④	⑤	⑤	⑤	⑤
Rust/Paint	❷	③	③	③	⑤	⑤	⑤	⑤
Steering	③	③	④	④	④	④	⑤	⑤
Suspension	③	③	④	④	④	④	⑤	⑤
Crash Safety								
Civic 2d	⑤	⑤	⑤	⑤	④	④	④	④
Civic 4d	❷	❷	❷	❷	③	⑤	④	④
CRX	⑤	⑤	—	—	—	—	—	—
Side Impact								
Civic 2d	—	—	—	—	—	—	—	❷
Civic 4d	—	—	—	—	—	—	③	③

HYUNDAI

Accent

Rating: Above Average (1995–98). Lots of standard features. **Maintenance/Repair costs**: Higher than average. Repairs are dealer dependent. **Parts**: Expensive and not easily found. Dealer servicing has been substandard in the past. Hyundai says that the timing chain should be replaced every 100,000 km.

Strengths and weaknesses: An upgraded Excel masquerading as an Accent, this front-drive, 4-cylinder sedan retains most of the Excel's underpinnings, while dropping the Mitsubishi powerplant in favor of a new home-grown 1.5L 4-cylinder. It's built better than the old Excel, though: upgraded, smoother-shifting automatic transmission; stiffer, better-performing suspension; stronger and quieter-running engine; optional dual airbags; and ABS.

Excels have always had a checkered reliability history, so the Accent has a lot of bad karma to overcome, such as mediocre body assembly and poorly applied paint. Past problem areas include the engine cooling system and cylinder head gaskets, transmission, wheel bearings, fuel system, and electrical components, as well as premature front brake wear and excessive noise when braking.

Dealer service bulletins: **1995**—The exhaust system has a tendency to release a rotten-egg odour. • Troubleshooting tips for locating and plugging interior water leaks. **1996**—Exhaust rattling. • Shift quality improvements. • Sticking headlight flasher switch.

Safety summary/Recalls: Horn controls may be hard to find in an emergency, rear head restraints appear to be too low to protect occupants, and rear seatbelt configuration complicates the installation of a child safety seat. **Recalls**: **1995**—The ECM wiring harness may short on vehicles equipped with a manual transaxle. **1995–97**—Corroded coil spring may break. **1996–97**—Faulty wiper motor.

Secret Warranties/Service Tips

1995–96—Harsh shifting may be fixed by installing an upgraded Transaxle Control Module (TCM). • Clutch drag may be caused by a restriction in the hydraulic line from grease used during the assembly of the clutch master assembly. **All models/years**—Apparent slow acceleration upon cold starts is dismissed as normal.

Accent Profile

	1995	1996	1997	1998
Cost Price ($)				
L	9,295	10,495	10,995	11,295
GL 4d	10,995	12,195	12,695	12,995
Used Values ($)				
L ↑	5,000	6,500	8,000	9,500
L ↓	4,000	5,000	6,500	8,000
GL 4d ↑	6,500	7,500	9,000	10,500
GL 4d ↓	5,000	6,500	7,500	9,500
Extended Warranty	Y	Y	Y	Y
Secret Warranty	N	N	N	N
Reliability	④	④	⑤	⑤
Crash Safety	—	③	③	③
Side Impact	—	—	—	③

Elantra

Rating: Above Average (1996–98); Average (1991–95). Way overpriced: between 1991 and 1996, the base MSRP rose a whopping $5,000. There's also a $1,000–$3,500 difference between the high-end and entry-level models. **Maintenance/Repair costs**: Higher than average. Repairs are dealer dependent. **Parts**: Higher-than-average cost, but not hard to find.

Strengths and weaknesses: This conservatively styled "high-end" sedan is only marginally larger than the Excel, but its overall reliability is much better. It's a credible alternative to the Toyota Corolla, Nissan Sentra, and Saturn, and the redesigned 1996–97 versions actually narrow the handling and performance gap with the Honda Civic. The 16-valve 1.6L 4-cylinder is smooth, efficient, and adequate when mated to the 5-speed manual transmission. It's not very quiet, however. The smooth ride causes excessive body lean when cornering, but overall handling is fairly good, due mainly to the Elantra's longer wheelbase and more sophisticated suspension.

The 4-speed automatic transmission robs the base engine of at least 10 horses. Brakes are adequate though sometimes difficult to modulate. Conservative styling makes the Elantra look a bit like an underfed Accord, but there's plenty of room for four average-sized occupants. Tall drivers might find rearward seat travel insufficient, making head room a bit tight.

1996–98 Elantras are the better buy due to their additional interior room, improved performance and handling, and quieter-running engine. Moreover, the Elantra was first offered in a wagon in 1996–97. Still, passing power with the automatic gearbox is perpetually unimpressive and the trunk's narrow opening makes for difficult loading.

Surprisingly for a Hyundai, owners report few serious defects. Nevertheless, be on the lookout for body deficiencies (fit, finish, and assembly), harsh shifting with the automatic transmission, oil leaks, and brake defects.

Dealer service bulletins: **1993**—Harsh shifting with the automatic transmission when accelerating or coming to a stop. • Excessive disc brake noise. • Engine has difficulty reaching recommended operating temperature (MPI fault code #21). • Low fuel pressure. • Oil leaks between the oil filter and mounting bracket. • Wheel cover discoloration. **1994**—Excessive front brake noise and premature wear. • Rear speaker whine. • Harsh automatic transmission engagement. • Slow windshield defrosting and defogging. **1995**—The exhaust system releases a rotten-egg odour. • Rear suspension squeaking noises. • An inaccurate fuel gauge. • Trunk water leaks and troubleshooting tips for locating and plugging other interior water leaks. **1996–97**—Automatic transmission won't engage

Overdrive. • Clutch pedal squeaking. • Tapping noise coming from the passenger-side dash panel/engine compartment area. • Exhaust system buzz. • Improved shifting into all gears. • Improved shifting into Reverse. • Clutch drag.

Safety summary/Recalls: **Recalls**: **1992–95**—Defective rear suspension trailing arm bolts. **1994–95**—Driver-side airbag could be defective or the warning light could illuminate unnecessarily. **1996–97**—A faulty wiper motor will be replaced.

Secret Warranties/Service Tips

1991–92—A harsh shift when coming to a stop or upon acceleration could be due to a misadjusted accelerator switch TCU. **1992**—Oil leaking from between the oil filter and mounting bracket could be caused by an overly wide mounting surface on the bracket. Correct this by replacing the bracket. **1992–94**—Hyundai has a field fix for manual transaxle gear clash/grind (DSB #9440-004). **1992–94**—The difficult-to-engage Reverse gear needs an upgraded part. **1994**—Rear speaker whine can be stopped by installing an improved noise reduction filter. **1996**—A cold exhaust system buzz can be silenced by installing a sub-muffler resonator. • Improved shifting into all gears can be accomplished by installing an upgraded transaxle control module (TCM). **1996–97**—A tip on eliminating clutch pedal squeaking is offered. • Tips on eliminating clutch drag are offered. **All models/years**—Hyundai has a new brake pad kit (#58101-28A00) that the company says will eliminate squeaks and squeals during light brake application. Hyundai also suggests that you replace the oil pump assembly if the engine rpm increases as the automatic transmission engages abruptly during a cold start.

Elantra Profile

	1991	1992	1993	1994	1995	1996	1997	1998
Cost Price ($)								
GL	10,595	10,795	11,295	11,795	12,295	13,495	13,995	14,295
GLS	11,795	11,995	12,795	13,695	14,195	16,745	17,245	17,545
Used Values ($)								
GL ↑	3,500	4,500	5,200	6,000	7,500	9,000	10,000	11,500
GL ↓	2,500	3,500	3,900	4,800	6,000	7,500	9,000	10,000
GLS ↑	4,000	5,200	6,000	7,000	8,500	11,000	12,500	14,000
GLS ↓	3,000	4,000	4,700	5,700	7,000	9,500	11,500	13,000
Extended Warranty	Y	Y	Y	Y	N	N	N	N
Secret Warranty	N	N	N	N	N	Y	N	N
Reliability	③	③	③	③	④	④	④	⑤
Air conditioning	③	③	③	④	④	④	⑤	⑤
Automatic transmission	❷	❷	❷	❷	❷	③	③	④
Body integrity	❷	❷	❷	❷	③	③	③	③
Braking system	❷	❷	③	③	③	③	③	③
Electrical system	❷	❷	❷	❷	③	③	③	④
Engines	③	④	⑤	⑤	⑤	⑤	⑤	⑤
Exhaust/Converter	③	③	④	⑤	⑤	⑤	⑤	⑤

Fuel system	③	③	③	③	④	④	⑤	⑤
Ignition system	③	③	③	④	④	③	③	④
Rust/Paint	③	④	④	④	④	⑤	⑤	⑤
Steering	④	④	④	⑤	⑤	⑤	⑤	⑤
Suspension	③	③	③	④	④	④	④	④
Crash Safety	—	❶	❶	❶	❶	③	③	③
Side Impact	—	—	—	—	—	—	—	③

Excel

Rating: Below Average (1992–94); Not Recommended (1986–91). **Maintenance/Repair costs**: Higher than average. Repairs are dealer dependent. To avoid costly engine repairs, check the engine timing belt every 2 years/40,000 miles. **Parts**: Expensive and not easily found.

Strengths and weaknesses: Replaced by the 1995 Accent, the Excel is a low-tech and low-quality economy car that was orphaned in 1995. Resale prices are low, but these cars are no bargain in the long run. Although overall comfort and handling are passable, the engine is distinctly short on power and the carburetor provides uneven throttle response, especially when the car is cold, resulting in difficult starts. Poor interior ventilation, with chronic window fogging and poor defrosting, is particularly irritating.

On early models, likely problem areas are defective constant velocity joints, water pumps, oil pan gaskets, oil pressure switches, front struts, and heat exchange under dash, as well as a leaking head gasket. Excels made after 1989 are noted for their noisy automatic transmissions and engines that fail to start when the weather turns cold or wet. Other problem areas are faulty radiator hoses, alternators, Hyundai radios, and wiper motors. Owners also complain of premature brake wear and temperamental carburetors, electrical problems, poor engine performance, and premature rusting due to poor-quality body parts and paint. Body construction is sloppy, giving rise to wind/water leaks, rattles, and breakage. Mufflers last a little over two years.

Dealer service bulletins: **1993**—Harsh shifting with the automatic transmission when accelerating or coming to a stop. • Excessive disc brake noise. • Engine has difficulty reaching recommended operating temperature (MPI fault code #21). • Low fuel pressure. **1994**—Tips for troubleshooting water leaks into the interior.

Safety summary/Recalls: **Recalls**: **1986**—A defective brake pedal cotter pin will be changed. **1986–87**—Excessive brake fade can be improved by installing upgraded metallic pads. **1986–89**—Cruise control operation is erratic. • A malfunctioning emission control system could cause an engine compartment fire. • Insufficient gear lubrication could cause drive wheels to lock. **1988–89**—A defective heater stem assembly could allow hot coolant leakage. **1990**—Front wheel hub nut lock washer could crack, causing a loss of vehicle control. **1990–94**—In frontal crashes, excessive fuel spillage could pose a fire hazard. **1994**—A short circuit in the crank-angle sensor could cause sudden stalling.

Secret Warranties/Service Tips

1990—A fifth-gear noise can be eliminated by replacing the third- and fourth-shift forks and synchronizer hub and sleeve. • Cranking with no spark may be due to a defective noise filter located between the ignition coil and tachometer. **1990–94**—Difficult-to-engage Reverse gear needs an upgraded part. **1991**—Stalling when shifting into gear immediately after starting a cold engine may be corrected by installing Hyundai's Cold Start Enrichment Kit (#39901-24Q00D). **1994**—A harsh downshift usually means that the accelerator switch is out of adjustment. • A rear suspension groaning noise heralds the need to install upgraded shock absorbers. • A rear suspension squeaking noise can be silenced by retorquing the rear suspension attachment nuts.

Excel Profile

	1988	1989	1990	1991	1992	1993	1994
Cost Price ($)							
Base	7,295	7,695	8,095	7,595	7,975	7,995	8,495
Used Values ($)							
Base ↑	1,200	1,500	1,500	2,200	2,500	3,500	5,000
Base ↓	800	1,000	1,200	1,800	2,000	3,000	4,000
Extended Warranty	Y	Y	Y	Y	Y	Y	Y
Secret Warranty	N	N	N	N	N	N	N
Reliability	❷	❷	❷	❷	❷	❷	❷
Air conditioning	❷	❷	❷	③	③	③	④
Body integrity	❷	❷	❷	❷	❷	❷	❷
Braking system	❷	❷	❷	❷	❷	❷	❷
Electrical system	❷	❷	❷	❷	❷	❷	❷
Engines	❶	❶	❷	③	③	③	③
Exhaust/Converter	❶	❷	③	③	③	③	③
Fuel system	❷	❷	❷	③	③	③	③
Ignition system	❷	❷	❷	③	③	③	③
Manual transmission	❷	❷	❷	❷	❷	❷	❷
- automatic	❷	③	③	③	④	④	④
Rust/Paint	❷	❷	❷	❷	❷	③	③
Steering	③	③	③	④	③	④	④
Suspension	③	③	③	③	③	③	③
Crash Safety	❷	❷	❷	❶	⑤	⑤	⑤

MAZDA

323, Protegé

Rating: Recommended (1996–98); Above Average (1991–95); Average (1985–90). If you can't find a reasonably priced 323 or Protegé, look for a Ford Escort or Tracer instead—they're basically Mazdas disguised as Fords. The redesigned 1995–96 Protegé offers fresh styling, a larger wheelbase, standard dual airbags, and a new 4-banger. Along with the 1997–98 versions, which were mostly carried over unchanged with a slightly restyled grille and headlights and interior refinements, they are excellent used-car buys. Plus, they should be plentiful at bargain prices as they come off their two- and three-year leases this year and next. **Maintenance/Repair costs**: Higher than average. Repairs are dealer dependent. To avoid costly engine repairs, check the engine timing belt every 2 years/40,000 km and replace it every 96,000 km ($300). **Parts**: Expensive and not easily found outside the dealer network.

Strengths and weaknesses: The GLC was replaced by the 323 in 1986. Both Mazdas are peppy performers with a manual transmission hooked to the base engine. The automatic gearbox, however, produces lethargic acceleration that makes highway passing a bit chancy. Handling and fuel economy are fairly good for a car design this old. However, overall durability is not as good as that of more recent Mazda designs, beginning with the 1991 323/Protegé, which was also sold as the Ford Escort. Catalytic converters plug up easily and other pollution-control components have been troublesome. Automatic transmission defects, air conditioner breakdowns, and engine oil leaks are also commonplace. Oil leaks in the power-steering pump may also be a problem. GLCs rust quickly, especially the rocker panels, door bottoms, hood, and rear hatch. A careful corrosion inspection is a must.

The fuel-injected 1.6L engine is a better performer than the 1.5L, but you also get excessive engine and exhaust noise. Stay away from the 3-speed automatic transmission. The car's small engine can't handle the extra burden without cutting fuel economy and performance. Both models are surprisingly roomy, but the Protegé's trunk is small for a sedan.

The 1985–90 models offer mediocre reliability. Owners report hard starting in cold weather, in addition to automatic transmission problems and electrical system failures. The engine camshaft assembly and belt pulley often need replacing around 80,000 miles. Clutch failure and exhaust-system rust-out are also common. Other areas of concern are constant-velocity-joint failures, rack-and-pinion steering wearout, and front brake wear. The front brakes wear quickly due to poor-quality brake pads and seizure of the calipers in their housings. Check for disc scoring on the front brakes. Stay away from models equipped with a turbocharger—few mechanics want to bother repairing it or hunting for parts. Many owners report premature paint peeling.

The 1991–95 models are a bit more reliable and reasonably priced; however, the 1996 and 1997 versions are *la crème de la crème.* When the 323 was dropped at the end of 1994, the Protegé became Mazda's least-costly model and underwent a major redesign the next year. It shares platforms with the Escort and Tracer, but keeps its own sheet metal, engine, and interior styling. Powered by a standard, fuel-efficient, 1.5L engine mated to a manual 5-speed transmission, the Protegé is one of the most responsive and roomiest small cars around.

Nevertheless, these cars aren't perfect and owners report problems with the front brakes (excessive noise and premature wear), weak rear defrosting, rough second-gear engagement, engine stalling, noisy suspension, AC failures, and body defects, including wind and water leaks into the interior, paint defects, and power mirror failures.

Dealer service bulletins: **1995**—A tapping, cracking noise that comes from the B- and C-pillars. • Excessive blower motor and brush noise. • Heater and AC unit noise after long storage. • No sound from audio. • A bump sound when opening the sunroof. • Detached sunroof tilt switch knob. • Power outside mirror glass vibration. • Fuse block cover and glove box hinge breakage. • A slightly off-centre steering wheel.

Safety summary/Recalls: **Recalls**: **1992–93**—Protegé may not meet regulations relative to child restraint tether anchorages. **1995 Protegé with 1.5L engine**—The engine valve springs are defective.

Secret Warranties/Service Tips

1990—Cold weather stalling usually requires the replacement of the ECU, which is under the emissions warranty. **1990–91**—Rough idle or vibration in drive may require changing the No. 1 and No. 4 engine mounts and the radiator lower mounts. **1990–92**—Noise coming from the front of the car when turning may be due to dirt accumulation in the top strut mount bushing. **1990–94**—Clutch squealing is fixed by installing an upgraded clutch cushioning plate. **1995–96**—A horn noise heard from the exhaust will be silenced with a special tail pipe tip furnished by Mazda.

323, Protegé Profile

	1991	1992	1993	1994	1995	1996	1997	1998
Cost Price ($)								
323	8,845	9,250	9,435	9,965	—	—	—	—
Protegé	11,655	12,425	12,675	13,265	13,370	13,895	14,685	14,675
Used Values ($)								
323 ↑	3,000	4,000	4,500	5,500	—	—	—	—
323 ↓	2,000	3,000	4,000	4,500	—	—	—	—
Protegé ↑	4,000	4,500	5,500	6,500	8,000	9,500	10,500	12,500
Protegé ↓	3,500	4,000	4,500	5,500	7,500	8,500	10,000	11,500

Extended Warranty	Y	N	N	N	N	N	N	N
Secret Warranty	N	N	N	N	N	N	N	N
Reliability	❷	③	④	④	④	④	④	⑤
Air conditioning	❷	❷	③	③	④	⑤	⑤	⑤
Body integrity	❶	❷	❷	❷	③	③	③	③
Braking system	❶	❷	❷	❷	③	③	③	③
Electrical system	❶	❷	③	③	③	④	④	④
Engines	③	③	③	④	④	④	④	④
Exhaust/Converter	❷	❷	❷	❷	③	⑤	⑤	⑤
Fuel system	❷	③	③	④	④	⑤	⑤	⑤
Ignition system	③	④	④	④	④	④	⑤	⑤
Manual transmission	④	④	④	④	④	④	⑤	⑤
- automatic	❷	❷	③	③	④	⑤	⑤	⑤
Rust/Paint	❷	❷	❷	③	③	③	③	④
Steering	③	③	③	④	④	④	④	④
Suspension	③	③	③	③	③	③	③	④
Crash Safety								
323	❷	❷	❷	❷	—	—	—	—
Protegé	⑤	⑤	⑤	⑤	—	❷	③	③

NISSAN

Sentra

Rating: Recommended (1995–98); Above Average (1991–94); Average (1988–90); Not Recommended (1983–87). The redesigned 1995 version offers fresh styling, a longer wheelbase, a peppier powerplant, standard dual airbags, and side-door beams. **Maintenance/Repair costs**: Higher than average on early models, but anybody can repair these cars. **Parts**: Reasonably priced and easily obtainable.

Strengths and weaknesses: Sentras aren't expensive to buy, they're relatively easy and inexpensive to repair, and they give good fuel economy. On the other hand, ride and handling are mediocre and build quality is spotty at best. Until 1991, mechanical and body components suffered from poor quality control, making these cars quite unreliable and sometimes expensive to repair. Clutches and exhaust systems were particularly problematic. The 1.6L engine is much more reliable, but even there the oil pressure switch may develop a leak that can lead to sudden oil loss and serious engine damage. Quality improved considerably with the 1991 version, yet the vehicle's base price rose only marginally, making these later model years bargain buys for consumers looking for a reliable "beater."

1991–94 Sentras are a bit peppier and handle better, although they have their deficiencies, too, such as the motorized two-point front seatbelts on the sedans. Other owner-reported problems:

faulty fuel tanks, leaking manual and automatic transmissions, a persistent rotten-egg smell, and noisy engine timing chains and front brakes. With the exception of electronic component failures, repairs are relatively simple to perform.

Redesigned for the 1995 model year, 1995–98 Sentra sedans are much improved, larger, and better-performing vehicles. The seatbelts are more comfortable, and dual airbags are a standard feature. Most owner complaints concern some stalling and hard starting; electrical glitches; premature brake wear and excessive brake noise; and accessories that malfunction. Body assembly is also targeted with some complaints of poor body fits, paint defects, and air and water leaks into the interior.

Dealer service bulletins: **1995**—AC compressor leaks or noise. • Excessive brake noise. • C-pillar finisher lifting. • Faulty fuel gauge. • Self-activating horn. • Poor driveability—Code 45. • Power door locks that self-activate during periods of high heat or humidity. • Power windows that won't roll up unless ignition key is cycled. • Front window misalignment. • Front window won't go completely down. • DSB #NTB95-052 addresses all the possible causes and remedies for wind noise intruding into the passenger compartment.

Safety summary/Recalls: **Recalls**: **1987–88**—These model years have been recalled because of cracked fuel tanks. **1987–91**—Takata seatbelts will be replaced. **1990–91**—Frayed front shoulder belts may lead to improper retraction and inadequate protection. **1991–92**—Valve vacuum hose may loosen. **1995**—ABS may be defective. **1995–98**—Dealer will install a water diversion seal to prevent water from entering the windshield wiper linkage.

Secret Warranties/Service Tips

1987–90—Manual transmission fluid leaks can be plugged with upgraded case bolts. **1989**—Install an AIV case assembly if the exhaust smells like rotten eggs. • If the trunk lid is hard to close or latch, install an upgraded latch and a softer weatherstrip. **1991–92**—Noisy front brakes can be silenced by installing upgraded, non-asbestos front disc brake pads (#41060-63Y90). • Timing chain rattle may be caused by insufficient oil in the chain tensioner. You can correct this by replacing the tensioner with a countermeasure part (#13070-53J03). **1991–93**—The manual transmission has no Reverse gear—install a Nissan upgrade kit. **1991–94**—Stiffer trunk torsion bars will help keep the trunk lid from falling. **1993–94**—Brake and steering wheel vibrations are most likely caused by excessive rotor thickness. **1994**—Door hinges may have received inadequate rust protection; Nissan will apply a sealer at no charge. **1995–96**—DSB #NTB96-001 gives lots of troubleshooting tips on finding and correcting various squeaks and rattles. • Nissan has a special kit to improve brake pedal feel, according to DSB #NTB96-041.

Sentra Profile

	1991	1992	1993	1994	1995	1996	1997	1998
Cost Price ($)								
Sentra	8,690	8,990	9,590	10,990	12,290	13,448	13,698	14,498
Used Values ($)								
Sentra ↑	2,500	3,000	3,500	4,500	6,00	8,000	10,000	12,000
Sentra ↓	2,000	2,500	3,000	4,000	5,000	7,000	8,500	11,000
Extended Warranty	N	N	N	N	N	N	N	N
Secret Warranty	N	N	N	N	N	Y	Y	N
Reliability	❷	③	④	④	④	⑤	⑤	⑤
Air conditioning	③	③	④	⑤	⑤	⑤	⑤	⑤
Body integrity	❷	❷	③	③	③	③	④	④
Braking system	❶	❷	❷	❷	❷	❷	③	④
Electrical system	❷	③	③	❷	❷	❷	③	④
Engines	③	③	④	④	④	⑤	⑤	⑤
Exhaust/Converter	❷	❷	③	③	④	⑤	⑤	⑤
Fuel system	❷	❷	③	③	③	③	③	④
Ignition system	❷	③	③	③	③	④	④	④
Manual transmission	④	⑤	⑤	⑤	⑤	⑤	⑤	⑤
- automatic	❷	④	④	④	④	⑤	⑤	⑤
Rust/Paint	❷	③	③	③	③	③	④	④
Steering	③	③	③	③	④	④	⑤	⑤
Suspension	③	④	④	④	④	④	⑤	⑤
Crash Safety	⑤	⑤	③	③	—	④	④	—

SUBARU

Chaser, DL, GL, Loyale, Impreza, XT

Rating: Recommended (1996–98); Average (1993–95); Below Average (1984–92). The earlier models aren't recommended because of poor-quality emissions components and the premature wearout of major mechanical systems (CV joints, steering, etc.). **Maintenance/Repair costs**: Higher than average, and 4X4 repairs must be carried out by a dealer. Only buy a Subaru if you must have AWD and you're confident you can get dependable service from your local Subaru dealer. **Parts**: Expensive and hard to find, especially for the discontinued Chaser. Emission components are often back-ordered for months, but cheap aftermarket components can be found outside the dealer network.

Strengths and weaknesses: 1987–92 are the years to avoid. Performance, handling, and ride are mediocre. Engine breakdowns and premature clutch and exhaust system wearout are commonplace. Early hatchbacks came with a weak and growly 1.6L flat 4-cylinder motor; later models have a 1.8L version of the same anemic engine. Expensive catalytic converters are often replaced at the owner's cost before the five-year emissions warranty has expired. Subaru will reimburse the cost if you raise a fuss.

On 1988 and later models, steering assemblies, CV joints, and front brakes are the main problem areas. These parts generally need replacing after three to five years and Subaru dealers charge the full rate for replacement.

In the early spring of 1994, the Impreza replaced the unpopular and aging Loyale. The Impreza, too, offered sluggish engine performance, jerky full-throttle downshifts, and mediocre fuel economy. Overall reliability was improved, however.

By 1995, when these small cars went AWD, overall quality control improved as well. Powertrain components are more durable and function more smoothly, and electronic components have fewer glitches. Rusting is less of a problem than with the earlier models, which are particularly susceptible to rapid rusting of the bumpers, door bottoms, rear hatch, and hood.

Safety summary/Recall:: 1994–95 Impreza—Inadvertent airbag deployment is a problem. Airbags could deploy after the underside of the car scrapes the road or if the car drives over a dip in the road, hits a pothole, or is stuck in a ditch in the snow. **Recalls**: **1985 XT**—Improperly installed bumpers will be fixed. **1985–87**—Corrosion of the rear suspension inner arms could affect the control of the vehicle. **1985–87 DL, GL, GT**—The rear suspension's inner arms could fail. **1987 DL, GL**—Faulty carburetor components could pose a fire hazard. **1988–91 XT6 and 1989–91 XT**—Faulty power steering will be repaired. **1989–90**—Automatic transmission may engage abruptly, causing a sudden lurch into Reverse. **1991 XT 4X4**—5-speed manual gearbox may leak, causing transmission to suddenly seize. **1988–1990 Loyale**—Takata seatbelts will be replaced. **1989–90 Loyale**—Automatic transmission may engage abruptly, causing a sudden lurch into Reverse. **1990 Loyale with 3AT transmission**—Vehicle may jump into gear from Park. **1990–93 Loyale**—5-speed manual transmission may suddenly seize.

Secret Warranties/Service Tips

1985–90—Correct rear gate door rattle with a plastic sheet buffer. **1987**—An overly rich choke condition during cold starts can be corrected by installing a modified auxiliary choke pull-off spring. **1990–91**—A knocking noise from the exhaust flex joint may require the replacement of an exhaust flange gasket with an upgraded gasket (#44022-GA 191). **1991 Loyale**—Weak, noisy AM reception can be corrected by installing a modified antenna feeder cable (#86324AA040). • Banging over bumps is likely caused by the struts, strut mounts, and brake cable clamps on trailing arms or by the rear defogger condenser hitting the quarter panel. • A popping noise heard when going over small bumps may be caused by the front stabilizer bar bushing clamps. **1991–92**—Ignition relay failure is the likely cause of no-starts. **1992–94**—The heater mode door actuator may be the culprit of an annoying clicking in the heater area.

Chaser, Impreza Profile

	1991	1992	1993	1994	1995	1996	1997	1998
Cost Price ($)								
Chaser	10,365	—	—	—	—	—	—	—
Impreza	—	—	—	12,995	17,995	—	—	—
Impreza 4X4	—	—	—	—	19,695	17,995	21,395	21,395
Used Values ($)								
Chaser ↑	3,000	—	—	—	—	—	—	—
Chaser ↓	2,500	—	—	—	—	—	—	—
Impreza ↑	—	—	—	7,000	9,000	—	—	—
Impreza ↓	—	—	—	6,000	8,000	—	—	—
Impreza 4X4 ↑	—	—	—	—	11,500	13,000	14,500	16,000
Impreza 4X4 ↓	—	—	—	—	10,500	12,000	13,500	15,000
Extended Warranty	Y	Y	Y	Y	Y	Y	Y	Y
Secret Warranty	N	N	N	N	N	N	N	N
Reliability	❷	③	③	③	④	④	⑤	⑤
Air conditioning	❷	③	④	④	⑤	⑤	⑤	⑤
Body integrity	❷	❷	③	③	③	③	③	④
Braking system	❷	❷	❷	③	③	③	④	④
Electrical system	❷	③	③	③	③	④	④	④
Engines	❷	③	③	③	④	⑤	⑤	⑤
Exhaust/Converter	❷	③	④	④	④	④	⑤	⑤
Fuel system	④	④	❷	❷	❷	④	④	⑤
Ignition system	❷	❷	❷	❷	④	④	④	⑤
Manual transmission	④	④	④	④	⑤	⑤	⑤	③
- automatic	③	③	③	③	③	③	③	③
Rust/Paint	❷	③	③	③	③	④	④	⑤
Steering	③	③	③	③	③	④	④	④
Suspension	③	③	④	④	④	④	④	④
Crash Safety								
Impreza	—	—	—	—	—	④	④	—

Loyale Profile

	1990	1991	1992	1993	1994
Cost Price ($)					
Loyale	12,549	13,050	11,899	11,255	—
4X4	13,749	14,299	12,909	13,193	14,995
Used Values ($)					
Loyale ↑	3,000	3,500	4,000	5,500	—
Loyale ↓	2,500	3,000	3,500	5,000	—
4X4 ↑	3,000	3,700	4,000	4,500	6,500
4X4 ↓	2,700	3,200	3,500	4,000	5,500
Extended Warranty	Y	Y	Y	Y	Y
Secret Warranty	N	N	N	N	N

Reliability	❷	❷	❷	❷	③
Air conditioning	❷	③	③	④	④
Body integrity	❷	❷	❷	❷	③
Braking system	❷	❷	③	③	③
Electrical system	❷	❷	③	③	③
Engines	❷	❷	❷	❷	③
Exhaust/Converter	❷	❷	❷	❷	③
Fuel system	③	④	④	④	④
Ignition system	❷	③	③	③	④
Manual transmission	④	④	④	⑤	⑤
- automatic	❷	❷	③	③	③
Rust/Paint	❷	❷	❷	③	④
Steering	③	③	③	③	④
Suspension	③	④	④	④	④
Crash Safety					
GL Loyale	❷	—	—	—	-

SUZUKI

Esteem

Rating: Recommended (1996–98). An incredibly slow rate of depreciation means that bargains will be rare. Nevertheless, both the base GL and upscale GLX come loaded with standard features that cost extra on other models. The GL, for example, comes with power steering, rear window defroster, remote trunk and fuel-filler door releases, tinted glass, and a fold-down rear seat (great for getting extra cargo space). GLX shoppers can look forward to standard ABS, power windows and power door locks, and a host of other interior refinements. Shop around for a better-made, second-series (made after March 1996) 1996 Esteem rather than a 1997, inasmuch as they're practically identical and a late-model 1996 should be much cheaper. **Maintenance/Repair costs**: Higher than average. Repairs must be carried out by a Geo, Chevrolet, or Suzuki dealer. **Parts**: Average cost, but some long waits reported.

Strengths and weaknesses: The Esteem, Suzuki's largest car, is a small four-door sedan that is a step up from the Swift. Smaller than the Honda Civic and Chrysler Neon, it has a fairly spacious interior, offering rear accommodation (for two full-sized adults) that is comparable to or better than most cars in its class. Suzuki's top-of-the-line econobox stands out with its European-styled body and large array of such standard features as air conditioning, a fold-down back seat, and remote trunk and fuel-door releases.

The Esteem has only been on the market for four years, but early reports indicate a high level of quality and dependability. In this respect, it competes well with rivals like the Chevy Cavalier, Ford Escort, and Honda Civic. However, some owners complain of premature front brake wear, noisy front brakes, and occasional electrical short circuits.

Safety summary/Recalls: **1998–99**—Defective engine block heaters.

Secret Warranties/Service Tips

1996–97—Uneven wear of the front disc brake pads can be corrected by modifying the upper bushing tolerance, says DSB #TS 5-03-04126.

Esteem Profile

	1996	1997	1998
Cost Price ($)			
GL	13,495	13,495	13,895
GLX	14,495	15,495	16,895
Used Values ($)			
GL ↑	9,500	10,000	11,500
GL ↓	8,500	9,500	10,000
GLX ↑	10,500	11,500	13,000
GLX ↓	9,000	10,500	12,500
Extended Warranty	**N**	**N**	**N**
Secret Warranty	**N**	**N**	**N**
Reliability	⑤	⑤	⑤

Note: The Esteem hasn't been crash-tested yet.

TOYOTA

Corolla

Rating: Recommended (1995–98); Above Average (1991–94); Average (1985–90). The 1995–96 models combine the best array of standard features, quality control, and "reasonable" (for a Toyota) used prices. Although the 1997 model was "decontented" (less soundproofing, fewer standard features, etc.), there has been little reduction in quality or performance. **Maintenance/Repair costs**: Lower than average, and repairs can be done anywhere. **Parts**: Reasonably priced and easily found.

Strengths and weaknesses: Corollas are economical, high-quality, dependable little cars, but age can take its toll, especially in eastern Canada where rust snacks on these cars. 1985–87 versions may carry the Toyota name and appear to be bargains at first glance, but they're likely to have serious rusting problems and need costly brake, steering, and suspension work. Stay away from the 1.8L diesel version; it lacks performance and parts aren't easy to find. 4X4 versions are also risky. Wiper pivot assemblies may seize due to corrosion. Front shocks on rear-drive models wear out more quickly than average. Exhaust parts aren't very durable.

Post-1987 models are much improved. The two-door models provide sporty performance and good fuel economy, especially when equipped with the 16-valve engine. The engines and drivetrains are exceptionally reliable. Front-drive sedans and five-door hatchbacks offer more room than their rear-drive counterparts. The base engine, however, lacks power and is especially deficient in low-end torque, making for agonizingly slow merging and passing on the highway. Owners report problems with premature front suspension strut and brake wear, brake vibration, faulty defrosting that allows the windows to fog up in winter, and rusting of body seams, especially door bottoms, side mirror mounts, trunk and hatchback lids, and wheel openings.

The 1990–94 Corolla's problems are limited to harsh automatic shifting, early front brake pad and strut/shock wearout, electrical glitches, ignition problems, and some interior squeaks and rattles. They do, however, still require regular valve adjustments to prevent serious engine problems. Less of a problem with later models, rusting is usually confined to the undercarriage and other areas where the moldings attach to sheet metal.

1995–98 models are relatively problem-free, except for some minor brake and electrical problems, in addition to occasional suspension vibrations and body trim imperfections.

Safety summary/Recalls: Some owners find the driver's foot room inadequate to safely operate the accelerator, brake, and clutch pedals. **1993–96**—Researchers are looking into 20 similar incidents where the turn signal failed after the hazard warning light activated. **Recalls**: **1993–94**—An unused harness connector for vehicle accessories located under the carpet may cause an electrical short circuit or fire. **1995–95**—Liquid spilled onto the console could make the airbags deploy. **1994**—Seatbelt anchor straps may be faulty. **1995**—Defective terminal could drain the battery or make it explode. Dealer will replace the battery.

Secret Warranties/Service Tips

1987–88—Engines that run on, surge, or have flat power spots may need an upgraded carburetor assembly, cold-mixture heater temperature switch, or engine sub-wire harness. **1993–96**—Toyota has upgraded the hazard switch to improve turn signal performance in cold climates. **1993–97**—Inoperative front-passenger power side window switch may be caused by lubricant from the wire harness contaminating the window switch contacts. **1994**—A loose rear seat bolster cover is a common problem, according to DSB #B094-005. • Windshield A-pillar wind noise can be stopped by modifying the molding lip. **1995–96**—To enhance the performance of the rear door glass, Toyota has upgraded the mounting channel rubber insert and offers it as a service part. **All models/years**—Improved disc brake pad kits are described in DSB #BR94-004. • Brake pulsation/vibration, another generic Toyota problem, is fully addressed in DSB #BR94-002, "Cause and Repair of Vibration and Pulsation." • Complaints of steering column noise may require the replacement of the steering column assembly, a repair covered under Toyota's base warranty.

Corolla Profile

	1991	1992	1993	1994	1995	1996	1997	1998
Cost Price ($)								
Base	11,948	12,788	14,598	14,798	15,628	13,508	13,968	14,928
Used Values ($)								
Base ↑	5,000	6,500	8,500	10,000	10,500	11,000	12,000	13,000
Base ↓	4,000	6,000	8,000	9,500	10,000	10,500	11,000	12,500
Extended Warranty	Y	N	N	N	N	N	N	N
Secret Warranty	N	N	N	Y	Y	Y	Y	Y
Reliability	③	③	③	④	④	⑤	⑤	⑤
Air conditioning	⑤	⑤	⑤	⑤	⑤	⑤	⑤	⑤
Body integrity	❷	❷	❷	❷	③	④	④	⑤
Braking system	❷	❷	❷	❷	③	③	④	④
Electrical system	❷	③	③	③	③	④	④	⑤
Engines	④	④	④	④	⑤	⑤	⑤	⑤
Exhaust/Converter	❷	❷	③	③	④	④	⑤	⑤
Fuel system	❷	③	③	③	③	③	⑤	⑤
Ignition system	❷	❷	③	③	④	④	⑤	⑤
Manual transmission	④	④	④	④	⑤	⑤	⑤	⑤
- automatic	③	③	④	④	⑤	⑤	⑤	⑤
Rust/Paint	❷	❷	❷	❷	③	④	⑤	⑤
Steering	③	④	④	④	④	⑤	⑤	⑤
Suspension	③	③	③	④	④	⑤	⑤	⑤
Crash Safety	❷	❷	⑤	⑤	④	④	④	④
Side Impact	—	—	—	—	—	—	③	③

Paseo

Rating: Recommended (1994–98); Above Average (1992–93). **Maintenance/Repair costs**: Lower than average. Repairs can be done anywhere. **Parts**: Reasonably priced and easily found.

Strengths and weaknesses: This baby Tercel's main advantages are a peppy 1.5L 4-cylinder engine, a smooth 5-speed manual transmission, good handling, a supple ride, great fuel economy, and above-average reliability. On the other hand, this light little sportster is quite vulnerable to side winds; there's lots of body lean in turns; there's plenty of engine, exhaust, and road noise; front head room and leg room are limited; and there is very little rear seat space.

Dealer service bulletins show that the 1993s may have defective Panasonic tape and CD players, and a radio hum at low volume caused by fuel pump interference. Later models have fewer reliability problems, except for some brake and drivetrain vibrations.

Safety summary/Recalls: Thick rear pillars reduce rear visibility. **Recalls**: N/A.

Secret Warranties/Service Tips

1992—Low-volume radio hum can be corrected by installing spacers (insulators) between the radio chassis and the printed circuit board. **1992–93**—Toyota will improve the shift "feel" on its automatic gearboxes by increasing the C1 accumulator control pressure. **1996**—Toyota has developed an upgraded thermostat to improve heater performance. **1997**—Fujitsu radios may not eject/accept CDs. **All models/years**—DSB #B0003-97 recommends the use of a new wind noise repair kit. • DSB #AC002-97 gives lots of troubleshooting tips on eliminating AC odours. • AM radio static is likely caused by a damaged power antenna or by poor grounding due to corrosion.

Paseo Profile

	1992	1993	1994	1995	1996	1997	1998
Cost Price ($)							
Base	13,338	14,398	14,698	16,878	17,21	17,608	15,998
Used Values ($)							
Base ↑	7,000	8,000	9,700	10,500	12,000	13,000	15,000
Base ↓	6,000	7,500	8,700	9,500	11,000	12,000	13,500
Extended Warranty	N	N	N	N	N	N	N
Secret Warranty	N	N	N	N	N	N	N
Reliability	④	④	⑤	⑤	⑤	⑤	⑤
Crash Safety	❷	—	—	—	—	④	—

Tercel

Rating: Recommended (1993–98); Above Average (1991–92); Average (1987–90). **Maintenance/Repair costs**: Inexpensive. Repairs can be done anywhere. **Parts**: Reasonably priced and easily obtainable.

Strengths and weaknesses: Don't buy a Toyota on reputation alone, because many early models (1985–90) can have serious braking, electrical, and rusting problems, and be overpriced to boot. As well, stay away from the troublesome 4X4 versions made from 1984 to 1987.

All Tercels should be checked for door panel and underbody rust damage. 1987–90 Tercels are rust-prone around the rear wheels and side mirror mounts, and along the bottoms of doors, hatches, and rear quarter-panels. Early models suffer from extensive corrosion of rear suspension components.

1987–88 Tercels give you more for your money with a restyled aero look, a better-performing multi-valve, an overhead-cam engine, the impressive performance of a 5-speed manual gearbox, and additional sound insulation. Their main shortcomings are insufficient power when merging into traffic or hill climbing (particularly when shifting from second to third gear with the automatic transmission), cruise control glitches, fuel system malfunctions, excessive carbon buildup on the engine intake valve, occasional air conditioner breakdowns, cracked front exhaust pipes, and exhaust system/catalytic converter rust-out. Tercels are also plagued by pulsating brakes that wear out much too quickly. Owners of the four-wheel drive wagon complain of manual transmission failures and the occasional bug in the transfer case (these repairs are *very* expensive).

The redesigned 1989–90 versions are roomier, better performing, and above average in quality and reliability (except for some paint peeling and surface rusting). The sunroof is a frill that cuts head room drastically and causes irritating water leaks and wind noise. Tercels have a great reputation for exemplary durability, but overall performance is not outstanding and the interior is cramped.

1991–94 Tercels are pretty reliable, but they're not perfect. They were the first to be fuel-injected, which makes for livelier and smoother acceleration, and the interior space feels much larger than it is. Owners report faulty clutch-sleeve cylinders, hard shifting with the automatic transmission, premature brake and suspension-component wearout, brake pulsation, defective CD players, leaking radiators, windshield whistling, and myriad squeaks and rattles.

Redesigned 1995–98 Tercels offer a bit more horsepower, standard dual airbags, side door beams, aero styling, and a redesigned interior. They continue, however, to have some minor brake, electrical system, suspension, and body/accessories problems.

Safety summary/Recalls: **Recalls**: **1980–82**—Rear control arm could deform from corrosion.

Secret Warranties/Service Tips

1989–90—Fix harsh shifting from second to third on automatics by installing a new rubber check ball (#35495-22020). **1994**—A whistling noise coming from the windshield requires a urethane sealant applied at key points. • A steering column that's noisy or has excessive free play may need an upgraded steering main-shaft bushing. **1995**—Troubleshooting windshield molding wind noise is covered in the following DSB #BO95-005.

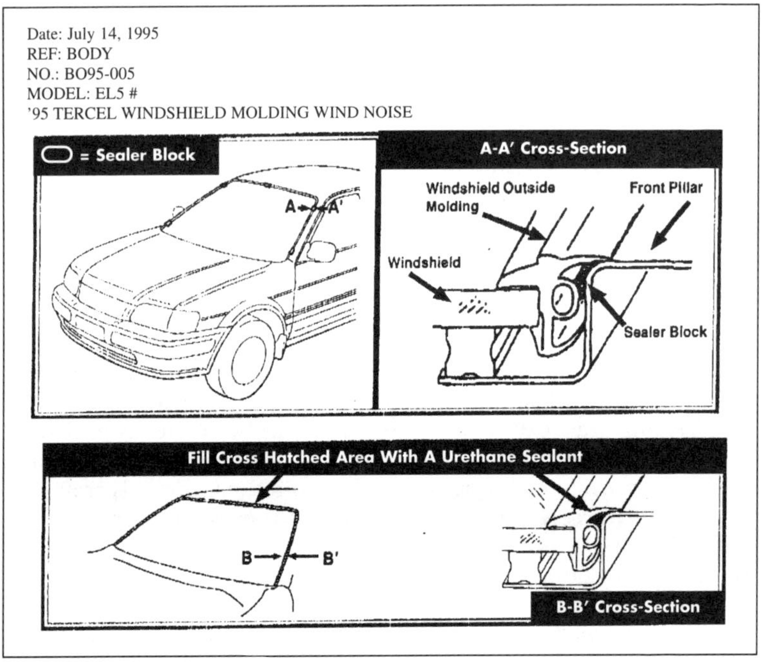

Put sealer blocks under the lip of the windshield moulding to silence an annoying whistling noise.

1995–96—Toyota has developed an upgraded thermostat to improve heater performance. • The company will also make available a longer passenger-side seatbelt. **1997**—Fujitsu radios may not eject/accept CDs. **All models/years**—DSB #B0003-97 recommends the use of a new wind noise repair kit. • Interior squeaks and rattles can be fixed with Toyota's kit (#08231-00801). • DSB #AC002-97 gives lots of troubleshooting tips on eliminating AC odours. • Older Toyotas with stalling problems should have the engine checked for excessive carbon buildup on the valves before any other repairs are done. • Improved disc brake pad kits are described in DSB #BR94-004. • Brake pulsation/vibration, another generic Toyota problem, is fully addressed in DSB #BR94-002, "Cause and Repair of Vibration and Pulsation." • A damaged power antenna or poor grounding due to corrosion are the most likely causes of AM radio static.

Tercel Profile

	1991	1992	1993	1994	1995	1996	1997	1998
Cost Price ($)								
Base	8,258	8,798	9,098	9,618	10,998	11,948	12,498	12,498
Used Values ($)								
Base ↑	3,000	3,500	4,000	5,000	6,000	7,000	8,000	10,500
Base ↓	2,500	3,000	3,500	4,000	5,000	6,000	7,000	9,500
Extended Warranty	Y	N	N	N	N	N	N	N
Secret Warranty	N	N	N	N	N	N	N	N

Reliability	③	③	③	④	⑤	⑤	⑤	⑤
Air conditioning	③	③	③	④	⑤	⑤	⑤	⑤
Body integrity	③	③	③	④	④	⑤	⑤	⑤
Braking system	❶	❷	❷	❷	③	③	③	④
Electrical system	③	③	③	④	⑤	⑤	⑤	⑤
Engines	❷	❷	❷	③	④	⑤	⑤	⑤
Exhaust/Converter	❷	③	③	③	④	④	⑤	⑤
Fuel system	③	③	③	④	⑤	⑤	⑤	⑤
Ignition system	④	④	④	④	⑤	⑤	⑤	⑤
- automatic	③	④	④	④	⑤	⑤	⑤	⑤
Manual transmission	④	④	④	④	⑤	⑤	⑤	⑤
Rust/Paint	❷	❷	❷	③	④	⑤	⑤	⑤
Steering	④	④	④	③	④	④	⑤	⑤
Suspension	③	③	③	③	③	③	④	⑤
Crash Safety	❷	❷	⑤	⑤	③	③	④	—
Side Impact	—	—	—	—	—	—	—	③

VOLKSWAGEN

Cabrio, Golf, Jetta

Rating: Above Average (1997–98); Average (1994–96); Below Average (1993); Not Recommended (1985–92). These overpriced vehicles age particularly badly, and since VW is more of a figurehead in Canada—run out of the States—don't expect many "goodwill" repairs. A Jetta is a Golf with a trunk; a Cabrio is a Golf without a roof. There was no 1994 model Cabrio. **Maintenance/Repair costs**: Higher than average. Repairs are dealer dependent. **Parts**: Expensive, but generally available from independent suppliers.

Strengths and weaknesses: On the positive side, these small Europeans are fun to drive and provide great fuel economy. The 1.8L gasoline engine is very peppy and the diesel engines are very reliable and good all-around performers. Both engines are easily started in cold weather. But here's the rub: Golfs and Jettas, like the failure-prone Rabbit they replaced, age badly. What you save in fuel, you lose in the car's high retail price carried over into the used-car market; and the ever-mounting maintenance costs as the vehicle gains years and mileage will easily wear you down.

Reliability is impressive—for the first three years. Then the brake components, fuel, and electrical systems start to self-destruct as your wallet gets lighter. Exhaust system components aren't very durable, body hardware and dashboard controls are fragile, and the paint often discolours and is easily chipped.

Although the 1990–93 models are a tad improved, Volkswagen still has terrible quality problems. Owners report electrical short circuits; heater/defroster resistor and motor failures; leaking transmission and stub axle seals; defective valve-pan gaskets, head gaskets, timing belts, steering assemblies, suspension components, alternator pulleys, brake and electrical systems. Body problems are legion, with air and water leaks, faulty catalytic converters, inoperative locks and latches, poor-quality body construction and paint, and cheap, easily broken accessories and trim items.

The redesigned 1994–96 models are a bit more reliable and safer. Nevertheless, problems disclosed in service bulletins for these model years include poor driveability, water leaks, trim defects, and premature rear tire wear. Owners report the following: electric door locks that take a long time to lock; paint that is easily nicked, chipped and marked; a variety of trim defects; premature rear tire wear; and poor-quality seat cushions. One owner of a 1996 Golf relates this "goodwill" experience with Volkswagen:

> Enclosed is a copy of a work order...to replace both front seat cushions—the second replacement for the driver's side since we purchased the vehicle in January 1996.
>
> The work order states "Goodwill Repair," as we are out of warranty. We feel this statement is rather misleading. Goodwill repair would have been if, when we brought the problem to the dealership's attention, they had offered to have the problem rectified with no hassle and without our having to resort to pressure from an outside source.
>
> The seat cushions have been replaced, but we feel they will only go again. Whether this is a flaw in the design of them or what, we can't say. The store manager seemed to think it was the way we "entered the vehicle" that caused the problem. As someone commented, entry was gained via the door!

For 1997, Golfs and Jettas equipped with the 116-hp 4-cylinder engine got a redesigned cylinder head that cuts engine noise. The Golf GTI VR6 rides lower, thanks to new shocks, springs, and anti-roll bars. The Cabrio Highline received standard AC, 14-inch alloy wheels, halogen driving lights, and leather upholstery. The base convertible lost its standard ABS and a few other goodies in an unsuccessful attempt to keep a lid on price.

Jettas provide slightly more comfort and better road performance than their Golf hatchback counterparts. The 1.6L 4-cylinder found on early Jettas was surprisingly peppy, and the diesel engine is very economical, although quite slow to accelerate. Diesels have a better overall reliability record than gasoline models and are popular as taxis. Jettas are far more reliable than Rabbits, but they, too, suffer from rapid body deterioration and some mechanical problems after their fourth year in service. For example, on post-1988 Jettas, starters often burn out because they are vulnerable to engine heat; as well, sunroofs leak, door locks jam, window cranks break, and windows bind. Owners also report engine head gasket leaks and water pump and heater core breakdowns. It's axiomatic that all diesels are slow to accelerate, but VW's fourth gear can't handle highway speeds above 90 km/h. Engine noise is deafening when shifting down from fourth gear.

Factory defects on 1990–96 Golfs and Jettas are so numerous they make these models very risky buys. Problems include automatic transmission, engine, suspension components, and catalytic converter failures; electrical short circuits; AC malfunctions; and fragile trim items. Body assembly and paint are second-class, leading to rattles and air leaks as the vehicles age.

All 1996–98 models are much more reliable, but, nevertheless, owners still report chronic automatic transmission, brake, and electrical system malfunctions problems, in addition to sub-par body construction and paint, leaky sunroofs, malfunctioning gauges and accessories, fragile locks and latches, and defective security systems.

Dealer service bulletins: **1995**—CD changer skipping or grounding. • Slow or inoperative central locking. • Clutch pedal noise and vibration. • Front end knock. • Wind noise at A-pillar door glass seal. • Exterior door locks freezing. • Release handle won't open hood. • Faulty instrument cluster control light. • Poor fuel system performance. • Pulsating or clicking noise from engine. • Shifting difficulties. • Rotten-egg odour from exhaust. • Sunroof noise and vibration. • Water leaks from sunroof area. • Noise from 5-speed transmission when cornering. • Vehicle drifts or pulls to one side. • Suggestions for correcting wheel/tire vibrations. • Skipping wiper blades. **1996**—Front-end knock. • Noise in rear centre console. • Vehicle drifts to one side. • Skipping wiper blades. **1997**—Anti-theft alarm sounds for no apparent reason and car won't start. • Automatic transmission fluid seeps from final drive breather vent. • Battery cable clamp won't tighten. • Frequent causes for inoperative window regulator motor. • Tips on eliminating front and rear door air leaks. • Excessive vibration or knocking of the shifter lever. • Transmission may not shift in or out of Reverse. • Troubleshooting windshield wiper failures.

Safety summary/Recalls: **Recalls**: **1983–92**—Corrosion may cause the hood to fly up. **1985**—Faulty brake master cylinder. • Front seatbelt retractors on two-door models may not lock properly. **1985 Jetta**—The plastic clip that holds the brake line may be a fire hazard. **1985–86**—Excessive fuel spillage could occur during a collision. **1985–87**—The engine may suddenly stall due to a seized fuel pump. **1987**—Loose lug nuts on alloy wheels could cause wheel separation. **1987–89**—An incorrectly contoured brake line could result in brake failure. **1987–92**—A de-icing kit will prevent sudden acceleration. **1988**—The seatbelt retractor pawl is weak. • The brake booster is faulty. **1988–91**—The engine preheating tube may interfere with braking. **1988–92**—VW will replace the fuel hose and install new spring-type hose clamps. **1990**—The power-steering pump bracket is faulty. **1991**—Too-short front brake hoses may rupture. **1993**—The windshield wiper motor may fail. **1985–91 Golf, Jetta**—Hot coolant could escape into the passenger compartment. **1990**—Omitting air ducts could cause brake fluid to overheat. **1993–95**—The jack could collapse during use. • Rear brake lines may leak from rubbing against the fuel tank; dealers will reroute and replace damaged lines. • Radiator fan motor could seize. **1993–96**—Hood may fly up. **1983–92 Cabrio**—Hood may fly up. **1985**—The fuel supply hose is faulty. **1987–90**—The fuel tank may leak. **1990**—The airbag harness wire is poorly located. **1990–91**—A bent water separator panel could cause sudden acceleration. **1990–92**—A faulty fuel hose retaining clamp could be a fire hazard. **1991**—Defective track control arms used to support front wheels cause wheel misalignment and steering pull to one side.

Secret Warranties/Service Tips

1988–94—Poor driveability may be caused by a deteriorated oxygen sensor wire shield or poor ground connection. **1990**—Hard winter starting may require the installation of a new high-energy ignition coil, high-tension wires, and spark plugs. **1992–94**—Poor engine performance or a rough idle may be due to a misrouted EVAP vacuum hose. **1993–94**—DSB #95-04 gives simple service tips for frozen door locks. **1994**—Water leaks into the engine bulkhead. **1996–97**—Tape player may have distorted sound or may snack on tapes. • A shifter that's hard to move side-to-side or won't go into Reverse may signal that the selector shaft is binding in the selector shaft housing bearing. **1997**—Erratic electrical functions may be caused by a loose ground at one of two grounding studs located under the battery tray. • If the transmission pops out of gear, check for a hairline crack on the selector shaft shift detent sleeve. • Buzzing noise from right-side air outlet may be caused by loose outlet mounting screws.

Cabrio, Golf, Jetta Profile

	1991	1992	1993	1994	1995	1996	1997	1998
Cost Price ($)								
Golf	10,270	10,710	10,710	12,600	12,995	14,325	14,690	16,765
Jetta	12,615	13,140	13,140	15,370	18,995	17,650	18,050	18,620
Cabrio	19,490	21,395	19,990	—	26,495	26,495	25,230	25,300
Used Values ($)								
Golf ↑	3,500	4,000	4,500	7,000	9,000	10,500	11,500	13,000
Golf ↓	2,500	3,500	4,000	6,500	8,500	10,000	11,000	12,500
Jetta ↑	4,500	5,500	6,500	8,500	12,000	14,500	16,000	17,000
Jetta ↓	3,500	5,000	5,500	7,500	11,000	13,500	15,000	16,000
Cabrio ↑	7,000	9,000	11,000	12,500	—	18,000	21,000	23,000
Cabrio ↓	6,000	8,000	10,500	12,000	—	17,000	20,000	22,000
Extended Warranty	Y	Y	Y	Y	Y	Y	N	N
Secret Warranty	N	N	N	N	N	N	N	N
Reliability	❶	❷	❷	❷	❷	③	④	④
Air conditioning	❷	❷	❷	❷	❷	③	⑤	⑤
Body integrity	❶	❶	❶	❶	❶	❷	❷	③
Braking system	❶	❶	❶	❷	❷	❷	③	③

Electrical system	❶	❶	❶	❶	❶	❷	❷	③
Engines	❷	❷	❷	❷	③	④	⑤	⑤
Exhaust/Converter	❷	❷	❷	③	③	④	④	⑤
Fuel system	❷	❷	❷	❷	③	③	③	④
Ignition system	❷	❷	❷	❷	❷	③	④	④
Manual transmission	④	④	④	⑤	⑤	⑤	⑤	⑤
- automatic	③	③	③	③	④	⑤	⑤	⑤
Rust/Paint	❷	❷	❷	❷	❷	③	④	⑤
Steering	③	③	③	③	④	⑤	⑤	⑤
Suspension	③	③	③	③	④	④	④	⑤
Crash Safety								
Golf	—	—	—	—	—	—	③	—
Jetta	—	—	—	—	③	③	③	—
Side Impact								
Jetta	—	—	—	—	—	—	—	③

MEDIUM CARS

A medium-sized car is a trade-off between size and fuel economy, offering more room and comfort but a bit less fuel economy (25–30 mpg) than a small car. These cars are popular because they combine the advantages of smaller cars with those of larger vehicles. As a result of their versatility and upsizing as well as downsizing throughout the years, these vehicles shade into both the small and the large car niches. The trunk is usually large enough to meet average baggage requirements, and the interior is spacious enough to meet the needs of the average family (seating four persons in comfort and five in a pinch). It's the best car for combined city and highway driving.

Recommended

Honda Accord (1992–98)
Toyota Camry (1994–98)
Volkswagen Passat (1998)

Above Average

Acura 2.2L CL, 3.0L CL (1998)
Acura Integra (1990–98)
Acura Vigor (1992–94)
GM Bonneville, Cutlass, Cutlass Supreme, Delta 88, Grand Prix, Intrigue, LeSabre, Lumina, Malibu, Monte Carlo, Regal (1984–87 rear-drives)
Honda Accord (1990–91)
Mazda 626, Cronos, MX-6, Mystère (1996–98)
Nissan Altima (1996–98)
Nissan Stanza (1990–92)
Subaru Legacy (1995–98 AWD)
Toyota Camry (1988–93)
Volkswagen Passat (1995–97)

Average

Acura 2.2L CL, 3.0L CL (1997)
Acura Integra (1986–89)
Chrysler/Eagle Acclaim, LeBaron, Spirit (1991–95)
Ford Contour, Mystique (1996–98)
Ford Sable, SHO, Taurus (1998)
GM Achieva, Calais, Grand Am, Skylark (1995–98)
GM Beretta, Corsica, Tempest (1993–96)
Honda Accord (1985–89)
Hyundai Sonata (1995–98)
Mazda 626, Cronos, MX-6, Mystère (1994–95)
Nissan Altima (1993–95)
Nissan Stanza (1988–89)
Subaru Legacy (1989–94)
Toyota Camry (1985–87)

Below Average

Chrysler/Eagle Breeze, Cirrus, Stratus (1995–98)
Ford Contour, Mystique (1995)
GM 6000, Celebrity, Century, Ciera (1998)
GM Achieva, Calais, Grand Am, Skylark (1988–94)
Volkswagen Corrado (1992–95)

Not Recommended

Chrysler/Eagle Acclaim, LeBaron, Spirit (1989–90)
Ford Sable, SHO, Taurus (1986–97)
Ford Tempo, Topaz (1985–94)
GM 6000, Celebrity, Century, Ciera (1982–97)
GM Beretta, Corsica, Tempest (1987–92)
GM Bonneville, Cutlass, Cutlass Supreme, Delta 88, Grand Prix, Intrigue, LeSabre, Lumina, Malibu, Monte Carlo, Regal (1988–98 front-drives)
Hyundai Sonata (1986–93)
Mazda 626, Cronos, MX-6, Mystère (1985–93)
Nissan Stanza (1985–87)
Volkswagen Corrado (1989–91)
Volkswagen Passat (1989–94)

1993 Ford Tempo: Along with its Topaz twin, this is Ford's quintessential lemon. Both cars were replaced none too soon in 1995 by the Contour and Mystique (first-year owners called it the "Mistake").

ACURA

2.2L CL, 3.0L CL

Rating: Above Average (1998); Average (1997). Overpriced and hard to find; be wary of the first-year models. **Maintenance/Repair costs**: Predicted to be higher than average. Repairs are dealer dependent. **Parts**: Expensive, once the warranty expires.

Strengths and weaknesses: The only difference between these two coupes is the 3.0L CL's larger engine, different wheels, and larger exhaust tip. Other vehicles worth considering, but with fewer standard features: BMW 318, Honda Accord, Lexus SC300, Nissan Maxima, and Toyota Camry.

The most distinctive features of the 2.2L CL and its 3.0L CL twin are the engine and wheels. These cars are both stylish, front-drive, five-passenger, American-built-and-designed luxury coupes. They have a flowing, slanted back end and no apparent trunk lock (a standard remote keyless entry system opens the trunk from the outside and a lever opens it from the inside). And while other Japanese automakers are taking content out of their vehicles, Acura has put content into the CL, making it one of the most feature-laden cars in its class.

Sure, we all know that the coupe's mechanicals and platform aren't that different from the Accord's, but when you add up all of its standard bells and whistles, you get a fully loaded small car that costs thousands of dollars less than such competing luxury coupes as the BMW 318 and the Lexus SC300. Consider this array of standard features: power windows, power mirrors, power moon roof, six-way power driver's seat, remote keyless entry system, ABS, leather-wrapped steering wheel, simulated wood trim, automatic climate control, dual airbags, tilt steering wheel, cruise control, CD player, and AM/FM stereo with six speakers.

Although no one would consider the CL a high performance car, it gets plenty of power from its quiet and smooth-running 3.0L 24-valve SOHC Variable Valve Timing and Lift Electronic Control (VTEC) V6, as well as from the Accord's 2.2L, single overhead cam 4-cylinder VTEC engine. The latter engine's 145 horses (same as the Accord's) take the CL from 0 to 100 km/h in a respectable nine seconds, but the engine works hard and noisily. Handling is better than average, thanks to the Accord's upgraded suspension, variable assisted steering, and 16-inch wheels, which are one inch larger than the Accord's (I told you there was a lot of Accord in the CL).

Now that I've whetted your appetite with all that's right about these little Acuras, let's look at some of the problems reported with the 1997 models. This is not a car for seating passengers in the rear. Back seat room is insufficient, unless the front seats are pushed all the way forward. And the rear windows don't roll down. Furthermore, owners have become so incensed at what they perceive as Acura's arrogant stonewalling of customer complaints that one owner, Michael Hos, set up his own Acura Lemon website to air Acura gripes and put pressure on the company. He got his money back after a few months and took down his website as part of the bargain. Nevertheless, he shows how others can do the same in Part Two of this guide.

Some of the common defects reported on the Hos website: faulty transmission control unit; transmission downshift problems; chronic brake rotor pulsation and other brake problems leading to resurfacing of brake rotors and replacement of brake pads, rotors, calipers, and springs; repeated front end realignments; door and wind noise leading to replacement of door; and sunroof won't stop at closed position, requiring replacement of sunroof switch and controller. Honda's 14-year free repair secret warranty applies to all 1996–97 Acuras, except for the type R models.

Safety summary/Recalls: **1997**—Complete brake failure; vehicle hit a wall. Chronic hesitation and stalling. Passenger side airbag replaced after sensor indicated it was defective. Leaky oil pan seals. Premature catalytic converter failure. Subframe out of alignment, causing vehicle to pull to one side. **1998**—When driving on a flat surdface at 45 km/h, or 1500 rpms, vehicle will jerk and pull for about 30 seconds. **Recalls**: **1998**—Defective shifter/parking pawl may allow vehicle to roll down an incline with shift lever ostensibly in Park. Dealer will put a collar on the pawl.

Secret Warranties/Service Tips

1997—Acura says it will replace any seatbelt's tongue stopper button for the life of the vehicle. • Remedy for front seat that won't slide forward or backward. • Fix for moonroof seal that sticks up. • Freeing up seatback adjustment lever • Silencing a dash pop or creak. • Fix for incorrect fuel guage and speedometer readings. • Correction for brake fluid leaking from the ABS modulator.

2.2L CL, 3.0L CL Profile

	1997	1998
Cost Price ($)		
2.2L CL	27,800	30,000
3.0L CL	30,650	34,000
Used Values ($)		
2.2L CL ↑	22,000	25,000
2.2L CL ↓	21,000	24,000
3.0L CL ↑	23,500	26,000
3.0L CL ↓	22,500	25,000
Extended Warranty	N	N
Secret Warranty	Y	N
Reliability	④	⑤
Crash Safety	—	—

Integra

Rating: Above Average, compromised by a stiff price for the Acura cachet and limited passenger room (1990–98); Average (1986–89). **Maintenance/Repair costs**: Higher than average. Repairs are dealer dependent. **Parts**: Expensive, but can be bought from cheaper independent Honda suppliers.

Strengths and weaknesses: A Honda spin-off, early Integras (1986–89) came with lots of standard equipment and are a pleasure to drive, especially when equipped with a manual transmission. The 4-speed automatic saps the base engine's power considerably. Engine and tire noise are intrusive at highway speeds. The car corners well and is more agile than later 1990–93 models. Its hard ride can be reduced a bit by changing the shocks and adding wide tires. The front seats are very comfortable, but they're set a bit low, and the side wheel wells leave little room for your feet. Rear seat room is very limited, especially on the three-door version.

A large number of 1989–90 Integras have faulty distributor igniters. When the igniter fails, the car stalls and it may be impossible to restart. The only remedy is to call a tow truck and have a garage replace the distributor.

Overall, assembly and component quality are good but not exceptional, as you can see from the "Secret Warranties/Service Tips" list. To avoid costly engine repairs, check the engine timing belt every 2 years/ 40,000 km.

For model years 1990–93, the high-revving 1.7L powerplant growls when pushed and lacks guts (read, torque) in the lower gears. The 1.8L engine is smoother running but delivers the same maximum horsepower as the 1.7L it replaced, until '94, when it gained 10 extra horses. Surprisingly, overall performance has been toned down and is seriously compromised by the 4-speed automatic gearbox. Interior design is more user friendly, with the front seating roomier than in previous years, but reduced rear seating is still best left to small children.

Mechanical reliability is impressive, but that's the case with most Hondas that sell for far less, and many mechanical components are so complex that self-service can pretty well be ruled out. The Integra's front brakes may require more attention than those of other Hondas. Surprisingly, what Integras give you in mechanical reliability they take away in poor quality control of body components and accessories. Water leaks, excessive wind noise, low-quality trim items, and plastic panels that deform easily are all commonplace. Owners also report severe steering shimmy, excessive brake noise, premature front brake pad wearout, and radio malfunctions.

In the 1994 model year, the Integra was dramatically restyled with a more aerodynamic profile and a few more horses were wrung out of the venerable 1.8L 4-banger through variable valve timing. The 1994–98 models also offer a smoother ride than previous versions. On the other hand, the powerful VTEC engine requires lots of shifting and interior room is still problematic. Overall, there are too few improvements to justify the high prices that late-model Integras command.

Owners report that steering wheel shimmy, fit and finish deficiencies, and malfunctioning accessories continue to be problematic on later models. Premature front brake wear is also an ongoing concern. Squeaks and rattles frequently crop up in the door panels and hatches, and the sedan's frameless windows often have sealing problems.

Safety summary/Recalls: **Recalls**: **1986–87**—An upgraded contact unit retainer needs to be attached to the front windshield wiper assembly. **1986–91**—Takata seatbelts will be replaced. **1987–88**—Acura will repair or replace defective heaters for free. **1994**—Automatic transmission retaining clip may show wrong gear.

Secret Warranties/Service Tips

1986–89—A faulty fuel pump check valve or fuel pressure regulator may be the cause of hard starts. Correct this problem by installing an improved fuel pressure regulator (#16740-PG7-663) or fuel pump (#16700-PG7-663). **1990–91**—DSB #91-015 is an excellent troubleshooting guide to the myriad squeaks and rattles in the dash, front doors, hatch, steering shaft, and sunroof. **1990–93**—Poor AM radio reception is likely caused by a poor ground connection between the antenna collar and the car body. **1992–97**—A defective seatbelt tongue stopper will be replaced free of charge with no ownership, time, or mileage limitations. **1994**—Rear hatch rattles are usually caused by a poorly adjusted striker. • Lots of troubleshooting bulletins are offered dealing with in-dash cellular phone problems. • Power door locks that cycle from locked to unlocked require a new power door lock control unit. **1994–95**—Rattling from a partially open window may be caused by excess clearance between the window guide pin and the centre sash guide or by the glass run channel having come out of the centre channel. These are also likely causes of moon roof chattering or shuddering. **1996–97**—In a settlement with the U.S. Environmental Protection Agency, Honda paid fines totaling $17.1 million and extended its emissions warranty on 1.6 million 1995–97 models to 14 years or 150,000 miles. This means that costly engine components and exhaust system parts, like catalytic converters, will be replaced free of charge, as long as the 14-year/150,000 mile limit hasn't been exceeded. Additionally, the automaker will provide a full engine check and emissions-related repairs at 50,000–75,000 miles and will give free tune-ups at 75,000–150,000 miles. It is estimated the free checkups, repairs, and tune-ups will cost Honda over $250 million. The story of the settlement was first reported on page 6 of the June 15, 1998, edition of *Automotive News*. Canadian owners may wish to use this settlement as leverage for free repairs in Canada, or use the full terms of the settlement when visiting the States. One thing is certain: neither Transport Canada nor Environment Canada is sufficiently enthused to render any assistance. **All models/years**—Severe and persistent steering wheel shimmy is likely due to an imbalanced wheel/tire/hub/rotor assembly.

Integra Profile

	1991	1992	1993	1994	1995	1996	1997	1998
Cost Price ($)								
LS	17,560	18,495	20,155	22,095	23,095	23,245	23,800	23,800
RS	14,800	15,495	16,250	17,655	18,595	18,795	19,500	21,000
Used Values ($)								
LS ↑	7,000	7,500	10,000	12,500	14,500	16,500	18,500	20,000
LS ↓	6,500	7,000	9,500	11,500	13,500	15,500	17,000	19,000
RS ↑	6,000	7,000	8,500	10,500	12,000	14,000	15,000	17,000
RS ↓	5,000	6,500	7,500	9,500	11,000	13,000	14,000	16,000
Extended Warranty	N	N	N	N	N	N	N	N
Secret Warranty	N	N	N	N	N	Y	Y	N
Reliability	④	④	⑤	⑤	⑤	⑤	⑤	⑤
Air conditioning	❷	③	④	④	⑤	⑤	⑤	⑤
Body integrity	❷	③	③	③	③	③	③	④

Braking system	❷	❷	❷	❷	❷	③	④	④
Electrical system	③	③	③	③	③	④	④	④
Engines	④	④	④	④	④	⑤	⑤	⑤
Exhaust/Converter	❷	❷	❷	③	④	④	⑤	⑤
Fuel system	③	③	③	④	④	⑤	⑤	⑤
Ignition system	❷	③	④	④	④	④	④	④
Manual transmission	⑤	⑤	⑤	⑤	⑤	⑤	⑤	⑤
- automatic	④	④	④	④	⑤	⑤	⑤	⑤
Rust/Paint	❷	③	③	③	③	③	④	④
Steering	③	③	④	④	④	④	④	⑤
Suspension	③	③	④	④	④	④	④	⑤
Crash Safety	⑤	⑤	❶	—	④	④	—	—

Vigor

Rating: Above Average (1992–94). **Maintenance/Repair costs**: Higher than average. Repairs are dealer dependent. **Parts**: Expensive, but most mechanical components can be bought more cheaply from independent suppliers. Accord parts can be used for most maintenance chores. Body panels and engine components, however, are more difficult to find.

Strengths and weaknesses: A spin-off of the Honda Accord sedan, the 5-cylinder Vigor went through only three model years (1992–94). Due to its high cost and the absence of a more economical, smoother, and easier-to-repair 6-cylinder powerplant, most buyers chose the cheaper Accord instead. This compact has power to spare, handles well, and has an impressive reliability/durability record. On the other hand, owners report the following problems: lots of engine, tire, and road noise intrude into the passenger compartment at highway speeds; the engine and transmission lack smoothness; and the brakes aren't easy to modulate. The suspension is firm but not harsh, rear leg room is limited, and there have been some complaints of fit and finish deficiencies. Vigors are also notoriously thirsty for premium fuel.

Safety summary/Recalls: **Recalls**: **1992**—Inadequate venting could allow moisture to build up in the distributor cap and prevent the engine from starting.

Secret Warranties/Service Tips

1992—DSB #92-034 goes into great detail on ways to silence the Vigor's many squeaks and rattles. **All models/years**—Excessive steering wheel shimmy is likely due to an imbalance of the wheel/tire/hub/rotor assembly in the front end.

Vigor Profile

	1992	1993	1994
Cost Price ($)			
LS	26,600	28,450	29,300
Used Values ($)			
LS ↑	10,000	13,000	16,000
LS ↓	9,000	12,000	15,000
Extended Warranty	N	N	N
Secret Warranty	N	N	N
Reliability	④	④	④
Crash Safety	❷	—	—

CHRYSLER/EAGLE

Acclaim, LeBaron, Spirit

Rating: Average only with an extended warranty (1991–95); Not Recommended (1989–90). These front-drives are a better choice than the Aries or Reliant (see "'Beaten' Beaters") due to their more recent vintage and more refined engines and suspension. 1995 was their last model year. Some early LeBaron GTC convertibles cost only about $2,000–$3,000 more than the standard LeBaron sedan. **Maintenance/Repair costs**: Lower than average. Repairs can be easily done by independent garages. The ease with which these cars can be repaired and the low cost of most replacement parts add to their attractiveness as relatively inexpensive five-passenger, fuel-efficient used cars. **Parts**: Reasonably priced, with good availability through independent suppliers.

Strengths and weaknesses: Although these cars have an outdated design, Chrysler improved their assembly quality and backed these cars with a strong base warranty that's transferable to second owners. The 100-hp 2.5L 4-cylinder engine and 3-speed automatic are models of ruggedness. Acceleration is adequate with the standard 4-cylinder engine and quite fast and smooth with the 150-hp Chrysler-bred 2.5L turbo.

The following are all prone to rapid wear: the cylinder head gaskets on the 2.5L engine, the front suspension, brakes, fuel system, and air conditioning. The Mitsubishi-built V6 has generated complaints about premature oil leaks and fuel system problems. The Ultradrive A604 4-speed automatic transmissions (renamed the 41TE in 1992) were found early in production to have poor shifting, internal fluid leakage leading to clutch burn-out, and sudden second-gear lockup. The problem first becomes apparent when the transmission tries to engage third and fourth gears simultaneously, causing either clutch failure or sticking in second gear. Between 1989 and July 1991, Chrysler tried 28 different changes to resolve the Ultradrive's clutch failures, second-gear lockup, and excessive shifting on hills. The 41TE automatic transmissions on 1992–96 models are also failure-prone.

The air conditioner switches on and off continuously. The following parts often need replacing: windshield wiper and interior fan motors; speedometer-sending unit and MAP sensor; and steering rack, bushings, and front stabilizer bar. Owners report that the steering wheel's vinyl coating rubs off on hands and clothing and the plastics used in the seats and dash area give off vapours that collect as a film on inside window surfaces.

All model years have fragile body hardware and sloppy body assembly. Many owners report water infiltration, especially around the windshield and into the trunk. Surface rust is common, especially around the windshield, door bottoms, and rear lip of the trunk lid.

LeBarons are a bit more high-tech, but overall road performance and reliability are just as disappointing. The ride is rough and vibrations and rattles are constant. Automatic transmission torque converters frequently malfunction. The front suspension is particularly trouble-prone, and electrical bugs and major computer module failures are common. Wheel bearings, fuel pumps, and rear window motors tend to wear out prematurely.

Owners of 1991–95 models report that oil leaks and oil pump failures are common; rear brakes are noisy; the heating, air conditioning, and ventilation systems often malfunction; and electrical components (notably electronic modules) aren't very durable. Paint and assembly quality are also below par, characterized by body hardware and trim that rust out, break, or fall off.

Safety summary/Recalls: In all three cars, the front wheels tend to lock prematurely when braking heavily. • According to CFAS, engine fires due to fuel and oil leaks from 1978–90 front-drive Chryslers have generated hundreds of complaints. • Rear brake lockup and master cylinder failures on 1978–92 models are also common, according to CFAS. **Recalls: 1989–90**—Engine oil leakage means a new engine valve cover gasket is needed. **1991**—Pin bolts on the front disc brake caliper guide might be too loose. • Both airbag front-impact sensors might be improperly mounted. • Outboard front seatbelts might not latch properly. **1992**—The coupling bolts on the steering column shaft could be faulty. **1994**—Seatbelt assembly could fail in an accident. **1990 LeBaron**—Airbag inflator module could be defective.

Secret Warranties/Service Tips

1989—Leaking fuel injectors are a common problem. • If the AC heater check valve freezes, install a new vacuum-control check valve (#5264270). • Models with 3.0L engines that lose power may need a new throttle body base gasket. • 3.0L engines that burn oil, stall, or lack power may have excessive sludge accumulation in the left bank (front) rocker cover baffle oil drain hole. • 2.5L turbo engines that lose power may need the Chrysler driveability kit (#4419460), or, during hot weather, a new vacuum connector (#5277577). **1989–90**—A604 automatic transmissions with excessive upshifting/downshifting require a new A604 controller (#4557585). • Rough idling might require the replacement of the EGR valve. **1989–91**—A604 automatic transmissions with 1–2 upshift shudder may need the 2–4 clutch replaced. **1989–93**—3.0L engines that burn oil or produce a smoky exhaust at idle can be fixed by installing snap rings on the exhaust valve guides and replacing all of the valve guide stems or the cylinder head. **1989–95**—Intake valve deposits are the likely cause of stalling, loss of power, hesitation, or hard starting. • Failure to go into Reverse is likely caused by faulty low/reverse piston. • Various tips are offered on correcting third-gear shuddering. **1990**—A604 automatic transmissions that limp or fail to shift require the replacement of the PRNDL and Neutral safety switch. **1990–91**—Surging or bucking at 50–90 km/h with an A413 or A670 automatic transmission can be fixed with driveability kit (#4419447). **1990–92**—Erratic idle speeds that occur after deceleration from a steady cruising speed can be corrected by replacing the idle air-control motor with a revised motor. **1990–94**—Harsh automatic shifts can be tamed by installing the following revised parts: kickdown, accumulator, reverse servo cushion springs, and accumulator piston. **1991**—Customer Satisfaction Notifications #499 and #521 set out Chrysler's commitment to correct, free of charge, engine oil pump failure. **1991–92**—Engines with a rough idle and stalling following a cold start may require a new Single Board Engine Controller (SBEC). **1991–93**—The serpentine belt may come off the pulley after driving the vehicle through snow. Install an upgraded shield, screw, and retainers. **1991–94**—Engines that stall following a cold start may need an upgraded Park/Neutral/start switch. **1991–95**—Poor AC performance while the AC blower continues to operate is likely due to the evaporator freezing. **1992**—If the heater and ventilation system changes to defrost mode during acceleration, trailer towing, or hill climbing, the installation of a revised vacuum check valve should cure the problem. • Long crank times, a rough idle, and hesitation may be corrected by replacing the intake manifold assembly. • A low-frequency moan or groan coming from the rear brakes can be reduced by installing an upgraded rear disc pad set (#4423667). **1992–93**—Rough idling after a cold start with 2.2L and 2.5L engines can be corrected by installing an upgraded powertrain control module (PCM). • Some 41TE transaxles may produce a buzzing noise when shifted into Reverse; this problem can be corrected by replacing the valve body assembly or valve body separator plate. • Deceleration shudder can be eliminated by replacing the powertrain control module with an upgraded version. **1992–95**—Tips are offered on eliminating AC odours and tracing likely causes of a rotten-egg exhaust smell. **1993**—Failure of the fuel pump check valve can cause start-up die-out, reduced power, or erratic shifting. **1993–94**—Acceleration shudder may be caused by automatic transmission front pump leakage. • Improve automatic shifting by installing an upgraded transmission control module. **1993–95**—Troubleshooting tips are offered concerning excessive brake noise. **1994–95**—Poor AC performance is likely due to defective compressor suction or discharge lines. **1995**—Transmission limp-in may be caused by an intermittent transaxle input speed-sensor signal. **All models/years**—Excessive effort required to disengage Park may mean a new Park rod assembly is required (#4431530). • Headlight condensation requires installation of vents in the headlight assemblies. • A rotten-egg odour coming from the exhaust is likely the result of a defective catalytic converter, which should be replaced for free under the emissions warranty.

Acclaim, LeBaron, Spirit Profile

	1989	**1990**	**1991**	**1992**	**1993**	**1994**	**1995**
Cost Price ($)							
Acclaim/Spirit	13,500	13,650	12,425	12,605	13,575	14,100	16,425
LeBaron Sedan	13,485	—	19,850	16,120	16,720	16,760	—
Convertible	20,065	21,785	20,395	21,715	22,400	22,995	23,615
Used Values ($)							
Acclaim/Spirit ↑	2,500	3,500	4,500	5,000	6,000	7,500	8,500
Acclaim/Spirit ↓	2,000	3,000	4,000	4,500	5,500	6,500	8,000
LeBaron Sedan ↑	3,500	—	5,500	6,000	7,500	8,500	—
LeBaron Sedan ↓	3,000	—	5,000	5,500	6,500	7,500	—
Convertible ↑	5,500	6,500	7,000	7,500	8,500	11,000	12,000
Convertible ↓	5,000	6,000	6,500	7,000	8,000	10,000	11,500
Extended Warranty	Y	Y	Y	Y	Y	Y	Y
Secret Warranty	N	N	Y	Y	Y	Y	Y
Reliability	❷	③	③	③	③	③	④
Air conditioning	❶	❶	❷	❷	❷	③	③
Body integrity	❶	❶	❷	❷	❷	❷	❷
Braking system	❶	❶	❷	❷	❷	❷	❷
Electrical system	❷	❷	❷	❷	③	③	③
Engines	❷	❷	③	③	④	④	④
Exhaust/Converter	③	❷	❷	④	⑤	⑤	⑤
Fuel system	❷	❷	③	③	③	④	⑤
Ignition system	❷	③	③	④	④	④	④
Manual transmission	—	—	—	—	—	—	—
Automatic	❷	❷	❷	❷	❷	❷	③
Rust/Paint	❶	❶	❷	❷	❷	❷	❷
Steering	❷	③	③	⑤	⑤	⑤	⑤
Suspension	❷	❷	③	③	③	④	⑤
Crash Safety							
Acclaim/Spirit	❶	❶	⑤	⑤	❶	❷	④

Breeze, Cirrus, Stratus

Rating: Below Average (1995–98). Stay away from any model carrying the anemic and failure-prone 4-cylinder engine. Also, be prepared to experience a number of nasty safety-related failures such as blown engine head gaskets, airbags not deploying, sudden acceleration, and loss of braking. **Maintenance/Repair costs**: Higher than average, but repairs aren't dealer dependent. **Parts**: Higher-than-average cost (independent suppliers sell for much less), but they are not hard to find. Nevertheless, don't even think about buying any one of these cars without a three- to five-year supplementary warranty.

Strengths and weaknesses: Roomy and stylish, well-appointed, comfortable, and smooth, the Chrysler Cirrus and Dodge Stratus are mid-sized sedan replacements for the LeBaron. The Breeze, launched as a 1996 model, is essentially a "decontented" version of the more expensive Cirrus and Stratus.

Most components have been used for some time on other Chrysler models, particularly the Neon subcompact and Avenger/Sebring sports coupes. Power is supplied by three engines: a 2.0L 4-cylinder engine, shared with the Neon; a 2.4L 4-banger; and the recommended optional 2.5L V6. Carrying Chrysler's "cab forward" design a step further up the evolutionary ladder, these cars have short rear decks, low noses, and massive sloping grilles. A wheelbase that's two inches longer than the Ford Taurus makes these cars comfortable for five occupants, with wide door openings and plenty of trunk space.

These cars aren't all that impressive, either from a performance or a quality control standpoint. Furthermore, owners have serious misgivings as to whether these vehicles can withstand a harsh climate. One radio producer relates the following "adventure" with his 1995 Cirrus:

> ...as I left a shopping mall parking lot, I heard a slight "pop." I instantly lost power steering. I had it towed to Triple Seven. They diagnosed the same problem that the university professor had: it appears Chrysler engineers made the sender hose from the power steering pump too long. In our severe climate, the line freezes, contracts, starves the power steering pump, and sends it into coronary arrest. And the professor and I weren't alone. Two other Cirruses with the same problem were seen Saturday at this dealer alone. Probably, this model was collapsing in large numbers across the West this last week in our record-breaking low temperatures....

Other problems reported by owners include the following: excessive road noise and vibrations, chronic automatic transmission failures, early and frequent engine head gasket failures (no doubt, part of the Neon engine legacy) and erratic engine operation, ABS malfunctions and sudden brake loss, paint delamination, electrical short circuits, weak headlights, AC glitches, water leaks into the trunk area and interior, easy-to-break trim items, lots of squeaks and rattles, and poorly located head restraints.

> ...the headrests were virtually useless, being 7" from the back of the head. I went to an upholstery shop and had them build up the headrests to 2" behind the head....

Dealer service bulletins: 1995—Inoperative AC system. • Misaligned windshield washer nozzles. • Faulty deck-lid and door latches. • Rattling exhaust system crossover pipes. • Faulty radios. **1996**—Won't start due to faulty neutral safety switch. • Single cylinder misfire. • Cold engine sag or rough idle. • A number of transmission problems and upgrades, including intermittent control module failures; reduced transmission limp-in default sensitivity; changes to lockup control system; and an upgraded Overdrive clutch hub. • Difficulty going into second gear or Reverse after a cold start. • Shuddering during upshifts or when torque converter is engaged. • Steering noise on right turns. • Vehicle drifts or leads at high speeds. • Cowl plenum water leaks. • Inoperable door glass. • Erratic windshield wiper delay intervals. • Exhaust cover pipe rattle, buzz, and moan. • Ratcheting sound when coming to a stop. • Headlight pattern improvement. • Interior window film buildup. **1997**—Troubleshooting tips are available for poor AC performance. • The powertrain "bumps" when AC engages. • The CD changer rattles. • A low frequency rumbling sound can come from the front of the car while highway cruising. • Transmission shudders. **1998**—Excessive cold crank time, start die-out or weak run-up. • Poor driveability with high-test fuel. • Delayed transaxle engagement. • Clunking or rattling steering wheel/column. • Cold-start power steering noise. • Intermittent front end popping noise. • Low-frequency front end rumble. • Inoperative right headlight high beam. • Seat adjuster rises when seat is unoccupied. • Broken or loose antenna mount.

Safety summary/Recalls: **1995–1996**—Frequent reports of engine fires. • Chronic engine oil leaks and oil galley plug failures. • Repeated engine head gasket failures, some resulting in engine compartment fires. • Timing belt failure after recall correction. • Chronic stalling or engine loss of power blamed on timing belt tensioner and pulley failure. • Engine sometimes loses power, then quickly accelerates. At other times, while at highway speeds, vehicle won't slow down when foot is taken off the gas pedal; instead, it speeds up. • Engine warning light often alight for no reason. • Reports of sudden acceleration in Drive and Reverse. • Owner claims that sudden acceleration is caused by a design flaw in cable to throttle body. • Airbags often fail to deploy. • Frequent ABS brake failures and prematurely worn rotors, calipers, and pads. • Rear brake failures also reported. • Faulty master cylinders are the cause of some early brake failures. • Floor mat can catch the steering shaft clamp and jam the steering column assembly, causing the steering to lock up. • While driving off the freeway at 100 km/h, steering tie rod came apart, resulting in complete steering loss. • Steering components worn out prematurely. • Main computer failed seven times. • Oxygen sensor prone to early failure and fluid leakage. • Alternator burned out. • Child was able to take parked vehicle out of gear without applying brake. • Vehicle jumped out of third gear. • Defective electrical switch causes transmission to stick in second gear. • Sudden transmission lockup. • Transmission won't engage or upshift to third or fourth gear. • Leaking transmission front pump seal. • Prematurely worn strut links and wheel bearings. • Heat is unevenly distributed, causing front and rear passengers to be cold. • Heater failures. • Defogger doesn't adequately defrost windshield. • AC fails to cool interior unless vehicle is traveling at high speed. • AC failures. • Dash reflection into windshield hampers view. • Inadequate headlight illumination. • Windshield wipers don't run fast enough to clear windshield in a heavy downpour. • Rear brake lights and brake switch failure. • Sloping hood design creates poor visibility for parking. • Fuel tank gauge indicates empty when tank is half full. • Trunk lid closes on its own; this gave one owner a mild concussion. • Plastic on seatbelt clinch bar is self-destructing. • Seatbacks collapsed when vehicle was rear-ended. • Seat frame and anchor broke as driver sat down. • Power window motor failures. • Door locks work intermittently. • Binding door hinges make for difficult closing. • Door handle design pinches fingers. • Key sticks in the ignition when vehicle is shut off. • Weak front speakers prevent radio balance. **1997**—Engine compartment fire ignited while car parked in garage. • Reports of sudden acceleration in Drive and Reverse. • Chronic stalling at highway speeds. • Vehicle rolled away while parked with shifter in Park position and keys pulled from ignition. • Shifter came off in hand when shifting. • Airbags fail to deploy. • Floor mat can catch the steering shaft clamp and jam the steering column assembly, causing the steering to lock up. • Frequent engine head gasket failures. • When the head gasket blew, one driver believes it overheated the transmission causing it to lock up. • It's common for the engine timing belt idler pulley to fail and damage the timing belt. • Sudden brake and power steering loss. • Brake pads and rotors fail prematurely. • Noisy brakes. • AC emits odour and white flakes through ventilation system, causing headaches, burning sensation, and congestion. • Noise coming from the high-pressure power steering line. • Missing part causes seatbelt to twist. • Seat buckle design is too short, causing difficulty in latching. • Driver-side shoulder belt failed to restrain driver in a collision. • Starter short caused fuse to blow. • Fuel sender for gauge often defective. • Inoperative power door lock motor. **1998**—Airbag exploded rather than inflated. • Airbags fail to deploy. • Frequent complaints of ABS brake failures; ABS brakes failed five times on one owner despite dealer attempts to correct the problem. • There is excessive brake noise whenever brakes are applied. • Floor mat jammed the steering column assembly causing the steering to lock up. • Automatic transmission floor console design throw from Drive to Reverse to Park is too long, resulting in consumer thinking vehicle is in Park when it's really in Reverse. • Floor shift indicator on the dash doesn't give a true reading of which gear is engaged. • Gear shift lever can be moved into Drive without putting foot on brakes to engage the transmission/brake interlock system. • High trunk lid makes it impossible to see directly behind the vehicle. **Recalls**: **1995 Cirrus, Stratus**—Following an NHTSA lawsuit, Chrysler has been ordered to replace the rear seatbelt anchor belts. **1995–96 Breeze, Cirrus, Stratus**—Rusting in the ABS unit could cause the car to jerk to one side when stopping. **1995–97**—Dealers may replace, free of charge, prematurely corroded ball joint components. **1995–98 Cirrus, Stratus**—Faulty ignition switches, console shifter, and cables may cause the vehicle to roll away or render the ignition-park interlock system inoperative. **1996–97**—The hood could fly up. **1996–97 Breeze with 2.4L engine**—An engine oil leak correction will, hopefully, eliminate fire hazard—Faulty ignition switches, console shifter, and cables may cause the vehicle to roll away or render the ignition-park interlock system inoperative.

Secret Warranties/Service Tips

1995–96—Water leaks into the passenger compartment from behind the door trim panel (see below). Correct the leakage by installing new door panel clips, door watershields, and additional tape to seal the watershield. • Front brake lining wears prematurely (see DSB #05-01-96 on following page).

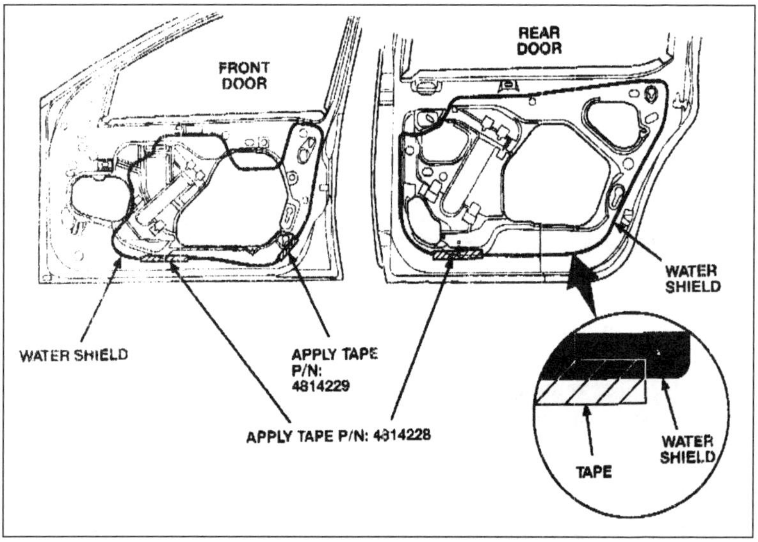

Water leakage from both doors is a common problem.

NO: 05-01-96
GROUP: Brakes
DATE: Jan. 26, 1996
SUBJECT: Premature Front Brake Lining Wear
MODELS: 1995-1996 (JA) Breeze/Cirrus/Stratus
SYMPTOM/CONDITION:
Grinding or squeaking noise from front brakes. This condition can result from premature front pad wear in city commuter type driving or commercial use where the brakes are operated at elevated temperatures for extended periods of time.
DIAGNOSIS:
Inspect the front disc brake lining, as described on page 5-234 of the 1996 Cirrus/Strauts Service Manual (Publication No. 8`-270-6121). If one or more of the front brake linings are excessively worn, perform the Repair Procedure.
PARTS REQUIRED:
1 4874548 Shoe Lining Set, Front Heavy Duty
REPAIR PROCEDURE:
This bulletin involves replacing the front disc brake linings with a heavy duty lining set.
NOTE:
This heavy duty front lining could generate customer vehicle brake applied noise and is not recommended for replacement on vehicles with front brake noise complaints that do not exhibit premature wear. Use PIN 4798885 brake lining for normal replacement.
1. Replace the front brake disc brake lining, P/N 4874548, as outlined in the Service Manual, page 5-24.
POLICY: Reimbursable within the provisions of the warranty.
TIME ALLOWANCE:
Labor Operation No: 05-70-24-90 0.5 hrs.
FAILURE CODE: P8 - New Part

Upgraded linings will reduce wear but increase noise.

1995–97—Troubleshooting tips are available to correct poor AC performance. • A powertrain "bump" when the AC engages is normal, according to Chrysler. • Transmission shudder could be caused by using the wrong transmission fluid. **1997**—Send a rattling CD changer back to the factory. • A low-frequency rumble heard while at highway cruising speed can be silenced by replacing the front hub bearing assemblies. **1997–98**—Excessive cold crank time, start die-out, or weak run-up may be corrected by replacing the powertrain control module (PCM) under warranty, according to DSB #18-18-98. • **All models/years**—Anecdotal reports confirm Chrysler has a 7-year/160,000 km secret warranty covering engine head gasket failures.

Breeze, Cirrus, Stratus Profile

	1995	1996	1997	1998
Cost Price ($)				
Breeze	—	18,200	18,865	19,505
Cirrus	22,115	23,235	24,125	—
LXi	24,555	25,695	26,465	26,465
Stratus	17,895	18,200	18,865	19,505
V6	19,750	20,100	24,060	24,475
Used Values ($)				
Breeze ↑	—	11,500	13,000	15,700
Breeze ↓	—	10,500	12,500	14,500
Cirrus ↑	12,000	14,000	15,500	—
Cirrus ↓	11,000	13,000	14,500	—
LXi ↑	13,000	15,000	17,000	19,000
LXi ↓	12,500	14,000	16,000	18,000
Stratus ↑	10,000	11,500	13,000	15,500
Stratus ↓	9,000	10,500	12,000	14,500

V6 ↑	11,000	12,500	15,000	18,000
V6 ↓	10,000	11,000	14,000	17,000
Extended Warranty	Y	Y	Y	Y
Secret Warranty	Y	Y	Y	Y
Reliability	❷	❷	❷	③
Crash Safety	③	③	③	③
Side Impact	—	—	③	③

FORD

Contour, Mystique

Rating: Average (1996–98); Below Average (1995). Mazda's 626 is a worthwhile alternative to the Contour and Mystique. It's a more stylish, highway-proven sedan with a better-than-average warranty and excellent parts supply, and it's been powered by a 2.0L and 2.5L V6 for the past several years. If you absolutely must have a Contour/Mystique, choose the much-improved (better seating, upgraded mechanicals) 1996 or a later version. For the latest reports on problems look at the Contour website listed in Appendix I. **Maintenance/Repair costs**: Higher than average. Most repairs are dealer dependent. **Parts**: Higher-than-average cost, and sometimes hard to find.

Strengths and weaknesses: These front-drive, mid-sized twin sedans (Contour has a more angular nose and different dashboard) are based on the European-designed Mondeo, which has met with respectable sales after many years on the market. The four-door, five-passenger Contour sells for a bit less than its practically identical Mercury counterpart.

The main advantages of the Contour and Mystique are exceptional handling and a powerful, limited-maintenance V6 engine. Their drawbacks are an unacceptably high base price; cramped rear seating; a wimpy, noisy 4-banger; and spotty quality control that is most evident with the electrical system, brakes, and poor body assembly.

Both vehicles are set on a wheelbase slightly larger than that of the Taurus and come with two engines and transmissions: a base 16-valve 125-hp 2.0L 4-cylinder and an optional 24-valve 170-hp 2.5L V6. Either engine may be hooked to a standard 5-speed transaxle or an optional 4-speed automatic. A smooth but firm ride and crisp handling are guaranteed by the standard MacPherson-strut front suspension, an anti-roll bar, and fully independent rear suspension. Other interesting standard features include dual airbags, adjustable head restraints, 60/40 split-fold rear seats, rear heater ducts, and a sophisticated air filtration system for the passenger compartment. Four-wheel disc brakes, a sport-tuned suspension, and high performance tires are standard features on the V6.

These cars have had lots of first-year problems. Owners report frequent computer module failures and a long wait for parts. One CompuServe member had the following to say:

> My '95 Mystique ran fine for six months, but has been in the dealer repair shop—still not fixed—for three weeks now. First the "over-lock" froze, so you couldn't shift out of Park—even with a foot on the brake. The dealer said an "electrical short circuit kept making a module fail." They told me the part was part of a recall that had not yet been announced. No sooner had they fixed that when the Overdrive on-off button on the automatic gearshift stopped working. That's still not fixed after 11 working days. First the dealer claimed they couldn't "locate the cause." They sent the car out to a transmission expert who found another "failed module." The current problem is that the replacement part is "much in demand," and they're "trying to locate one."

Dealer service bulletins: **1995**—Fluid leaking from the transaxle pump seal. • Reduced power due to throttle plate icing. • Reduced steering assist during wet weather. • Insufficient AC cooling or excessive clutch end gap. • Delayed transmission engagement and shift errors, or no forward engagement or third gear. • Brake roughness upon application. • PCV tube wear-through. • Excessive vibration upon acceleration. • Noise upon hard acceleration or when turning. • Faulty speedometer. • Stuck fuel-filler door. • Water dripping on floor. **1996**—Stalling or hard starts. • Grinding or clashing noise when shifting into third gear. • CD4E transaxle vent right-hand differential leak. • Fluid seeping from CD4E transaxle vent. • CD4E transaxle makes a whistle noise when in Park. • Harsh ride or rubbing noise from rear of vehicle. • Inoperative fuel-filler door. • Fog/film on windshield/interior glass. • Musty and mildew-type odours. • Water dripping on floor from AC condensation leak. • **2.5L engine**—Squealing noise from water pump. • Stall or hooting noise from engine compartment. **1997**—Air blows out the defroster ducts only. • AC has a musty odour. • Transaxle fluid seepage. • Harsh transmission shifting. • Chirping or squeaking blower motor. • Fog/film on windshield/interior glass. • Front-end accessory drivebelt slippage. • Parking brakes stick or bind. • Poor carpet fit. • Tips on preventing brake vibration. • Wind noise around doors. • Stall, hooting, or moosing noise from engine compartment. • Stall and/or exhaust sulfur smell.

Safety summary/Recalls: NHTSA has opened an official probe into front-suspension coil spring failures on the 1995 Contour, and headlight switch failures on 1996 models. **Recalls**: **1995**—The rear door window will be replaced. • The addition of a ground strap prevents an electrostatic charge from building up and igniting fuel vapours when refueling. • The hardware that attaches the outboard ends of the front seatbelts to the front seat frames may be cracked or fractured. • The passenger airbag may not inflate properly. • The fuel-filler pipe opening reinforcement may leak fuel. Tank replacement includes a new fuel filter and a FREE FILL UP! The last item in the recall procedure involves filling up the tank (at Ford's expense) in order to confirm that there are no leaks at the tank-to-filler-neck seal. **1995–96**—Faulty traction control throttle cables may prevent the engine from returning to idle. **1996–97**—Natural gas–equipped vehicles may explode in a collision. **1996–98**—The floor shift indicator may give an incorrect reading.

Secret Warranties/Service Tips

1995–97—Stall and/or exhaust sulfur smell requires a revised power control module. • Stall, hooting, or moosing noise from engine compartment can be fixed by replacing the air intake duct, idle air resonator, and idle air hose with a revised duct and resonator assembly (#F6RZ-9B659-CA). **1995–98**—Air that blows out the defroster ducts only may signal that the defrost actuator door linkage has become disconnected from the crank. • Transaxle fluid seepage can be corrected by servicing with a remote vent kit or by replacing the main control cover. • Parking brakes that stick or bind need a parking brake cable service kit. • Front-end accessory drivebelt slippage can be corrected by installing an upgraded FEAD belt, steel idler pulley, and splash shield kit.

Contour, Mystique Profile

	1995	1996	1997	1998
Cost Price ($)				
Contour GL	16,895	17,595	18,395	18,495
LX V6	20,995	19,855	20,810	20,910
Used Values ($)				
Contour GL ↑	10,000	11,000	13,000	14,500
Contour GL ↓	9,000	10,000	12,500	13,500
LX V6 ↑	11,500	12,500	14,500	16,000
LX V6 ↓	10,000	11,000	13,500	15,500
Extended Warranty	Y	Y	Y	Y
Secret Warranty	Y	Y	Y	Y
Reliability	❷	③	③	④
Crash Safety	⑤	⑤	⑤	⑤
Side Impact	—	—	③	③

Sable, SHO, Taurus

Rating: Average (1998); Not Recommended (1986–97). An extended powertrain warranty is a prerequisite for any Sable or Taurus. The SHO has fewer powertrain defects; however, it shares most of the other generic problems common to the Sable and Taurus.

Sables and Tauruses are aging badly. Serious 3.8L engine head gasket failures, automatic transmission failures, and chronic paint/rust problems are the main reasons the rating has been downgraded this year. I've recommended these cars in the past because Chrysler's and GM's defects were more serious and more frequent. Plus, Ford had set up a number of effective "goodwill" programs to compensate owners for powertrain failures once the warranty had expired. Unfortunately, much of that "goodwill" dried up during the past several years, prompting my meeting with Ford officials early this year to re-prime the "goodwill" pump. I am encouraged by Ford's commitment to me that it will be more generous in settling customers' extra-warranty complaints (see details below); however, I'll maintain my cautious ratings of the Sable and Taurus until I'm convinced all Ford owners are being treated fairly—not just *Lemon-Aid* readers, or owners whose names I pass onto the company.

Incidentally, for those owners who wonder how I can stop recommending model years I once recommended, let me be clear: as vehicles age, their ratings always change to reflect new information from owners, service bulletins, etc., relating to durability and the automaker's warranty performance. I warn shoppers of the changes in subsequent issues of *Lemon-Aid* or in updates to my website, *www.lemonaidcars.com.* But that's not enough. Throughout the year, I also lobby automakers to compensate out-of-warranty owners through formal "goodwill" programs or on an individual case-by-case basis. This way, I assist my readers in avoiding a bad purchase and provide an additional means to get compensation if that purchase has already been made. **Maintenance/Repair costs**: Higher than average, but repairs aren't dealer dependent. **Parts**: Average cost (independent suppliers sell for much less) and very easy to find, except for the SHO's Yamaha engine, which is practically indestructible anyhow.

Strengths and weaknesses: Although they lack pickup with the standard 4-cylinder engine, these mid-sized sedans are competent family cars, offering lots of interior room, nice handling, and plenty of safety and convenience features. The best combination of high performance engine and transmission for all driving conditions: the 3.8L V6 hooked to a 4-speed for the family sedan, and the Yamaha powerplant harnessed to a manual gearbox on the high performance SHO.

SHO

The Taurus SHO (Super High Output) sedan, debuting in 1989, carries a Yamaha 24-valve 3.0L V6 with 220 horsepower, a stiff, performance-oriented suspension, and 5-speed manual transmission. As of 1993, a 4-speed automatic transmission became available. In mid-1996, a redesigned SHO debuted with a standard Yamaha 32-valve V8. Unfortunately, the manual transmission was dropped at that time, a move that has turned off most die-hard performance enthusiasts.

The SHO is an impressive high performance car that is apparently better built than regular-production Sable and Taurus versions. In fact, early SHOs escaped Ford's powertrain problems because of their reliance upon Yamaha for their engines and the wide use of a manual gearbox.

Is the SHO a good buy? The answer is yes, but only if you're willing to spend big bucks for your performance thrills. SHOs hold their value well and give impressive performance. On the other hand, they're hard to find (only 10,000 units a year are sold, and only three percent of Taurus buyers opt for a SHO), they're expensive, and engine repairs may become a problem if Ford abandons the model. The SHO's price is likely to remain stable, or even drop somewhat, inasmuch as most buyers feel the vehicle is already overpriced and they are becoming increasingly wary of Ford's reputation for poor quality control.

Reliability in steep decline

Since 1986, Sables and Tauruses (joined by the Lincoln Continental in 1991) have been the object of multiple safety-related recalls, secret warranties, and service adjustments. They've distinguished themselves for being unsafe, unreliable, and costly to maintain and repair (see "Dead Ford" and "Ford Taurus, Sable Automatic Trnasmission Victims" website links in Appendix I). Their deficiencies fall into the following major categories: the powertrain (engines and transmissions), brakes, fuel and air conditioning system, and body.

Automatic transmission failures

Automatic transmissions have always been failure-prone, function erratically, and are slow to shift—an annoying drawback if you need to rock the car out of a snowbank, and fairly dangerous if you need to pull onto a busy roadway. These problems are caused principally by a cracked aluminum forward clutch piston, although dozens of other causes, including major hardware and software components, have been linked to the above failures. Breakdowns usually occur after three years of use, around the 80,000–120,000 km mark, and can cost $2,500 to repair at the dealer, or half that much at an independent garage.

When Ford modified the automatic transmission in 1991 for easier shifting, it also made it less reliable (Chrysler had a similar experience with its A604 automatic transmission). In fact, *Consumer Reports*' annual member survey shows that almost 20 percent of owners of 1991 Tauruses, Sables, and Continentals experienced a serious transmission failure—four times the average for 1991 vehicles. Ford subsequently extended the warranty on these transmissions for up to six years under a special Owner Notification Program (ONP). Although it's too late for a full refund for 1991–92 models, you should strive for a satisfactory pro rata adjustment; if not successful, challenge Ford's refusal before the small claims court. Obviously 1993 and later models should be eligible for 100 percent repair reimbursements.

Why are we so hard on Ford? Two reasons. First, the automaker has routinely denied that it considers the normal life span for an automatic transmission to be at least 6 years/100,000 *miles*, judging by Ford's original 1991 ONP parameters. Second, the company's own internal bulletins shout out the fact that Ford has known about the transmission's premature failures and erratic shifting since 1986! Yet, the company kept using the cheaper, though failure-prone, aluminum forward clutch piston—which has been principally responsible for the above failures—for over a decade.

Let's take a look at what Ford's bulletins disclose, as background and as arguing points with Ford or in your small claims filing:

Detroit, We Have a Problem! (1986–95)

Article No
94-24-7
11/28/94
TRANSAXLE - AXOD, AXOD-E, AX4S - FORWARD/REVERSE ENGAGEMENT CONCERN - REVISED FORWARD CLUTCH PISTON
FORD:
1986-95 TAURUS
1993-95 TAURUS SHO
LINCOLN-MERCURY:
1986-95 SABLE
1988-94 CONTINENTAL
LIGHT TRUCK:
1995 WINDSTAR
ISSUE:
The forward clutch piston may crack on its outside diameter, seal groove or apply wall (bottom of piston). This condition could allow internal clutch leakage resulting in engagement concerns.
STEEL CLUTCH PISTON PART APPLICATIONS
F4DZ-7A262-A F4DZ-7A262-B
3.0L TAURUS
3.0L SABLE
3.8L TAURUS
3.8L SABLE
3.8L CONTINENTAL
3.8L WINDSTAR 3.2L TAURUS SHO
ACTION:
Use the chart for proper application if replacement of the forward clutch piston is necessary.
NOTE:
PREVIOUS TSB ARTICLES 91-5-7, 92-1-4, 92-26A-6 AND 93-9-11 SHOULD NOT BE USED DUE TO REVISED PARTS LISTED IN THIS ISSUE.
Use a magnet to verify the forward clutch piston being installed is the new steel piston. The magnet will adhere to the steel piston and not to the aluminum piston.
Do not use any service stock of aluminum forward clutch pistons. Return stock to parts depot for credit.

PART NUMBER	PART NAME	CLASS
F4DZ-7A262-A	Forward Clutch Piston	A
F4DZ-7A262-B	Forward Clutch Piston (SHO)	C

SUPERCEDES: 91-5-7, 92-1-4, 92-26A-6, 93-9-11

This first bulletin updates others going back several years and lists 1986–95 as the model years affected. Note how Ford clearly blames the aluminum piston as the culprit causing what the automaker understates as "engagement concerns."

Independent Garages Confirm Ford's Mistake (1986–95)

SUBJECT: Forward Piston Change
APPLICATION: Ford
DATE: 1995
Forward Piston Change

- No forward or reverse engagement.
- Delayed forward and/or reverse engagement.
- Shifts out of gear when coming to a stop.

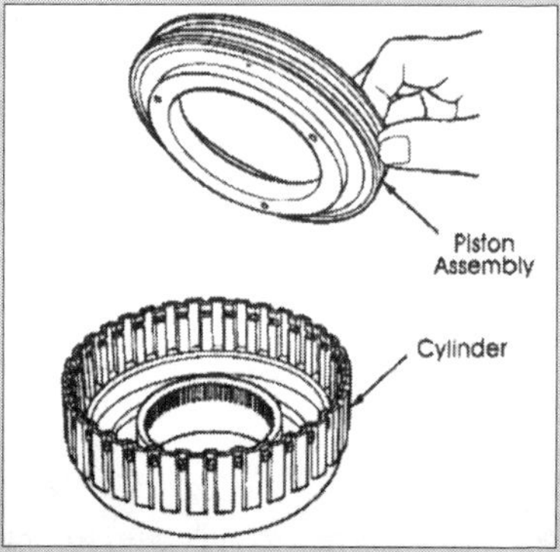

All of these complaints can be attributed to a cracked or broken forward clutch piston. There have been three different versions of the aluminum piston in this location (the original plus two updates). Problems with cracking still persist. A steel version of this part has been released that should prevent this from happening. The Ford part number is F4DZ-7A262-A. The aluminum piston should always be replaced with the steel piston.

Note
The part number for the SHO forward piston is F4DZ-7A262-B.

The Automatic Transmission Rebuilders Association (ATRA) prepared this explanation and illustration to help thousands of independent U.S. and Canadian shops to understand and repair Ford's piston defect. It basically confirms what Ford says, in that the aluminum piston is the most likely cause of transmission breakdowns. Unlike Ford, however, the ATRA bulletin lists three specific symptoms related to the piston failure.

Surprise! Upgraded Steel Piston Fails, Too (1994–98)

Article No.
98-3-7
02/16/98
TRANSAXLE - AX4N - INTERMITTENT NEUTRAL CONDITION - NO FORWARD OR REVERSE MOVEMENT - VEHICLES BUILT THROUGH 2/1/98
FORD:
1994-98 TAURUS
LINCOLN-MERCURY:
1994-98 SABLE
1995-98 CONTINENTAL

ISSUE:
Some vehicles may experience an intermittent Neutral condition after driving and coming to a stop. This may be caused by the bonded seal on the forward clutch piston intermittently not sealing during the 3-2 downshift.
ACTION:
Replace the forward clutch piston with a revised Forward Clutch Piston (F8DZ-7A262-AB). Refer to the following Service Procedure for details.
SERVICE PROCEDURE
Clean and reseal the transaxle completely including replacement of the forward clutch piston with revised forward clutch piston and replace the forward clutch plates if darkened or discolored from heat. Refer to the appropriate Continental Service Manual, Section 07-01, or the appropriate Taurus/Sable Service Manual, Section 07-01B, for details.
Be sure to check end clearance on all three (3) select fit thrust washers (# 16, # 8, 1.02-1.50 mm (0.040-0.059")). Be sure to clean and inspect the main control (pump and valve body) and servos. Prior to returning vehicle to customer recheck fluid level at operating temperature.

PART NUMBER	PART NAME
F5DZJ153-AA	Seal And Gasket kit
F8DZ-7A262-AB	Forward Clutch Piston
F8DZ-7B164-AC	Forward Clutch Plates - Friction (4)
F2DZ-7B442-A	Forward Clutch Plates - Steel (4)

Apparently, the move to a steel piston didn't make Ford's transmissions more reliable or durable, as the above bulletin clearly demonstrates. Hence, my ratings downgrade for Ford's latest models.

3.8L engine failures

Ford's other major powertrain problem is the 3.8L engine's chronic head gasket failures. Symptoms include engine overheating; poor engine performance; and a thin film deposited on the inside of the windshield, thus cutting down night driving visibility. Repairs range $700–$1,500, depending upon what other damage has occurred from overheating. Left untreated, the failure can "cook" your engine, requiring $3,000–$4,000 in repairs.

I've spent the last 2-days at the Ford dealership. The president of the dealership finally agreed to cover the FULL cost of the head-gasket repair (see my email for details) on my '93 Taurus. They also gave me a courtesy car for the time my car's going to be in the shop (sometime early next week). I had been over the warranty (their bulletin stated the warranty ended Dec. 31,98) by 7-days!!

Regards,
Ernst

p.s: Personally, I want to thank-you for your interest in my situation and for "taking" up my case!!

There have been hundreds of cases reported to me of head gasket failures after three to five years or 60,000 km of use. Although these engine repairs are covered by Ford's 98M01 "goodwill" engine warranty up to five years or 100,000 km (see page 75 for Ford's ONP letter), many Sable, Taurus, and Windstar owners have had their claims rejected because their vehicles fell outside of the 1994–95 limit set out in the above warranty extension. This makes no sense whatsoever, especially since Ford's own bulletin below traces the problem back to the 1993 models.

3.8L V6 Engines Blow Their Tops (1993–95)

Article No.
98-4-9
03/02/98
COOLING SYSTEM - OVERHEATING AND/OR LOSS OF COOLANT - 3.8L VEHICLES
FORD:
1988-96 TAURUS
LINCOLN-MERCURY:
1988-94 CONTINENTAL
1988-95 SABLE
LIGHT TRUCK:
1996 WINDSTAR
ISSUE:
Coolant may leak from the head gaskets and/or the vehicle may overheat. There may also be concerns of reduced heater output due to low coolant levels. This may be caused by insufficient sealing of the head gaskets.
ACTION:
Replace the head gaskets and head bolts. The revised head gaskets and bolts provide improved sealing capability and higher clamping force between the cylinder head and block. Refer to the following Service Procedure for details.

PART NUMBER	PART NAME
F5PZ-6051-AA	Head Gasket And Bolt Kit (One Side)

OTHER APPLICABLE ARTICLES: 91-1-8, 94-10-10
WARRANTY STATUS: Eligible Under The Provisions Of Bumper To Bumper Warranty Coverage

OPERATION	DESCRIPTION	TIME
980409A	Verify Coolant Loss	0.3 Hr.
980409B	Replace Cylinder Head Gaskets - Taurus/Sable	8.0 Hrs.
980409C	Replace Cylinder Head Gaskets - Windstar	6.8 Hrs.
980409D	Replace Cylinder Head Gaskets - Continental	8.0 Hrs.

DEALER CODING

BASIC PART NO.	CONDITION CODE
6051	68

OASIS CODES: 208000, 402000, 490000, 499000

Interestingly, the above head gasket failures have also been reported with 3.8L V6 engines equipping Ford's Mustang.

Paint delamination

Over the past decade, there have been frequent complaints of paint delamination, peeling, and premature rusting. One owner of a 1990 Sable found pinpoint rust spots during the first year of ownership and had the entire car repainted at Ford's expense. Unfortunately, the problem didn't go away. The owner reports the following:

> I have recently noticed continued paint defects causing the car to rust prematurely, specifically under the front edge of the hood....I maintain that the sealer and paint were improperly applied when the car was manufactured and that this is a defect that the Ford Motor Company should correct.

That *Lemon-Aid* reader is right, and that's why Ford is the target of multiple class action paint lawsuits in the States. The automaker is settling most Canadian small claims court cases, although class actions may still be imminent north of the border.

Article No. 93-8-4
April 14, 1993
PAINT - EXTERIOR CLEARCOAT "MICROCHECKING," HAZING OR PEELING

FORD:
1983–93 THUNDERBIRD
1984–93 TEMPO
1985–93 ESCORT
1986 LTD
1986–93 TAURUS
1989–93 CROWN VICTORIA, MUSTANG, PROBE

LIGHT TRUCK:
1983–93 RANGER
1986–93 AEROSTAR
1987–90 BRONCO II
1988–93 F SUPER DUTY, F-47
1991–93 ECONOLINE, EXPLORER
1992–93 F-150-350 SERIES
1993 VILLAGER

LINCOLN-MERCURY:
1983–93 COUGAR
1984–92 MARK VII
1984–93 TOPAZ
1985–89 TRACER
1986 MARQUIS
1986–93 SABLE
1988–93 CONTINENTAL, TOWN CAR
1989–93 GRAND MARQUIS
1991–93 TRACER
1993 MARK VIII

ISSUE: The clearcoat layer of the basecoat/clearcoat paint system may "microcheck" (crack and erode), turn white, flake or peel off vehicle. This condition is noticeable on the horizontal surfaces only.
ACTION: Inspect the vehicle and if repair is necessary, refer to the following procedure for service details. This is a wet on wet procedure. Sanding is not required after the seal coat is applied.

Paint delamination is a problem that goes back to the 1983 model year, according to Ford's DSB #93-8-4. Other internal documents show the company saved up to $16 (U.S.) on each vehicle where it skipped the primer surfacer.

Ford's warranty performance

In December of 1998, I advised Ford of Canada President Bobbie Gaunt that the automaker's handling of owners' complaints was unsatisfactory and requested a meeting with her officials to extend the warranty on powertrain problems and improve its customer claims handling procedures.

November 9, 1998

Bobbie Gaunt
President and CEO
Ford Motor Company of Canada Limited
The Canadian Road
Oakville, Ontario
L6J 5E4

Dear Ms. Gaunt,

I am very much concerned over the frequent complaints I have received from Ford Taurus, Sable, and Windstar owners relative to the premature failure of their 3.8L V6 engines and chronic automatic transmission malfunctions.

Having monitored these problems over the past several years and given your company the benefit of the doubt in recommending the above-mentioned vehicles in my *Lemon-Aid* guides, I was convinced your customers would be compensated and the vehicles would be fixed through service campaigns.

Unfortunately, this does not appear to be the case. On the contrary, it appears Ford has taken an unusual hard-line approach.

A perusal of Ford service bulletins and owner complaints show expensive automatic transmission breakdowns (often due to a defective aluminum forward piston, used throughout your entire model lineup since 1985) and engine headgasket failures are still quite common. Furthermore, the 98M01 5-year customer service campaign for 3.8L engines, announced last June, following my first airing of this problem, doesn't compensate owners who fall outside of the program's restrictive model-year parameters ('94 and '95 owners, only), even though your own internal service bulletins show the failures date back to 1988!

As a former Member of Parliament responsible for consumer issues, and thirty-year Canadian consumer advocate, I appeal to your sense of fairness and ask that you extend the warranty retroactively to cover the above transmission and engine defects 100%, up to seven years, without a mileage or prior ownership limitation.

There are many precedents at Ford where similar retroactive compensation has been given to owners. Over two decades ago, I first asked your predecessor, Roy Bennett, to compensate owners for thousands of rust-cankered Fords. Following the formation of Rusty Ford Owners Associations in every province, he agreed, complaints subsided, and Ford gained market share.

I am hopeful we can resolve this issue, as well.

My only regret is that many of the Ford staffers I've enjoyed working with in the past, like Larry Johnson in Service and Tony Fredo in Public Relations, have now retired.

What a sense of timing.

Awaiting your earliest, convenient reply,

Phil Edmonston, *Lemon-Aid*
lemonaid@earthlink.net

After receiving the above letter, Ford of Canada President Bobbie Gaunt quickly set into motion a series of meetings between her staff and me that culminated, three months later, in the company accepting previously rejected claims, upgrading its customer assistance efforts, and extending on a case-by-case basis the warranty relating to Sable, Taurus, and Windstar engine and transmission failures.

Most all of my original 75 complaints submitted to Ford have been settled since my original letter was sent. As I see it, Ford is presently paying 50–75 percent on claims that go back to the 1993 models and haven't run up more than 160,000 kilometers. Other claimants who have been rejected a second time are using the small claims courts and Ford's bulletins in order to get more substantial refunds.

> Hi Phil:
>
> Early in the Fall of 98 you helped me with some information regarding a transmission failure in my '94 Ford Taurus wagon. Ford Canada covered half the cost of replacement although the warranty had expired. That still left me with a $1500 bill, which I resented paying on a car with less than 60,000 km. So I sent Ford a letter threatening to get it back through small claims. For several months the small claims forms have sat on my desk but I have never got around to filing a claim. Recently I received a call from Ford Canada asking for further information about my case. They mentioned that they were following up on some cases you had raised. After some back and forth, they have proposed covering an additional 25% of the full cost, which would reduce the damage to me to $750. Sounds to me like a reasonable settlement, especially considering I have delayed court action. Thanks again for your help.
>
> M.S.

Just a word about Ford's customer assistance. In early January 1999, I phoned on behalf of a Ford owner, and I immediately felt I had descended into one of Dante's circles of Hell.

The 1993 Ford Probe owner had called me with the following story. While he was driving, the car's spring suddenly broke and gouged out the front tire, throwing the Probe into a ditch. He knew that some Sables and Tauruses had been recalled for a similar problem and felt Ford should refund him the cost of the tire. His first call was met with the response that he should simply file an insurance claim with his insurer. When he asked to speak with a supervisor, he was given a contact name. When he attempted to speak with her, he was told that she didn't take customer calls. This owner wasn't all that angry before he called Ford—but after the above exchange, he was fit to be tied.

When I called Ford several hours later, I was told that it was true the supervisor never took customer calls, and that the customer who asked for the supervisor's number "NEVER SAID HE ACTUALLY WANTED TO SPEAK WITH HER." I was also informed that customers who have accidents should file an insurance claim, not bother Ford.

Having been advised at an earlier meeting that day with Ford officials that all claims involving accidents or parts failures that affect safety are handled exclusively by Ford's Product Liability section, I requested the customer assistant send the owner's complaint to both the Product Liability VP and the Manager of the Ford

Customer Assistance office. She agreed it would be done through email. Nothing was ever received by either party.

Many of the customer complaints I had submitted earlier to Ford related to how the company lost owner files, didn't read owner letters, wouldn't call them back, and generally mishandled their claims. In my last meeting with the Customer Assistance Manager and the Product Liability VP, they both expressed their displeasure at how some complaints had been handled in the past, and pledged that steps had been taken and other steps would be taken to ensure that customers who write or phone are treated in a polite and efficient manner. Incidentally, Mr. Howell, owner of the '93 Probe, got most of his money back—three months later.

In looking over the past six months, I have seen my Ford complaint emails and letters taper off and I'm getting lots of positive feedback from owners whose claims are being accepted on the first submission and from those who were previously rejected and are now getting refunds of up to 75 percent.

I hope this turnaround continues. However, I am convinced Ford has an uphill battle because it contracts out its customer assistance to independent companies, whereas a company like Chrysler, with its own owner assistance staffers, can act and react more quickly. Nevertheless, I'm monitoring customer feedback for both automakers, and plan to revisit the question in the near future.

In the meantime, anyone seeking assistance should call Ford's toll-free number: 1-800-565-3673 (FORD). I have been assured each owner will have his or her claim thoroughly reviewed. Please let me know how you were treated.

Dear Phil,

First and foremost, please let me extend to you a very grateful thank you for your personal intervention to Ford of Canada on my behalf regarding the problem with my 1993 Ford Taurus. I was pleasantly surprised when you took time out of your busy schedule to personally telephone me (twice) to provide me with advice regarding this situation.

As you know my car had an engine failure due to a defective head gasket. Ford would not provide any compensation, claiming that there was no warranty on a 1993 3.8L engine, only 1993 and 1994 engines were covered. After discussing this briefly with you and being continually and completely stonewalled by the local Ford dealership, I reluctantly authorized my mechanic to proceed with the installation of a used replacement engine as we appeared to be running out of options and urgently required a means of transportation (a copy of the $1800 invoice is attached). Armed with your DSBs on the Ford Taurus (which the local Ford dealership initially refused to acknowledge), I continued my arguments but to no avail. After a month or so, due to apathy or discouragement on my part, I almost gave up hope on obtaining any kind of compensation from Ford.

That is until one day, I received a call from Beth Hannan, who I believe is a customer representative with Ford of Canada. Ms. Hannan explained to me that that she had received a telephone call from Phil Edmonston inquiring about my situation. Ms. Hannan suggested that I discuss the situation further with my local dealership. I proceeded, once again to the local dealership at which coincidently, Ford Canada's Dealer Operations Manager for Customer Services-Ontario, a Mr. Dean Rowely, happened to be on site.

I showed Mr. Rowely the documentation I had with respect to my engine repair/replacement (ie. invoices, DSBs as well as other conspicuous papers, which I referenced either your name and/or *Lemon-Aid*). Upon review of the documentation, Mr. Rowely did ask if I had indeed corresponded with "Phil," to which I replied that I had. It appears that these people quake at the thought of having to deal with the wrath of Phil Edmonston!

After some discussion about the circumstances associated with my vehicle, Mr. Rowely offered to cover 75% of the repair/replacement cost ($1352.40) associated with my Taurus engine. After conferring with you by fax (following which you responded back to me by phone), I accepted Ford's offer.

I've printed many letters in Lemon-Aid *from Ford owners (Chrysler Caravan owners, too) who have gotten refunds not to pat myself on the back or castigate Ford or Chrysler, but rather to encourage readers to fight back when they're asked to pay for a defect that's clearly factory related.*

Other problems

The 4-cylinder engine is a dog that no amount of servicing can change. It's slow, noisy, prone to stalling and surging, and actually consumes more gas than the V6.

The 3.0L 6-cylinder is noted for engine head bolt failures and piston scuffing, and is characterized by hard starting, stalling, excessive engine noise, and poor fuel economy. Transmission cooler lines leak and often lead to the unnecessary repair or replacement of the transmission, as this vacationing owner of a 1993 Taurus was told the following:

> ...
>
> While in Florida a month ago the local Ford dealer plugged a tester in our car and announced that our problem with the transmission could only be fixed by a new one—at an estimated $2,800 US. Another garage checked the colour of the transmission fluid and came to the same conclusion...a new transmission, but the estimate was lower, about $1,800 US. They added oil which temporarily fixed the problem. On our long drive home, towing a boat, we kept the fluid level topped up.
>
> On arrival, we visited our local CTC station in Port Hope, and discovered that a transmission cooling line had been leaking. The two mechanics who listened while I described the problem, immediately guessed what it would likely be. It turns out they have seen a lot these rusted lines on Fords. The fix cost $150, because the radiator had to be removed to get at the line. In fairness to the mechanics in Florida, perhaps this rusting problem is limited to climates where salt is used on roads.
>
> ...

Other things to look out for: blown heater hoses, malfunctioning fuel gauge sending units, and brakes that need constant attention—in front, they're noisy, pulsate excessively, tend to wear out prematurely, require a great deal of pedal effort, and are hard to modulate. Master cylinders need replacing around 100,000 km.

Although these sedans and wagons were redesigned in 1992, quality wasn't improved that much. In fact, only the latest 1996 redesign seems to have tackled some of the cars' quality shortcomings. Owners are keeping their fingers crossed though, because the reduction in quality complaints may simply be caused by the vehicles' low mileage.

1988–95 models continue to have defective ignition modules, oxygen sensors, and fuel pumps, causing rough running, chronic stalling, hard starting, and electrical system short circuits. Other problem areas include the following: defective engine mounts; an automatic transmission that is slow to downshift, hunts for Overdrive, and gives jerky performance; air conditioners that are failure-prone and can cost up to $1,000 to fix; malfunctioning heaters that are slow to warm up and don't direct enough heat to the floor (particularly on the passenger side); a defective heater core that costs big bills to replace (buy from an independent supplier); and noisy, prematurely worn rack-and-pinion steering assemblies. Front suspension components also wear out quickly.

Electrical components, like windshield wipers, fuel pumps, and the rear defroster, interfere with radio reception. The automatic antenna often sticks, electric windows short-circuit, power door locks fail, and the electronic dash gives inaccurate readings. Owners report that electrical short circuits—which illuminate the Check Engine light and cause flickering lights and engine surging—are frequently misdiagnosed. Customers end up paying for the unnecessary replacement of the alternator, voltage regulator, or battery, in addition to unnecessary tune-ups. The speedometer is noisy and often inaccurate in cold weather.

Body/trim items are fragile on all cars (did somebody mention door handles?). Paint adherence is particularly poor on plastic components, weld joints, and the underside—even with mud guards. Owners also report that water leaks into the trunk through the taillight assembly and that 1986–95 versions produce an annoying sound of fuel sloshing when accelerating or stopping.

1996–98 models were radically redesigned with a totally new, more rounded styling that turned off as many buyers as it turned on. Other changes included upgraded engines, new electronic controls for the LX, and a revamped, oval dash panel. Other improvements: better handling and ride quality, more effective soundproofing, and some transmission refinements (beginning with the 1997 models, but ineffective). Despite these changes, however, owners still report serious safety-related deficiencies (see "Safety summary/Recalls") and other performance-related problems like the non-upgraded engines being noisy, slow and hard to start; the automatic transmission shifting erratically, or not at all; engine head gaskets still failing; and an assortment of rattles, buzzes, whines, and moans (see following 1998 model bulletin summary).

DSB SUMMARY
$15 SUMMARY, $5 EACH BULLETIN
$3 FAX CHARGE ONLY ON BULLETIN ORDERS
1-954-563-2448
2805 E. OAKLAND PARK BLVD., SUITE 211,
FT. LAUDERDALE, FLA. 33306

1.	JUN-98	A/C-FILTERING REFRIGERANT AFTER COMPRESSOR REPLACEMENT
2.	JUN-98	A/C-IDENTIFYING NON-FORD APPROVED REFRIGERANTS
3.	MAY-98	A/C-NO TEMPERATURE CONTROL
4.	MAR-98	A/T, BLUE SERVICE TAGGED TRANSAXLES, AX4S
5.	JUL-98	A/T-FLUSHING OIL COOLER AND LINES-SERVICE TIP
6.	MAY-98	A/T-NO 4TH GEAR AND TROUBLE CODES STORED, AX4N
7.	MAR-98	A/T-TORQUE CONVERTER CLUTCH NOT ENGAGING/CODES STORED
8.	OCT-98	ADHESIVE BONDING NOT RECOMMENDED FOR BODY PANEL REPAIR
9.	APR-98	AUTOMATIC TRANSMISSION FLUID USAGE CHARTS
10.	FEB-98	AX4S/AX4N NEW TRANSAXLE FLUID
11.	SEP-98	BODY-SOLVENTS REMOVE SAFETY CERTIFICATION LABEL PRINT
12.	MAR-98	BRAKE HUB-MOUNT ROTOR MACHINING EQUIPMENT
13.	SEP-98	BRAKES-FRONT DISC BRAKE ROTOR PROTECTOR REMOVAL
14.	AUG-98	BUMPER-REMOVAL OF TRANSEAL PROTECTIVE COATING
15.	OCT-98	BUZZ/RATTLE NOISE-LOOSE CATALYST OR MUFFLER HEAT SHIELD
16.	MAR-98	COOLING SYSTEM CONCERNS, 3.0L GASOLINE/ETHANOL
17.	AUG-98	EVAPORATIVE EMISSION CANISTER COMPONENTS AVAILABILITY
18.	MAY-98	FUEL INJECTORS, NON-WARRANTY REIMBURSEMENT TESTING
19.	JAN-98	FUEL PUMP REPLACEMENT PROCEDURE
20.	OCT-98	HARD START/LONG CRANK TIME, 3.0L, 3.4L
21.	FEB-98	INTERMITTENT NEUTRAL CONDITION, NO FORWARD/REVERSE, AX4N
22.	SEP-98	LOCKS-REPLACE/REPAIR LOCK COMPONENTS-INDIVIDUAL LOCKS
23.	JUL-98	MOANING NOISE FROM POWER STEERING DURING IDLE, 3.0L, 4V
24.	SEP-98	NO START/NO CRANK-BATTERY DISCHARGED/CURRENT DRAIN
25.	APR-98	OIL LEAKS-NEW SILICONE GASKET AND SEALANT
26.	MAR-98	PAINT ANTI-CHIP STONE ABRASION UNDERCOAT SERVICE TIP
27.	SEP-98	PAINT MATERIALS FOR AFTER WARRANTY AND WARRANTY REPAIRS
28.	SEP-98	PASSIVE ANTI-THEFT SYSTEM (PATS)-DIAGNOSTIC SERVICE TIPS
29.	DEC-97	PROPER DOOR PANEL REMOVAL PROCEDURE-SERVICE TIP
30.	DEC-97	RATTLING NOISE DURING ACCELERATION, 3.0L
31.	FEB-98	REPAIRS TO VINYL COVERED SURFACES-SERVICE TIP
32.	SEP-98	SEAT-LEATHER CLEANING PROCEDURE-SERVICE TIP
33.	SEP-98	STEERING WHEEL NIBBLE FELT WHEN DRIVING AT 45 AND 75 MPH
34.	FEB-98	SURFACE DEFECT REMOVAL WITHOUT REPAINTING-SERVICE TIP
35.	MAR-98	TIRE COMPANY TELEPHONE NUMBERS - CUSTOMER ASSISTANCE
36.	DEC-97	TROUBLE CODES P0442, P0455-NO DRIVEABILITY CONCERNS
37.	APR-98	WHINING/BUZZING NOISE THROUGH RADIO SPEAKER
38.	SEP-98	WINDNOISE AROUND SIDE DOORS-SERVICE TIP
39.	JUN-98	WIRE HARNESS TERMINAL REPAIR AND WIRE SPLICE KIT

Dealer service bulletins: **1992–95**—Hesitation, no-start, reduced power, stalling, no-crank due to solenoid corrosion.

Article No. 94-25-3
12/26/94
NO CRANK — POSSIBLE CORROSION AT STARTER SOLENOID — VEHICLES WITH 3.0L AND 3.8L ENGINES
FORD:
1992—95 TAURUS, THUNDERBIRD
LINCOLN-MERCURY:
1992–94 CONTINENTAL
199—95 COUGAR, SABLE
LIGHT TRUCK:
1996 WINDSTAR
ISSUE: Some vehicles may exhibit a "No-Crank" condition due to corrosion at the starter solenoid connector.
ACTION: Replace the corroded connector with a service wiring and connector assembly. Applying di-electric grease, along with replacing the wiring and service connector, to the terminal connector will aid in corrosion protection.

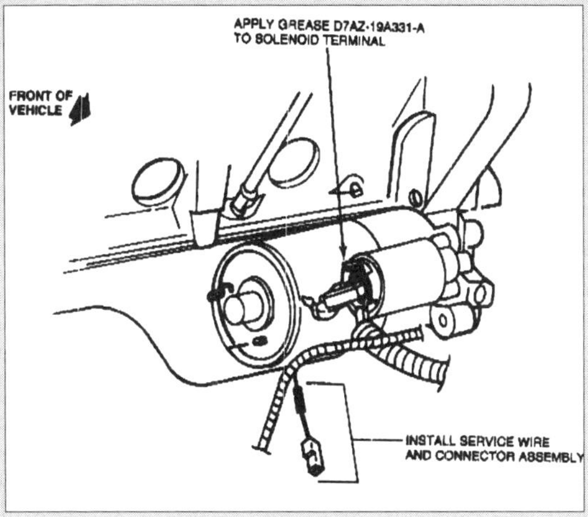

This is one inexpensive-to-repair cause of no-starts on 1992–96 cars and minivans equipped with 3.0L and 3.8L engines. Check it out before authorizing other costly repairs.

1993—Poorly performing air conditioning systems caused by a slipping clutch at high ambient temperatures. • Growling AC FX-15 compressor. • Fuel odours in the passenger compartment. • Noisy power-steering units. • Inoperative power door locks. • Under-hood squeaks, chirps, and knocks; wind noise coming from front door windows. • Intermittent long cranks or no-starts. • A service engine light that has a mind of its own and may go on for no apparent reason. **1994**—A faulty 3.8L engine rocker arm assembly may be the cause of squeaking, chirping, and knocking noises. • A rough idle, hesitation, and excessive fuel consumption. • Defective fuel pumps often produce extraneous noise in radio speakers. • Faulty electric rear window defrosters. • Inadequate AC operation caused by a faulty cold engine lockout switch and hose assembly. **1995**—Delayed transmission engagement and shift errors, harsh shifts, no

3–4 shift, erratic shifts, Forward/Reverse gear malfunctions, and transaxle click when in Reverse. • Insufficient AC cooling or excessive clutch end gap; AC compressor has moans, chirps, or squeaks coming from the blower motor at low speeds. • Fuel pump buzz/whine heard through the radio speaker. • Faulty temperature gauge. • No-crank caused by a corroded starter solenoid (carried over several model years). • Brake roughness upon application and a clacking/thumping noise when braking. • Premature inner-edge wear on the rear tires. • A fuel tank sloshing noise. **1996**—No-start or stall. • Stall or hard start after one- to four-hour soak • Cold engine hesitation/stumble. • Harsh automatic transmission shifting. • Click from transmission when going into Reverse. • Transaxle driveline noises. • Case breakage at rear planet support. • Grunt or groan noise during steering wheel return. • Reduction in power-steering assist. • Door hinge correction. • Fog/film on windshield/interior glass. • Musty and mildewy odours. • Dead battery diagnosis. • Blower motor noise. • Acceleration or deceleration clunk noise. • Wind noise at A-pillar at highway speeds. • A-pillar creaking noise. • Front suspension creak/groan. • Hard starting, long crank, stalling. • Rear headliner sag. • Door handle malfunctions. • High idle, surge, stall, harsh transmission engagement. • Loose catalyst or muffler heat shields. • Click noise when going into Reverse. • AM static. • Warm weather stalling. • Thump, clunk, or chuckle noise from front end. • Troubleshooting driveline noises. • Water leaking onto passenger compartment floor. • Excessive wind noise. **1997**—Harsh automatic shifting. • Chronic dead battery. • Excessive blower motor noise. • Acceleration or deceleration clunk. • Front suspension clunk. • A-pillar creaking noise. • Front end accessory drive belt (FEAD) may slip during wet conditions, causing a reduction in steering power assist. • Steering wheel grunt or honk noise. • Loose catalyst or heat shields. • AM band static. • Stall or surging with automatic transmission engagement. • Water leaks onto passenger floor area.

Safety summary/Recalls: **1993**—Front coil springs may fracture causing the suspension to collapse. • Windshield wiper failures. **1995**—Sudden windshield shattering. • AC failures. • Headlight failures. **1995–96**—Fuel pump failures. • ABS failures. • Airbag fails to deploy or is accidentally deployed. • Transmission slips out of Park. • Sudden acceleration. • Stalling. • Engine compartment fires. • Defective door locks. **1996**—Chronic stalling. • Sudden acceleration. • Cruise control won't slow vehicle on slopes. • Left front wheel may separate from car. • Frequent ABS brake failures. • Airbag failed to deploy. • Inadvertent airbag deployment. • Loss of steering when it rains. • Loss of power steering. • Frequent engine head gasket failures. • Transmission fails or shifts erratically. • Dash reflects into the windshield. • Faulty door lock switch. • Defective heating/defrosting system causes excessive windshield fogging. **1997**—Sudden acceleration.

• Chronic stalling. • Transmission jumps from Park to Reverse. • Sudden steering loss. • Wheels fly off. • Frequent ABS brake failures. • Several engine head gasket failures. • Chronic transmission failures. **1998**—Accelerator and brake pedals are too close. • Vehicle won't slow when accelerator pedal is released. • Faulty cruise control won't slow vehicle down. • Automatic transmission malfunctions. • Chronic brake failures. • Defective rotor and wiring assembly caused ABS failure. • Loss of steering when steering belt pulley and pump failed. • Defective rack and pinion steering spring yoke. • Sudden steering lockup. • Trunk lid fell on owner's head due to defective torsion bar. • Trunk light burned garment in the trunk. • Faulty headlights. • Headlights don't give enough light to the sides. • Daylight running lights flicker due to defective module. • Dashboard reflects upon the windshield, causing reduced visibility. • Heater system failed. • Driver's seatbelt won't retract or lock into position. • Hatchback window suddenly exploded while vehicle was parked. **1999**—Frequent complaints of sudden acceleration or high idle. • ABS brakes locked up when applied and vehicle suddenly accelerated. • Front passenger's seatbelt won't retract or lock into position. • Seatbelt broke. • No-start caused by fuel pump failure. • Faulty remote keyless entry. • Defective windshield wiper motor. • In the morning and evening, the light tan dashboard reflects upon the windshield, causing reduced visibility. **Recalls**: **1986**—Ignition key can be removed when the ignition switch isn't locked. • Faulty cooling fan motor resistor may cause the air conditioner to malfunction. • Misrouted battery wire may lead to premature radiator leakage. **1986 wagons**—The right-quarter tinted window was improperly tempered. **1986–87**—Rear windows may break suddenly. • A faulty spring-lock fuel line coupling is a fire hazard. **1986–91**—The front brake rotors may snap as a result of corrosion. **1986–92**—The rear storage compartment is a hazard to children because it can't be opened from the inside. Owners should deactivate the slam-down latching mechanism with a screwdriver and then get a dealer to install a replacement mechanism. **1986–93**—Vehicles may have detached body mounts at the rear corners of the car's subframe. This defect could allow the subframe to drop and make steering difficult. The mounts will be inspected and a reinforcement plate installed with new attaching bolts. **1986–95 Sable, Taurus**—Subframe engine and transmission mounts could fail due to excessive corrosion that could cause the subframe to drop and make steering difficult. **1987**—Lower steering shaft may separate. • A defective rear spindle assembly could separate and cause loss of vehicle control. **1988 3.0L**—The air conditioner compressor shaft seal is faulty. • The power-steering pump pulley may fail causing loss of power steering and other accessories. **1988–89**—A misrouted power seat switch could cause an electrical fire. **1988–90**—Engine mount failure could lead to engine surges, a stuck throttle, or power-steering hose failures. **1989–90**—End release seatbelt buckles may not latch properly. **1991–95**—Cruise control units and throttle control cable are faulty. **1992**—The inner tie-rod may collapse suddenly. The son of a West Coast owner of a 1992 Taurus relates this incident:

> The right inner tie-rod, a piece of the suspension critical to the steering and thus safety of my 1992 Taurus, broke while my father was attempting to make a right turn from a stop sign. The car lost all steering control and the front wheels were seized. Fortunately, the car was barely moving, and no collision occurred...I can tolerate a radio or AC failure on a "medium aged" car but critical safety components should have a longer service life designed into them. I hope you can inform all Taurus/Sable owners of the inherent dangers lurking in their steering system. And for those cars still under warranty, specify that the part be thoroughly inspected by first removing the rubber protecting boot....

1992 wagons—Children could lock themselves in the footwell area or storage area. • The liftgate could open while the vehicle is in motion. **1992–95**—The engine cooling fan may freeze and cause the cooling fan motor to overheat, in turn causing wiring damage and sparking a fire (3.0L and 3.8L engines only). • The throttle can stick and not return to idle if water enters the throttle cable area and freezes (3.8L engines only). **1993**—The following is not a safety problem in accordance with U.S. Federal Regulation 573 ; however, it is deemed a safety improvement campaign by the NHTSA. Ford is providing an extended warranty through the year 2001 for replacement of front springs due to fracture. The front coil springs can fracture as a result of corrosion in combination with small cracks in the springs. The front tire could deflate due to a broken front coil spring contacting the tire, increasing the risk of a vehicle crash. Dealers will install a spring catcher bracket which will prevent a fractured spring from contacting a tire. Ford has decided to provide an extended warranty through the year 2001 for replacement of front springs due to fracture. Owners can contact Ford at 1-800-392-3673. **1993 ABS**—Rear-drive controllers were installed in error. **1995**—Brake master cylinder may be defective. **1996**—Although apparently in Park, the vehicle can roll away. • The brake fluid indicator may malfunction. • Fuel may leak from a faulty fuel pressure regulator. • The brake system fluid level indicator lamp switch may be faulty. **1996 SHO**—Faulty fuel pressure regulator could cause a fire. **1996–97 Sable, Taurus**—The transmission may not engage; PRNDL may give a false reading, allowing vehicle to move although it appears to be in Park. • Defective fuel rail may deliver fuel to the injectors at more than 43 psi; it may cause chronic stalling under low-speed deceleration or acceleration. • Transmission fluid leakage could cause a fire. **1996–98**—Dealers will replace the dash insulator retainer clip so it doesn't interfere with the accelerator cable or pedal. **1998**—On vehicles equipped with manual seat tracks, the front seatbelt buckle attaching stud may be defective. **1998–99**

Sable, Taurus with manual seats—Dealers will replace the seatbelt buckle mounting bracket and stud assembly.

Secret Warranties/Service Tips

1986–89—If the accessories frequently cut out, install a new ignition switch wire harness. **1986–90**—Extended or no 3–4 shift may require a reassembled direct clutch piston and spring retainer. • No shifts, harsh shifts, or extended shifts may be due to faulty oil pump body and valve body check balls. • AC evaporator water leaks onto the carpet require a new core and seal assembly (#E9DZ-19860-A). • Poor AM radio reception may be caused by interference from the heated windshield system. **1986–91**—Poor forward shifting may require a new clutch piston. • Engine knocking at idle may require the installation of a new, thicker thrust plate to reduce camshaft end play. **1986–92**—A buzz or rattle from the exhaust system may be caused by a loose heat shield catalyst. **1986–94**—A squeak or chirp coming from the blower motor can be stopped by installing an upgraded blower motor with improved brush-to-commutator friction. • A rear suspension clunk or rattle when a wagon goes over a bump may be caused by a loose rear tension strut. • A speaker whine or buzz caused by the fuel pump can be stopped by installing an electronic noise RFI filter. **1986–95**—A cracked forward clutch piston may cause Forward/Reverse problems. Install the improved clutch piston and ask Ford to cover part of the cost inasmuch as their bulletins confirm it's a design defect. • Premature wear on the inner edges of the rear tires calls for the installation of Ford's rear suspension adjustable camber kit. **1988–89**—Harsh downshifts may require a new 1990 level pump and main control assembly. **1988–92**—Cold hesitation when accelerating, rough idle, long crank times, and stalling may all signal the need to clean out excessive intake valve deposits. **1989–93**—A persistent fuel odour in the interior when the AC is running signals the need to install a new auxiliary vapour-tube service kit and relocate the vapour tube near the rear bumper. **1990–93**—Noise coming from the power-steering pump may be caused by air in the system; purge the system. **1991–93**—Growling from the FX-15 AC compressor can be eliminated by installing a new compressor rubber damped disc and hub assembly. **1991–95**—A sloshing noise from the fuel tank when accelerating or stopping requires the installation of an upgraded tank. Cost may be covered under the emissions warranty. **1992**—A 3.0L engine that stalls or idles roughly after a cold start may require a new EEC IV processor. **1992–95**—A corroded solenoid may be the cause of starter failures. **1993**—Free replacement through the year 2001 of front coil springs that fracture or may fracture due to excessive corrosion (see "Safety summary/Recalls"). **1993–94**—An inoperative AC blower probably needs an improved cold engine lockout switch and hose assembly. • Stalling or hard starts in high ambient temperatures or high altitudes may be due to fuel tank contamination, which causes damage to the fuel pump. Ford paid for a fuel tank flush and a new fuel pump/sender and in-line fuel filter until May 31, 1997, under Service Program 94B48. **1993–97**—If the front end accessory drive belt (FEAD) slips during wet conditions, it can cause a reduction in steering power assist; Ford suggests the belt be replaced. **1994–95**—A thumping or clanking noise heard from the front brakes signals the need to service the front disc brake rotors. **1994–98**—Forward clutch piston returns as the most likely suspect in malfunctioning automatic transmissions. **1996–97**—An acceleration or deceleration clunk is likely caused by the rear lower subframe isolators allowing movement between the mounts and the subframe. • A front suspension clunk may signal premature sway bar wear. • Harsh automatic shifting 1–2 may be caused by a malfunctioning electronic pressure control or the main control valves sticking in the valve body (see following).

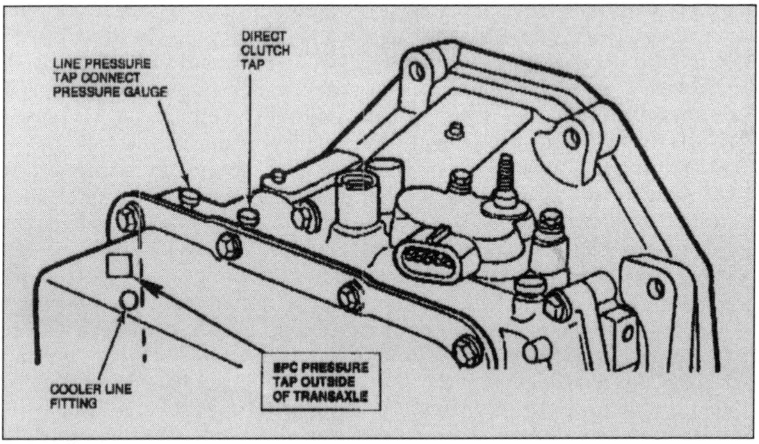

1996–98—Water leaking onto the passenger floor area is likely caused by insufficient sealing of the cabin air filter to the cowl inlet. **1997**—Stalling or surging of 3.0L engines when shifting may signal the need to reprogram the power control module (PCM). **All models/years**—A rotten-egg odour coming from the exhaust probably means that you have a faulty catalytic converter; replacement may be covered under the emissions warranty. • Vehicles equipped with 3.0L engines where piston scuffing is evident (hard starting, stalling, excessive engine noise, poor fuel economy) may be eligible for free engine repairs or replacement on a case-by-case basis. On other engines, Ford has been repairing premature engine head gasket, ring, and valve wear for free when the emissions warranty applies. • Two other components that frequently benefit from Ford "goodwill" warranty extensions are fuel pumps and computer modules. If Ford balks at refunding your money, apply the emissions warranty for a full or partial refund. • Paint delamination, fading, peeling, hazing, and "microchecking" are also candidates for secret warranty compensation (see page 112 for details on claiming a refund).

Sable, SHO, Taurus Profile

	1991	1992	1993	1994	1995	1996	1997	1998
Cost Price ($)								
Sable GS	17,895	18,995	19,470	20,995	21,195	22,595	23,595	24,395
LS Wagon	22,096	23,550	23,570	24,196	24,195	25,496	26,596	25,096
Taurus GL	16,495	17,495	18,195	19,295	20,695	22,195	23,195	—
Taurus LX	20,095	21,195	22,595	23,495	23,895	25,095	26,195	23,295
L Wagon	17,595	18,895	—	—	—	—	—	—
GL Wagon	—	20,250	18,196	19,296	20,636	22,196	23,196	—
SHO	26,695	28,595	30,095	30,095	31,095	32,430	32,695	37,795

Used Values ($)								
Sable GS ↑	4,000	5,000	7,000	8,500	11,000	13,500	15,000	18,000
Sable GS ↓	3,500	4,500	6,000	8,000	10,000	12,500	14,000	17,000
LS Wagon ↑	5,500	6,500	8,500	12,000	13,000	15,000	17,000	19,500
LS Wagon ↓	4,500	5,500	7,500	10,000	12,000	14,000	16,000	19,000
Taurus GL ↑	4,000	5,500	7,000	8,000	11,000	13,000	14,500	17,500
Taurus GL ↓	3,000	4,000	5,500	7,500	10,000	12,500	13,500	16,700
Taurus LX ↑	5,000	6,000	8,000	9,500	12,500	14,500	16,500	18,500
Taurus LX ↓	4,000	5,000	7,000	8,500	12,000	14,000	16,000	17,500
L Wagon ↑	—	4,500	6,000	—	—	—	—	—
L Wagon ↓	—	4,000	5,000	—	—	—	—	—
GL Wagon ↑	—	5,500	7,500	9,500	12,000	13,500	15,500	—
GL Wagon ↓	—	5,000	6,500	9,000	11,000	12,500	14,000	—
SHO ↑	6,000	8,000	8,500	11,000	14,000	17,000	21,500	27,000
SHO ↓	4,500	6,500	7,000	9,000	12,000	14,500	19,000	26,000
Extended Warranty	Y	Y	Y	Y	Y	Y	Y	Y
Secret Warranty	Y	Y	Y	Y	Y	Y	Y	Y
Reliability	❶	❶	❷	❷	❷	❷	❷	③
Air conditioning	❶	❶	❶	❷	❷	③	③	③
Automatic transmission	❶	❶	❷	❷	❷	❷	❷	③
Body integrity	❷	❷	❷	❷	❷	❷	③	③
Braking system	❷	❷	❷	❷	❷	❷	❷	③
Engines	❷	❷	❷	❷	❷	❷	❷	③
Exhaust/Converter	❷	❷	❷	❷	③	③	③	④
Fuel system	❷	❷	❷	❷	❷	❷	③	③
Ignition system	③	③	③	③	③	③	④	④
Rust/Paint	❷	❷	❷	❷	❷	❷	❷	③
Steering	❷	❷	❷	❷	③	③	③	④
Suspension	❷	❷	❷	③	④	⑤	⑤	⑤
Crash Safety	❶	⑤	⑤	❶	❶	④	④	④
Side Impact	—	—	—	—	—	—	③	③

Note: The low frontal crash ratings represent severe leg trauma.

Tempo, Topaz

Rating: Not Recommended (1985–94). Be wary of the all-wheel drive's low price—it's no bargain at any price. The V6 is the powerplant of choice, but it's found only on 1992 and later versions. In 1994, these cars were replaced by the 1995 Contour and Mystique. **Maintenance/Repair costs**: Higher than average. Repairs aren't dealer dependent, but troubleshooting usually costs an arm and a leg. **Parts**: Higher-than-average cost, but can be bought for much less from independent suppliers.

Strengths and weaknesses: This front-drive compact's strong points are its attractive, rounded styling; smooth, quiet V6, which works well with the automatic gearbox; and a roomy interior. On the downside, these cars are under-powered with the base 2.3L 4-cylinder engine, fuel thirsty when coupled with an automatic transmission, and failure-prone. They are riddled with design and manufacturing bugs. The engine, transmission, electrical systems, electronic modules, fuel pump, power steering, suspension components, and cruise control all tend to fail prematurely. The 2.0L Mazda diesel engine is unreliable and doesn't deliver traditional diesel durability.

The 2.3L gas engine isn't much better. Cylinder head gaskets tend to leak and the engine's cooling, fuel, and ignition systems are plagued by a multitude of breakdowns.

For all models, stalling and hard starting are often caused by a malfunctioning catalytic converter. If the car still won't start, mechanics advise owners to tap the solenoid switch behind the battery. The starter motor is weak and the oil pan gasket tends to leak. There are many complaints of prematurely worn front axles and leaking seals. The air conditioning fails frequently and is expensive to repair. Heater noise often signals the need to change the heater motor. Power-steering rack seals deteriorate quickly. Suspension components such as tie-rod ends and strut bearings need replacement almost annually. Shocks, for example, last barely 20,000 km. Front brakes wear out almost as quickly, and rotors are easily damaged. Even though these cars have been dropped, parts are plentiful; but they cost more than the North American average.

Body components are substandard and poorly assembled on all models and all years, with peeling paint and premature rusting being the main offenders. The front door seal tears every time the seatbelt doesn't retract properly, the car's air dam often works loose, and the hood cable-release mechanism tends to jam. Radio reception is mediocre. The 4X4 models are particularly unreliable, with most of the complaints centring on driveline deficiencies.

Safety summary/Recalls: **1991–94**—Engine cooling-fan motor may overheat and catch fire. • Owners report that their cars' chronic stalling places them constantly at risk. One owner of a 1991 Tempo says:

> The car will stall in all weather conditions and in all driving conditions. The engine failure is most apparent when decelerating or stopping after exiting from highways. The car frequently, and dangerously, stalls in the midst of left-hand turns. There is absolutely no warning. The car engine dies and there is occasionally some difficulty re-igniting the engine.

Other owners report the opposite problem with their 1993 Topaz—the car may suddenly accelerate. **Recalls**: **1985**—Rear-suspension control-arm bolts may fail. • 2.3L engine stalling, hesitation, and hard starting is due to a faulty ignition module. **1985–86**—Manual shift lever may accidentally slip into Neutral gear. **1986**—Sudden acceleration on cars with 2.3L engines may be due to a defective electronic control module. **1987**—Again, a faulty 2.3L engine electronic control module may cause high idling, loss of power, and stalling. **1987 Tempo**—Stainless-steel lug nuts may cause stud failure/wheel separation. **1988**—Faulty throttle sensor could cause unintended acceleration. **1988–93**—The ignition switch could experience an internal short circuit, creating the potential for overheating, smoke, and possibly fire in the steering column area of the vehicle. **1989–94**—Cars with 2.3L engines may suffer a frozen PCV system when operated at sustained highway speeds for extended periods in extremely cold weather. • Oil forced out through the dipstick tube or past some other gasket or seal area of the engine could pose a fire hazard. **1990**—Fan motor could overheat and cause engine damage. **1992–94**—Engine cooling-fan motor may overheat and cause a fire.

Secret Warranties/Service Tips

1983–90—Buzzing or humming coming from the fuel pump when the engine is shut off, a low battery, and hard starts or no-starts signal the need to install an upgraded fuel pump relay (#F19Z-9345-A). **1984–88**—2.3L engines with Duraspark II ignition may have starting problems; correct with a new module (#D9UZ-12A199-A). • Hard starting/stalling is caused by defective TFI modules that link the vehicle's distributor to its microprocessor and signal the plugs. • Coolant and oil leaks that plague 2.3L and 2.5L HSC engines can be stopped by installing an improved cylinder head gasket (#E83Z-6051-A). **1985–94**—An exhaust buzz or rattle can be fixed by installing new clamps to secure the heat shield attachments. • The in-tank fuel pump is the likely cause of all that radio static you hear; squelch the noise by installing an electronic noise RFI filter (#F1PZ-18B925-A). **1986–94**—Ford offers an upgraded wiper motor service kit for wiper motor malfunctions. **1987–94**—Loss of Reverse gear in cold weather may be due to defective inner and outer Reverse clutch piston seals. • A rusty catalytic converter inlet pipe flange may be replaced free of charge with a stainless-steel flange under the emissions warranty. **1989–90**—Owner Notification #B90: cooling fans on AC may loosen; replace with upgraded fan (#F03Z-8600-B). **1990**—Loss of AC cooling may be due to a cracked spring lock connector. Put in a new discharge manifold and discharge hose. **1990–93**—Noise heard from the power-steering pump may be caused by air in the system. **1992–93**—An inoperative air conditioner may be due to a faulty cooling fan relay (#E93Z-8Z658-A). **1992–94**—If the idle speed fluctuates excessively in cold weather, you may need to change the powertrain control module (PCM). **All models/years**—If the transmission seems erratic in shifting from second gear to third gear, or downshifting from third to second, a newly designed governor spring will have to be installed. • A rotten-egg odour coming from the exhaust is the result of a malfunctioning catalytic converter. • Three components that frequently benefit from Ford "goodwill" warranty extensions are catalytic converters, fuel pumps, and computer modules. If Ford balks at refunding your money, apply the emissions warranty for a full or partial refund. Regarding paint delamination, fading, peeling, hazing, and "microchecking," see page 112 for details on claiming a refund.

Tempo, Topaz Profile

	1987	1988	1989	1990	1991	1992	1993	1994
Cost Price ($)								
L	9,520	9,794	10,695	11,695	10,695	—	—	—
GL	10,087	10,697	11,536	12,547	11,395	11,295	9,995	10,995
LX	11,814	12,208	12,751	13,497	12,295	12,995	10,775	16,195
4X4	—	—	14,864	15,607	14,195	—	—	—
Used Values ($)								
L ↑	800	1,500	2,000	2,500	3,000	—	—	—
L ↓	500	1,000	1,500	2,000	2,500	—	—	—
GL ↑	1,300	1,500	2,500	3,000	3,500	4,000	4,500	5,500
GL ↓	800	1,300	2,000	2,500	3,000	3,500	4,000	5,000
LX ↑	1,800	2,100	2,700	3,500	4,500	5,500	6,200	7,500
LX ↓	1,500	1,500	1,900	2,400	3,300	4,500	4,900	6,200
4X4 ↑	—	—	2,500	3,000	3,500	—	—	—
4X4 ↓	—	—	2,000	2,500	3,000	—	—	—
Extended Warranty	Y	Y	Y	Y	Y	Y	Y	Y
Secret Warranty	N	N	N	Y	Y	Y	Y	Y
Reliability	❶	❶	❶	❶	❷	❷	❷	❷
Air conditioning	❶	❷	❷	❷	❷	❷	❷	❷
Body integrity	❷	❷	❷	❷	❷	❷	③	③
Braking system	❶	❶	❷	❷	❷	❷	❷	❷
Electrical system	❷	❷	❶	❶	❷	❷	❷	❷
Engines	❶	❶	❶	❶	❷	❷	③	③
Exhaust/Converter	❶	❶	❷	❷	❷	③	③	③
Fuel system	❶	❶	❶	❶	❷	❷	❷	❷
Ignition system	❶	❶	❶	❷	❷	❷	③	③
Manual transmission	④	④	④	④	④	④	⑤	⑤
- automatic	❷	❷	❷	❷	❷	③	③	③
Rust/Paint	❶	❷	❷	❷	❷	③	③	④
Steering	❶	❶	❷	❷	③	③	③	③
Suspension	❶	❶	❶	❶	❷	❷	❷	③
Crash Safety	—	—	—	❶	⑤	⑤	⑤	❶

GENERAL MOTORS

6000, Celebrity, Century, Ciera

Rating: Below Average (1998); Not Recommended (1982–97). The same failure-prone components have been used year after year and have included the discontinued Celebrity, as well. The 1996 Century isn't in the same league as the more expensive, revised 1997 version. The 1997 Century moved over to the W platform that also serves the Chevrolet Lumina, Pontiac Grand Prix, and 1998 Oldsmobile Intrigue. You'll probably find a cheaper 1996 Ciera, but you won't have the important mechanical and body upgrades offered by its 1997 replacement, the Oldsmobile Cutlass, a new mid-sized sedan similar to the new Malibu. Be careful not to confuse the new Cutlass with the Cutlass Supreme, a 10-year-old model that was replaced by the Intrigue, which is equipped like the Century. **Maintenance/Repair costs**: Higher than average, but repairs aren't dealer dependent. **Parts**: Higher-than-average cost (independent suppliers sell for much less), but not hard to find. Nevertheless, don't even think about buying one of these front-drives without a three- to five-year supplementary warranty.

Strengths and weaknesses: The A-body line, long a mainstay in GM's family sedan market, has disappeared. This is good news, because these cars are outclassed by the competition and are in desperate need of high-quality components and fresher styling. Overall quality has improved somewhat since the introduction of these cars in 1982, but with the arrival of the Ford Taurus in 1986 and better-quality Japanese imports, these derivatives of the X-bodies aren't really in the running.

Nevertheless, these cars were consistently popular with fleet buyers and car rental agencies because they were useful as comfortable family sedans and wagons. Handling and other aspects of road performance varied considerably depending on the suspension and powertrain chosen.

1988–96 models are particularly unreliable. The 2.5L 4-cylinder engine suffers from engine-block cracking and a host of other serious defects. The 2.8L V6 engine hasn't been durable either; it suffers from premature camshaft wear and leaky gaskets and seals, especially the intake manifold gasket. The 3-speed automatic transmission is weak and the 4-speed automatic frequently malfunctions. Temperamental and expensive-to-replace fuel systems (including the in-tank fuel pump) afflict all models/years, causing chronic stalling, hard starting, and poor fuel economy (use the emissions warranty to get compensation). Fuel system diagnosis and repair for the 3.0L V6 are difficult, and the electronic controls are often defective. Air conditioners frequently malfunction and the cooling system is prone to leaks.

Prematurely worn power-steering assemblies are particularly commonplace. Brakes are weak and need frequent attention due to premature wear and dangerously rapid corrosion, front brake rotors warp easily, excessive pulsation is common, and rear drums often lock up. Shock absorbers and springs wear out quickly. Rear wheel alignment should be checked often. Electric door locks frequently malfunction. Premature and extensive surface rust—due to poor paint application, delamination, and defective materials—is common for all years. Far more disturbing are scattered reports of severe undercarriage/suspension rusting, possibly making the vehicles unsafe to drive and costing lots of money to correct, as the following owner of a 1990 Century relates:

> Recently I was doing an oil change on my car and I noticed a small divot in the engine cradle (or sub frame). I poked at it and put my finger right through it! I discovered that the cradle was rotted on both sides near the idler arm. The car is only 8 years old and has only 112,000 kms. on it. I have had it into two collision repair places and they both said they have never seen a rotted engine cradle. One man has been in the business 25 years!

The 1997 Century received a complete make-over that includes the following: gobs of room and trunk space (rivaling that of the Taurus, Concorde, Accord, and Camry); sleeker styling; a much quieter interior; and an upgraded, standard ABS system that produces minimal pedal pulsation. Engine noise was also reduced. Other new features include a starter that has an anti-grind feature in case you turn the key with the engine running, upgraded door seals, steering-wheel-mounted radio controls, and additional heating ducts for rear passengers. On the downside, you can get only the 160-hp 3.1L V6 engine; the new Century's speed-dependent power steering is too light and vague; and its suspension and handling are more tuned to comfort than performance.

Dealer service bulletins: 1993 models equipped with 3.3L V6 engines—Converter seal leaks on 4T60E automatic transmissions. • Loss of power. • Unexpected downshifts. • Defective cruise control. • Stalling when decelerating. • Extended cranking. • Faulty Reverse gear on 3T40 automatic transmissions. • Additionally, the Century may experience brake drag and loss of power. **1994**—Fuel pumps may fail prematurely. • A front-end squeak may require the replacement of the exhaust manifold pipe springs with dampers. • Excessive noise or vibrations can be caused by defective rear transmission mounts. • Engine wiring harnesses could melt. • Exhaust moan. • Temperature gauges could give false readings. **1995**—The cruise control fails to engage. • Oil leaks at the rocker cover. • Grinding/growling when in Park on an incline. • Popping, banging, and rattling upon hard acceleration. • Engine tick/rattle upon cold start-up. • Cold-start stall, hesitation, and sag; cold rough idle. • Cold-start rattle with the 4T60E automatic transmission. • Brake vibration and/or pedal pulsation. • Low voltage reading or dim lights at idle. • Exhaust manifold seal squeaks and exhaust boom/moan at idle. • Loose windshield garnish molding. • Wind rush at top of doors. • Poor paint application and rust spots. **1996**—Second-gear starts; poor 1–3 shifting. • Vibration at high speeds. • Steering column noise. • Air conditioning odours. • Radio frequency interference diagnosis. **1997 Century**—Air temperature from HVAC outlets doesn't change. • Rear brake clicking or squealing. • Door glass creaking noise. • Insufficient heater performance on passenger side floor area. • Intermittent Neutral/loss of Drive at highway speeds. • Loose airbag housing door on passenger side dash. • Low voltage reading/dim lights at idle. • Pop, groan, or moan from rear of vehicle. • Remote keyless entry range is shorter than expected. • Slow window operation. • Front brake squealing, grinding noise. • Wiper arms/blades park at incorrect positions.

Safety summary/Recalls: **1996**—Vehicle suddenly accelerated on its own. • Cruise control speed increases upon descending a hill. • Sudden brake loss. • Chronic stalling. • Oxygen sensor failures believed to be cause of stalling problems. • Steering radius is too large, and steering response is sluggish. • Dash reflection on windshield causes poor visibility. • Cannot read clock in daylight. • Airbag assembly on steering wheel blocks view of instrument panel. • Back windows often shatter. • Frequent battery failures. • Fuel pump failures. • Transmission failures; gear shift lever fell off in driver's hand. • Front door power motors failed. **1997**—Sudden brake loss. • Car moved forward when put into Reverse. • Battery exploded twice. • Poor design of magnetic variable steering results in difficult handling with vehicle swaying at 60 km/h and struts contributing to instability. • Sway bar links failure. • Horn buttons difficult to access and depress due to their small size; must take eyes off the road. • Fuel tank warning system activates prematurely. • Dash reflection on windshield causes poor visibility. • Poor headlight design makes for poor visibility, as well. • Defroster system button breaks easily. • Windshield wipers fail frequently. • Defective airbag cover. • Power door lock failures. • Driver's seatbelt won't fully retract. **1998**—Airbags failed to deploy. • Idle surge after releasing brake due to faulty oxygen sensor. • Leaking lower intake manifold. • Engine oil pan leakage. • Chronic stalling. • Transmission shifts erratically. • Transmission hard to put into Reverse; faulty gear shift lever. • Sudden loss of electrical power. • Climate control switch failures. • Dash reflection on windshield causes poor visibility. • Excessive brake vibrations. • Headlights provide poor visibility. • Headlight switch failure. • Windshield wiper arm failures. • Water leaks into trunk. • Drivers report that the head restraints don't stay up. • Horn buttons hard to access in an emergency because of the airbag located directly under the horn and the radio control on the steering wheel. • Horn blows on its own when car is not running. **Recalls**: **1983–87**—The fuel system on the four-cylinder engine may leak fuel, creating a fire hazard. • Some wagons were recalled due to poor braking performance. **1985**—Throttle on V6 engines may stick open, causing sudden acceleration. • Misrouted clutch cable could cause fluid leakage from the brake master cylinder. **1986**—Defective headlight switch can cause headlights to flicker or fail. **1988**—GM will repair a leaking fuel feed hose on vehicles with 2.8L engines. • Dealer will inspect and repair the front suspension. • The lower arm bracket could develop cracks and cause the disengagement of the lower control arm, resulting in steering loss. **1989–90 Celebrity, 1989–91 6000, 1989–96 Ciera**—Rear seat anchor bolts don't meet federal load standards. **1990–91**—A short circuit in the six-way power seat or power recliner could set the seat on fire. **1992**—Automatic 4-speed transmissions were recalled because they performed poorly or remained in Reverse while the indicator showed Neutral. • Wagons may have erratically operating interior lights and suffer a sudden tailgate opening. **1993**—Dealers will install a redesigned intake manifold gasket at no charge on vehicles with rough-running 2.2L engines. • Right front brake hoses are defective. **1986 Century**—Cars with 2.8L engines could suffer an under-hood fire due to misrouted wiring. **1989–96**—Rear seat anchor bolts don't meet federal load standards. **1994 Century**—Sudden acceleration may occur when the 3.1L primary accelerator spring binds. • Water leakage into the power door lock may cause a short-circuit fire.

Secret Warranties/Service Tips

1984–90—Frequent loss of Drive with 440-T4 transmissions is likely caused by a maladjusted 1–2 band stop unit. **1986–90**—Vehicles equipped with the 3T40 automatic transmission may experience slippage in manual Low or Reverse gears. Install service package #8628222, which includes a Low/Reverse clutch release spring (#8664961) and a clutch retainer and snap ring (#656/657). **1987–88**—Poor engine performance troubleshooting shortcuts are detailed in DSB #88-T-47B. **1987–90**—A rattle or buzz from the instrument panel may require a new, upgraded brake booster check valve (#18012017). **1989–93**—Vehicles equipped with a 3300 or 3800 engine that experience stalling upon deceleration or hard starts may need a new air control motor. **1990**—Door lock rods that fall off need an upgraded inside handle to lock the rod. **1994**—Loss of Drive or erratic shifts may be caused by an intermittent short to ground on the A or B shift solenoid or an electrical short circuit in the transaxle. • A front-end clunking noise when driving over rough roads may require the repositioning of the diagonal radiator support braces. **1994–95**—DSB #43-81-29 troubleshoots cruise controls that fail to engage. **1995–97**—Intermittent Neutral/loss of Drive at highway speeds can be fixed by replacing the control valve body assembly. **1997**—Rear brake clicking or squealing may be caused by a maladjusted park brake cable. • Insufficient heater performance on the passenger-side floor area can be fixed by installing a new I/P insulator panel and bracket. **1991–92 Ciera**—Uncomfortable front seat shoulder belts will be replaced at no charge. **All models/years**—A rotten-egg odour coming from the exhaust is probably the result of a malfunctioning catalytic converter; replacement cost may be covered by the emissions warranty. **3.8L V6 engines**—Low oil pressure is likely caused by a failure-prone oil pump. A temporary remedy is to avoid low-viscosity oils and use 10W-40 in the winter and 20W-50 for summer driving. • THM 44C-T4 automatic transaxles with V6 engines are particularly failure prone, due to pinched or kinked vacuum lines that cause oil starvation. • **All models/years**—Paint delamination, peeling, or fading (see pages 76, 113).

6000, Century, Ciera Profile

	1991	1992	1993	1994	1995	1996	1997	1998
Cost Price ($)								
6000	15,798	—	—	—	—	—	—	
Century	17,498	17,798	17,298	20,398	23,080	23,820	24,545	25,215
Ciera S/SL	16,198	16,398	17,298	20,598	22,960	23,625	—	—
Used Values ($)								
6000 ↑	3,500	—	—	—	—	—	—	—
6000 ↓	3,000	—	—	—	—	—	—	—
Century ↑	4,000	5,000	7,500	9,000	11,000	13,000	17,000	20,000
Century ↓	3,500	4,500	6,500	8,000	10,000	12,000	16,000	18,500
Ciera S/SL ↑	3,500	4,500	6,000	8,500	10,500	12,500	—	—
Ciera S/SL ↓	3,000	4,000	5,500	7,500	9,500	11,500	—	—
Extended Warranty	Y	Y	Y	Y	Y	Y	Y	Y
Secret Warranty	Y	Y	Y	Y	Y	Y	Y	Y
Reliability	❷	❷	❷	❷	❷	❷	③	③
Air conditioning	❷	❷	❷	❷	③	③	③	③
Automatic transmission	❷	❷	❷	❷	❷	❷	❷	③
Body integrity	❶	❶	❶	❶	❶	❷	❷	❷
Braking system	❷	❷	❷	❷	❷	❷	❷	❷
Electrical system	❷	❷	❷	❷	❷	❷	❷	③
Engines	❷	❷	❷	❷	③	③	③	③
Exhaust/Converter	❷	❷	③	③	④	⑤	⑤	⑤
Fuel system	❷	❷	❷	❷	❷	③	③	③
Ignition system	❷	❷	❷	❷	❷	③	④	④
Rust/Paint	❶	❶	❶	❶	❶	❷	❷	❷
Steering	❷	❷	❷	❷	③	③	③	④
Suspension	❷	❷	❷	❷	❷	❷	③	④
Crash Safety								
Century 4d	❷	❷	❶	❷	④	④	—	—
Ciera	—	—	❶	—	—	—	—	—
Side Impact								
Century 4d	—	—	—	—	—	—	—	③

Achieva, Calais, Grand Am, Skylark

Rating: Average (1995–98); Below Average (1988–94). The 1992 Achieva replaced the Calais; except for styling, this Achieva is practically identical to the others. There is little to recommend in these cars. More reliable, reasonably priced family haulers and sporty sedans are available from other domestic automakers, principally Ford. **Maintenance/Repair costs**: Higher than average. Repairs aren't dealer dependent. **Parts**: Higher-than-average cost, but can be bought for much less from independent suppliers.

Strengths and weaknesses: These cars come with a standard 150-hp Quad SOHC engine, a 5-speed transaxle, and ABS. The basic front-wheel drive platform continues to be a refined version of the Sunbird and Cavalier J-body. These cars are too cramped to be family sedans (rear entry/exit can be difficult), too sedate for sporty coupe status, and too ordinary for inclusion in the luxury car ranks.

In their basic form, these cars are unreliable, unspectacular, and provide barely adequate performance. An upgraded and more reliable 3.1L V6 powerplant gives you only five more horses than the base four-banger. There's been a lot of hype about the Quad 4 16-valve engine, available with all models, but little of this translates into benefits for the average driver. A multi-valve motor produces more power than a standard engine, but always at higher rpms and with a fuel penalty and excess engine noise. The Quad 4 is rougher than most multi-valve engines when revved to cruising speed and so does little to encourage drivers to get the maximum power from it.

These cars ride and handle fairly well but share chassis components with the failure-prone J-bodies. This explains why engine, transmission, brake, and electronic problems are similar. Water leaks and body squeaks and rattles are so abundant that GM has published a six-page troubleshooting DSB that pinpoints the noises and lists fixes (see "Secret Warranties/Service Tips").

The 2.5L 4-cylinder engine doesn't provide much power and has a poor reliability record. Avoid the Quad 4 and 3.0L V6 engines with SFI (sequential fuel injection) because of their frequent breakdowns and difficult servicing. If you have a blown head gasket or other problems with the Quad 4 engine, keep in mind that GM has a 7-year/100,000 km secret warranty covering its free repair or replacement; ask for a pro rata refund for the repair, even though the warranty coverage may have expired (again, see "Secret Warranties/Service Tips"). Poor engine cooling and fuel system malfunctions are common; diagnosis and repair are more complicated than average, however. The engine computer on V6 models has a high failure rate, and the oil pressure switch often malfunctions. The electrical system is gremlin plagued. Seals in the power-steering rack deteriorate rapidly. Front brake discs need more frequent replacement than average. Paint defects are common; consequently, surface rust may occur sooner than expected.

Dealer service bulletins: By the way, DSB #43-1007A gives an exhaustive review of all the possible causes and remedies for body squeaks and rattles afflicting 1992–95 models. **1993 Achievas and other models equipped with 2.3L engines or 3.1L V6 engines—** They won't continue running after a cold start. • The converter seal leaks on 4T60E automatic transmissions. • Front door windows may be hard to roll up and seal poorly. • Intermittent electrical problems. • Stalling when decelerating. • Hard starting. • Faulty Reverse gear on 3T40 automatic transmissions. • Rear brake squawk. **1994—** V6-equipped models may experience excessive second-gear vibration caused by a defective hub shaft bearing and sleeve assembly. • Loss of engine coolant likely caused by faulty surge tank caps. •

Power steering may lead or pull. • A faulty automatic transmission fluid-level indicator could give false readings. • A poorly running 2.3L Quad 4 engine can often be traced to a moisture-contaminated or corroded ECM connector; correction requires the resealing of the engine harness pass-through. **1995**—Oil leaks at the rocker cover. • Engine tick/rattle upon cold start-up. • Cold-start stall or tip-in hesitation. • Grinding/growling when in Park on an incline. • AC cut-off after an extended idle. • A cold-start rattle with the 4T60E automatic transmission. • Longer-life front brake linings. • Brake vibration and/or pedal pulsation. • Rear wheel brake drag. • Inoperative washer pump. • A low voltage reading or dim lights at idle. • A flickering "Check Oil"/"Check Gauges" light. • Water entering into spoiler (Grand AM). • Poor paint application and rust spots. • Right rear-quarter window wind noise and water leaks. • **1996**—Second-gear starts, poor 1–3 shifting. • Driveability problems, whistling noise, or reduced fuel economy. • Engine overheating/faulty cooling fan. • Excessive engine roll. • Air conditioning odours. • Front seatbelt webbing twists and won't retract. • Fuel-filler door won't open. • Radio frequency interference diagnosis. • Wind noise at front-door outer-belt sealing strip. • Water entering rear compartment at taillight area. • Water leaking into taillight harness. **1997**—Cold-start rattle noise is normal, according to GM, and no fix is planned. • Inoperative power door locks. • Noisy instrument panel. • Intermittent loss of Drive at highway speeds. • No-starts. • Engine popping noises. • Power window malfunctions. • Front suspension squawk noise.

Safety summary/Recalls: **1995**—Fire ignited in the engine compartment while vehicle was on the road. • Sudden acceleration due to weak pedal return spring. • Chronic stalling. • Airbag failed to deploy. • Malfunctioning airbag causes horn to suddenly go off. • Driver's seatbelt failed to restrain driver. • Frequent brake failures and extended stopping distance. • Power-steering fluid leak due to high pressure hose chaffing by the two fuel-injector tabs. • Brake caliper seizure damages pads and rotors. • Transmission jumps out of gear. • Shoulder belt rides across driver's neck. • Driver's side door handle popped out. • Erratic fuel gauge operation. • Headlights suddenly shut off. • Passenger seat not anchored securely. **1996**—Inadvertent airbag deployment. • Airbag failed to deploy. • Sudden acceleration and stalling. • When putting car in Reverse, it suddenly accelerated forward. • Left wheel came off after the stud that holds the wheel unbolted from the wheel. • Oxygen sensor failures cause Check Engine light to come on. • Frequent brake failures. • Excessive brake noise. • Defective master cylinder, drums, pads, and rotors. • Engine, transmission, and AC failures. • Seatbelt sticks into driver's side or will not fasten properly. • Insufficient insulation under steering column allows draft to come in under left lower dash area. • Headlight failure. • Water leaks into the trunk. •

Cracking around the outside edge of all tires. • Passenger power door lock and window lock do not work properly. **1997**—Many reports of vehicle first losing power and then suddenly accelerating. • Accelerator cable snapped, causing pedal to go to the floor. • While vehicle was being driving, the hood suddenly flipped backwards, hitting the windshield. • Windshield reflection obstructs vision. • Premature brake replacements. • Seatbelt improperly fitted. • Design flaw allows wheels to rub against front fender. • Mirrors aren't adjustable enough to see other vehicles, and seatbacks are too high for some drivers to see over. • Driver-side bucket seat isn't anchored properly; rocks from side to side. • Intermittent windshield wiper failures. • Headlights sometimes cut out. **1998**—Airbag failed to deploy. • Premature front brake pad wearout and warped rotors. • Seatbelts tend to twist. **Recalls**: **1985**—A faulty throttle return spring on 2.5L engines could lead to sudden acceleration. **1985–86**—The door pillar could crack. **1986–1990**—Headlights could operate erratically. **1987**—The fuel feed or return hose could leak. **1991**—Power windows could short-circuit and remain in the down position or start an electrical fire. **1994**—Weak fuel tank welds could create a fire hazard in a rear-end collision. **1996**—Front or rear hazard warning lights could be faulty. • The airbag could deploy behind the instrument panel. • A loose steering column bolt could cause loss of steering.

Secret Warranties/Service Tips

1991–95—Front brake linings can be made to last longer by replacing the front brake pads with a new 8100 lining compound (#18022600) **1992–93**—A front-end engine knock troubleshooting chart is found in DSB #306001. **1992–94**—Corrosion of the ECM connectors can lead to a host of driveability problems (DSB #338109A). • DSB #431007 is an excellent troubleshooting guide to finding and correcting squeaks and rattles; it contains six charts that show noise sources and remedies. **1994**—Gear whine with the 4T60E automatic transaxle can be stopped by replacing the final drive and updating the PCM calibration. • Insufficient AC cooling may be due to a leak at the low-charge primary-port seal. • Loss of Drive or erratic shifts may be caused by an intermittent short to ground on the A or B shift solenoid or an electrical short circuit in the transaxle. • A front-end clunking noise when driving over rough roads may require the repositioning of the diagonal radiator support braces. **1990–93 2.3L Quad 4 engine** —A notoriously bad engine. Head gasket leaks are covered under GM's 7-year/160,000 km secret warranty—oops, I mean Special Policy (see following document).

Dear (Buick/Oldsmobile/Pontiac) Owner:

As the owner of a 1992 or 1993 (Buick Skylark/Oldsmobile Achieva/Pontiac Grand Am) equipped with a 2.3L Quad OHC (RPO Code L40, VIN Engine Code 3) engine, your satisfaction with our product is of utmost concern to us. Your vehicle was provided with a new vehicle warranty which covers certain parts of your vehicle for a specified period. These warranties are of considerable value to you if you should experience problems with your vehicle.

Condition: This letter is intended to make you aware that some 1992 and 1993 (Buick Skylark/Oldsmobile Achieva/Pontiac Grand Am) models with 2.3L Quad OHC (RPO Code L40, VIN Engine Code 3) engines may develop a cracking condition of the cylinder head that allows coolant to leak from the cylinder head. Early evidence of this would be a loss of coolant in the coolant reservoir and coolant deposits on the exterior of the cylinder head. These may be combined with the odor of coolant from the engine compartment.

Action: (Buick/Oldsmobile/Pontiac) Division is therefore taking the following action: We are providing owners with special coverage. If the above mentioned condition occurs within seven (7) years of the date the vehicle was originally placed in service or 100,000 miles (160,000 km), whichever occurs first, the condition will be repaired for you at no charge.
This is not a recall campaign. Do not take your vehicle to your (Buick/Oldsmobile/Pontiac) Dealer/Retailer as a result of this letter unless you believe that your vehicle has the condition as described above.

Reimbursement: (Statement for all states except as shown below.)
If you have already paid for some or all of the cost to have the cylinder head replaced and your mileage and in service time was less than seven (7) years and 100,000 miles (160,000 km), you should contact your (Buick/Oldsmobile/Pontiac) Dealer/Retailer Service Department to seek reimbursement. Please provide your retailer with your original paid receipts or invoices verifying the repair, the amount charged, proof of payment, and the date of payment of those charges by November 1, 1996.

Yikes! This is one important free engine fix—just the labour alone is nine hours. Tell your friends before the seven-year limit expires.

1988–89—Power loss in cold weather can be corrected by installing ventilation kit (#12339306). • A whistle or whine coming from the engine is likely caused by a noisy oil pump; replace it with an improved oil pump (#22538689). • The engine wiring harness may have been cut by constant rubbing on the alternator. This will cause a rough-running engine, inoperative air conditioner/alternator, the illumination of the Check Engine light, failure of the ECM module, and excessive white smoke (DSB #88-8-11). **1992–97**—Inoperative power door locks may need an upgraded external bumper on the actuator arm. **1995–97**—Intermittent loss of drive at highway speeds may require the replacement of the control valve body assembly. • Engine popping noises can be silenced by tightening the torque strut mount bolts. **1997**—No-starts may be due to an improperly routed and pinched wire from the generator to the wiring harness. **2.5L engine**—Excessive oil consumption may be caused by one or more damaged intake valve guides. • Hard starting and engine pinging can be fixed by the installation of a new PROM module (#16121217), DSB #88-6E-11. • Frequent engine overheating is caused by a defective thermostat; replace it with an upgraded part (#3059793). **1995–97 Skylark**—Intermittent Neutral/loss of Drive at highway speeds can be fixed by replacing the control valve body assembly. **All models/years**—A rotten-egg odour coming from the exhaust may be the result of a malfunctioning catalytic converter—possibly covered by the emissions warranty. Stand your ground if GM or the dealer claims you must pay. • Paint delamination, peeling, or fading (see pages 76, 113).

Achieva, Calais, Grand Am, Skylark Profile

	1991	1992	1993	1994	1995	1996	1997	1998
Cost Price ($)								
Achieva S	—	15,098	15,298	16,698	18,485	19,925	20,735	21,200
Calais S	14,298	—	—	—	—	—	—	—
Grand Am	13,698	13,698	14,898	15,798	17,365	18,000	19,035	19,610
Skylark	14,598	16,350	15,298	16,398	19,035	20,035	21,220	22,965
Used Values ($)								
Achieva S ↑	—	4,500	5,500	7,500	9,000	12,000	14,000	17,000
Achieva S ↓	—	4,000	4,500	6,500	8,000	11,000	13,000	16,000
Calais S ↑	3,500	—	—	—	—	—	—	—
Calais S ↓	3,000	—	—	—	—	—	—	—
Grand Am ↑	4,500	5,000	6,000	7,500	10,000	11,500	13,500	15,000
Grand Am ↓	3,500	4,500	5,000	6,500	9,500	10,500	12,000	14,000
Skylark ↑	5,500	6,000	7,000	8,500	10,500	12,500	15,000	17,000
Skylark ↓	4,500	5,500	6,500	7,500	10,000	11,000	13,500	16,000
Extended Warranty	Y	Y	Y	Y	Y	Y	Y	Y
Secret Warranty	Y	Y	Y	Y	Y	Y	Y	Y
Reliability	❷	❷	❷	❷	❷	③	④	④
Air conditioning	❷	❷	③	③	④	④	④	④
Body integrity	❷	❷	❷	❷	❷	❷	❷	❷
Braking system	❶	❶	❷	❷	❷	③	③	③
Electrical system	❶	❷	❷	❷	❷	❷	❷	❷
Engines	❷	❷	❷	❷	③	③	③	③
Exhaust/Converter	❶	❷	③	③	③	④	⑤	⑤
Fuel system	❷	❷	❷	❷	❷	③	③	③
Ignition system	❷	❷	③	③	③	③	③	④
Manual transmission	③	③	③	④	⑤	⑤	⑤	⑤
- automatic	❷	❷	❷	❷	❷	❷	❷	④
Rust/Paint	❶	❶	❷	❷	❷	③	③	③
Steering	❷	❷	❷	③	④	⑤	⑤	⑤
Suspension	③	③	③	⑤	⑤	⑤	⑤	⑤
Crash Safety								
Achieva 2d	—	❷	❷	—	④	—	—	—
Achieva 4d	—	—	❶	—	④	—	⑤	—
Calais 4d	⑤	—	—	—	—	—	—	—
Grand Am 2d	⑤	—	—	—	—	—	⑤	—
Grand Am 4d	—	—	❶	—	—	④	—	—
Skylark 4d	—	—	❶	—	—	—	⑤	—
Side Impact								
Achieva 4d	—	—	—	—	—	—	❶	❶
Grand Am 2d	—	—	—	—	—	—	❶	❶
Skylark 4d	—	—	—	—	—	—	❶	❶

Beretta, Corsica, Tempest

Rating: Average buy, if you can adjust to the brakes' poor performance (1993–96); Not Recommended (1987–92). The Corsica was the entry-level model, followed in price by the Beretta. Best bet is an ABS-equipped and V6-powered version available 1992–96. The trio was dropped in 1996. **Maintenance/Repair costs**: Higher than average. Repairs aren't dealer dependent. **Parts**: Higher-than-average cost, but can be bought for much less from independent suppliers.

Strengths and weaknesses: These reasonably priced, roomy compacts came with standard anti-lock brakes as of the 1992 model year. Overall road performance is unimpressive, however, and emergency handling is below par. From a quality standpoint, these compact coupes, sedans, and hatchbacks are almost as unreliable as the Skyhawk, Firenza, J2000, and 2000 they replaced. The anemic, failure-prone, and expensive-to-repair 2.0L 4-cylinder engine with its faulty computer modules is a major disappointment. You need to shift into low gear on modest inclines. Owners report frequent no-starts and stalling with all engine variations, mainly due to temperamental electronic modules, particularly in conjunction with multi-port fuel-injection systems. Fuel system and ignition glitches are legion, and cruise control operation is erratic. Servicing is difficult because of the tight engine compartment. The V6 engine is seriously weakened by air conditioning and the automatic transmission's poor quality control. The lockup on the automatic engages and disengages constantly. The 5-speed manual performs better, but it too is unreliable, being handicapped by long clutch-pedal travel and a tendency to stall at light throttle. Owners report some instances where the clutch has shifted into Reverse rather than first gear. Steering is vague and components are unreliable. Braking is terrible and the front brakes rust and wear out quickly. The standard suspension offers poor ride control on bumpy roads, produces excessive body roll in turns, and is characterized by imprecise handling, especially at highway speeds. The sport suspension option (standard on some models) offers better handling and a firm, more comfortable ride. Shock absorbers often begin leaking before 50,000 km, and the replacement of the suspension struts is sometimes an annual affair. Electrical components often short circuit. Owners report that the windshield wiper motor fails frequently.

1990–96 models also have more than their share of performance problems and factory-related defects. Braking is still scary, even with ABS, and cornering is a white-knuckle affair. The 2.2L engines have a "piston scuffing" problem and the Quad 4 engine's head gaskets fail prematurely. Hard starting, stalling, and engine surging are common, and AC operation is erratic.

Body panels are poorly assembled, paint delaminates and peels away, and rattles are a constant companion. GM announced in the

January 18, 1993, edition of *Automotive News* that Corsica and Beretta paint peeling would be covered for six years under a special extended warranty. This change in GM's policy was first communicated to American dealers on October 16, 1992, in a series of letters and bulletins sent to each division's dealers.

Amazingly, GM weasels out of paying many claims for paint repairs by ignoring the *Automotive News* article or by pretending that the extended warranty never existed. Owners who won't take no for an answer and turn to small claims court usually get their cars repainted for free (see "Secret Warranties/Service Tips").

Safety summary/Recalls: There have been many reports of rear-wheel lockup and sudden brake loss on non-ABS equipped models, and ABS brake performance hasn't been impressive. **Recalls**: **1987–88**—The hood could fly up. • Replace broken front-door hinges with upgraded hinges (#10092242-3). **1988–89**—Front shoulder belt retractors could fail in a collision. **1989**—Dealers will install a front seatbelt latch plate and buckle for free. • The fuel tank could leak. • Dealers will replace both front seat frames if found defective. • The port fuel injector on 2.8L engines will be replaced for free based on the emissions warranty. **1994–95**—The reinforcement panel is missing from the right-side rocker assembly. **1989 Corsica**—A steel wheel fracture could cause wheel separation. **1991**—A loose steering wheel nut may cause steering wheel separation. **1992**—The brakelight switch is faulty. **1992–93**—Dealers will install a redesigned intake manifold gasket on vehicles with rough-running 2.2L engines. **1990 models with 2.3L Quad 4 engines**—Faulty ignition coils that cause engine misfiring may be replaced under an emissions recall campaign. **1990–94**—Vehicles may have a cracked fuel-hose feed, causing a fuel leak.

Secret Warranties/Service Tips

1987–91—Vehicles equipped with the 3T40 automatic transmission may experience slippage in manual Low or Reverse. Install Service Package #8628222; it includes a Low/Reverse clutch release spring (#8664961) and a clutch retainer and snap ring (#656/657). **1988**—Erratic performance of the window motor may call for a replacement brush package (#22094719). • 2.0L engine valve-train noise can be reduced by adjusting or replacing the rocker arms. • Lack of heat in the rear may require upgraded floor heater outlets. **1988–89**—Product Campaign #89C06 provides for the rewiring of a faulty 2.0L engine coolant switch. • 2.0L engines that fail to run, or stall in cold weather, probably need a new PROM. **1988–95**—A binding or popping noise coming from the front door glass when the glass is rolled down means that the regulator arm stabilizer plate is maladjusted. **1989**—Product Campaign #89C16 provides for the replacement of the 5-speed manual transmission. **1989–91**—A power-steering shudder, moan, or vibration signals the need for a "tuned" power-steering return hose and/or high-expansion pressure hose between the steering pump and gear. **1990**—Engine piston scuffing may cause cold engine knock. GM DSB #90-433-6A recommends a partial engine replacement. **1990–91**—No-starts, stalling, and rough running may be caused by a DIS ignition wiring short circuit. •

Poor starting may be caused by the spring in early starter drives compressing too easily. Install an upgraded starter motor drive assembly (#10473700). • Engine overheating or poor AC performance may be due to an inoperative engine cooling fan. **1990–92**—Poor braking may be due to excessive corrosion of the front disc brake caliper bolt bore. **1991–94**—A crunching noise coming from the right side of the dash when turning the steering wheel can be silenced by installing a new steering grommet on the right side. • Water leaking onto the right front carpet from a gap between the air inlet screen and windshield can be stopped by applying a urethane sealing strip. **1992**—Chafing of the engine harness wires can cause hard starting, engine surging, stalling in gear, the display of the "Service Engine Soon" warning, and an inoperative temperature gauge. • Engine hesitation or roughness, particularly at idle, may be corrected by installing a new lower intake manifold gasket (#10103647). **1992–93**—No Reverse or slipping in Reverse can be corrected by installing an upgraded Low/Reverse clutch return spring and spiral retaining ring. **1993–96**—A front suspension squawk noise heard when passing over small bumps can be silenced by using a special GM service kit (see following bulletin).

File In Section: 3 Steering/Suspension
Bulletin No.: 73-33-01
Date: April 1997

Subject:
Squawk Noise Coming From Front Suspension or Engine Compartment Area
(Install Ultra High Molecular Tape)

Models:
1993–97 Buick Skylark
1993–94 Chevrolet Cavalier
1993–96 Chevrolet Beretta, Corsica
1993–97 Oldsmobile Achieva
1993–94 Pontiac Sunbird
1993–97 Pontiac Grand Am

Condition
Some owners may comment on a "squawk" noise coming from the front suspension or engine compartment area. This noise may be more noticeable over small bumps, entering parking lots, and including any irregular road surfaces. This condition may also be more noticeable in cold weather conditions.

Cause
The squawk noise may be caused by the rubber stabilizer bushing material bleeding through the Teflon/Polyester sock (on later models) and coming into contact with the stabilizer shaft. On earlier models, the squawk may be caused from friction when the stabilizer bushing is grabbing and releasing the stabilizer shaft.

Correction
If the above conditions exist, perform the following repair:

1. Remove front stabilizer bushing clamps and bushings. Refer to Section 3C of the Service Manual.
2. Inspect stabilizer bushings for excessive wear, replace if necessary.
3. Use crocus cloth (or equivalent) to sand the stabilizer shaft where the stabilizer bushings contact the stabilizer shaft. Sand all rough corrosion thoroughly.
4. Install UHM (Ultra High Molecular) tape (provided in kit, P/N 22602686) into stabilizer bushing; adhesive side of tape should be installed onto the bushing. Install UHM tape to the opening of the bushing (the slit) as well as the circumference (this will secure the tape in place). Kit provides enough tape for two bushings (one vehicle).

1994—Insufficient AC cooling may be due to a leak at the low-charge primary port seal. • Automatic transaxle gear whine may be eliminated by installing a redesigned transaxle final drive assembly. • A squeaking noise heard when going over bumps, accelerating, or shifting can be stopped by replacing the exhaust manifold pipe seal. • Harsh automatic transmission upshifts can be corrected by installing an upgraded accumulator valve in the control valve body. • Loss of Drive or erratic shifts may be caused by an intermittent short to ground on the A or B shift solenoid or an electrical short circuit in the transaxle. **1994–96**—If the accelerator pedal is difficult to depress or if the accelerator cable separates, install an upgraded accelerator cable and clip. • Chronic engine miss or a loss of power can be caused by a micro-arcing corroded spark plug wire or ignition coil. **1990–91 2.3L Quad 4 engine**—Head gasket leaks were once covered by a secret warranty extension. The first sign of trouble is a loss of power caused by combustion gases mixing with coolant. This is followed by a cloud of steam and coolant loss through the exhaust system. If these warnings are ignored, cylinder bore scoring, a warped cylinder head, and piston seizure will likely result as the engine continues to overheat. **1995–96 Beretta and Corsica**—Intermittent Neutral/loss of Drive at highway speeds can be fixed by replacing the control valve body assembly. **All models/years**—A rotten-egg odour coming from the exhaust is probably caused by a malfunctioning catalytic converter, which may be covered by GM's emissions warranty. • Paint delamination, peeling, or fading (see pages 76, 113).

Beretta, Corsica, Tempest Profile

	1989	1990	1991	1992	1993	1994	1995	1996
Cost Price ($)								
Beretta	13,098	14,099	12,698	12,998	12,999	15,300	16,950	17,275
Beretta GT	14,998	16,498	15,398	14,898	14,998	—	—	—
Corsica LT	12,298	12,298	12,098	12,698	12,998	15,698	17,365	17,685
Tempest	12,798	13,500	12,396	—	—	—	—	—
Used Values ($)								
Beretta ↑	1,700	2,500	3,000	4,000	5,000	6,500	8,000	8,500
Beretta ↓	1,500	2,000	2,500	3,500	4,500	6,000	7,500	9,000
Beretta GT ↑	3,000	3,500	4,000	5,000	—	—	—	—
Beretta GT ↓	2,500	3,000	3,500	4,500	—	—	—	—
Corsica LT ↑	3,000	3,500	4,000	5,000	6,000	7,500	9,000	11,000
Corsica LT ↓	2,500	3,500	4,000	5,000	7,000	8,500	8,500	10,500
Tempest ↑	2,500	3,000	4,000	—	—	—	—	—
Tempest ↓	2,000	2,500	3,000	—	—	—	—	—
Extended Warranty	Y	Y	Y	Y	Y	Y	Y	Y
Secret Warranty	N	N	Y	Y	Y	Y	Y	Y
Reliability	❶	❷	❷	❷	❷	❷	❷	❷
Air conditioning	❷	❷	❷	❷	❷	❷	❷	③
Body integrity	❶	❶	❶	❶	❶	❷	❷	❷
Braking system	❶	❶	❶	❶	❶	❷	❷	❷
Electrical system	❶	❶	❷	❷	❷	❷	③	③
Engines	❶	❶	❷	❷	❷	❷	③	③
Exhaust/Converter	❷	❷	❷	❷	③	③	③	④
Fuel system	❶	❶	❷	❷	③	③	③	④
Ignition system	❶	❷	❷	❷	③	③	③	④
Manual transmission	❷	❷	③	③	③	④	④	⑤

- automatic	❷	❷	❷	❷	❷	❷	❷	❷
Rust/Paint	❶	❷	❷	❷	❷	❷	❷	③
Steering	❶	❷	❷	③	③	④	④	⑤
Suspension	❶	❷	❷	③	③	④	④	⑤
Crash Safety								
Corsica 4d	—	❷	⑤	⑤	❶	❶	③	③
Beretta 2d	—	❶	⑤	⑤	⑤	—	—	—

Bonneville, Cutlass, Cutlass Supreme, Delta 88, Grand Prix, Intrigue, LeSabre, Lumina, Malibu, Monte Carlo, Regal

Rating: Front-drives are Not Recommended (1988–98); rear-drives, if you can find them, are Above Average (1984–87). Although these GM models are generally classed as medium-sized cars, some of them move in and out of the large car class as well. **Maintenance/Repair costs**: Higher than average, but repairs aren't dealer dependent. **Parts**: Higher-than-average cost (independent suppliers sell for much less), but not hard to find. Nevertheless, don't even think about buying one of the front-drives without a three- to five-year supplementary warranty.

Strengths and weaknesses: Body assembly on all models is notoriously poor and is no doubt one of the main reasons why GM has lost so much market share over the past decade. Premature paint peeling and rusting, water and dust leaks into the trunk, squeaks and rattles, and wind and road noise are all too common. Accessories are also plagued by problems, with defective radios, power antennas, door locks, cruise control, and alarm systems leading the pack.

Rear-drives: The rear-drives are competent and comfortable cars, but they definitely point to a time when handling wasn't a priority and fuel economy was unimportant. Their overall reliability isn't impressive, but at least repairs are easy, defects are obvious, and any independent garage can service them. Models equipped with diesel engines or with the turbocharged gas V6 should be approached with extreme caution. These cars have a higher-than-average incidence of repairs, but parts are inexpensive and all mechanical work is very easy to perform.

Original-equipment shock absorbers and springs aren't durable, and electrical malfunctions increase proportionally with extra equipment. The AC module and condenser and wheel bearings (incredibly expensive) also have short life spans. The 4-speed automatic transmission available in later models isn't reliable. Surface rust caused by poor paint quality and application is common. The rear edge of trunk lids, roof areas above doors, and the windshield and windshield posts rust through easily.

Front-drives: The front-drives are a different breed of car: less reliable and more expensive to repair, with a considerable number of mechanical and electrical deficiencies directly related to their front-wheel drive configuration. Nevertheless, acceleration is adequate, fuel economy is good, and they're better at handling than their rear-drive cousins—except in emergencies, when their non-ABS brakes lock up and directional stability is the first to go. The front-drive's many design and manufacturing weaknesses make for unimpressive high-speed performance, mediocre interior comfort, a poor reliability record, and expensive maintenance costs. That's why most fleets and police agencies use rear-drives when they can get them. They've seen the rear-drive's safety and operating cost advantages.

These aren't driver-friendly cars. Many models have a dash that's replete with confusing push-buttons and gauges that are washed out in sunlight. At other times, there are retro touches, like the Intrigue's dash-mounted ignition, that just simply seem out of place. The keyless entry system often fails, the radio's memory is frequently forgetful, and the fuel light comes on when the tank is below the "1/2" fuel-level mark. The electronic climate control frequently malfunctions and owners report that warm air doesn't reach the driver-side heating vents. Servicing, especially for the electronic engine controls, is complicated and expensive.

Other major problem areas: the engine, automatic transmission, leaking and malfunctioning AC systems—due mainly to defective AC modules—faulty electronic modules, rack-and-pinion steering failure, bursting steering hoses on 1991 models, weak shocks, excessive front brake pad wear, warping rotors, and seizure of the rear brake calipers, rear brake/wheel lockup, myriad electrical failures requiring replacement of the computer module (a $500–$750 repair if the emissions warranty has expired), leaking oil pan, and suspension struts.

Owners report that 3.8L engines won't continue running after a cold start, the exhaust system booms, 3T40 automatic transmissions may have faulty Reverse gears, and the instrument panel may pop or creak.

Intrigue

The 1998 Oldsmobile Intrigue is last year's replacement for the Cutlass Supreme and represents the most refined iteration of the W-body shared by the Century, Grand Prix, Lumina, and Regal. It's more luxurious than the Lumina and performs as well as the Accord, Camry, and Maxima. Its rigid chassis has fewer shakes and rattles than found on GM's other models, and its 3.8L engine provides lots of low-end grunt but lacks the top-end power that makes the Japanese competition so much fun to toss around.

Lumina and Monte Carlo

The Lumina and Monte Carlo are popular two- and four-door versions of Chevy's "large" mid-sized cars, featuring standard dual airbags, ABS, and 160-hp V6 power. The Monte Carlo was formerly sold as the

Lumina Z34. Powertrain enhancements have increased horsepower and fuel efficiency. Each car has been given a slightly different appearance and a distinct "personality." A 3.1L V6 is the standard engine, while a 3.8L V6 equips the more upscale versions.

Same old generic front-drive shortcomings as mentioned earlier. In spite of some noise reduction progress, body construction is still below par—with loose door panel mouldings, poorly fitted door fabric, and misaligned panels. Other common problems: fuel pump whistling, frequent stalling, vague steering, premature paint peeling on the hood and trunk, heavy accumulation of hard-to-remove brake dust inside the honeycomb-design wheels, and front tires that scrape the fenders when the wheel is turned. Despite its own recent redesign, the 3.1L engine isn't entirely problem-free. Electronic fuel-injection systems and engine controls have created many problems for GM owners. The 4-speed automatic transmission still has bugs. The front brakes wear quickly, as do the MacPherson struts and shock absorbers. Steering assemblies tend to fail prematurely. The electrical system is temperamental. The sunroof motor is failure-prone. Owners report water leaks from the front windshield. Front-end squeaks may require the replacement of the exhaust manifold pipe springs with dampers.

The Lumina's engine is buzzy and anemic, giving out only 140 hp with the V6; the instruments and steering column shake when the car is travelling over uneven road surfaces; and lots of road and wind noise comes through the side windows thanks to the inadequately soundproofed chassis. Seating isn't very comfortable due to the lack of support caused by low-density foam, knees-in-your-face low seating, and the ramrod-straight rear backrest. The ride is acceptable with a light load, but when fully loaded, the car's back end sags and the ride deteriorates.

Malibu and Cutlass

These two front-drive, medium-sized sedans are slotted in between the Cavalier and Lumina in both size and price. This niche was once filled by the long-discontinued Corsica, Beretta, Tempest, Celebrity and Ciera. Malibu and Cutlass are boringly styled cars that use a more rigid body structure to cut down on noise and improve handling. Standard mechanicals include a 2.4L twin-cam 4-cylinder engine or an optional 3.1L V6. There's plenty of passenger and luggage space. Although headroom is tight, the Malibu can carry three rear passengers and gives much more legroom than either the Cavalier or Lumina.

Other points to consider: the base 4-cylinder is loud, handling isn't on par with the Japanese competition, there's lots of body lean in turns, outside mirrors are too small, there's no traction control, and the ignition switch is mounted on the dash (a throwback to your dad's Oldsmobile).

In addition to the generic front-drive problems listed above, owners also report: fuel-injector deposits cause chronic stalling, poor idling, or hard starts; excessive vibration occurs at any speed; transmission doesn't lock when the key is in the accessory position; steering is very loose; and the high-beam light switch is faulty.

Dealer service bulletins: **1993–94**—3.8L engine stalls after a cold start. • Converter seal leaks on 4T60E automatic transmissions. • AC hisses from the instrument panel and it becomes too warm on extended idle. • Poor heat distribution. • Door speakers buzz. **1993–94 LeSabre**—Remedies for stalling and hard starts. • Whistling when the AC is in recirculate mode. • Rattling when car passes over bumps. **1994 Delta 88, Bonneville**—3T40 automatic transmissions have faulty Reverse gears. **All 1995 models**—A knocking noise from the accessory drive belt tensioner. • Continuous spark knock. • Excessive oil consumption and oil leaks at the rocker cover. • Coolant leak near the throttle body. • Grinding/growling when in Park on an incline and a cold-start rattle with the 4T60E automatic transmission. • Harsh 1–2 upshifts. • Brake vibration and/or pedal pulsation. • Low voltage reading or dim lights at idle. • Headlights/parking lights remain on. • An upgraded wire protector shield is needed to correct a noisy steering column. • Squeak/creak from rear of vehicle, whistle noise from the heater/AC unit, and excessive radio static. • Door window rattle, popping noise during moderate braking or acceleration, front suspension pop noise, and rear-strut-related squeaking/thumping noise. • Left rear door binds. • Poor paint application and rust spots. • Frequent reports of wind noise affecting the 1990–95 Cutlass, Grand Prix, and Regal and the 1988–94 Lumina have led to the publication of DSB #53-15-16, which outlines the causes of and remedies for persistent wind noise. **Cutlass Supreme**—Snapping noise from door window, and exhaust manifold seal squeak. • Door lock/power-window switches binding. **Regal**—Left rear door binds when opening. **Delta 88**—Inadvertent low fuel chime. **Bonneville**—Headrest cover splits, loosens, or separates. **1996 Supreme**—Stalls at low rpm and high loads, extended crank time. • Front suspension popping. • HVAC blower motor noise/vibration. • Loose left side instrument panel access cover. • Inoperative antenna, power mirrors, and door locks. **All 1996 models**—Second-gear starts, poor 1–3 shifting. • Steering column noise. • Air conditioning odours and diagnosis of AC noises. • Whistle noise from HVAC. • Diagnosis and correction of fluttering, popping, ticking, and clunking noises. • Popping noise from the front of the vehicle when turning. • Radio frequency interference diagnosis. • Rear door rattle when closing. • Excessive wrinkles in seat cushion trim. • Condensation on exterior light. **1997**—AC flutter or moan. • Cold-start rattle. Engine cranks but will not run. • Engine oil leak at oil pan sealing flange and rear of engine near flywheel cover. • Engine oil level indicates over-full. • Excessive vibration of electrochromic mirror. • High beams are intermittent. • Inoperative power door locks. • Intermittent AC noise. • Intermittent Neutral/loss of Drive at highway speeds. • Instrument panel buzz and rattle when the brakes are applied. • Popping or thump noise from the left rear of vehicle is normal, according to GM. • Transmission gear whine at 40–70 km/h.

Safety summary/Recalls: **1986–88**—Brakes either fail or the pedal goes to the floor when depressed. **1988–93**—Corroded brake calipers may cause brake failures. **1990**—ABS is tough to modulate; it doesn't always engage quickly enough or it's sometimes too sensitive. **1990–91**—Wheels may crack and fracture, leading to wheel failure and loss of control. **1991–92**—Damaged fuel lines. **1992**—A poorly aligned PRNDL indicator may allow shifts from Park to Reverse while the engine is running. • There are reports of the driver's seat breaking from its moorings. **1994**—Windshield wiper failures. • When heater is turned on, it emits a strong, irritating odour. **1997**—Airbag failed to deploy. • Seating design forces driver to sit too close to airbag mechanism in the steering column. • Chronic stalling. • Throttle sticks. • Frequent cruise control failures. • Inadequate braking; ABS brake failures are common. • Emergency brake doesn't work properly; it won't remain locked. • Backfires caused by defective ECM. • Premature suspension strut failures. • Vehicle bottoms out with four or more passengers aboard. • Ventilation system emitted fumes that made occupants ill. • Frequent AC compressor failures. • Seatbelt didn't restrain passenger sufficiently in an accident. • Inoperative rear seat buckles. • Rear seatbelts are too short to secure a child seat or large person. • Plastic part of buckle came off when trying to buckle up. • Rear seatbelt design forces user to sit on buckles; they are too close to the seat and difficult to fasten. • Turn signal lever won't return to neutral position. • Excessive wind noise comes in around the door openings. • Many reports that door locks continually engage and disengage while driving. • Power window motor failure. • Outside rear view mirrors positioned too far back for a clear view. • Dash reflection onto windshield cuts visibility. • Windshield wipers operate erratically. • Water collects in the headlight lenses, causing them to fog or malfunction. • Headlights often dim for no apparent reason. • Stop lights and taillights aren't very durable. **1998**—Airbags failed to deploy. • Sudden acceleration; faulty fuel pressure regulator suspected. • Power seat puts occupant too close to airbag. • Headrest can't be raised high enough for someone over six feet tall. • Engine hesitates when accelerating. • Many reports of vehicle suddenly stalling in traffic. • Ignition coil failure also causes engine to stall and backfire. • Flexible hose line from fuel pump rests against sharp metal edge of the heat shield. • Trunk popped open while driving. • Excessive wind noise enters the interior. **Recalls**: **1988–91**—Faulty front shoulder belt guide loop. **1989–90**—Faulty brake lights. **1992**—4-speed automatic transmissions that slip in Reverse, lock in Reverse when the indicator shows Neutral, or generally perform poorly. **1993**—The seat back may suddenly recline. **1996–97**—

Backfire can break upper intake manifold, making car hard to start and possibly starting a fire. **1997**—Seatbelt may not latch properly. **1990 Lumina**—Cracks can develop in the Kelsey Hayes steel wheels. **1996 Lumina, Regal**—Brake line could rub against the transaxle mounting bracket. • **1996 Lumina, Monte Carlo**—Improperly installed brake booster. **1991 Regal**—The fuel-feed hose on vehicles with 3.8L V6 engines may leak fuel; dealer will install a new fuel-feed hose. **1994–95 Regal**—Improperly installed brake hoses could leak fluid and cause partial brake loss. **1995 Regal**—Steering could fail if the bolts on the steering column support bracket aren't tightened. **1990 models with 2.3L Quad 4 engine**—Faulty ignition coils that may cause engine misfiring will be replaced under an emissions recall campaign. **1991 Delta 88, 1992 LeSabre, 1992 Bonneville**—Parking brakes may not hold well enough, allowing the car to roll when the brake is on. **1994 Cutlass Supreme**—Improperly installed brake hoses could leak fluid and cause partial brake loss. **1994–95**—Washer wiper may malfunction. **1995**—Seatbelt anchor could fracture in a crash. If cracks appear in the centre rear seatbelt anchor, the dealer must replace the entire centre rear seatbelt system. **1994–95 Olds 88**—Headlight switch may not work. **1988–90 Cutlass, Grand Prix, Regal**—Cracks can develop in the Kelsey Hayes steel wheels. **1992–93 LeSabre, Park Avenue, 88, 98, and Bonneville with 4T60-E automatic transmission**—Oil cooler line can leak transmission fluid, posing a fire hazard.

Secret Warranties/Service Tips

1980–89 front-drives—First gear malfunctions with the THM 125C transmission may require new forward clutch piston seals (#8631986). **1984–89**—No third gear with the automatic 440-T4 transmission means that a new thrust bearing assembly should be installed. **1985–89**—Wind noise around the doors can be cured by using several kits mentioned in DSB #89-286-10. **1986–87**—Serious stalling problems can usually be traced to a defective PROM module or a malfunctioning TCC solenoid. **1986–88**—Models with anti-lock brakes may have defective hoses. • 3.8L V6 engines have a history of low oil pressure caused by a failure-prone oil pump. A temporary remedy is to avoid low viscosity oils and use 10W-40 in the winter and 20W-50 in the summer. **1987**—Frequent engine stalling may require that a different PROM be put in the ECM (DSB #87-6-17A). **1987–89**—If the engine constantly stalls and won't restart, consider replacing the fuel pump and installing a fuel-sender kit. **1988**—Poor engine performance diagnostic shortcuts are detailed in DSB #88-T-47B. **1988–89**—Poor FM reception can be improved by installing an RFI suppresser harness (#25027405). • If your door hinge breaks, GM will replace it for free up to ten years under a special policy. • Rear suspension thud noise may be reduced by replacing the strut assemblies. **1988–93**—A vehicle equipped with a 3300 or 3800 engine that stalls when decelerating or is hard to start may need a new air control motor (IAC). **1989**—Constant stalling with the 2.8L engine may be fixed by installing a new service MEMCAL (a computer module that regulates powertrain and emissions). **1989–90**—Hard cold starts may require a new MEMCAL. **1990**—Sunroof failures may be due to static electricity blowing the sunroof's electric module; install a plastic button switch. **1990–92**—A vehicle equipped with a 3.1L or 3.4L engine that is hard to start when cold or that chronically stalls may need an engine calibration or an upgraded MEMCAL module. **1991–94**—Loss of Drive or erratic shifts may be caused by an intermittent short to ground on the A or B shift solenoid or an electrical short circuit in the transaxle. • Harsh automatic transmission upshifts can be corrected by installing an upgraded accumulator valve in the control valve body. **1992–94**—A front-end engine knock troubleshooting chart is found in DSB #306001. • Water leaking from the doors into the passenger compartment has a number of causes and remedies, according to DSB #431003. **1993–94**—Knocking from the accessory drive belt tensioner requires an upgraded replacement. • Owners who complain of automatic transmission low-speed miss, hesitation, chuggle, or skip may find relief with an improved MEMCAL module. **1995–96**—Wind noise around front and rear doors; diagnosis and repair. **1995–97**—Intermittent Neutral/loss of Drive at highway speeds can be fixed by replacing the control valve body assembly. **1991–92 LeSabre**—Premature bore corrosion of the front brake caliper bolt can be fixed by changing the rubber bushings and honing out the bores. **1991–94**—A scraping noise or increased effort required to open the front doors can be fixed by bending the door's lower check ear. • Water leaking from the doors into the passenger compartment has a number of causes and remedies (DSB #431003). **1995–97**—Intermittent Neutral/loss of Drive at highway speeds can be fixed by replacing the control valve body assembly (see following DSB).

File In Section: 7 - Transmission
Bulletin No.: 67-71-64
Date: February 1997
Subject:
Intermittent Neutral or Loss of Drive at Highway Speeds or from Fourth Gear
(Replace Control Valve Body Assembly)
Models:
1995–96 Buick Skylark, Regal Century, Park Avenue, Riviera, LeSabre
1997 Buick Skylark, Regal, Century, LeSabre
1995 Cadillac DeVille
1995–96 Chevrolet Beretta, Corsica, Lumina APV, Lumina, Monte Carlo 1997 Chevrolet Lumina, Monte Carlo, Venture
1995–96 Oldsmobile Cutlass Ciera, Cutlass Cruiser, LSS, Ninety Eight, Ninety Eight Regency, Eighty Eight, Achieva, Silhouette, Cutlass Supreme
1997 Oldsmobile Eighty Eight, Achieva, Cutlass Supreme, LSS, Silhouette
1995–97 Pontiac Bonneville, Grand Am, Trans Sport, Grand Prix with HYDRA-MATIC 4T60-E Transaxle (RPO M13)
Condition
Some owners may comment about an intermittent, neutral condition while driving at highway speeds or intermittent neutral from fourth gear.
Cause
The 3–2 Manual Downshift valve may be sticking intermittently. When the 3–2 downshift valve sticks, this allows the 2–3 shift valve to float in its bore either exhausting 2ND, 3RD and 4TH clutch oil or sending D-4 oil into the AUX input clutch feed. Exhausting the 3RD and 4TH clutch oil could cause an intermittent loss of drive. Sending D-4 oil into the input circuit could cause a potential tie up condition by engaging the input clutch and 4TH clutch.
Correction
Inspect the transmission fluid and pan for signs of distressed clutches or any abnormal amounts of debris. Remove and disassemble the 1–2 and 2–3 accumulator housing assembly. Inspect the accumulator housing and cover for any signs of scoring, sediment, or abnormalities that could generate debris. If any of the above conditions are found, make the necessary repairs to those items. Then inspect the 2–3 (357) and 3 2 (356) valve line up. If the valves are scored, replace the control valve body assembly.

This major transmission repair should be covered by GM's base warranty or a "goodwill" warranty since the bulletin clearly shows it's a factory-related defect.

• Transmission gear whine at 40–70 km/h means the final drive assembly may have to be replaced. **1990 Lumina**—Vehicles equipped with a 3.1L engine may exhibit piston scuffing, which produces cold engine knock at low ambient temperatures. GM recommends a partial engine replacement rather than changing the pistons. **1990–91**—Late transaxle upshifts may be fixed by resetting the TV cable. **1994**—Excessive brake-pedal effort when cold can be fixed by installing upgraded brake pads. **1994–95**—A steering wheel clicking or scrubbing noise when turning can be fixed by installing an upgraded wire protector shield. **1990–91 Lumina and Regal models with 3.1L V6 engine**—If the starter makes a grinding noise or won't engage the engine, the fault may be a weak starter spring. Replace the starter-motor drive assembly with #10473700. • A defective sending unit will cause a display of high oil pressure or erratic readings. Replace the old pressure sender with #25605389. • Excessive brake-pedal effort can be corrected by replacing the original pads with modified parts. **1990–92**—Vehicles equipped with the 3T40 automatic transmission may experience slippage in manual Low or Reverse. Install service package #8628222, which includes a Low/Reverse clutch release spring (#8664961) and clutch retainer and nap ring (#656/657). **1991–92 Lumina**—A delayed shift between Drive and Reverse is likely caused by a rolled or cut input-clutch-piston outer seal. **1991–94**—Loss of Drive or erratic shifts may be caused by an intermittent short to ground on the A or B shift solenoid or an electrical short circuit in the transaxle. **1992–93**—No Reverse or slipping in Reverse can be corrected by installing an upgraded Low/Reverse clutch return spring and spiral retaining ring. **1991–94 Regal**—Loss of Drive or erratic shifts may be caused by an intermittent short to ground on the A or B shift solenoid or an electrical short circuit in the transaxle. **1992**—Front-door wind noise and water leaks can be fixed by replacing the run channel retainer and adding sealer between the retainer and the door frame. **1997 Regal, Grand Prix**—Insufficient heater performance on passenger-side floor area can be fixed by installing a new I/P insulator panel and bracket. **1990–91 2.3L Quad 4 engine**—Head gasket leaks are a problem that was once covered by a secret warranty extension. The first sign of trouble is a loss of power caused by combustion gases mixing with coolant. This is followed by a cloud of steam and coolant loss through the exhaust system. If these warnings are ignored, cylinder-bore scoring, a warped cylinder head, and piston seizure will likely result as the engine continues to overheat. Up to 7 years/150,000 km, GM has replaced at no charge the head gasket with an improved version. **All models/years**—A rotten-egg odour coming from the exhaust is probably caused by a malfunctioning catalytic converter (covered by the emissions warranty). • Paint delamination, peeling, or fading (see pages 76, 113).

Bonneville, Cutlass, Cutlass Supreme, Delta 88, Grand Prix, Intrigue, LeSabre, Lumina, Malibu, Monte Carlo, Regal Profile

	1991	1992	1993	1994	1995	1996	1997	1998
Cost Price ($)								
Bonneville	20,098	22,898	23,498	25,298	28,175	29,440	31,175	33,255
Cutlass GL	—	—	—	—	—	—	—	23,140
Cutlass Supreme	18,798	18,398	18,858	21,398	24,310	25,285	26,355	—
Delta 88	21,298	22,250	23,098	25,598	28,320	30,190	32,185	32,950
Grand Prix	18,098	17,598	18,498	22,098	24,555	23,940	26,305	26,035

Intrigue	—	—	—	—	—	—	—	27,998
LeSabre	21,298	23,198	23,997	25,498	28,235	29,560	32,370	33,100
Lumina	16,098	15,998	17,798	19,898	20,730	21,455	22,340	22,980
Malibu	—	—	—	—	—	—	19,995	20,595
Monte Carlo	—	—	—	—	22,578	23,625	24,275	24,895
Regal	18,998	19,398	20,006	21,898	20,060	25,035	27,795	28,410
Used Values ($)								
Bonneville ↑	6,500	7,500	9,500	12,000	14,000	17,000	20,000	24,000
Bonneville ↓	5,500	6,500	8,500	10,500	13,000	15,500	17,500	22,500
Cutlass GL ↑	—	—	—	—	—	—	—	20,000
Cutlass GL ↓	—	—	—	—	—	—	—	18,500
Cutlass Supreme ↑	5,000	6,000	7,000	9,000	11,500	13,000	16,000	—
Cutlass Supreme ↓	4,500	5,500	6,500	8,000	10,500	11,500	14,500	—
Delta 88 ↑	5,500	7,000	8,500	12,00	14,000	16,000	19,500	21,000
Delta 88 ↓	4,500	6,000	7,500	10,000	12,500	14,500	17,000	19,500
Grand Prix ↑	4,500	5,500	7,000	8,500	11,000	13,000	18,500	21,500
Grand Prix ↓	4,000	5,000	6,000	7,500	10,00	11,500	16,000	20,000
Intrigue ↑	—	—	—	—	—	—	—	21,500
Intrigue ↓	—	—	—	—	—	—	—	20,000
LeSabre ↑	5,500	7,500	9,500	12,500	14,500	17,500	21,500	25,000
LeSabre ↓	5,000	6,500	8,500	11,000	13,000	16,000	20,000	24,000
Lumina ↑	4,500	5,000	6,500	8,000	10,500	12,000	14,500	16,500
Lumina ↓	4,000	4,500	5,000	7,000	9,500	11,000	13,000	15,500
Malibu ↑	—	—	—	—	—	—	14,000	16,000
Malibu ↓	—	—	—	—	—	—	13,000	15,000
Monte Carlo ↑	—	—	—	—	11,500	14,000	16,000	19,000
Monte Carlo ↓	—	—	—	—	10,500	13,000	15,000	18,000
Regal ↑	5,500	6,500	8,000	10,500	12,500	14,000	19,000	22,000
Regal ↓	4,500	5,500	7,000	9,500	11,000	13,000	17,500	21,000
Extended Warranty	Y	Y	Y	Y	Y	Y	Y	Y
Secret Warranty	Y	Y	Y	Y	Y	Y	Y	Y
Reliability	❶	❶	❶	❷	❷	③	③	③
Air conditioning	❷	❷	❷	❷	③	③	③	③
Automatic transmission	❶	❶	❶	❶	❶	❷	③	③
Body integrity	❶	❶	❶	❶	❶	❷	❷	③
Braking system	❶	❶	❶	❷	❷	❷	❷	③
Electrical system	❶	❶	❶	❷	❷	③	③	③
Engines	❷	❷	❷	❷	③	③	③	④
Exhaust/Converter	❷	❷	❷	③	③	④	④	④
Fuel system	❷	❷	❷	❷	③	③	③	④
Ignition system	❷	❷	③	③	③	③	③	④
Rust/Paint	❶	❶	❶	❶	❶	❶	❷	③
Steering	❷	❷	❷	❷	③	③	③	③
Suspension	❷	❷	❷	❷	③	③	③	③
Crash Safety								
Bonneville 4d	⑤	⑤	❶	❶	⑤	—	⑤	⑤
Cutlass 4d	—	—	—	—	—	—	④	④
Cutlass Supreme 2d	❷	❷	❷	❷	—	—	—	—
Cutlass Supreme 4d	❶	❶	❶	❶	—	—	—	—
Delta 88 2d	⑤	⑤	⑤	—	—	—	—	—
Delta 88 4d	❷	❷	❶	❶	—	—	—	—

Grand Prix 2d	❷	❷	❷	❶	④	④	④	—
Grand Prix 4d	❶	❶	❶	❶	—	—	—	—
Intrigue	—	—	—	—	—	—	—	④
LeSabre 2d	❷	❷	❷	❷	—	—	—	—
LeSabre 4d	—	—	—	—	—	—	④	④
Lumina 4d	⑤	⑤	❷	❷	—	⑤	⑤	④
Malibu	—	—	—	—	—	—	④	④
Monte Carlo	—	—	—	—	④	—	—	—
Monte Carlo 2d	—	—	—	—	—	④	—	—
Regal 2d	⑤	⑤	—	—	—	—	—	—
Regal 4d	—	—	⑤	❶	—	—	—	—
Side Impact								
Bonneville 4d	—	—	—	—	—	—	—	③
Cutlass 4d	—	—	—	—	—	—	❶	❶
Intrigue	—	—	—	—	—	—	—	③
Lumina	—	—	—	—	—	—	④	④
Malibu	—	—	—	—	—	—	❶	❶
Regal 4d	—	—	—	—	—	—	—	③

HONDA

Accord

Rating: Recommended (1992–98); Above Average (1990–91); Average (1985–89). Fast and nimble without a V6, this is the car of choice in the compact sedan class for drivers who want maximum fuel economy and comfort along with lots of space for grocery hauling and occasional highway cruising. Surprisingly, the Accord built during the past few years has racked up an unusually large number of safety complaints, including reports of sudden acceleration, brake failures, and airbag malfunctions that cause the devices to go off when they shouldn't, and not deploy when they should. **Maintenance/Repair costs**: Lower than average. Repairs aren't dealer dependent. **Parts**: Higher-than-average cost, but can be bought for much less from independent suppliers. Owners report that Honda frequently charges for parts that should be replaced for free under the emissions warranty and that those charges are far in excess of what other automakers ask.

Strengths and weaknesses: The Accord doesn't really excel in any particular area; it's just very, very good at everything. It's smooth, quiet, mannerly, and competent, with outstanding fit and finish, inside and out. Every time Honda redesigned the line it not only caught up with the latest advances but went slightly ahead. Strong points are comfort, fit and finish, ergonomics, impressive assembly quality, reliability, and driveability. With the optional 16-valve 4-cylinder engine or V6, the Accord is one of the most versatile compacts you can find. It offers something for everyone, and its high resale value means there's no way you can lose money buying one.

Despite all the foregoing praise, this hasn't always been a great car. During the '80s, Accords were beset with severe premature rusting, frequent engine camshaft and crankshaft failures, and severe front brake problems. Engines leaked or burned oil and blew their cylinder head gaskets easily and carbureted models suffered from driveability problems through 1986.

Between 1986 and 1989, the brakes, automatic transmission (particularly the 2–4 clutch assembly), rack-and-pinion steering, suspension (coils are practically biodegradable), and electrical system became the major problem areas. Also, water pumps and alternators need replacing about every three years. Rapid front brake wear and frequent brake rotor replacements are common. The automatic transmission shifts a bit harshly upon hard acceleration. Shock absorbers go soft quickly, and replacement prices are often less at independent suppliers. To avoid costly engine repairs, check the engine timing belt every 2 years/40,000 km.

Early Accords were surprisingly vulnerable to paint chipping, flaking, and premature surface rust. If left untreated, sheet metal perforations develop unusually quickly. Especially vulnerable spots are front fender seams, door bottoms, and areas surrounding side view mirrors and door handles, rocker panels, wheel openings, windshield posts, front cowls, and trunk and hatchback lids.

1990–93 models got more room (stepping up to the mid-sized car niche) and additional power through a new quieter 2.2L 4-cylinder engine. Nevertheless, rear seating is still inadequate, the added weight saps the car's performance, and the automatic transmission shifts harshly at times. Owners report prematurely worn automatic transmissions, constant velocity joints, and power-steering assemblies. Poor quality control in the choice of body trim and assembly leads to numerous air and water leaks.

Redesigned again for the 1994 model year, the Accord continues to add interior room and other refinements. However, the addition of the V6 powerplant in the 1995 model year gives the Accord plenty of power in reserve without the high rpms. The automatic transmission still works poorly with the 4-banger, producing acceleration times that are far from impressive, and owners still complain of excessive road noise and tire whine. Nevertheless, no significant reliability problems have been reported with the latest redesign.

Confidential dealer service bulletins show that the 1994–98 models are susceptible to AC malfunctions, engine oil leaks, transmission glitches, power steering pump leaks, windows falling off their channels, and numerous air and water leaks. Usually, these problems are simple to repair and Honda customer relations staff are helpful; however, Honda staffers may be getting a bit too arrogant in their dealings with the public. One *Lemon-Aid* reader wrote the following:

> We returned the car to the dealer several times (at least six) to have the doors adjusted to reduce wind noise affecting my 1994 Accord LX sedan. They were finally able to reduce the noise a small amount. A letter written to Honda explaining our dissatisfaction with the car resulted in a response letter with very definite "screw you" overtones.

Dealer service bulletins: **1994**—Buzzing when the turn signals are activated. • A clunking noise from the door glass, and instrument panel creaking. • Damaged door handle seals. • Faulty fuel-filler doors. • Rattling moon roof deflectors. • Outside mirror wind noise. • Power door locks that unlock themselves. • Difficulty in closing the trunk lid. • Lots of water leaks and wind noise. **1995**—Noise from the front passenger's footwell, exhaust system, and shoulder belt anchors; rear shelf will buzz. • The dash panel and clutch pedal may creak. • A screeching noise occurs when the driver's window is lowered. • A wind whistle emanates from the top of the windshield. • Heater control indicators may not light. • DSB #95-017 shows which rear brake pads produce less noise and which ones last the longest. **1996**—Instrument panel creaking. • Poor fit of wheel centre cap. • Screeching noise when lowering the driver's window. • Seatbelt is slow to retract. **1997**—AC won't blow cold air (we're talking secret warranty here). • Oil seepage from the engine block. • Torque converter won't lock up. • Fifth gear grinds during upshift. • Leak from the power steering pump. • Rear wheel bearing noise. • Static when adjusting the radio volume. • Front door glass comes out of run channel. • Cracking paint on passenger's side airbag cover. • Missing alloy wheel centre cap. **1998**—Trunk spoiler damages paint. • Brake system indicator stays on. • Creak from the rear shelf area, headliner, windshield, and rear window. • Front ABS wheel sensor harness rubs against wheel. • Rattle from rear stabilizer bar. • Water leaks from the rear doors. • Wind noise from the top of the windshield.

Safety summary/Recalls: **1995**—Rear seatbelt buckle has insufficient slack, preventing buckle from latching. • Airbag warning light stays on. • Excessive windshield glare. • AC failure. • Headlight failure. • Cruise control malfunctions. • Premature front/rear brake wear. **1995–96**—Faulty power windows. • Brake failures/lockup. • Airbag failed to deploy or accidentally deployed. • Injury from airbag. • Sudden acceleration, stalling. • Passenger-side seatbacks won't stay upright. **1996**—Front passenger seatbelt locks up. • The door lock design gives a boost to thieves. • The Check Engine light is always on. • The defroster could be faulty. • Faulty power door locks. • The steering column separates from the shaft. **1997**—Airbags failed to deploy. • Inadvertent airbag deployment. • Fire caused by faulty wiring harness. • Sudden acceleration while braking. • Cruise control doesn't accelerate properly and won't downshift the transmission. • Transmission shifted into Reverse and vehicle moved forward.

• ABS brakes lock up. • Sudden brake failure. • Location of the oil filter allows oil to leak onto the exhaust system and catalytic converter. • Faulty brake master cylinder. • All four front brake pads cracked right down the centre. • Power-steering fluid leakage. • Oil plug fell off into the oil pan and sprayed oil everywhere. • Left side seatbelt fails to retract. • Seatbelt tightened and locked up; occupant had to cut belt. • Seatbelt continually ratchets tighter. • When sun visor is opened, it blocks driver's vision due to its large size. • Check Engine light stays on continually. • Defroster fails to defrost side windows, and actually causes them to fog up. • Power windows fail to operate properly in cold weather. • Door continually out of adjustment. **1998**—Sudden acceleration while vehicle stopped in traffic. • Sudden acceleration occurred when vehicle hit from the rear. • Chronic stalling. • ABS brake light comes on continually. • Gas and brake pedals are too close together and often get pressed at the same time. • Airbag failed to deploy. • Frequent brake failures. • Sudden brake lockup. • Brake master cylinder failures. • Floormat bunches under the brake pedal. • Engine oil leakage. • Power-steering fluid leakage, causing sudden loss of steering control. • Vehicle rolled back when parked. • Automatic transmission gears disengage and make a loud noise when engaging. • Transmission fails to engage at slow speeds. • Transmission fails to fully lock up in Overdrive. • Transmission hunts for the right gear. • Clutch pedal failure. • Automatic transmission parking mechanism failure. • Due to design of dashboard lights, it's hard to read odometer, digital clock, and radio indicator. • Can't see high beam indicator light in the daytime. • Light tan dash reflects too much sunlight into the eyes. • Instrument panel lights are too bright at night and can't be dimmed enough. • To activate horn, driver must remove hand from steering wheel. • Fuel gauge shows two-thirds when the gas tank is full, or indicates an empty tank with warning light on while five gallons remain in the tank. • Poor seatbelt design allows for belt to wrap around the release lever and get stuck, or causes seatback to suddenly recline. • Seatbelts get trapped underneath the seatback electric switch. • Seatbelts ratchet too tight, trapping occupants. • Rear passenger-side door won't unlock. • Sunroofs and headliners often need replacing. **Recalls**: **1985**—A defective electric control unit could cause the airbags to deploy outside of a crash situation. **1986–1991**—Takata seatbelts will be replaced. **1990–91**—Defective crank pulleys connected to the drive generator and power-steering pump could fail, preventing the battery from recharging. **1991**—The power window malfunctions. • **1991–93**—Faulty rear seatbelts won't pull out if the car is parked on a steep incline. **1994**—Faulty tire stems may lead to sudden air loss. **1995–97 with 4-cylinder engines (excluding DX version)**—Improperly routed air conditioning wires may short and cause fires or other electrical malfunctions; dealers must install protective plastic tubing on the harnesses, reroute the wires, and replace damaged harnesses. **1998**—Vehicle may roll away with transmission lever in "Park."

Secret Warranties/Service Tips

1988–93—A creaking sound coming from the window regulator can be corrected by installing an upgraded regulator spiral spring. **1990**—Delay after shifting into Drive is corrected by adjusting the cable (#87-040). • A right rear suspension clunking noise can be fixed with a new spring silencer tube. • An interior roaring noise can be silenced by installing blind body plugs (#95550-15000) in the door rocker panels. • If the front inside door handle doesn't work, check for a loose or broken actuator rod clip. • A moaning sound heard when the steering wheel is turned may mean that the steering-pump outlet-hose orifice has slipped out of position or that the outlet hose is faulty. • Water leaks behind the dashboard require sealing near the windshield locating blocks and frame panel seams in the cowl. • Whistling from the front of the car is likely caused by poor hood sealing. **1990–91**—A faulty automatic transmission countershaft nut was once replaced free under a product update program. • A new distributor body and kit will help cars that have a starting problem. **1990–93**—Poor AM reception or a popping noise from the speakers is likely due to poor ground connection between the antenna collar and car body. **1992**—A faulty oil pressure switch was once replaced for free under a product update program. **1994**—Damaged door handle seals were once replaced for free under a "goodwill" warranty; try for a partial refund. **1994–97**—If the AC doesn't blow cold air, Honda will replace both the evaporator and the receiver/dryer free of charge. • When operating a manual or power-assisted front window, the rear edge of the glass comes out of the channel. • Oil seepage from the engine block requires sealing and the installation of a new exhaust manifold bracket. (The following three bulletins give more details on the above 1994–97 problems.)

97-031
Applies To: 1994-97 Accord - ALL
1994-97 Civic - ALL
1994-95 del Sol - ALL
August 4, 1997
A/C Does Not Blow Cold Air
SYMPTOM
The air conditioning system does not blow cold air.
PROBABLE CAUSE
The evaporator tubes corrode from ocean salt spray, causing pinholes that allow refrigerant to leak. This problem is limited to areas where a hot, humid climate is combined with ocean air (Florida, Hawaii, Puerto Rico, or Gulf Coast areas).
CORRECTIVE ACTION
Replace only the evaporator if the vehicle comes in with some refrigerant in the high side line. Replace both the evaporator and the receiver/dryer if the vehicle comes in without any refrigerant in the high side line.

MODEL	EVAPORATOR	RECEIVER/DRYER	PAG REFRIGERANT OIL (120 ml) (for ordering only, do not put on warranty claim)	O-RINGS
Accord	P/N 80210-8V1-A11, H/C 4335725 (Use this number only if you are located in Florida, Hawaii, Puerto Rico, or theGulf Coast areas. Otherwise, use the part number in your parts catalog.)	V6 P/N 80351-SV7-A11, H/C 4602330	V6 P/N 38897-P13-A01AH, H/C 5023627	
		All except V6 P/N 80351-SV1-A11, H/C 4276267	1994 EX P/N 38897-P0A-A01AH, H/C 5172663	Evaporator (one of each) P/N 80871-ST7-000, H/C 4256103
			1994 LX, DX & All 1995-97 P/N 38897-PR8-A01AH, H/C 5023635	P/N 80873-ST7-000, H/C 4256129
1994-95 Civic, del Sol	P/N 80210-SR1-A12, H/C 43233242	P/N 80351-ST7-A11, H/C 4255964	P/N 38897-P13-A01AH, H/C 5023627	Receiver/Dryer (two required) P/N 80873-ST7-000, H/C 4256129
1996-97 Civic	P/N 80210-SR1-A12, H/C 43233242	P/N 80351-S01-A01, H/C 4809620	P/N 38897-P13-A01AH, H/C 5023627	

PARTS INFORMATION
WARRANTY CLAIM INFORMATION
In warranty:
The normal warranty applies.
Failed part: P/N 80210-ST7-A21, H/C 4327656
Defect code: 011
Contention code: B02
Skill Level: Repair Technician
Out of warranty:
Any repair performed after warranty expiration may he eligible for goodwill consideration by the District Service Manager or your Zone Office. You must request consideration, and get a decision, before starting work.

This secret warranty pays for major AC components plus two hours of labour needed to get the job done. It also establishes that AC failures are routinely covered by Honda up to four years. Smart owners of other models should argue that any failure of the AC during that time, whether caused by salty air, road salt, or any other defect, falls within these "goodwill" parameters. Repairs required after four years of use should get partial refunds.

97-060
September 22, 1997
Applies To: 1994-97 Accord 4-door - All
Front Door Glass Comes Out of Run Channel
SYMPTOM
When operating a front window (power or manual), the rear edge of the glass comes out of the B-pillar run channel.
PROBABLE CAUSE
The window regulator is out of adjustment.
CORRECTIVE ACTION
Adjust the window regulator, the front channel, and the glass.
WARRANTY CLAIM INFORMATION
In warranty:
The normal warranty applies.

Failed part:	P/N 72250-SV4-A11, H/C 4272282
Defect code:	030
Contention code:	B01
Skill level:	Repair Technician

Out of warranty:
Any repair performed after warranty expiration may be eligible for goodwill consideration by the District Service Manager or your
Zone Office. You must request consideration, and get a decision, before starting work.

This job also falls under Honda's "goodwill" policy.

98-042
May 12, 1998
Applies To: 1994 - 97 Accord - DX, LX, and SE models with L4 engine
Oil Seepage From the Engine Block
SYMPTOM
Oil seepage from the front of the engine at the boss for the exhaust manifold bracket.
CORRECTIVE ACTION
Seal the cracked boss on the engine block, and install a new exhaust manifold bracket onto the A/C compressor bracket.
NOTE:
On DX models without A/C, an A/C compressor bracket must be installed. See PARTS INFORMATION.
WARRANTY CLAIM INFORMATION
In warranty:
The normal warranty applies.

Operation number:	110107
Flat rate time:	2.4 hours
Failed part:	P/N 10002-P0A-A01
H/C 4511465	
Defect code:	060
Contention code:	B06
Template ID:	98-042A (DX w/o A/C)
98-042B (All others)	
98-042C (LX w/o ABS)	
Skill level:	Repair Technician

Out of warranty:
Any repair performed after warranty expiration may be eligible for goodwill consideration by the District Service Manager or your Zone Office. You must request consideration, and get a decision, before starting work.

More "goodwill," we hope.

• Power-steering pump fluid leakage requires a new O-ring, which is also eligible for "goodwill" consideration, says Honda. **1995**—A whistling or howling noise coming from the top of the windshield can be silenced by applying sealant under the upper windshield molding. • Honda will supply an exhaust buzz silencing kit for free on a case-by-case basis. • The Accord's noise problems will be fixed for free only if the dealer makes the request to Honda. **1995–96**—A creaking noise coming from the instrument panel can be silenced through a variety of measures outlined in a series of Honda bulletins. **1996–97**—In a settlement with the U.S. Environmental Protection Agency, Honda paid fines totaling $17.1 million and extended its emissions warranty on 1.6 million 1995–97 models to 14 years or 150,000 miles. This means that costly engine components and exhaust system parts, like catalytic converters, will be replaced free of charge as long as the 14-year/150,000 mile limit hasn't been exceeded. Additionally, the automaker will provide a full engine check and emissions-related repairs at 50,000–75,000 miles and will give free tune-ups at 75,000–150,000 miles. (See Honda Civic rating for full details.) **All models/years**—Steering wheel shimmy is a frequent problem taken care of in DSB #94-025.

Accord Profile

	1991	1992	1993	1994	1995	1996	1997	1998
Cost Price ($)								
LX	15,095	15,690	17,195	18,695	19,795	20,295	20,995	23,800
EXi/EX	17,975	18,695	20,545	21,495	22,595	22,995	23,495	26,800
Used Values ($)								
LX ↑	5,500	6,500	8,500	9,500	12,000	14,000	15,500	18,000
LX ↓	5,000	6,000	7,500	9,000	11,000	13,000	14,500	17,000
EXi/EX ↑	7,000	8,500	10,000	12,000	14,500	17,000	18,500	22,000
EXi/EX ↓	6,500	7,500	9,000	11,000	13,500	15,500	17,500	21,000
Extended Warranty	N	N	N	N	N	N	N	N
Secret Warranty	N	N	N	Y	Y	Y	Y	N
Reliability	③	④	⑤	⑤	⑤	⑤	⑤	⑤
Air conditioning	③	③	③	③	③	④	④	④
Body integrity	❷	❷	❷	❷	③	③	③	④
Braking system	❷	❷	❷	❷	❷	③	③	④
Electrical system	❷	③	③	③	④	⑤	⑤	⑤
Engines	④	④	④	⑤	⑤	⑤	⑤	⑤
Exhaust/Converter	❷	❷	❷	④	④	⑤	⑤	⑤
Fuel system	④	④	④	⑤	⑤	⑤	⑤	⑤
Ignition system	③	③	④	⑤	⑤	⑤	⑤	⑤
Manual transmission	⑤	⑤	⑤	⑤	⑤	⑤	⑤	⑤
- automatic	④	④	④	④	⑤	⑤	⑤	⑤
Rust/Paint	③	③	④	④	④	④	④	④
Steering	❷	❷	③	③	③	④	⑤	⑤
Suspension	③	③	③	④	⑤	⑤	⑤	⑤
Crash Safety								
2d	—	—	—	—	—	—	—	④
4d	⑤	⑤	⑤	❷	④	④	④	④
LX wagon	⑤	—	❷	—	—	—	—	—
SE	—	—	❶	—	—	—	—	—
Side Impact								
4d	—	—	—	—	—	—	❷	④

HYUNDAI

Sonata

Rating: Average (1995–98); Not Recommended (1986–93). The 1994 model year was skipped. These cars aren't rated very high because of their poor quality control. Nevertheless, they haven't registered one-tenth the number of safety complaints as has the higher-rated Honda Accord. If you must have a Sonata, I suggest you buy a 1996–98 version, plan to keep it at least five years to shake off the depreciation, and put some of the savings on the purchase price into a comprehensive supplementary warranty to protect yourself during that period. **Maintenance/Repair costs**: Higher than average. Repairs aren't dealer dependent. **Parts**: Higher-than-average cost, and often back-ordered.

Strengths and weaknesses: This mid-sized front-drive sedan was built under Mitsubishi licensing, but its overall reliability isn't anywhere near as good as what you'll find with Mitsubishi's cars and trucks. Acceleration is impressive with the manual gearbox, and only passable with the automatic. Handling and performance are also fairly good, although emergency handling isn't confidence-inspiring, particularly due to the imprecise steering and excessive lean when cornering. As with other Hyundai models, the automatic transmission performs erratically, the engine is noisy, and reliability is a problem: it's way below average for the 1989–93 models; the 1995–98 models are moderately improved (remember, Hyundai skipped the 1994s).

Redesigned for the 1995 model year, the car got additional interior room, more horsepower, and an upgraded automatic transmission. Nevertheless, acceleration with an automatic is still below average with the 4-banger, and the automatic gearbox still downshifts slowly. 1995–97 Sonatas came with more standard features, like air conditioning, power steering, and a 2.0L 16-valve 4-cylinder engine.

Consumer Reports magazine reports that owners of 1990–92 Sonatas had 120 percent more complaints than the average car owner. Sonatas have had body assembly and preparation deficiencies that show that Hyundai doesn't yet have a firm grip on quality control. For example, the sloppily applied paint pits like an orange peel, and the sun visors and headliner look cheap and fragile. If the experience of Stellar and Pony owners is any guide, Sonata owners will notice serious deficiencies by the third and fourth year.

Hyundai dealer service bulletins indicate that the Sonata's automatic transmission could exhibit what Hyundai describes as "shift shock," as well as delayed shifting. Problems reported by owners in the past: poor engine performance (hard starting, poor idling, stalling); the engine runs hot, and when you're stopped at a traffic light, it shakes like a boiling kettle; #3 spark plug often needs replacing or cleaning; rough engine rattle; high oil consumption (one litre every two to three months); excessive front brake pulsation and premature wear; steering defects (when the steering wheel is turned to either extreme, it makes a sound like metal cracking); cruise control malfunctions and electrical short circuits; battery life of only 18 months; malfunctioning lights; radio failures; falling interior roof liner; faulty hood locks; rotten-egg smell coming from the catalytic converter; broken muffler; faulty resonator; defective exhaust pipe; poor door and window sealing (water leaking into the interior when the car is washed); premature paint peeling; and rusting.

The 1995–98 Sonatas haven't elicited as many quality control and safety complaints as previous versions. However, engines, automatic transmissions, brakes, airbags, and electrical system components still top the list of parts most vulnerable to premature failure or malfunctioning.

Safety summary/Recalls: There are reports of electrical fires for the 1989–90 models. • Emergency handling leaves a lot to be desired. **1996**—Tire fell off of car. • Battery blew up while car was idling. • Airbag light is continually lit. • Brake malfunctions. • Chronic stalling, particularly when AC is engaged. • Frequent engine valve cover gasket failures. • Poor quality spark plugs cause sluggish acceleration. • Check Engine light is lit continually. • Frequent automatic transmission failures. • Water leaks from the front of the car. **1997**—Airbags fail to deploy. • Frequent engine failures. • Transmission gear shift will not stay in place. • Power window switch failed. **1998**—Frequent inadvertent airbag deployment. • Sudden transmission failure while driving. **Recalls**: **1986–91**—Part of the wiring harness connecting the alternator to the battery could detach, create a short circuit, and cause a loss of electrical power or an engine fire. **1989–90**—Fuel leaks into engine compartment. • Hood could fly up. **1989–93**—The motorized shoulder belt could malfunction. **1995**—There is the potential of loss of rear spring support. **1996–97**—The wiper motor is faulty.

Secret Warranties/Service Tips

1989–90—Vibrations caused by excessive hub and brake disc runout require a new hub, machining of the disc, or new brake discs. • Cold-start stalling with the 2.4L engine can be fixed by installing start enrichment kit #39901-326000D. • Engine oil drain-plug leaks are likely caused by a faulty gasket or incorrect gasket installation. **1990**—Difficult shifts into fourth gear require a new transmission restrict ball assembly. **1992**—Oil leaking between the oil filter and mounting bracket could be due to an overly wide mounting surface on the bracket; correct this leakage by replacing the bracket. **1992–93**—Hyundai has a field fix for manual transaxle gear clash/grind (DSB #9440-004). **1992–95**—Difficult-to-engage Reverse gear needs an upgraded part. **1995–96**—Harsh shifting might be fixed by installing an upgraded Transaxle Control Module (TCM). **1995–97**—Lots more tips are offered on getting the automatic transmission to shift properly. **All models/years**—Harsh shifting when coming to a stop or upon initial acceleration is likely caused by a maladjusted accelerator pedal switch TCU. • A faulty air exhaust plug could cause harsh shifting into second and fourth gears on vehicles with automatic transmission. • Brake pedal pulsation can be corrected by installing upgraded front discs and pads (see following bulletin).

Number: 96-50-004
Date: AUGUST, 1996
Model: SONATA (ALL)
Subject: BRAKE PEDAL PULSATION: DIAGNOSIS AND REPAIR
DESCRIPTION: Some Sonata vehicles may experience brake pedal pulsation during light brake application if brake disc thickness variation exceeds specifications. A pulsating brake pedal condition does not affect the stopping distance of the vehicle. If a road test has confirmed that the vehicle has this condition, the front brake discs and pads should be replaced.
SERVICE PROCEDURE:

1. Remove front brake discs and pads, following the procedure outlined in the shop manual.
2. Discard brake discs, pads, and return springs.

Part Name	Part Number	
	Previous	Current
Disc-FR Brake	51712-33001	51712-33001
Pad Kit-FR Disc Brake	58101-35A10	58101-34A10

Sonata Profile

	1990	1991	1992	1993	1995	1996	1997	1998
Cost Price ($)								
Base	13,198	12,595	13,095	13,495	15,595	16,595	16,995	17,495
Used Values ($)								
Base ↑	3,000	3,500	4,000	5,000	7,500	8,000	11,000	13,500
Base ↓	2,500	3,000	3,500	4,500	7,000	7,5000	10,000	12,500
Extended Warranty	Y	Y	Y	Y	Y	Y	Y	Y
Secret Warranty	N	N	N	N	N	N	N	N
Reliability	❷	❷	❷	❷	③	③	③	③
Air conditioning	❷	❷	③	③	③	③	④	④
Body integrity	❷	❷	❷	❷	❷	❷	❷	③
Braking system	❷	❷	❷	❷	❷	③	③	③
Electrical system	❷	❷	❷	❷	❷	④	④	④
Engines	③	③	③	③	④	④	④	⑤
Exhaust/Converter	❷	❷	③	④	④	⑤	⑤	⑤
Fuel system	③	③	③	⑤	⑤	④	⑤	⑤
Ignition system	③	④	④	④	⑤	⑤	⑤	⑤
Manual transmission	④	④	④	③	③	④	⑤	⑤
- automatic	❷	❷	❷	❷	❷	③	④	④
Rust/Paint	❷	❷	❷	❷	③	③	③	④
Steering	③	③	④	④	④	⑤	⑤	⑤
Suspension	❷	❷	③	④	④	⑤	⑤	⑤
Crash Safety	—	—	—	—	③	③	③	③
Side Impact	—	—	—	—	—	—	❶	❶

MAZDA

626, Cronos, MX-6, Mystère

Rating: Above Average (1996–98); Average (1994–95); Not Recommended (1985–93). 1997 was the last model year for the MX-6 and Probe, its Ford twin. **Maintenance/Repair costs**: Higher than average. Repairs aren't dealer dependent. Mazda suggests changing the engine timing chain after 100,000 km. **Parts**: High cost of parts is confirmed by CAA's surveys, which show that parts for the 626 cost twice as much as parts for other cars in its class. Try independent suppliers to keep costs down.

Strengths and weaknesses: Although far from being high-performance vehicles, these cars ride and handle fairly well and still manage to accommodate four people in comfort. The 1988–92 versions incorporated a third-generation redesign that added a bit more horsepower to the 4-banger. Apart from that improvement, these cars are still easy riding, fairly responsive, and not hard on gas. On the downside, the automatic transmission downshifts roughly, the power steering is imprecise, and the car leans a lot in turns.

Four-wheel steering was part of the sedan's equipment in 1988, and it was added exclusively to the MX-6 a year later. Wise buyers should pass over this option and look instead for anti-lock brakes and airbags on 1992 LG and GT versions. The manual transmission is a better choice because the automatic robs the engine of much-needed horsepower, as is the case with most cars this size.

A mid-sport and mid-compact hybrid, the MX-6, Mystère is a coupe version of the 626. It has a more sophisticated suspension, more horsepower, and better steering response than its sedan alter ego. The 1993 model gained a base 2.5L 165-hp V6 powerplant. Overall reliability and durability are on par with the 626.

Through 1995, 626s have sub-par body construction, electrical system glitches, brakes that wear out prematurely, and automatic transmissions that shift poorly and are prone to premature failure. Expect jerky downshifts when the 4-cylinder is at full throttle. Shocks and struts (MacPherson) are expensive to replace (especially when the model is equipped with the electronic adjustment feature). The electronically controlled shock absorbers haven't been durable and they, too, cost a lot to replace. Body problems include door and hatch locks that often freeze up, headliner rattles, and the metal surrounding the rear wheel wells is prone to rust perforation, as are hood, trunk, and door seams. The paint seems particularly prone to chipping. The underbody and suspension components on cars older than five years should be examined carefully for corrosion damage. The exhaust system will rarely last more than two years, and wheel bearings fail repeatedly within the same period.

You will want to buy a 1996–98 626 for the improved performance, handling, and overall reliability, plus reasonable fuel economy (premium grade, though).

Dealer service bulletins: **1993**—Door glass rattles. • Vibrating hood. • Water leaking into the trunk. • Engine misses during hard cornering. **1994**—Air conditioning defects. • Head gasket leaks. • Automatic transmission malfunctions. • Excessive engine hydraulic lifter noise. • Brake vibration and premature wearout of the front brakes. • Steering gear could produce a clunking noise or be off-centre. • Door glass pulls out of its track. • Inside rear view mirrors fall off the windshield. • Driver-side power seat might not work. • Lots of wind noise around the side mirrors and A-pillars. **1995**—Camshaft friction gear noises. • 3–4 shift hunt. • Door side moulding detaches in cold weather. • Expansion valve whistling noise. Front strut squeaks on turns. • Heater and AC unit noise after long storage. • Loose, rattling sunroof and outer door handles. • Creak/rattle noise from passenger-side wiper. • Steering wheel slightly off-centre. **1996**—3–4 shift hunt. • Front strut squeaks on turns.

Safety summary/Recalls:: **1995–96**—Inadvertent airbag deployments. **1997 626 ES**—Head restraints are too low. One North Vancouver neurologist says the 626's head restraints are set too low and cannot extend to a safe level. He says there is an additional two inches required for a six-foot-tall occupant. When informed of his assessment, the dealer replied that Mazda "cannot help you with your problem." The doctor maintains his Mazda 626 cannot be safely operated by a driver over five feet and ten inches in height. He concludes: "As a result of my occupation, I see many motor vehicle accident neck injuries and have a keen interest in making my new vehicle, and those of others, safe." **Recalls**: **1986**—The throttle could stick open due to a defective nylon rotor. **1986–87**—An ignition switch failure will result in faulty wipers, washer, engine fan, heater blower, and air conditioner compressor. **1988**—Automatic shoulder belt could break. • The floor mat could interfere with the gas pedal. • Frost could accumulate in the throttle body. • A rear brake shoe could separate from a wheel cylinder piston. • Band flexing could cause the fuel tank to leak. **1988–91**—Mazda will replace original door handles with new handles that are more durable so that door won't fly open. **1988–89 MX-6**—Takata seatbelts will be replaced. **1995**—Airbags malfunction. **1995–96 626, MX-6**—Inadvertent airbag deployment when passing over bumps in the road. Sensors will be reprogrammed. **1997 626, MX-6**—Tensioner spring may break and get caught in the engine timing belt, stalling the engine; dealers will replace the tensioner. **1997 626**—Inadvertent airbag deployment when passing over bumps in the road. **1998**—Faulty computer module causes stalling.

Secret Warranties/Service Tips

1993-94—Freezing door and hatch lock cylinders are addressed in DSB #021/94. • Headliner rattles can be fixed by using Mazda's fastener kit. • The driver-side power seat might not work if the wiring harness touches the seat frame. • A clunking noise coming from the steering gear is caused by excessive backlash in the steering gear assembly. **1995-96**—A 3–4 shift hunt is probably caused by failure in the 3–4 shift solenoid hydraulic circuit. • Front strut squeaks on turns could be caused by interference between the upper seat spring and the strut dust cover or between the dust cover and the rubber bump stopper. **All models/years**—Non-turbo models that idle roughly after a warm restart could have fuel vapourizing in the distribution pipe due to high under-hood temperatures; to eliminate this problem, the emission control unit has been redesigned to allow higher fuel pressure at lower temperatures (DSB #023/87R). • Excessive rear brake squealing can be reduced with improved brake pads (DSB #015/89-11). • Excessive vibrations felt in the brake pedal, steering wheel, floor, or seat when applying the brakes can be fixed by installing a redesigned brake assembly.

626, Cronos, MX-6, Mystère Profile

	1991	1992	1993	1994	1995	1996	1997	1998
Cost Price ($)								
626/Cronos	15,215	15,595	17,395	18,725	19,365	19,995	19,995	20,140
MX-6/Mystère	15,540	15,825	19,375	20,695	21,835	22,780	23,325	—
Used Values ($)								
626/Cronos ↑	5,000	6,000	7,500	9,500	11,000	13,500	15,000	17,000
626/Cronos ↓	4,000	5,000	7,000	9,000	10,500	12,000	14,000	16,000
MX-6/Mystère ↑	6,000	7,500	9,000	10,500	11,500	14,500	16,000	—
MX-6/Mystère ↓	5,000	7,000	8,500	9,500	10,500	14,000	15,000	—
Extended Warranty	Y	Y	Y	Y	Y	N	N	N
Secret Warranty	N	N	N	N	N	N	N	N
Reliability	❷	❷	❷	❷	❷	③	④	④
Air conditioning	❷	❷	③	③	③	③	④	④
Body integrity	❷	❷	❷	❷	❷	③	③	④
Braking system	❷	❷	❷	❷	❷	③	③	③
Electrical system	❷	❷	❷	❷	③	③	③	④
Engines	④	④	④	⑤	⑤	③	③	④
Exhaust/Converter	⑤	⑤	⑤	③	④	⑤	⑤	⑤
Fuel system	④	④	③	④	④	④	⑤	⑤
Ignition system	④	④	④	④	⑤	⑤	⑤	⑤
Manual transmission	④	④	④	④	⑤	⑤	⑤	⑤
- automatic	❷	❷	❷	❷	❷	③	④	④
Rust/Paint	❷	❷	❷	❷	❷	③	③	④
Steering	③	③	③	④	④	⑤	⑤	⑤
Suspension	❷	❷	❷	③	③	④	④	④
Crash Safety								
626 4d	—	—	❶	❶	④	④	④	—
Side Impact								
626 4d	—	—	—	—	—	—	❷	③

NISSAN

Altima

Rating: Above Average (1996–98); Average (1993–95). The four-cylinder engine barely provides the necessary versatility needed to match the competition: the Toyota Camry, Mazda 626, updated Ford Probe, or larger Ford Sable and Taurus. Although the SE gives the sportiest performance, the less-expensive GXE is the better deal from a price/quality standpoint. **Maintenance/Repair costs**: Higher than average. Repairs are dealer dependent. **Parts**: Owners complain of parts shortages, and the CAA reports that parts are more expensive than those for most other cars in this class.

Strengths and weaknesses: The Altima's wheelbase is a couple of inches longer than the Stanza's, and the car is touted by Nissan as a mid-size, even though its interior dimensions put it in the compact league. The small cabin seats only four, and rear seat access is difficult to master due to the slanted roof pillars, inward-curving door frames, and narrow clearance.

The base engine gives average acceleration and fuel economy. Good manoeuvrability around town. No reliability problems reported with the 16-valve powerplant or transmission. Uncluttered underhood layout makes servicing easy. Good body assembly.

With its noisy and rough engine performance, this car cries out for a V6 like the one used in the Maxima. The 4-banger has insufficient top-end torque and gets buzzier the more it's pushed. In order to get the automatic to downshift for passing, for example, you have to practically stomp on the accelerator. The 5-speed manual transmission is sloppy. The Altima's sporty handling is way overrated; there's excessive body roll and front-end plow in hard cornering, tires squeal at moderate speeds, and steering isn't as precise or responsive as befits a car with performance pretensions. In spite of the car's independent suspension, it gives a busy, uncomfortable ride that's punishing over bumps. Lots of engine, road, and tire noise.

Dealer service bulletins: **1993**—A creaking, squeaking, or tapping noise coming from the window and door areas. • Water leaks. • Electrical system, heater, air conditioning, and defrosting malfunctions. • Body/trim deficiencies. **1994**—No Reverse (automatic). • Excessive brake and steering wheel vibration. • Inaccurate fuel gauges. • Rear suspension noise. **1995**—Front cover oil leakage. • No or low line pressure rise (control valve). • Poor driveability—Code 45. • AC compressor leaks or is noisy. • Troubleshooting excessive front and rear brake noise. • Some windshield cracks are covered under warranty.

Safety summary/Recalls: **1997**—Fire ignited in the engine compartment while vehicle was parked. • Fire started by fuse box in passenger compartment. • Sudden acceleration. • Frequent reports that the airbag failed to deploy. • Poor braking performance. • Poor design causes electronic control unit failure. • Defective automatic transmission solenoid. • Windshield wiper fails periodically. • Water leaks into trunk area causing premature rusting. • Automatic door lock failure. • Seat and shoulder belts lock up and don't retract. **1998**—Stalling when accelerating. • Rear seats won't lock upright. • Windows rattle excessively. • Wheel cover failure. **Recalls**: **1993–94**—Throttle cable may not return to idle. **1995**—Automatic shift lever plate can break, causing unintended vehicle movement. • Brake hose may leak. **1996**—AC refrigerant leaks. • Leaking AC evaporator drain hose causes ECM damage. **1997**—Defective seatbelt buckles.

Secret Warranties/Service Tips

1993—Centre caps on alloy wheels may not fit properly; order upgraded caps (#40315-D9000). • An exhaust rattle during acceleration can be resolved by installing an upgraded muffler-exhaust assembly (#203000-IE860). • Inadequate defrosting on vehicles made before February 1993 can be resolved by following Nissan's suggestions in its #NTB93-059 bulletin. • #NTB93-086, a lengthy Nissan bulletin sent out May 25, 1993, troubleshoots the Altima's most common water leaks into the interior. • Poor AC performance can be corrected by installing a thermal control unit (TCU). • An inoperative AC and heater may have a kinked heater/water valve cable that has damaged the air mix door actuator. **1993–94**—No Reverse gear with the manual transmission: install a Nissan upgrade kit. • Fuel gauge is inaccurate or it may be difficult to fill a fuel tank fixed with upgraded parts. **1993–96**—DSB #NTB96-046 gives lots of useful tips for troubleshooting squeaks and rattles.

Altima Profile

	1993	1994	1995	1996	1997	1998
Cost Price ($)						
XE	16,491	17,690	18,990	20,598	20,798	19,398
GXE	18,690	19,990	21,690	23,298	23,498	21,398
SE	21,190	22,490	24,190	25,798	25,998	24,398
Used Values ($)						
XE ↑	8,000	9,500	11,500	13,500	15,000	17,000
XE ↓	6,500	7,500	10,500	12,500	14,500	16,500
GXE ↑	8,500	9,500	13,000	14,500	16,500	18,500
GXE ↓	8,000	8,500	12,000	13,500	15,500	17,500
SE ↑	10,500	12,500	14,500	16,000	18,000	20,000
SE ↓	9,500	11,500	13,500	15,500	17,500	19,000
Extended Warranty	**N**	**N**	**N**	**N**	**N**	**N**
Secret Warranty	**N**	**N**	**N**	**N**	**N**	**N**
Reliability	③	③	④	④	④	⑤
Air conditioning	❷	③	④	④	④	⑤

Body integrity	❷	❷	③	③	③	③
Braking system	❷	❷	③	③	③	③
Electrical system	❷	❷	❷	③	③	③
Engines	④	⑤	④	⑤	⑤	⑤
Exhaust/Converter	③	④	④	⑤	⑤	⑤
Fuel system	③	③	④	⑤	⑤	⑤
Ignition system	④	⑤	⑤	⑤	⑤	⑤
Manual transmission	④	⑤	⑤	⑤	⑤	⑤
– automatic	④	⑤	⑤	⑤	⑤	⑤
Rust/Paint	❷	❷	③	③	③	③
Steering	③	③	③	④	⑤	③
Suspension	③	③	③	③	③	③
Crash Safety	⑤	⑤	—	④	④	③
Side Impact	—	—	—	—	—	③

Stanza

Rating: Above Average (1990–92); Average (1988–89); Not Recommended (1985–87). The Stanza was replaced by the Altima in 1993. **Maintenance/Repair costs**: Lower than average. Repairs aren't dealer dependent. **Parts**: Average cost, but can be bought for much less from independent suppliers.

Strengths and weaknesses: The 1988–89 Stanzas were mini Maximas in that they were redesigned along luxury Maxima lines. Unfortunately, they were under-powered, the automatic transmission was reluctant to downshift, the engine was noisy, and the base suspension allowed the car to bounce about. This is not the car for a full passenger or cargo load.

1990–92 Stanzas are roomy, reasonably priced, four-passenger compacts that offer peppy performance, more responsive steering, nimble handling, and good fuel economy. Overall reliability has been fairly good during the past seven years. Except for some road noise, suspension thumps, starting difficulties, transmission malfunctions, and a biodegradable exhaust system, no major problems have been reported on 1988–92 Stanzas.

Owners complain that the automatic transmission vibrates annoyingly when the car is idling in gear, clutches are noisy, and steering wheel vibrations are frequent. Power antenna malfunctioning, premature front brake wear, and surface rust problems are common for all years. Especially prone to rust perforation are wheel openings, the front edge of the hood, the rear hatch, and door bottoms.

Safety summary/Recalls: **Recalls**: **1986 Wagon 2X4**—Cover the fuel-filler pipe for increased accident protection.

Secret Warranties/Service Tips

1982–88—Improved brake pad material will cut brake noise and add to pad durability. Upgraded semi-metallic pads carry the Hitachi HP12 FE designation. **1987–89**—Trunk lid torsion bars won't fall off if you install a torsion bar spacer (#84449-D4060). **1989–92**—Clutch shudder and steering wheel vibration can be eliminated by checking brake rotor thickness or installing an upgraded pressure plate release lever. **1990**—Starting in D4 or delayed downshifting requires a new valve body separator plate. **1990–91**—Starting difficulties can often be traced to an ECU connector not fully seated in the ECU. **1990–92**—Clutch whine or screech in cold temperatures requires the installation of an upgraded clutch disc. **All models/years**—A rotten-egg smell could be caused by a defective catalytic converter; have it replaced free of charge if the emissions warranty still applies.

Stanza Profile

	1987	1988	1989	1990	1991	1992
Cost Price ($)						
Base	13,687	14,288	15,089	15,690	15,290	15,290
Used Values ($)						
Base ↑	1,800	2,500	3,000	3,500	4,500	5,000
Base ↓	1,500	2,000	2,500	3,000	4,000	4,500
Extended Warranty	Y	Y	Y	Y	N	N
Secret Warranty	N	N	N	N	N	N
Reliability	❷	③	③	③	④	④
Air conditioning	③	④	④	④	⑤	⑤
Body integrity	❷	❷	③	③	③	③
Braking system	❷	❷	③	③	③	③
Electrical system	❷	❷	❷	③	③	③
Engines	❷	③	④	④	④	④
Exhaust/Converter	❷	❷	❷	❷	③	③
Fuel system	❷	③	③	③	③	③
Ignition system	③	③	③	③	③	③
Manual transmission	③	④	④	④	④	④
- automatic	❷	③	③	③	③	③
Rust/Paint	❷	❷	③	③	③	③
Steering	❷	③	③	④	④	④
Suspension	❷	③	③	③	③	④
Crash Safety						
Stanza 4d	—	—	—	❶	⑤	⑤

SUBARU

Legacy

Rating: Only the AWD models are Above Average (1995–98); Average (1989–94).The 1995–98 models are way overpriced considering they don't offer much more than AWD. Before buying a front-drive Subaru, first consider what Honda, Mazda, or Toyota have to offer. The only reason to buy a Subaru is for its 4X4 capability, and on most used models you'll have to pay a $1,500–$2,000 premium to get it. If it's not what you need for your basic driving requirements, pick a car with a conventional drivetrain. **Maintenance/Repair costs**: Higher than average. Repairs are dealer dependent. **Parts**: Higher-than-average cost can't be attenuated through purchasing from independent suppliers.

Strengths and weaknesses: First launched in 1989 as front-drives, these compacts are a bit slow off the mark. The 5-speed is a bit notchy, and the automatic gearbox is slow to downshift and has difficulty staying in Overdrive. Early Legacys are noisy, fuel-thirsty cars with bland styling that masks their solid, dependable AWD performance. Actually, the availability of a proven 4-wheel drive powertrain in a compact family sedan and wagon makes these cars appealing for special use. In spite of their reputation for dependability, though, Subarus are not trouble-free—engine, clutch, turbo, and driveline defects are common on the early models through to the 1994 models.

The redesigned 1995–98 models have sleeker styling, additional interior room, a bit more horsepower with the base engine, and a new 2.5L 4-cylinder driving the 1996 AWD GT and Lsi. The Outback, a Legacy Madison Avenue spin-off, was transformed into a sport-utility wagon with a taller roof. Even with the improvements noted above, acceleration is still only passable (if you don't mind the loud engine), but handling and ride are remarkably good. Overall, it's a competent car with few faults and only one outstanding feature—AWD. If you don't need that feature, there are better-performing, less-expensive vehicles available. Another word of warning: *Consumer Reports* says that the Outback upgrade isn't worth the extra money. They point out that Subaru's ads are misleading when they show Paul Hogan zipping around off-road. The publication also reports that the Subaru's handling isn't as competent as that of the regular Legacy (something to do with the higher centre of gravity, no doubt).

Subaru's overall product lineup for 1997 marked a return to the company's four-wheel drive roots with the repackaging of its Legacy and Impreza 4X4 lineup as Outbacks (half of all Legacys sold are Outbacks). A Legacy 2.5L GT all-wheel drive sporting sedan, or wagon variant, also joined the group that year. In addition to these redesignated models—and the squeezing out of a bit more horsepower from its limited range of engines—Subaru continued to tap the sport-utility craze through lower prices for its AWD and ABS options and by offering a greater variety of AWD vehicles.

Are Subarus good buys? Devotees will insist that there's no other car as versatile, and the manufacturer is justifiably proud of the high J.D. Power owner satisfaction rating earned in the U.S. On the other hand, I have found that a Subaru can be a good used-car choice only if its AWD feature is essential for your driving needs, it's been maintained carefully, the body hasn't begun to rust, and dealer servicing is easily available. This last point is crucial because these cars are very dependent on the dealer network for parts and servicing.

Through 1995, premature exhaust system rust-out and automatic transmission (front seals, especially) and clutch breakdowns are the more common complaints. Brakes require frequent attention. Shock absorbers, constant velocity joints, and catalytic converters often wear out prematurely. Other problems that appear over many model years include starter and ignition relay failures and front-end suspension noises. Subarus are rust-prone: fenders, door bottoms, rocker panels, wheel openings, bumpers and supports, rear quarter-panels, tailgates, trunk lids, and hoods are particularly vulnerable. Additionally, the underbody and chassis components should be examined very carefully for corrosion damage.

1996–98 Legacys are much improved, with few problems except for some minor brake, electrical, clutch, and body integrity shortcomings.

Safety summary/Recalls: **1996**—Sudden acceleration. • When the vehicle is being driven, transmission may suddenly jump out of Drive into Neutral. • Inadvertent airbag deployment. • Airbags failed to deploy. • ABS brakes locked up. • Brake failures. • ABS performs poorly on snow and ice. • Floor carpet prevented brake pedal application. • Complete engine failure at 18,000 km. • Chronic stalling. • Computer sensor control unit failure. • Excessive shaking at highway speeds. • Windshield wiper bolt failure. • Alternator failures while driving. • Left turn signal fails intermittently. • Climate control button sticks. • AC seizure. • O-ring failure causes AC to leak freon. • Bridgestone tires frequently blow out. • Keyless entry failed due to pinched wire in driver's door. • New-design headlights give poor illumination. **1997**—Airbags deployed but failed to inflate. • Airbags failed to deploy. • Sudden acceleration. • Cruise control failed to disengage when brakes were applied. • Igniter failed, allowing unburned fuel to flow into catalytic converter. • AC blew fumes into interior, causing driver to black out. • Stalling caused by igniter failure. • Complete engine failure due to defective valves and pistons. • Sudden loss of steering. • Poor braking performance; ABS brakes frequently fail or lock up. • Front brake pad failure. • Premature wearout of all brake components. • Transmission surges when cold, or shifts into Neutral at low speed or when descending a small hill. •

Transmission failures. • Frequent electronic control unit failures. • Rear seatbelts are too long to properly secure child safety seat, and the locking mechanism doesn't lock properly. • Brake and engine lights continually on. • Three alternators replaced by one owner. • Shorted hazard switch drained battery. • Alternator belt snapped, causing battery and brake warning lights to come on and making car hard to steer. • Alternators frequently quit while vehicle is under power. **1998**—Many instances where the airbags failed to deploy. • Oil leak from oil filter seam caused fire. • Sudden brake loss after linings, calipers, and master cylinder had been replaced. • Cruise control failed to disengage when brakes were applied. • Excessive shaking at highway speeds. • Seatback collapsed when vehicle was rear-ended. • Subaru told car owner that tendency to pull to the right was a design feature. **Recalls**: **1989–93**—Frost could build up in the manual transmission dipstick vent, causing transmission oil leakage and eventual wheel lockup. **1990–91**—The defroster is faulty. • Help! Door might not open from the inside. • Automatic transmission could engage abruptly, causing a sudden lurch into Reverse. **1990–93 4X4**—Cold weather or high humidity could cause the manual transmission to seize. **1993**—The top of the fuel tank could leak. **1998**—Due to poor welds, ignition keys can stick, shift levers and linkages can break, and shift levers can move; dealers will replace the automatic transmission shift lever assembly.

Secret Warranties/Service Tips

1990–91—The defroster is faulty. • Front door malfunction could make it impossible to open the door from the inside. • The 4EAT automatic transmission Park gear might not engage properly. **1990–93 AWD**—The manual gearbox could leak, causing the transmission to seize. **1993**—Legacys could have a headliner droop, which is addressed in a March 1994 DSB. **1994**—Torque converter squeaking is addressed in a June 1993 DSB. **1995**—Tips are provided on silencing excessive front strut noise and engine oil pump leaks. **1995–96**—If the antenna won't fully retract, Subaru suggests cleaning the antenna mast and replacing the dress nut. **1995–97**—Troubleshooting tips on a sticking anti-lock brake relay are offered. **1997**—Troubleshooting tips are offered on transfer clutch binding and/or bucking on turns. **All models/years**—A rotten-egg smell could be caused by a defective catalytic converter. It will be replaced, after a bit of arguing, free of charge under the emissions warranty.

Legacy Profile

	1991	1992	1993	1994	1995	1996	1997	1998
Cost Price ($)								
Legacy	16,482	16,888	16,888	19,995	17,995	23,195	—	—
Legacy 4X4	18,042	18,969	19,126	22,695	24,695	25,195	26,495	25,695
Used Values ($)								
Legacy ↑	4,500	5,500	8,000	9,500	10,500	14,500	—	—
Legacy ↓	3,500	5,000	7,500	9,000	9,500	13,500	—	—
Legacy 4X4 ↑	6,000	7,500	9,000	10,500	13,500	15,000	20,000	23,000
Legacy 4X4 ↓	5,000	6,500	8,500	10,000	12,500	14,000	19,000	22,000

Extended Warranty	Y	Y	Y	Y	Y	Y	Y	Y
Secret Warranty	N	N	N	N	N	N	N	N
Reliability	❷	③	❷	❷	③	④	④	④
Air conditioning	❷	❷	③	③	④	⑤	⑤	⑤
Body integrity	❷	❷	❷	❷	❷	③	③	③
Braking system	❷	❷	❷	❷	❷	③	③	③
Electrical system	❷	❷	❷	❷	③	③	③	④
Engines	③	④	④	⑤	⑤	⑤	⑤	⑤
Exhaust/Converter	❶	❶	❷	❷	③	④	⑤	⑤
Fuel system	③	③	④	④	④	④	⑤	⑤
Ignition system	❷	❷	③	③	④	⑤	⑤	⑤
Manual transmission	③	③	③	④	⑤	⑤	⑤	⑤
- automatic	❷	❷	③	③	③	③	⑤	⑤
Rust/Paint	❷	❷	❷	③	③	③	④	⑤
Steering	③	③	④	④	⑤	⑤	⑤	⑤
Suspension	③	③	③	③	⑤	⑤	⑤	⑤
Crash Safety								
Legacy 4d	❶	❶	❶	❶	④	④	④	④
Side Impact	—	—	—	—	—	—	—	③

TOYOTA

Camry

Rating: Recommended (1994–98); Above Average (1988–93); Average (1985–87). 1996 was the wagon's last model year. For the past four years, Camrys have elicited an unusual number of safety complaints that are carried over from one model year to the next. The complaints include: severe wandering at highway speeds; sudden acceleration; engine-compartment fires; brake failures; poor headlight illumination; and transmission interlock failures, which allow a parked vehicle to roll away. **Maintenance/Repair costs**: Higher than average, but repairs aren't dealer dependent. **Parts**: Parts are more expensive than for most other cars in this class (alternator and ignition module, for example). Parts availability is excellent.

Strengths and weaknesses: Safety complaints, aside, the Camry is an excellent family-car buy because of its spacious, comfortable interior, good fuel economy, and impressive reliability and durability.

1987–93 models have few problems, although they're far from perfect. Main areas of concern are failure-prone cylinder head gaskets; suspension and electrical system failures; defective starter drive and ring gear; leaking low pressure and high pressure power-steering lines and outer CV boots that split, causing grease to leak; premature brake wear; and some paint peeling and rusting. Mufflers last only two years on earlier models, and sunroofs are rattle-prone.

Persistent problems with all Toyota vehicles are premature brake wear, excessive noise, and vibrations. Stung by consumer criticism that these problems haven't been fixed for over a decade and that owners are charged for useless repairs, Toyota published a "Brake Repair" service bulletin (POL94-18) in October 1994 for its American dealers, which defines those repairs that will be done under warranty. Toyota states that premature brake wear and noise will be fixed under warranty for the first 12 months/12,500 miles, and that vibrations will be attended to, under warranty, up to 3 years/36,000 miles. Canadians should use these parameters as their guide in requesting "goodwill" repairs.

Front suspension bushings wear out quickly, leading to clunking and squeaking noises when going over bumps or when stopping quickly. There's also the so-called Camry chop (exceptionally rough rides when passing over uneven roadways) reported by owners of 1992–94 models. Cruise control fails frequently on all years. Owners of the 1992 Camry have reported that a chronic drone noise, along with a vibration felt from the floor and the gas pedal, occurs mostly when the automatic transmission changes from second to third gear at 1,800–2,000 rpm.

1994–98 Camrys aren't perfect: owners report premature brake failures, faulty window regulators, smelly ACs, and myriad rattles, clunks, and groans that seem to come from everywhere. There is also an annoying surging and shuddering when decelerating, which appears to be more common with the 1995–96 models. It seems as though it's a sticking throttle position sensor, but mechanics say that the problem is intrinsic to the way the engine/transmission computer module is calibrated. Other deficiencies reported by owners: brake vibrations, premature brake pad wear, AC malfunctions, and defective automatic transmissions that slip out of gear when parked. 1994–98 model body problems include excessive wind noise coming from the front windshield, back doors, and sunroof. Trim items rust and fall off, door handles pull away, and mufflers have a short life span. No reports of rust perforation problems, but weak spots are door bottoms, rear wheel openings, and trunk and hatchback edges. There are complaints concerning premature rusting on cars painted white. Toyota generally corrects these rust/paint deficiencies for free.

Totally redesigned, the 1997 Camry is taller, longer, wider, more powerful, and cheaper, in both a literal and a figurative sense. Gone are the coupe and station wagon variants. The wheelbase was extended by two inches, giving backseat passengers more room. Other changes: it's powered by a base 2.2L 133-hp 16-valve 4-cylinder engine (taken from the Celica)and an optional 3.0L 24-valve V6 that unleashes 194 horses. Either engine will be mated to a 5-speed manual or an electronically controlled 4-speed automatic. ABS and traction control are standard on all V6-equipped Camrys, rear seats have shoulder belts for the middle passenger, brighter low beam lights, optional heated mirrors, more cup holders, a sunglasses holder, and an additional power port in the centre console.

"Decontenting" hit Toyota's 1997 lineup hard, resulting in many changes that cheapened the Camry through less sound insulation (when more was needed, I'd say) and less durable trim and accessories. Although Toyota admits to engine head gasket leaks for the first time (see "Secret Warranties/Service Tips"), reliability doesn't appear to have been affected, notwithstanding a sharp increase in noise complaints. The changes include: less expensive S-rated tires on models with 4-cylinder engines, cheaper heating/ventilation system components, no more assist handles for front occupants, no more chrome trim around the windshield, one door seal instead of three (greater chance for wind and water leaks), fewer airbag sensors, an LCD odometer, a distributorless ignition with the 4-cylinder, and a windshield-embedded antenna. Owners report the 1997 models have limited rear visibility (due to the side pillars and high trunk lid), less steering "feel," and more squeaks and rattles than previous versions.

Dealer service bulletins issued so far for the 1998 Camry show fewer trim problems and no engine head gasket alert, but noise concerns relating to the suspension, steering, and power front seat still remain (see following bulletin).

STEERING
ST004-98
August 21, 1998
Title:
STEERING RACK HOUSING BUSHING NOISE
Models:
'97 & '98 Camry
Introduction

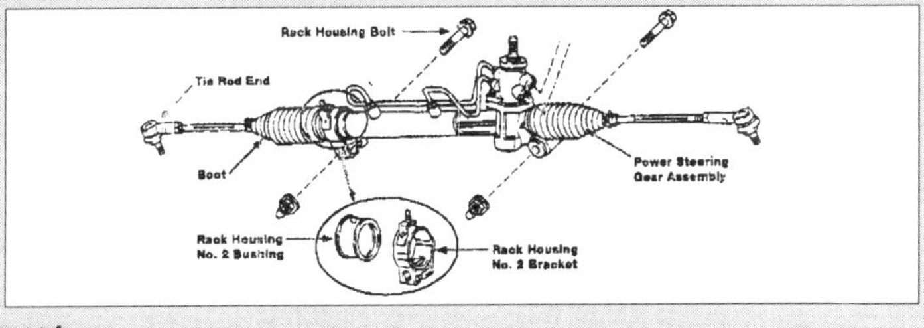

Parts Information

PREVIOUS PART NUMBER	NEW PART NUMBER	PART NAME
45517-06020		
45517-06030	45517-33021	Steering Rack Housing Bushing No. 2

OPCODE	DESCRIPTION	TIME	OPN	T1	T2
ST8002	Replace Steering Rack No. 2 Bushing	1.0	45517-06030	91	89

Warranty Information
Applicable Warranty*:
This repair is covered under the Toyota Basic Warranty. This warranty is in effect for 36 months or 36,000 miles, whichever occurs first, from the vehicle's in-service date.
Warranty application is limited to correction of a problem based upon a customer's complaint.
Repair Procedure
If the customer encounters noise in cold weather when turning at very low speed (2-3 m.p.h.), replace the Rack Housing Bushing with the updated part.

Toyota won't give you the upgraded steering component unless you ask for it.

Dealer service bulletins: **1994**—Improved durability of Overdrive clutch bearing. • Rear brake squeak. • Rear suspension noise. • Steering column clicking noise. • New seals to prevent water leakage into interior. • Water pump seal improvements. • Wind noise from window. • AC evaporator odour. • Brake pad kits. Cause and repair of brake vibration and pulsation. • Front brake noise. • Improved power window regulator. • Moon roof panel wind noise. • **1995**—Front brake noise. • AC evaporator odour. • Moon roof panel wind noise. • Improved power window regulator. • Rear brake squeak. • Rear suspension noise. **1996**—Front brake noise. • AC evaporator odour. • Charcoal canister humming noise. • Rear window rattle. • Rear brake squeak. • Rear suspension noise. • Moon roof panel wind noise. **1997**—Engine head gasket coolant leak. • AC evaporator odour. • Charcoal canister humming noise. • Difficulties with moon roof operation. • Exterior rear view mirror improvement. • Steering rack housing bushing noise. • Front shoulder belt anchor buzzes. • Front suspension groans. • Suspension rattle and popping. Tailpipe contact with heat shield. • Headliner buzzes or rattles. • Moon roof rattles. • Manual front seat movement/noise. • Power front seat chattering. • Radio volume control too sensitive. • Rubbing noise from door trim. • Seat cover loose at lower rear corners. • Seat movement field fix procedure. • Armrest bum improvement. • CD player won't accept/eject CDs. • Fuel door operation improvement. • Wind noise repair kit.

Safety summary/Recalls: Owners report that the Dunlop D60 A2 tire is a poor wet-weather performer. **1987–91**—Leaking fuel tanks. **1995**—Injury from airbag. • Premature front/rear brake wear. • Defective, poor performance (wet) Goodyear Invicta tires. • Windshield reflects dashboard image. **1995–96**—Brake failure. • Noisy, vibrating brakes. • Premature front and rear brake wear. • Excessive engine noise. • Transmission lever can slip from Drive to Neutral. • Airbag fails to deploy or is accidentally deployed. • Sudden acceleration, stalling. • Passenger-side seatbacks won't stay upright. • Passenger seatbelts over-retract. • Defective radio antenna. • Taillight and turn signal bulbs frequently burn out. • **1996**—Vehicle wanders over road. • Door bottom/undercarriage rusting. • Airbag warning light and Check Engine light always on. • Windshield film buildup. • Window water leaks. **1997**—Engine compartment fire following ABS brake failure. • Airbags failed to deploy. • Violent deployment of airbag caused death, during an accident. • Sudden acceleration when brakes were applied. • Tendency for car to wander all over the roadway. • Steering wheel suddenly locked up, causing an accident. • Many reports of transmission interlock system failures. • Many other reports, probably related to the interlock system, that vehicle was put in Park and keys taken out of ignition and car then proceeded to roll away. • Century infant seat won't fit in the centre of rear seat. • Seatbelts continually ratchet tighter; rear seatbelt was strangling child, who had to be cut free. • Plastic part fell behind the dash and lodged behind the brake pedal arm, causing an accident. • Frequent brake failures. • Premature brake pad and caliper wear or failure. • Loud grinding brake noise when braking. • Excessive noise coming from underneath the car at highway speeds. • Vehicle sits too low, has minimal ground clearance. • Driver's knee can hit the steering wheel adjuster lever making steering wheel go up and down. On several occasions, steering wheel suddenly tilted all the way up while on the highway. • Very poor headlight illumination; headlight safety cap cover design cuts visibility severely; there's a blind spot on the driver-side headlights. • Low beam lights aren't bright enough. • Dash indicator lights are too small and low in intensity. • Inoperative rear window defroster. • Driver's seatback rocks back and forth. • Power door lock relay failure. • Car locks and unlocks on its own. • Fuel door doesn't open fully when lever is pulled. • Rear windshield exploded while car was parked overnight. **1998**—Several reports of engine fires. • Vehicle stalled, oil light came on, and fire ignited in engine compartment. • Sudden acceleration. • Many reports that airbags failed to deploy. • Frequent complaints that vehicle wanders at highway speeds (50 km/h) and is difficult to control in a crosswind. • Overly soft suspension allows the chassis to scrape the roadway when passing over a small bump. • Frequent ABS brake failures. • Excessive brake noise and extended stopping distances. • Transmission gear shift lever went from Neutral to Drive without pressing button. • Airbag service indicator light stays on. • Engine malfunction light stays on. • Inadequate night illumination from headlights; low beam halogen headlights don't carry very far; dark spot cast from left headlight results in poor visibility; metal deflector inside the concealed headlights blocks out all light beyond 30 feet. • Electrical system failure; running lights won't shut off. • Lock design allows for occupants to be temporarily locked in vehicle if someone gets out before them and locks the doors. • Power door locks fail intermittently. • Front restraints lock up when vehicle is parked on an incline. • Sunvisors are too small to block the sun, and cut visibility. • Back windshield shattered. • Fumes from inside the vehicle fog up the windshield. • Gas tank makes sloshing noise when applied. • General Tire wears excessively on the inside tread. • Frequent complaints of moon roof leaks, which may cause electrical short. • Doors have to be slammed shut. **Recalls: 1987**—Liquid spilled on console could short out the automatic seatbelt motor. **1987–89**—Front seatbelt guides that stay retracted will be changed. **1987–90**—Malfunctioning power door locks. **1988 station wagon**—Original equipment jack may collapse. **1991**—Dealer will replace faulty electrical components in Fujitsu Ten radios to eliminate the chance of fire. **1994**—Steering wheel may disengage from the steering column or shaft assembly. One owner of a 1992 Camry reports that her car suddenly went out of control. She found that the left rear tire was turned to a 45 degree angle: the front control arm that holds the tire in alignment had completely broken off at the tire end of the arm. **1996**—Taillight assembly lacks sufficient heat resistance. **1997**—Ignition key can be removed when vehicle not in Park. • Brake failure may occur due to moisture freezing in the brake vacuum hose.

Secret Warranties/Service Tips

1987–90—Engine ping, surging, or jerk can be fixed with an upgraded ECU. **1987–91**—A front inner shoulder belt guide is available to keep the belt away from the neck and face. This free accessory is covered under the seatbelt warranty. **1988**—Harsh 2–3 shifts may require the replacement of worn valve body rubber check balls. **1988–89**—Hard starting may require a new cold-start time switch. **1989–90**—A rattling sunroof will be silenced with an upgraded sliding mechanism and new cables and shoes. • Wind noise coming from the upper windshield molding requires that sealant be injected underneath the molding. **1992**—Noisy rear brakes can be silenced by installing upgraded brake pads identified by the letter "N" stamped on the part number label. **1992–93**—Upgraded brake rods will reduce front brake groan/grind. **1993**—Under a special program, Toyota has paid for the repair of AC units that fail to cool due to a faulty expansion valve. Ask for partial compensation. **1993–96**—Suspension squeaks and groans are addressed in DSB #SU95-003. **1994**—A steering column clicking noise calls for the replacement of the steering main shaft assembly and steering column tube assembly. • Rear window wind noise can be stopped by replacing the front centring-type bolt with a non-centring-type bolt and a washer. **1995–96**—Use upgraded brake pad material to eliminate brake groaning, according to the following DSB #BR002-96.

REF.: BRAKES
NO.: BROO2-96
DATE: MARCH 29, 1996
MODEL: CAM & AVL
CAMRY & AVALON FRONT BRAKE GROAN NOISE
To reduce front brake groan noise on Camrys and Avalons, the brake pad material has been changed.
IDENTIFICATION OF NEW PARTS:

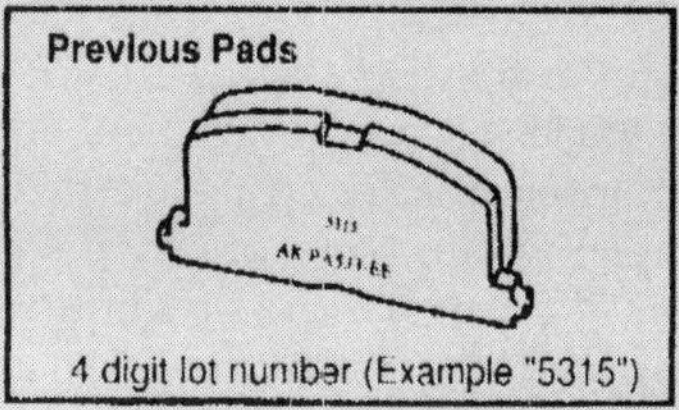

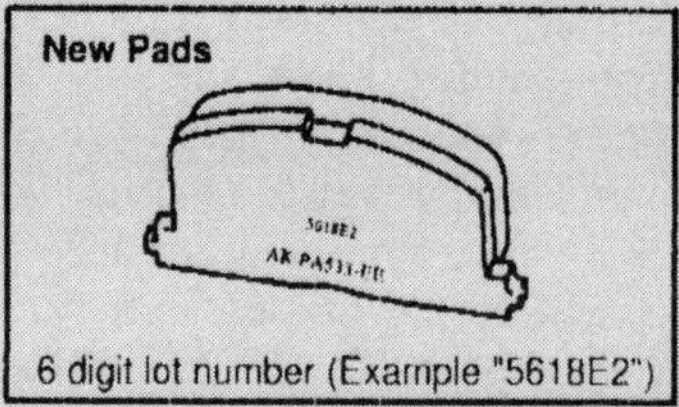

1996–97—A charcoal canister humming noise can be silenced by installing an upgraded vacuum hose. **1997**—Head gasket leaks are covered by a special Toyota program that is applied only if the customer complains (see the following bulletin and illustration).

ENGINE
EG008-98
July 10, 1998
Title:
5S-FE COOLANT LEAK FROM HEAD GASKET
Models:
'97 Camry
Introduction
To improve head gasket sealing on 5S-FE engines, follow the specified procedure for applicable conditions listed.

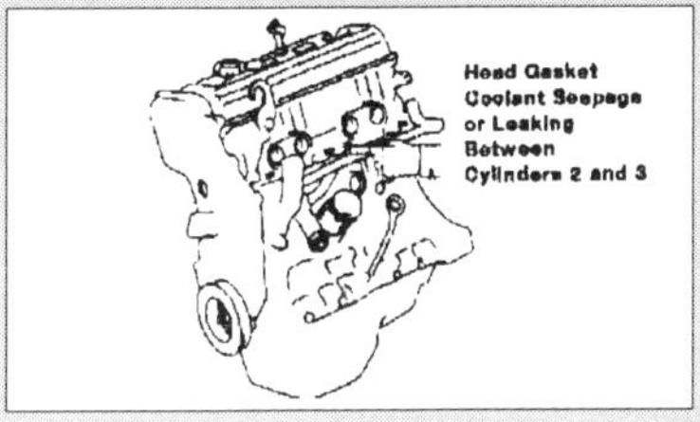

A. Slight coolant seepage or residue at head gasket area on left side of engine block (see area indicated in illustration).
B. Coolant visible leaking down left side of cylinder block originating from area indicated in illustration.

OPCODE	VWC	DESCRIPTION	TIME	OPN	T1	T2
110011	Q	R & R Cylinder Head	11.4	11115-74XX0*	64	13
110051	Q	R & R Cylinder Head Gasket	5.4	11115-74XX0*	64	13

* Where xx=12 if metal or 08 if composite

Warranty Information
Applicable Warranty*:
This repair is covered under the Toyota Powertrain Warranty. This warranty is in effect for 60 months or 60,000 miles, whichever occurs first, from the vehicle's in-service date.
* Warranty application is limited to correction of a problem based upon a customer's complaint.

Is a less-durable V6 engine the price to pay for Toyota's cheapening of its products in 1997? It's too early to tell. Nevertheless, this kind of problem hasn't been seen since the automaker extended the head gasket warranty to eight years for the identical defect on its 1988–94 V6-equipped sport-utilities and pickups. Remember, this 5- to11-hour job won't be done unless the customer complains.

• A front suspension groan can be fixed by replacing the front spring bumper under Toyota's three-year warranty. • If the driver's seat "rocks," Toyota has an upgraded assembly that will secure the seat. **All models/years**—A decade-old brake pulsation/vibration problem is fully described and corrective measures are detailed in DSB #BR94-002, issued February 7, 1994. Sometimes only the parts are covered; the owner has to pay for labor. • To reduce front brake squeaks on ABS-equipped vehicles, ask the dealer to install new, upgraded rotors (#43517-32020). • Owner feedback over the last decade plus dealer service managers who wish to remain anonymous tell me that Toyota has a secret warranty that will pay for replacing front disc brake components that wear out before 2 years/40,000 km. If you're denied this coverage, threaten small claims court action. • A wind noise repair kit is now available.

Camry Profile

	1991	1992	1993	1994	1995	1996	1997	1998
Cost Price ($)								
Base Coupe	—	—	—	19,238	19,998	20,488	—	—
Base Sedan	16,048	17,948	18,778	20,138	20,638	21,138	21,178	21,348
LE	18,648	20,998	23,058	24,918	24,718	25,458	25,268	26,508
4X4	19,488	—	—	—	—	—	—	—
Wagon V6	23,568	26,008	27,668	29,708	31,278	32,178	—	—
Used Values ($)								
Base Coupe ↑	—	—	—	11,500	13,000	14,000	—	—
Base Coupe ↓	—	—	—	10,500	12,000	13,500	—	—
Base Sedan ↑	7,500	10,000	11,000	12,500	13,000	14,500	16,500	19,500
Base Sedan ↓	6,500	9,000	10,000	11,500	12,500	13,500	15,500	18,500
LE ↑	9,000	11,000	12,000	14,000	16,000	18,000	20,000	23,000
LE ↓	8,000	10,000	11,000	13,000	15,000	17,500	19,000	22,000
4X4 ↑	6,500	—	—	—	—	—	—	—
4X4 ↓	5,000	—	—	—	—	—	—	—
Wagon V6 ↑	10,000	12,500	13,000	14,500	16,500	19,000	—	—
Wagon V6 ↓	9,000	11,000	12,500	13,000	15,500	18,000	—	—
Extended Warranty	N	N	N	N	N	N	N	N
Secret Warranty	N	N	N	N	N	Y	Y	Y
Reliability	⑤	⑤	⑤	⑤	⑤	⑤	⑤	⑤
Air conditioning	④	④	❷	⑤	⑤	⑤	⑤	⑤
Body integrity	❷	❷	❷	❷	③	③	③	③
Braking system	❷	❷	❷	❷	❷	❷	❷	③
Electrical system	③	③	③	③	③	③	③	④
Engines	③	④	④	④	⑤	⑤	③	⑤
Exhaust/Converter	❷	❷	④	④	④	⑤	⑤	⑤
Fuel system	③	③	③	④	④	③	④	⑤
Ignition system	③	③	③	④	④	④	⑤	⑤
Manual transmission	④	④	④	⑤	⑤	⑤	⑤	⑤
- automatic	③	③	③	③	③	③	③	③
Rust/Paint	③	③	③	④	④	④	⑤	⑤
Steering	③	③	③	③	③	⑤	⑤	⑤
Suspension	③	③	④	④	④	⑤	⑤	⑤
Crash Safety	—	⑤	⑤	⑤	④	④	④	④
Side Impact	—	—	—	—	—	—	③	③

VOLKSWAGEN

Corrado

Rating: Below Average (1992–95); Not Recommended (1989–91) **Maintenance/Repair costs**: Way higher than average, and only a VW dealer can repair these cars. **Parts**: Frequently back-ordered; outrageously high cost for parts confirmed by CAA. Owners complain that the alloy wheels are easily damaged and can cost over $450 to replace. Improved wheels cost over $900 each.

Strengths and weaknesses: The Corrado is an attractive, mid-sized, two-door coupe with a comfortable interior for the driver and front passenger. It gives good all-round performance, with the accent on smooth acceleration, a firm but not harsh ride, and excellent handling with little body roll.

The manual transmission's long shift throw is annoying and rear passenger room is quite limited. The following systems can be troublesome: heating, defrosting, brake, electrical, and fuel. Although the V6 engine runs quietly, there's lots of road/wind noise intrusion into the passenger compartment.

Dealer service bulletins: **1993**—Excessive diagonal tire wear. • Excessive brake noise. • Faulty windshield wipers. **1994**—AC expansion valve noise. • An erratic idle. • Abnormal rear tire wear due to the rear axle having too much positive toe. • CD changer could skip or be grounded. • Difficulty shifting the manual transmission.

Safety summary/Recalls: Rear visibility is seriously compromised by the high tail and spoiler. • One Montreal owner of a 1993 Corrado reports that his car suddenly fishtailed out of control while rounding a curve at a moderate speed. **1992**—NHTSA is looking into reports of heater cores that rupture and spew hot coolant into the passenger compartment. **Recalls**: **1989–90**—Hot water could leak onto driver's feet and steam up the vehicle's interior.

Secret Warranties/Service Tips

1990–91—Abnormal engine noise that occurs when the engine is warm can be corrected by installing an upgraded crankshaft (#026-105-101E). • If the radiator fan stays on high speed when the ignition is off and thus discharges the battery, replace the radiator fan high-speed relay with part number #321-919-505A. **1991–93**—AC expansion valve noises require the installation of an upgraded expansion valve. **1990–95**—On Corrados equipped with an automatic transmission, an engine that won't start could have a loose contact in the ECM power supply relay; if so, replace the power supply relay. **1992–94**—Poor 2.8L engine performance or a rough idle could be due to a misrouted EVAP vacuum hose or an improperly routed positive crankcase ventilation hose that causes a vacuum leak. **1993**—If the ABS warning light won't go out, a faulty start switch/lock is the likely culprit. **1993–94**—AC expansion valve noise can be silenced with an upgraded expansion valve. • Troubleshooting tips are offered for insufficient AC cooling. • If the ABS light won't go off, you might have a faulty ignition switch/lock. **All models/years**—Poor driveability (hard starts, surging, stalling, low fuel economy) could be due to a deteriorated oxygen sensor (02S) wire shield or an improper oxygen sensor wire shield ground; ask that it be replaced under the emissions warranty. • Manual transmissions that are difficult to shift in or out of gear could require the installation of a clutch disc with a nickel-plated hub.

Corrado Profile

	1990	1991	1992	1993	1994	1995
Cost Price ($)						
Base	26,350	24,930	27,600	29,175	29,050	30,995
Used Values ($)						
Base ↑	7,500	9,500	11,500	13,000	15,500	18,500
Base ↓	6,500	8,500	10,000	12,000	14,500	17,500
Extended Warranty	Y	Y	Y	Y	Y	Y
Secret Warranty	N	N	N	N	N	N
Reliability	❷	❷	③	③	③	③
Air conditioning	③	③	④	④	④	④
Body integrity	❷	❷	❷	③	④	④
Braking system	❷	❷	❷	❷	❷	③
Electrical system	❷	❷	❷	❷	❷	③
Engines	❷	❷	③	③	⑤	⑤
Exhaust/Converter	③	⑤	⑤	⑤	⑤	⑤
Fuel system	❷	❷	❷	③	③	④
Ignition system	③	③	③	④	④	⑤
Manual transmission	⑤	⑤	⑤	⑤	⑤	⑤
- automatic	❷	❷	❷	❷	❷	❷
Rust/Paint	❷	❷	❷	❷	④	⑤
Steering	⑤	⑤	⑤	⑤	⑤	⑤
Suspension	③	③	③	④	④	⑤

Note: No crash tests have been carried out on the Corrado.

Passat

Rating: Recommended (1998); Above Average (1995–97); Not Recommended (1989–94). Be wary of any Passat sold as a 1994; all 1994s were carried over unchanged from the 1993 model year. Don't buy any Passat without a comprehensive extended warranty. If you must have a Passat, wait until the Audi spin-off 1998 Passats come off their two-year leases. The 1998 Passat is much improved over previous models. **Maintenance/Repair costs**: Much higher than average. Most major repairs are dealer dependent. **Parts**: Parts and service are more expensive than average and dealers complain that VW support is spotty.

Strengths and weaknesses: This front-drive compact sedan and wagon uses a standard 2.0L engine and other mechanical parts borrowed from the Golf, Jetta, and Corrado. Its long wheelbase and squat appearance give the Passat a massive, solid feeling, while its styling makes it look sleek and clean. As with most European imports, it comes fairly well appointed.

As far as overall performance goes, the Passat is no slouch. The multivalve 4-cylinder engine is adequate and its handling is superior to that of most of the competition. The 2.8L V6 provides lots of power when revved and is the engine that works best with automatic transmission.

The redesigned 1995 version came with standard dual airbags, a restyled interior, rear headrests, a softened suspension, and a much-improved crashworthiness rating. The 1998 Passat, however, is practically an entirely different car, based upon the Audi A4 and A6, and offers much better performance and reliability.

Passats are infamous for transmission malfunctions that are costly to repair. Even when they're operating as they should, Passat's manual and automatic gearboxes leave a lot to be desired. For example, the 5-speed manual transmission gear ranges are too far apart: there's an enormous gap between third and fourth gear, and the 4-speed automatic shifts poorly with the 4-banger. Also, owners report problems with the front brakes, MacPherson struts, and fuel and electrical systems as the car ages. Interior trim and controls are fragile.

Competent servicing and parts aren't easy to find away from the larger cities, and many of the above-mentioned deficiencies can cost you an arm *and* a leg to repair.

Dealer service bulletins: **1993**—AC expansion valve noise. • Faulty ABS warning lights. • Excessive diagonal tire wear. • Frequent windshield wiper failures. **1995**—Skipping CD changer. • Instrument cluster loss of memory. • Poor fuel system performance. • Excessive rear tire wear. • Drifting, pulling to one side. • Skipping wiper blades. • Engine won't start. **1996–97**—Transmission fluid seepage. • Shifter is hard to move or won't go into Reverse. • Knocking/vibrating shift lever. • Transmission pops out of gear. • Vehicle will not move into any forward gear. • Eliminating musty odour from AC. • Malfunctioning CD player. **1998**—Eliminating musty odour from AC.

Safety summary/Recalls: 1996—Many complaints that transmission slips out of third gear into Neutral. • Vehicle suddenly accelerated forward as lever was put into Reverse. • Intermittent stalling due to electronic control module failure. • Check Engine light constantly goes on and off. • Engine valve cover gasket failures. • Plastic shroud on top of engine rubs against the fuel line. • Leaking windshield and door seals. • Premature wheel bearing failure on the driver's side. • Door lock failure allows door to open while under power, or makes doors difficult to open. • Instrument cluster wiring harness failures. • Repeated trunk switch failures. • Chronic trunk water leaks. • Premature brake pad wearout. • All dash gauges suddenly stop working. **1997**—When manual transmission lever is put into Reverse, it often goes into first gear instead. • Sudden steering wheel lockup. • Cooling fan control module failure. • Rear window defroster power button works only when held in the ON position. • Door handle failures. • Frequent window regulator failures. **1998**—Engine head gasket leaks. • Erratic transmission performance. • Gas tank can't be filled without the pump shutting off repeatedly. • AC recirculation switch failure. • Serious blind spots caused by the small size and narrow view of the three mirrors. • Front seats move back and forth when braking or accelerating. • Total electrical system failure. **Recalls**: **1989–90**—Hot water could leak onto driver's feet and steam up the vehicle's interior. **1993**—Axle could separate from struts, causing loss of vehicle control. **1993–95 VR6 engine**—Faulty radiator fan motor could cause engine to overheat and stall.

Secret Warranties/Service Tips

1990–91—If the radiator fan stays on high speed with the ignition off and thus runs down the battery, replace the fan's high-speed relay with part number #321-919-505A. **1990–95**—On Passats equipped with an automatic transmission, an engine that won't start could have a loose contact in the ECM power supply relay; fix this by replacing the power supply relay. **1991–93**—AC expansion valve noises require the installation of an upgraded expansion valve. • Poor driveability could be caused by magnetic interference due to a deteriorated oxygen sensor wire shield or improper oxygen sensor wire shield ground connection. **1992–94**—Poor 2.8L engine performance or a rough idle could be due to a misrouted EVAP vacuum hose or an improperly routed positive crankcase ventilation hose, which will cause a vacuum leak. **1993**—If the ABS warning light won't go out, a faulty start switch/lock is the likely culprit. **1995–96**—Troubleshooting tips are offered on automatic transmission fluid seepage. **1996–97**—If the transmission pops out of gear, check for a hairline crack on the selector shaft shift detent sleeve. **All models/years**—Vehicles that won't move into any forward gear could have broken retaining lugs for selector plugs of B2/K1, which causes the selector valve to partially protrude or fall out of the valve body.

Passat Profile

	1991	1992	1993	1995	1996	1997	1998
Cost Price ($)							
Base	18,995	19,790	21,525	25,870	27,230	28,620	28,450
Used Values ($)							
Base ↑	5,500	7,000	9,500	14,000	17,000	22,000	24,500
Base ↓	5,000	6,500	8,500	13,000	16,000	21,000	23,500
Extended Warranty	Y	Y	Y	Y	Y	Y	Y
Secret Warranty	N	N	N	N	N	N	N
Reliability	③	③	③	③	④	④	⑤
Air conditioning	❷	❷	③	④	④	④	④
Automatic transmission	❶	❶	❶	❷	❷	③	③
Body integrity	❷	❷	❷	❷	③	④	④
Braking system	❷	❷	❷	❷	❷	③	③
Electrical system	❷	❷	❷	❷	③	③	③
Engines	⑤	⑤	⑤	⑤	⑤	⑤	⑤
Exhaust/Converter	⑤	⑤	⑤	⑤	⑤	⑤	⑤
Fuel system	❷	❷	❷	③	④	⑤	⑤
Ignition system	❷	❷	❷	③	④	⑤	⑤
Rust/Paint	③	③	⑤	⑤	⑤	⑤	⑤
Steering	③	③	③	③	④	④	⑤
Suspension	③	③	③	③	④	④	⑤
Crash Safety	❷	❷	❶	④	④	④	—

LARGE CARS/WAGONS

For motorists who can write off relatively high gasoline consumption, maintenance, and insurance premiums, these are excellent cars for extensive highway driving.

The term "large car" is relative. It once designated vehicles that had a wheelbase of more than 114 inches and that weighed about 3,200 lb. But now that the automakers have shortened most of the wheelbases of their large cars, reduced their weight, and switched to front-wheel drive, traditional definitions of "large" may no longer be accurate indicators of a car's size. Some large cars, like GM's Caprice and Roadmaster, bucked this trend, however, and remained long and heavy.

Owners have to pay a premium for these vehicles, which usually come fully loaded with performance and convenience features, but they are happy to do so because these vehicles offer considerable comfort and stability at high speeds. Large cars also depreciate slowly, can seat six adults comfortably, and are ideal for motoring vacations.

These vehicles generally incur less damage from front, rear, and side collisions, although recent U.S. government crash tests show that some smaller cars absorb frontal crash just as well.

Recommended

Ford Crown Victoria, Grand Marquis (1997–98)

Above Average

Ford Cougar, Thunderbird (1995–97)

Ford Crown Victoria, Grand Marquis (1994–96)

GM Caprice, Custom Cruiser, Impala SS, Roadmaster, Wagon (1995–96)

Average

Ford Crown Victoria, Grand Marquis (1984–93)

GM Caprice, Custom Cruiser, Impala SS, Roadmaster, Wagon (1994)

GM Olds 98, Regency, Park Avenue (1997–98)

Below Average

Chrysler Concorde, Intrepid, LHS, New Yorker, Vision (1995–98)

Ford Cougar, Thunderbird (1985–94)

GM Olds 98, Regency, Park Avenue (1991–96)

Not Recommended

Chrysler Concorde, Intrepid, LHS, New Yorker, Vision (1993–94)

Chrysler Dynasty, Fifth Avenue, Imperial, New Yorker (1985–93)

GM Caprice, Custom Cruiser, Impala SS, Roadmaster, Wagon (1985–93)

GM Olds 98, Regency, Park Avenue (1985–90)

Station wagons (full-sized)

If passenger and cargo space and carlike handling are what you want, a large station wagon may not be the answer—a used minivan, van, light truck, or compact wagon can fill the same need for less cost and will probably still be around a decade from now. Popular (though troublesome) wagons—like the Caprice and Roadmaster (both axed in 1996)—in which you could cram a Little League team are an endangered species, losing out to the van and minivan craze. (See the *Lemon-Aid New 4X4s, Vans and Trucks* and *Lemon-Aid Used 4X4s, Vans and Trucks* guides.)

Some disadvantages of large station wagons: difficulty of keeping the interior heated in winter, atrocious gas consumption, sloppy handling, and poor rear visibility. Exterior road noise is also a frequent problem, since the vehicle's interior has a tendency to amplify normal road noise. Rear hatches tend to be rust-prone. Crash safety is variable.

No full-sized station wagons are recommended.

CHRYSLER

Concorde, Intrepid, LHS, New Yorker, Vision

Rating: Below Average (1995–98); Not Recommended (1993–94). If you want maximum passenger room, choose the 1995–96 LHS or New Yorker. They're about five inches longer and use the larger engine. Whichever vehicle you're considering buying, before paying a cent, make sure you get an extended warranty and take a test drive at night to assess the efficacy of the headlights. **Maintenance/Repair costs**: Higher than average, but most repairs aren't dealer dependent. **Parts**: Higher-than-average cost (independent suppliers sell for much less), but not hard to find. Only vehicles that still carry Chrysler's seven-year powertrain and body warranty are worthy of any consideration.

Strengths and weaknesses: These sleek-styled front-wheel drives are Chrysler's mid-size flag bearers. They're roomy, fuel efficient, and highly manoeuvrable. The base engine is a 3.3L 153-hp 6-banger, but 70 percent of buyers chose the 3.5L for its 61 extra horses. Both engines provide plenty of low-end torque and acceleration; they blow the Camry and Accord away with a 0–60 time of 8.9 seconds. This advantage is lost somewhat when traversing hilly terrain: the smaller V6 powerplant strains to keep up. You'll find good or better handling and steering response than in the Sable and Taurus, and the independent suspension maximizes control, reduces body roll, and provides lots of suspension travel so that you don't get bumped around too much on rough roads.

With all these positives, why aren't these cars recommended? Simple: they're unsafe, unreliable, and cheaply made. One 1994 Intrepid owner's comments posted on the Internet sum it up: "The lack of overall quality and dangerous headlights made me dump the car after only 15 months' use."

A perusal of dealer service bulletins and comments from car rental agencies and other owners tell me that these cars continue to have many serious safety- and performance-related shortcomings that can no longer be explained away as "first series" teething problems that all new cars experience.

The automatic climate control system operates erratically, blowing cold air when it's set for warm, and warm air when it's set for cool. The instrument panel must be removed before servicing the ventilation system because the AC ducts are moulded into the plastic panel.

The shift console needs lighting, the radio and climate controls are too small, the trunk release is hidden in the glove compartment, the hood release is on the floor, the rear view mirror is too narrow, and the fuel-filler door needs a lock. The trunk has a high deck lid, making for difficult loading and unloading, and there's no inside access by folding down the rear seat, as in the Camry.

Owner reports confirm that there are chronic problems with leaking 3.3L engine head gaskets and noisy lifters that wear out prematurely around 27,000 km. Water pumps often self-destruct and take the engine timing chain along with them (a $1,200 repair). Other common complaints: a poor fit between the exhaust manifold and the engine block often results in oil leaks and noisy engine operation; if you wash your car during extremely cold weather, the resulting ice buildup and leaking deck lid seals can easily damage the heater fan motor.

AC malfunctions and failures abound. The problem has become so prevalent that Chrysler has a little-known warranty extension paying for the replacement of the evaporator up to seven years.

The 4-speed LE42 automatic transmission is a spin-off of Chrysler's failure-prone A604 version—and owner reports show it to be troublesome as well. Owners tell of chronic glitches in the computerized transmission's shift timing and computer malfunctions, which result in driveability problems (stalling, hard starts, and surging).

Body problems abound, with lots of interior noise, uneven fit and finish, poor-quality trim items that break or fall off easily, exposed screw heads, faulty door hinges that make the doors rattle and hard to open, distorted windshields, windows that come off their tracks or are misaligned and poorly sealed, power window motor failures, and steering wheel noise when the car is turning.

Other body defects specifically addressed in dealer service bulletins include the following: noisy front suspension; glue oozing out at the backlight and windshield moulding; water collecting in the park/turn lights; noisy rear upper strut mounts; water draining into the trunk when the deck lid is raised; and water entering into the AC/heater housing.

Dealer service bulletins: **1993**—AC operates erratically (covered by a secret warranty). • Automatic temperature control (ATC) works poorly in warm weather. • An ominous clicking noise comes from the passenger compartment, and the 3.3L engine emits a ticking sound when cold. • Hard starting and long crank times when the engine is hot. • Ignition noise on the AM band. • Malfunctioning AC ducts, which blow different temperatures of air. **1993–95**—Upgraded spark plugs will improve cold starting; ask for part #56027275. **1994**—Acceleration shudder. • Faulty engine timing belts. • The transmission wiring harness bracket could break, transmission wires could short-circuit, and the floor shifter knob could stick. • The radio often "locks up." **1995**—Glitches in the computerized transmission's shift timing cause driveability problems (stalling, hard starts, and surging). Amazingly, the 4-speed LE42 automatic transmission—a spin-off of Chrysler's failure-prone A604—appears to be just as problem-plagued as its predecessor. • AC refrigerant leaks. • Heater/AC housing leaks water and there's moisture in headlights. • Excessive engine noise is likely caused by carbon buildup on the top of the pistons. • Engine mount rattles. • Transmission clicks and clunks. • Fuel line rattles caused by faulty fuel rail assembly. • Squeaking front or rear brakes require upgraded brake linings. • A-pillar wind noise. • Rattling C-post appliqué or poor fit of the appliqué to the back glass. • Front hub clicks and clunks. • B-pillar and rear spring rattles. • Upper strut mount squeaks. • Excessive road noise from the front wheels and rear seat require reduced tire pressure and the addition of foam/sealer insulation to the front upper load beam or the C-pillar. **1996**—3.3L/3.5L lower engine oil leaks. • Reduced transmission limp-in default sensitivity. • Shuddering on upshift or whenever the torque converter is engaged. • Upgraded Overdrive clutch hub. • Front suspension clunking or rattling. • Metallic knocking from front of vehicle when passing over bumps. • High-speed windshield washer spray knockdown. • Trunk water leaks (see "Secret Warranties/Service Tips"). • Polycast wheel centre falls off. • Cup holder improvements. **1997**—Tips on troubleshooting noisy brakes. • AC evaporator leaks. • AC suction line failure. • Warped or poor-fitting passenger airbag door. • A metallic popping noise might be heard coming from the front of the vehicle. • Sticking sunroof or sunshade. • Rear drum brake ticking. • Water ingestion into heater/AC housing.

Upscale LHS and New Yorker versions haven't escaped Chrysler's notorious poor-quality body components and sloppy assembly. In addition to the above-mentioned owner and bulletin-related problems, additional service bulletins for these cars indicate that they're likely to have their own persistent problems—such as faulty fuel pumps causing stalling, reduced power, or erratic transmission shifting; radio lockups; water leaks coming from the heater/AC housing; and moisture in the headlights. Bulletins also address how to silence a noisy AC compressor, a squeaking or creaking noise coming from the rear window when travelling at slow speeds over rough roads, and a high-pitched whistling noise caused by a defective idle air control motor.

Safety summary/Recalls: **1993–95 models with 3.5L engines—** Faulty fuel-injector components could cause an engine compartment fire. • Headlights are too dim for safe motoring, and some owners report that the headlights sometimes cut out completely. This problem also affects post-1995 versions. Defrosting is inadequate on some 1993s, allowing ice and moisture to collect at the base of the windshield. Chrysler has a fix for these two problems that requires the installation of a new headlight lens and small foam pads into the defroster outlet ducts. • Both ABS and non-ABS brakes perform poorly, resulting in excessively long stopping distances—more so than for other cars in this class, as the following Port Coquitlam owner's email confirms:

> I recently had an accident in my 1997 Intrepid with ABS brakes. On bone dry pavement I was unable to do more than slow the car down. Lots of room to stop. Both feet on the pedal pushing like hell! No tire skid and no ABS ratcheting. Air bags in my face and $7500 damage. I've been trying to find others that have had the same experience....
>
> Brian Hyndman
> Port Coquitlam, BC
> "Centurion" <cent@wave.home.com>

• The overhead digital panel is distracting and forces you to take your eyes from the road. • The emergency brake pedal catches pant cuffs and shoelaces as you enter or exit the vehicle. • A high rear window sill obstructs rear visibility. • These cars have safety-related peculiarities that you wouldn't believe. The owner of a 1993 Intrepid recounts his unforgettable experience:

> I started having trouble getting the key out of the ignition after I shut the car off. This happened on a Saturday afternoon. I called the dealer and was told I was pretty much out of luck until a mechanic was available on Monday. If I had been smart, I would have left the keys in the ignition and prayed some stupid thief would steal the car. I finally had to call a friend who works at Transport Canada's Road Safety department here in Ottawa. He had the answer—you have to smack the gearshift lever from right to left so the internals under the

> button on the gearshift lever pop out. Lo and behold, this worked. Nifty feature on a $27,000-plus car. I still have to do this to this day, especially during colder weather....The power remote locks would not work and the driver's side door could not even be unlocked with the key (some rod was broken). The only way I could get into the car was to unlock the passenger door with the key and crawl across the front seat! Who is the brainiac who decided the location of the horn button? The horn is impossible to locate in an emergency.

1994–97—The NHTSA is looking into reports that the front suspension may collapse causing the driver to lose control of the vehicle. This government agency knows of 26 complaints and 105 warranty claims for 1994 years, and another 49 complaints covering 1995–97 years that used an upgraded suspension. Broken welds or cracks where the lower control arm attaches to the front cradle may be the cause of the failures. **1996**—Sudden acceleration when starting out in Drive. • Airbags fail to deploy. • Driver's pant leg gets caught on the parking brake assembly upon entering the vehicle. • Many reports that while the vehicle is being driven at 100 km/h, there's a sudden loss of power, the engine shuts down, power steering and brakes become inoperative, and the Check Engine light comes on. • High pressure fuel lines between the two cylinder heads leak gas on the engine. • When shifting into Reverse from Park, automatic transmission acts as if it's in Neutral, and engine races when accelerator is pressed. • ABS brake failures. • Brake rotors rust prematurely, warp easily, and pads have to be changed every 15,000 km. • Vehicle shakes violently when ABS brakes are applied. • Inadequate headlight illumination. • Dashboard reflection on the windshield hampers visibility. • Horn buttons difficult to access in emergency situations. • Stuck right rear door lock won't allow door to open. • Door hinges don't hold door open securely, causing injury to occupants when door closes unexpectedly. • Frequent water pump, AC, and battery failures. **1997**—Fire ignited in trunk. • Fuel line hoses disconnected from the engine, spilling fuel into the engine area. • ABS brake failures. • Airbags failed to deploy. • Engine hesitates and surges. • Vehicle decelerates while driving. • Premature wearout of front brake pads and rotors (rotors warp or become rust-pitted). • Excessive brake noise. • Several reports that Goodyear Eagle tires tend to hydroplane. • Steering linkage failure causes vehicle to wander all over the roadway. • Power steering failure—extremely hard to turn. • Many reports of sudden transmission failures, many due to cracked transmission casings. • Transmission fluid leakage caused by defective transmission casing bolt. • Seat backrest failure. • Frequent windshield replacements due to distortion—particularly annoying at night when combined with poor headlight illumination. • Intermittent turn signal operation. • Flawed interior door panels. • Frequent instrument panel malfunctions: lights, gauges, AC all go out at once. • Cigarette lighter shoots out so hard it shoots under driver's seat—while red hot. **1998**—Several reports of sudden acceleration when shifting into Reverse. • When relay fails, it causes a harsh downshift to second gear while at highway speeds. • Airbags failed to deploy. • Rear brakes improperly adjusted by Chrysler result in overloaded front brakes and warped front brake rotors. • Windshield distortions. **Recalls**: **1993**—Dealers will reroute the wiring harness to prevent shorting. **1993–97 Concorde, Intrepid, LHS, and Vision with 3.5L V6 engines and all 1993 models with 3.3L V6 engines**—A faulty O-ring in the engines' fuel-injection system may harden and let fuel escape. **1994**—Faulty transmission wiring might cause the car to start when not in the Park position. **1993 Intrepid and Vision**—Defective lower control arm washers could cause loss of steering.

Secret Warranties/Service Tips

1993—Failure of the fuel pump check valve could cause start-up die-out, reduced power, or erratic shifting. • Doors that are hard to open or close or that make a snapping sound likely need four-check door straps, which Chrysler will provide free of charge. • More importantly, AC evaporator failures will be covered under a special seven-year program, according to Chrysler service honchos. Although limited to 1993s, it's logical (to me, at least) that this warranty extension sets up a benchmark that owners of other more recent models can rely upon to get their ACs repaired under warranty (even if the evaporator isn't the cause of the failure). **1993–94**—Acceleration shudder could be caused by automatic transmission front pump leakage. • A-pillar wind noise requires sealing the upper load beam and A-pillar or sealing the roof rail body seam. • AC belt rollover requires the installation of a revised AC belt and idler pulley. • Exchange the base body control module (BCM) for an upgraded version if dash instruments and gauges suddenly quit working. • Upgraded disc brake linings will help reduce front or rear squeaking noises. • AM radio static (poor reception of distant stations) can be corrected by installing a supplemental engine-suppression strap to the left side of the engine. • Engine mount rattles signal the need to replace the engine mounts (always replace in pairs). • Cold-start piston knocking noise can be eliminated by replacing the piston and connecting rod assembly. • No-starts, poor engine performance, and loud noises when attempting to start can be corrected by installing a snubber over the timing-belt tensioner plunger (this applies to the 3.5L engine only). • Dealers will exchange small head restraints for larger ones free of charge. • A heater/AC housing that leaks water into the passenger compartment can be plugged by enlarging the right plenum drain hole. • Rear disc brake noise can be stopped by installing upgraded rear disc brake adapters. • Rear glass rattling is likely caused by a loose backlight. • Excessive rear road noise can be reduced by indexing the spring to the upper strut mount; in cases where indexing doesn't work, you may have to replace the rear upper strut mounts with revised mounts. • A transmission buzz or rattle can be stopped by replacing the transfer chain snubber and attaching screws. • If water leaks or dust accumulates inside the trunk, try sealing the quarter panel to the outer wheel house panel seam. **1993–97**—The blower motor could seize or freeze from water seeping into the heater/AC housing. • AC evaporator leaks caused by premature corrosion can be prevented by installing a cowl plenum screen; Chrysler says this is a warranty repair.

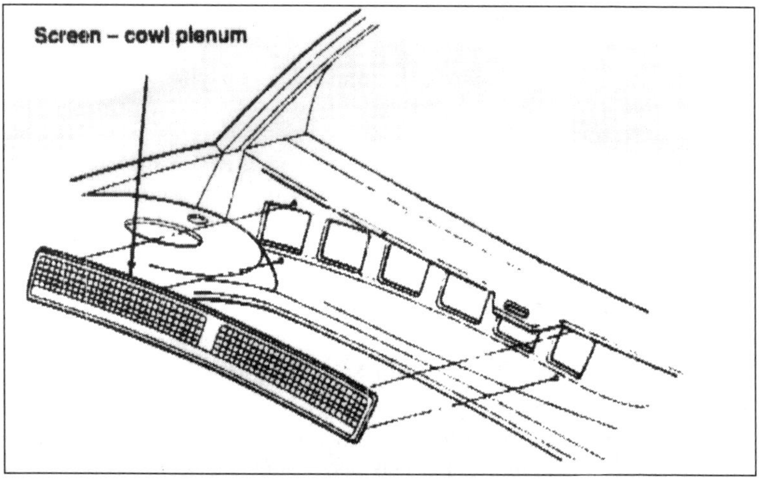

Both the screen and its installation are covered by Chrysler's warranty.

1994—Harsh, erratic, or delayed transmission shifts can be corrected by replacing the throttle position sensor (TPS) with a revised part. **1994–95**—The intermittent or total loss of air conditioning can be corrected by installing a revised AC pressure transducer. **1995–96**—More troubleshooting tips are offered on diagnosing and fixing trunk water leaks.

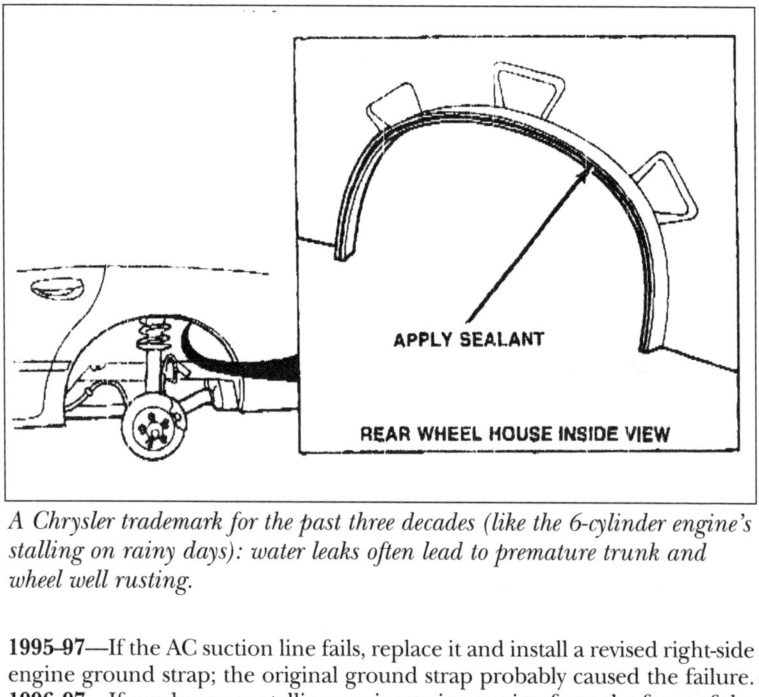

A Chrysler trademark for the past three decades (like the 6-cylinder engine's stalling on rainy days): water leaks often lead to premature trunk and wheel well rusting.

1995–97—If the AC suction line fails, replace it and install a revised right-side engine ground strap; the original ground strap probably caused the failure. **1996–97**—If you hear a metallic popping noise coming from the front of the vehicle when you accelerate from a stop, install two new upper and two new lower cradle mounting isolators. • Rear drum brake ticking can be silenced by burnishing the rear brakes. **All models/years**—A rotten-egg odour coming from the exhaust is probably caused by a malfunctioning catalytic converter; this is covered by Chrysler's original warranty *and* the emissions warranty. Don't take no for an answer. The same advice goes for all the squeaks and rattles and the water and wind leaks that afflict these vehicles. Don't let Chrysler or the dealer pawn these problems off as maintenance items. They're all factory related and should be covered for at least five years.

Concorde, Intrepid, LHS, New Yorker, Vision Profile

	1993	1994	1995	1996	1997	1998
Cost Price ($)						
Concorde	20,425	22,590	24,420	26,005	26,815	26,915
Intrepid	17,555	19,170	21,010	22,980	24,055	24,395
LHS	—	35,020	37,625	38,420	40,500	40,500
New Yorker	—	29,990	32,080	34,490	—	—
Vision	18,450	21,485	22,910	23,770	24,775	—
Used Values ($)						
Concorde ↑	9,500	11,000	12,500	15,000	17,000	20,000
Concorde ↓	8,500	10,000	11,500	13,500	16,000	18,000
Intrepid ↑	7,500	9,000	11,000	13,000	15,000	18,500
Intrepid ↓	6,500	8,000	10,000	12,000	14,000	15,500
LHS ↑	—	13,000	16,000	18,500	23,000	28,000
LHS ↓	—	11,500	15,000	17,500	22,000	26,000
New Yorker ↑	—	13,000	14,000	16,500	—	—
New Yorker ↓	—	12,000	13,100	15,500	—	—
Vision ↑	7,500	9,500	11,000	12,500	14,500	—
Vision ↓	7,000	8,500	10,000	11,500	13,500	—
Extended Warranty	Y	Y	Y	Y	Y	Y
Secret Warranty	Y	Y	Y	Y	Y	Y
Reliability	❶	❶	❶	❷	❷	❷
Air conditioning	❶	❶	❶	❷	❷	③
Automatic transmission	❶	❶	❶	❷	③	③
Body integrity	❶	❶	❶	❷	❷	❷
Braking system	❷	❷	❷	❷	❷	③
Electrical system	❶	❶	❷	❷	❷	③
Engines	③	③	③	④	④	④
Exhaust/Converter	③	④	④	④	④	④
Fuel system	③	③	③	③	④	④
Ignition system	❷	❷	③	④	④	④
Rust/Paint	③	④	④	④	④	④
Steering	③	④	⑤	⑤	⑤	⑤
Suspension	❷	❷	❷	③	③	④
Crash Safety						
Concorde	❷	❷	—	—	④	—
Intrepid	❷	❷	④	④	④	—
LHS	—	❷	—	—	④	—
New Yorker	—	❷	④	④	—	—
Vision	❷	❷	—	—	④	—

Side Impact						
Concorde	❶	—	—	—	④	—
Intrepid	❶	❶	④	④	④	—
LHS	—	❷	—	—	④	—
New Yorker	❷	❶	④	④	—	—
Vision	❶	—	—	—	④	—

Note: All these vehicles are practically identical and should have similar crashworthiness scores, even though not every model was tested each year.

Dynasty, Fifth Avenue, Imperial, New Yorker

Rating: Not Recommended (1985–93). The 1994 and later LHS New Yorker and New Yorker are completely different cars from previous New Yorkers and are rated later on. Even though the top-of-the-line Imperial has a rapid depreciation rate, it's no bargain at any price. **Maintenance/Repair costs**: Higher than average; repairs aren't dealer dependent. **Parts**: Higher-than-average cost, but can be bought for much less from independent suppliers. Body parts are often back-ordered. Imperial body parts are rare and costly.

Strengths and weaknesses: These relatively fuel-efficient four-door sedans handle and ride like the average mid-sized car popular two decades ago. Compared to large rear-drives equipped with V6 or V8 engines, they can't tow as much weight, their unit-body construction makes for a noisier ride, and long-term reliability is way below average.

The base 2.5L engines are a major source of mechanical woes, and when turbocharged, they can cost you a small fortune to repair. Owners complain of substandard piston rings, faulty oil seals, and a failure-prone timing-chain mechanism that can lead to severe engine damage unless it's checked and adjusted frequently.

The Mitsubishi 3.0L V6 engine, offered since 1988, has generated complaints concerning oil leaks, electronic malfunctions, and fuel system problems. Chrysler's 3.3L V6, available as of the 1990 model year, is more reliable. The electronic 4-speed automatic has elicited a number of complaints about erratic and noisy shifting. The A604 automatic transmission is a nightmare, and Chrysler is replacing it by the ton, paying the warranty deductible and compensating consumers for consequential damages.

Front suspension components and brakes wear out quickly. The electrical system is bug-plagued; avoid the electronic dashboard. Air conditioning malfunctioning occurs all the time, accompanied by the evaporator freezing up. Windshield wiper motors short-circuit, and often the motor is replaced unnecessarily. Owners report erratic horn operation and leaky radiators. Knobs and levers break very easily. Front seats lack lower-back support, and the middle passenger is punished by the hard edges of the split seat and folded armrest. Mediocre heating (when the Dynasty accelerates, there isn't enough engine vacuum to operate the actuators that maintain a consistent heat range); there aren't any ducts to distribute warm air to the rear. When going over bumps, these cars shake and rattle like old taxicabs. Many owners report water infiltration, especially around the windshield and into the trunk. Surface rust is common, especially around windshields, door bottoms, and the rear lip on trunk lids.

Safety summary/Recalls: **1993**—Sudden acceleration while in Reverse or forward gear; brakes wouldn't respond. • Inadvertent airbag deployment. • Airbag failed to deploy. • Short drivers feel uneasy having to sit so close to the airbag housing. • Airbag warning light stays lit for no reason. • When started, vehicle leaps forward. • Defective fuel-injectors cause vehicle to jerk and lunge. • Engine serpentine belt and belt tensioner failures, causing complete loss of power. • Premature head gasket failures and engine mount breakage. • Chronic stalling at any speed. • Fuel pumps often need replacing and may be part of the cause for stalling. • Frequent transmission failures. • When weather turns cold, transmission won't shift properly. • Faulty ECM (computer module) makes the automatic transmission downshift at irregular intervals. • When travelling at high speeds, with cruise control engaged, automatic transmission may suddenly downshift from fourth into second gear. • Transmission cooling lines often leak. • Transmission whines and growls. • Sudden steering wheel lockup when driving. • Steering column fails to lock when key is removed. • Power steering belt tensioner failures. • Brake failures continued after ABS recall repairs carried out. • Brake master cylinder leaks fluid. • Noisy brakes. • Shock absorbers failed • Leaky MacPherson struts. • Chronic battery failures. • Courtesy light mounted above rear view mirror can interfere with driver's vision. • Gas line from tank to fuel filter rubs on framing, thus wearing a hole in the line. **Recalls**: **1989–90 Dynasty/all 1990s**—The dealer will install a free by-pass valve to prevent automatic transmission damage in cold weather. **1990–91 models with ABS**—The high-pressure hose could leak. **1991**—The front outboard seatbelt latch is faulty. **1992**—The coupling bolts for the steering column shaft are faulty. **1990 Fifth Avenue**—The airbag could be defective on cars with gray interior. **1991 Fifth Avenue**—Pin bolts for the front disc brake caliper guide could be too loose. • A short circuit in the heater blower motor could cause a fire in the cowl area.

Secret Warranties/Service Tips

1988–93—3.0L engines that burn oil or produce a smoky exhaust at idle can be fixed by installing snap rings on the exhaust valve guides and replacing all of the valve guide stems or the cylinder head. **1990–93**—Harsh automatic shifts can be tamed by installing the following revised parts: kickdown, accumulator, reverse servo cushion springs, and accumulator piston. **1991–92**—Engines with a rough idle and stalling following a cold start could require a new single board engine controller (SBEC). **1991–93**—Poor AC performance while the AC blower continues to operate is likely due to the evaporator freezing; the evaporator freezes when the powertrain control module (PCM) isn't properly disengaging the AC clutch via the relay. • An engine that stalls following a cold start might need an upgraded Park/Neutral/start switch. • The serpentine belt could come off the pulley after driving through snow; install an upgraded shield, screw, and retainers to overcome this problem. **1992**—Long crank times, a rough idle, and hesitation can be corrected by replacing the intake manifold assembly or ECT and sensor connector. **1992–93**—Eliminate a deceleration shudder by replacing the PCM with an upgraded version. • Rough idling after a cold start with 2.2L and 2.5L engines can be corrected by installing an upgraded PCM. **1993**—Acceleration shudder could be caused by a leaking automatic transmission front pump. • For improved automatic shifting, install an upgraded transmission control module. • Fuel pump check valve failure can cause start-up die-out, reduced power, or erratic shifting. **All models/years**—Defective valve springs on the 3.3L V6 engine will be replaced free of charge (DSB R#466). • There is excessive air conditioner noise with 3.3L engine (DSB #24-8-89). • Most model years with the notoriously failure-prone A604 automatic transmission suffer clutch slippage (DSB #21-09-90). Ask for a repair refund for this or any other transmission failure that occurs within 7 years/160,000 km. • A rotten-egg odour coming from the exhaust is probably caused by a malfunctioning catalytic converter and might be covered under the emissions warranty.

Dynasty, Fifth Avenue, Imperial, New Yorker Profile

	1988	1989	1990	1991	1992	1993
Cost Price ($)						
Dynasty	15,000	15,865	16,715	15,575	16,505	16,910
Fifth Avenue	25,300	26,235	25,180	—	—	—
Imperial	—	—	38,200	34,415	35,915	35,915
New Yorker	21,000	21,000	25,180	25,600	21,910	21,755
Used Values ($)						
Dynasty ↑	2,000	2,500	4,000	4,500	5,500	6,500
Dynasty ↓	1,500	1,800	3,500	4,000	5,000	5,500
Fifth Avenue ↑	4,000	5,000	6,000	—	—	—
Fifth Avenue ↓	3,000	4,500	5,000	—	—	—
Imperial ↑	—	—	6,500	7,500	10,000	11,500
Imperial ↓	—	—	6,000	7,000	9,000	10,500
New Yorker ↑	2,500	3,500	5,000	7,000	7,000	8,000
New Yorker ↓	2,000	3,000	4,000	6,000	6,500	7,000
Extended Warranty	Y	Y	Y	Y	Y	Y
Secret Warranty	N	N	N	Y	Y	Y
Reliability	❶	❶	❶	❶	❶	❶
Air conditioning	❶	❶	❶	❶	❷	❷
Automatic transmission	❷	❷	❶	❶	❶	❶
Body integrity	❷	❷	❷	❷	❷	❷
Braking system	❶	❶	❶	❷	❷	③
Electrical system	❷	❷	❷	❷	❷	③
Engines	❶	❶	❶	❷	❷	❷

Exhaust/Converter	❷	❷	❷	⑤	⑤	⑤
Fuel system	❶	❶	③	③	③	③
Ignition system	❶	❶	❷	❷	❷	③
Rust/Paint	❷	❷	❷	③	③	③
Steering	❶	❶	❶	❷	❷	③
Suspension	❶	❶	❶	❷	③	③
Crash Safety						
Dynasty	❶	—	—	—	—	❷
Imperial	—	—	❷	❷	—	—
New Yorker	❶	—	—	—	—	❶

FORD

Cougar, Thunderbird

Rating: Above Average (1995–97); Below Average (1985–94). **Maintenance/Repair costs**: About average, and repairs aren't dealer dependent. **Parts**: Higher-than-average cost (independent suppliers sell for much less), and not hard to find. Despite the fact that 1997 is the last model year for both models, parts should remain plentiful.

Strengths and weaknesses: These are no-surprise, average-performing, two-door luxury cars that have changed little over the years. Nevertheless, they offer more performance and greater reliability than GM rear-drives and most of the Big-Three-produced front-drives. Handling and ride are far from perfect, though, with considerable body lean and rear-end instability when taking curves at moderate speed.

Overall reliability of these models has been average, as long as you stay away from the turbocharged 4-cylinder engine. True, it offers lots of power, but excessive noise and expensive repairs are the price you pay. Front suspension components wear out quickly, as do power-steering rack seals. Owners of recent models have complained of ignition module defects, electrical system bugs, premature front brake repairs, steering pump hoses that burst repeatedly (one Winnipeg owner of a 1994 Cougar wrote that he replaced the hose twice in the same year), erratic transmission performance, early AC failures, defective engine intake manifolds, excessive vibrations when driving, numerous squeaks and rattles, faulty heater fans, and failure-prone power window regulators.

Dealer service bulletins: **1993**—Poor-performing AC systems are caused by a slipping clutch at high ambient temperatures; AC cooling may be insufficient at idle; and the AC compressor may moan from idle to 1,500 rpm. • Power-steering units may be noisy and power door locks may not work. **1994**—Faulty 3.8L engine rocker arm assemblies may cause squeaking, chirping, and knocking. • Rough idle, hesitation, excessive fuel consumption, and poor heater output are likely caused by a thermostat sticking in an open position or opening before it should. • The transmission shudders under light to moderate acceleration. • There may be extraneous noise in radio speakers, caused by a malfunctioning fuel pump. • Electric rear window defrosters may fail prematurely and the floor ducts tend to leak water. • Moon roofs are also plagued by excessive wind noise and buffeting. **1995**—Exhaust moan or vibration. • Inefficient AC cooling or excessive clutch end gap. • Intermittent loss of torque at 3–4 upshift. • No-crank due to starter solenoid corrosion. • Faulty outside temperature display. • Water intrusion of the MLP/TR sensor. • Brake roughness upon application and a clacking/thumping noise when braking. • Exhaust moan or vibration, fuel pump buzz/whine heard through the radio speaker, and rocker arm noise. • Warranty coverage for a bowed antenna. • Faulty temperature gauge. **1996**—Stalling or hard starts. • Squeal or hoot noise from the engine compartment. • Transmission valve body cross leaks. • Power steering requires increased effort at low speeds. • Chatter during turns. • Growling noise from the steering column while vehicle is turning. • Roughness during braking. • Fuel pump buzz/whine continues to be heard through radio speaker. • Musty and mildew-type odours. • Tape won't eject from cassette player. • Window won't go to full up position. • Water drips on floor from AC evaporator core. • Water leaks from the cowl vent screen. **1997**—AC emits musty odour. • Erratic or prolonged 1–2 shift. • Door latch stuck open. • Fog/film on windshield/interior glass. • Loose catalyst or muffler heat shields. • Shudder or vibration while in third or fourth gear. • Chirping or squeaking blower motor. • Radio AM static while driving. • Water in air inlet elbow. • Vibration while driving at highway speeds. • Inoperative remote keyless entry.

Safety summary/Recalls: **1992–93**—NHTSA is probing allegations that the headlight switches may be failure-prone. **1996**—Vehicle caught fire after being parked for 13 hours. • Airbags failed to deploy. • Airbag indicator light comes on for no reason. • Right rear wheel came off. • During inclement weather, design of engine allows water to saturate the air filter, causing engine to die out. • PCM chip defect causes engine to cut out and the Check Engine light to come on. • Malfunctioning crankshaft and oxygen sensors may cause chronic stalling and no-starts. • Cracked intake manifold allows coolant leakage. • Frequent motor mount failures. • Transmissions are noisy, won't shift properly, and frequently won't shift at all. • Excessive driveshaft vibrations. • Power steering works poorly at low speeds and sometimes cuts out completely when cruising on the highway. • Heater and AC compressor failures. • Defective AC fan switch. • Frequent reports of sudden brake failures, front brake rotor warpage, and noisy brakes. • Brake pedal sinks below the accelerator pedal level, causing driver to depress the accelerator. • Hood struts are too weak to keep hood in open position. • Crooked steering wheel replaced. • Power window regulator failures. • Electric door locks are failure-prone. • Doors fit poorly. • Headlights require constant adjustment and don't provide as much illumination as with other model years. • Seatbelt doesn't release properly. • Defective sunroof guides, and the gasket barely lasts two years. • Sun visor hits driver in the head when it's adjusted. • Speedometer and gauges work erratically. **1997**—Sudden acceleration while stopped at a traffic light. • Left vehicle in Park with engine running and it suddenly lurched into Reverse. • Car suddenly pulled to the left when braking, and steering locked up. • Excessive vibrations while cruising that increase in intensity when braking. • Premature front brake wear. • Headrests can't be raised high enough to protect the head. • Seat won't latch. • Brake caliper moves in its bracket. • Transmission fluid leakage. • Water leaks onto passenger-side floor due to missing wiring harness plug. • Dash reflection on the windshield obstructs visibility. • Dealer can't correct windshield washer spray to prevent it from spraying the rear window. • Multiple Firestone tire failures. **Recalls**: **1988–93**—The ignition switch could experience an internal short circuit, creating the potential for overheating, smoke, and possibly fire in the steering column. **1989**—Rear suspension wheel knuckles could fracture. • There is excessive brake pedal travel. **1990–91**—Wiper motor nuts may loosen or fall off. • **1996**—The automatic transmission may disengage or fail to engage when shift lever is moved to Park. • Driver's door may not sustain the specified load when it's in the secondary latch position. • The semi-automatic temperature control blower may malfunction.

Secret Warranties/Service Tips

1985–92—A buzz or rattle from the exhaust system may be caused by a loose heat shield catalyst. **1989–94**—Water dripping from the floor ducts when the AC is working requires a relocated evaporator core. **1992–95**—A corroded solenoid may be the cause of starter failures. **1994**—Automatic transmissions with delayed or no forward engagement, or a higher engine rpm than expected when coming to a stop, are covered in DSB #94-26-9. • A no-crank condition in cold weather may be due to water freezing in the starter solenoid. • Hesitation or stumble in vehicles equipped with a 3.8L engine may be fixed by installing an upgraded PCM that allows for low-grade fuel. **1994–95**—A thumping or clacking noise heard from the front brakes signals the need to machine the front disc brake rotors; Ford will pay for this repair under its base warranty. **1994–97**—An erratic or prolonged 1–2 shift can be cured by replacing the cast aluminum piston with a one-piece stamped steel piston that has bonded lip seals and by replacing the top accumulator spring. These upgraded parts will increase the transmission's durability, according to Ford. **All models/years**—Ford's "goodwill" warranty extensions cover fuel pumps and computer modules that govern engine, fuel-injection, and transmission functions. If Ford balks at refunding your money for a faulty computer module, say that you wish to have the 5-year/80,000 km emissions warranty applied by either the company or the courts. There's nothing like a small claims court action to focus Ford's attention. The same advice applies if you notice a rotten-egg odour coming from the exhaust. It's likely the result of a malfunctioning catalytic converter—not a passenger's poor hygiene. • Paint delamination, fading, peeling, hazing, and "microchecking" (see page 112 for details on claiming a refund).

Cougar, Thunderbird Profile

	1990	1991	1992	1993	1994	1995	1996	1997
Cost Price ($)								
Cougar	20,052	18,295	19,395	19,650	21,395	22,095	23,495	24,995
T-bird	18,670	17,495	18,795	19,695	21,895	22,995	23,595	25,095
Used Values ($)								
Cougar ↑	4,000	5,000	5,500	7,500	9,500	12,000	14,000	15,500
Cougar ↓	3,500	4,500	5,000	6,500	8,500	10,500	12,500	14,500
T-bird ↑	3,500	4,500	5,000	7,000	9,000	11,500	13,500	15,000
T-bird ↓	3,000	4,000	4,500	6,000	8,000	10,000	12,000	14,000
Extended Warranty	Y	Y	Y	Y	Y	Y	N	N
Secret Warranty	Y	Y	Y	Y	Y	Y	Y	Y
Reliability	❷	❷	❷	❷	❷	③	④	④
Air conditioning	❷	❷	❷	❷	❷	③	③	④
Automatic transmission	❷	❷	❷	❷	❷	❷	③	③
Body integrity	❷	❷	❷	❷	❷	③	③	③
Braking system	❷	❷	❷	❷	❷	❷	❷	③
Electrical system	❷	❷	❷	❷	❷	❷	③	③
Engines	③	③	④	④	④	⑤	⑤	⑤
Exhaust/Converter	❷	③	③	③	④	⑤	⑤	⑤
Fuel system	❷	❷	③	③	③	③	③	③
Ignition system	③	③	③	④	④	⑤	⑤	⑤
Rust/Paint	❶	❶	❷	❷	❷	❷	❷	③
Steering	❷	❷	❷	③	③	③	③	⑤
Suspension	❷	❷	③	③	④	④	④	⑤
Crash Safety	—	⑤	⑤	❶	❷	⑤	⑤	⑤
Side Impact	—	—	—	—	—	—	—	③

Note: The 1993–94 models' low crash scores represent severe leg trauma.

Crown Victoria, Grand Marquis

Rating: Recommended (1997–98); Above Average (1994–96); Average (1984–93). Overall, these cars aren't as reliable as Japanese luxury vehicles, but they're the best of the domestic crop when it comes to price, power, performance, and overall comfort. Don't waste your money buying a 1997 or 1998 version if you can find a low-mileage 1996 version. It'll cost much less and give you the same features. The Marquis is a slightly more luxurious version that costs more but gives little of consequence for the extra expense. **Maintenance/Repair costs**: Average, but some AC and electronic repairs can be carried out only by Ford dealers. **Parts**: Higher-than-average cost (independent suppliers sell for much less), but not hard to find.

Strengths and weaknesses: Built in Ontario, these cars are especially suited to people who need lots of room or who prefer the safety blanket provided by road-hugging, gas-guzzling weight. Handling is mediocre, but it's about average for cars this size. Both the 4.6L and 5.0L V8s provide adequate though sometimes sluggish power, with most of their torque found in the lower gear ranges. The fuel pump, sender, fuel filter, and fuel hose assemblies are failure-prone. There are many complaints of EEC IV ignition module malfunctions that cause hard starting and frequent stalling. Brakes, shock absorbers, and springs wear out more quickly than they should. Furthermore, there is such a high number of safety-related complaints concerning brake and fuel lines, suspension, and steering components that an undercarriage inspection is a prerequisite to buying any model year.

Dealer service bulletins: **1993**—Poor-performing AC systems caused by a slipping clutch at high ambient temperatures. • Ticking, pinging, or popping when the AC clutch cycles, and a noisy FX-15 AC compressor. • Premature front brake rotor wear and vibrations when braking. • Noisy power-steering units. • Inoperative or malfunctioning cellular phones. • Power door locks that may not work. **1994**—Transmission shudder under light to moderate acceleration. • Radio speaker noise caused by a faulty fuel pump. • Defective electric rear window defrosters. **1995**—Delayed transmission engagement and shift errors, loose transmission connector, intermittent loss of torque at 3–4 upshift, irregular or no-torque converter operation, and shifts to Neutral at heavy throttle. • Insufficient AC cooling or excessive clutch end gap. • Water intrusion of the MLP/TR sensor. • A clacking/thumping noise when braking, and brake roughness upon application. • A fuel pump buzz/whine continues to be heard through the radio speaker. **1996**—Stalling or hard starts. • Delayed 1–2 shift. • Transmission valve body cross leaks. • Chatter during turns. • Fog/film on windshield/interior glass. • Musty and mildew-type odours. • Right front door window requires much effort to roll up. **1997**—ABS brakes may activate on their own or produce a grinding, pulsing, fluttering effect on the brake pedal. • AC emits musty odour. • Air rush and/or flutter noise from centre register. • Erratic or prolonged 1–2 shift. • Spark knock during acceleration. • Door latch sticks open. • Driver-side seat cushion is uncomfortable, sags. • Fog/film on windshield/interior glass. • Front door panel crack on rear edge. • Loose catalyst or muffler heat shields. • Wind noise around doors. • Shudder or vibration while in third or fourth gear. • Steering wheel noise or vibration. • Suspension leans to right side. • Axle whine at 100 km/h.

Safety summary/Recalls: One Marquis owner writes:

> Fuel is delivered from the gas tank to the engine via a pressurized fuel line. There is a gas return line that brings fuel back from the engine to the gas tank. At the location of the wheel well, the return line is installed slightly higher than the fuel line. Because of this, the return line is exposed to intermittent rubbing from the chassis. Since the fuel line is metal, it will not wear right away. In my car, it took six years, but now the line is leaking gasoline. I don't know whether all Marquis cars have this problem; however, the Alberta Motor Association mechanic, the Canadian Tire mechanic and the City Ford dealer mechanic all indicated that the car came from the factory this way. There's a serious risk of both fire and explosion while driving the car.

• The Saskatoon Police Department reports that four 1992–93 vehicles have had incidents of unintended sudden acceleration, and on two occasions the brakes failed • Ford has received 19 complaints of reduced power-steering assist during high-speed manoeuvres with 1992–93 Crown Victoria police cars. Lawsuits have been filed and the NHTSA is also involved. • There are reports of hoods suddenly flying up and shattering the windshield while 1992–93 vehicles are in motion. • There are also complaints of poor traction on ice, a hard-to-see shift indicator, and the instrument panel washing out in sunlight. **1996**—Fire caused by a short in the electrical system. • Airbag failed to deploy. • ABS brake failures. • Engine surges without warning or shuts down when brakes are applied. • Cruise control won't disengage when brakes are applied. • Oil pan/gasket leaks. • Many reports that the left frame bracket broke at weld, causing vehicle to pull sharply to the right when brakes were applied. • Inadequate lubrication caused the right lower ball joint to separate, causing an accident. • Other reports of ball joint and lower control arm failures say that the vehicle's front wheel assembly is also affected. • Police department inspectors report their vehicles have shown excessive play in the Pitman arms. • Brake lines are routed too close to the body and chafe excessively. • Brake pedal went to the floor due to failure of the stop switch and clip on pedal. • Frequent replacement of front brake components. • Steering locks up, will not return, or is so loose it won't steer vehicle. • Cracked intake manifolds cause loss of coolant. • Frequent catalytic converter failures. • Inoperative rear window defroster. • Seatbelt buckle doesn't stay latched. • Seatbelts failed to restrain occupants during a collision. **1997**—Fire caused by a faulty fuel line. • Other engine compartment fires reported, cause unknown. • Sudden acceleration due to a design flaw of the throttle linkage and bracket. • Door mats can shift under the gas or brake pedals, causing them to jam. • Vehicle left in Park position with engine on slipped into Reverse. • Many reports of brake failures caused by brake line rubbing against body components. • Collapsed body mounts allow undercarriage to rub flat spots on steel brake lines, causing brake failure. • Steering wheel lockup while driving. • Front suspension ball joint failure after having recall campaign correction done. • Lower passenger-side control arm fell off when vehicle was in first gear. • Sharp door edges have injured three people. • Engine intake manifold failures that cause anti-freeze to leak into the engine compartment. • Front doors are too large and difficult to manoeuvre around. **1998**—NHTSA investigators are looking into reports that the inertia fuel shutoff switch operates when it shouldn't, stalling the vehicle. • Vehicle caught fire in the engine compartment while parked in garage overnight. • Sudden acceleration after vehicle stalled and was restarted. • Airbags failed to deploy. • Sudden loss of power, stalling. • Traction control engages for no reason, causing loss of power and control. • Steering too sensitive when changing lanes, easy to lose control. • Dome light switch is poorly designed; it can only be activated by the driver due to its location. • Loss of lighting caused by sudden electrical system failure. • Rubber hose leading from the fuel tank is easily hit when going over a bump or pothole. **Recalls**: **1984**—Seatbelt anchors may not meet federal regulations. **1987**—Faulty fuel-injection tube assembly may cause fuel leakage and create a fire hazard. **1987–88 station wagons with dual-facing rear seats**—Automatic seatbelt retractors are faulty. **1988–89**—A faulty ignition module may cause a fire. **1991**—The car may roll away even though the automatic transmission lever says that it's in Park. **1992–93**—A short circuit could cause a front seat fire. • On non-ABS-equipped vehicles, an abraded rear brake line can cause loss of brake fluid and loss of braking power. **1992–97**—Hoods may fly open on taxis and police cars. **1995**—Fuel could leak from the tank and pose a fire hazard; dealer will replace the fuel-filler pipe seat. • Faulty headlight and power window circuit breakers may unexpectedly turn the headlights off. • Rear seatbelt attachments may be faulty. • Airbag may not inflate properly and the end cap could separate. **1996**—Driver's door may not sustain the specified load when in the secondary latch position. **1996–99**—Prematurely worn ball joint could break, causing loss of steering. **1995–96 fleet cars**—Pitman arm corrosion can cause abnormal wear. **1994 Grand Marquis**—The metal cylinder that holds the airbag can be projected into the passenger area as the airbag deploys.

Secret Warranties/Service Tips

1987–90—Excessive oil consumption is likely caused by leaking gaskets, poor sealing of the lower intake manifold, defective intake and exhaust valve stem seals, or worn piston rings. Install new guide-mounted valve stem seals for a more positive fit and install new piston rings for improved oil control. **1989–90**—Program B89 provides for the inspection and replacement of the Overdrive gear and installation of a repair kit (#E9AZ-7L22B-A). • No Overdrive or an extended 3–4 shift may require a new Overdrive band, transmission separator plate kit, overhaul kit, and reverse drum. • Harsh or rough shifting may be caused by sticking control valves. • Poor

AM radio reception due to ignition static can be improved by securing the antenna ground connection. **1990**—Excessive transmission noise, delayed shifts, or no engagements may be due to metal particles from the thrust washer that have plugged the filter or burnt clutch plates. • Install a new "service only" EEC IV processor to correct driveline clunk when the throttle is closed. • If the AC blower won't change speed, consider installing a new switch assembly. If the blower sometimes cuts out, look for a loose connection at the variable-speed blower control. • If the front seats move in their tracks or make noise, Ford will install new seat tracks and a memory track on the driver's side free of charge. • A sluggish speed-control response can be remedied by installing a new EEC IV processor. • A loose or wobbly steering wheel requires putting in a new bearing tolerance ring service kit (#F0DZ-3L539-D). **1990–91**—Fuel pump whining can be reduced by adding tank insulation material. **1990–93**—Brake pedal and steering vibration when braking can be reduced by installing improved brake rotors (#F1VY-1125-A) and linings (#F3AZ-2001-A). **1992**—Premature brake wear and seizure due to overheating can be corrected by installing air scoops to cool the brakes. **1992–94**—A moaning or loud noise coming from the engine compartment of cars equipped with the 4.6L engine can be corrected by installing a new air-idle by-pass tube and resonator assembly. • Automatic transmissions with delayed or no forward engagement, or a higher-than-expected engine rpm when coming to a stop, are covered in DSB #94-26-9. **1992–97**—An erratic or prolonged 1–2 shift can be cured by replacing the cast aluminum piston with a one-piece stamped steel piston that has bonded lip seals and by replacing the top accumulator spring. These upgraded parts will increase the transmission's durability, according to Ford. **1995–97**—ABS brakes that activate on their own or produce a grinding, pulsing, fluttering effect on the brake pedal probably need upgraded wiring connectors at the ABS sensors.

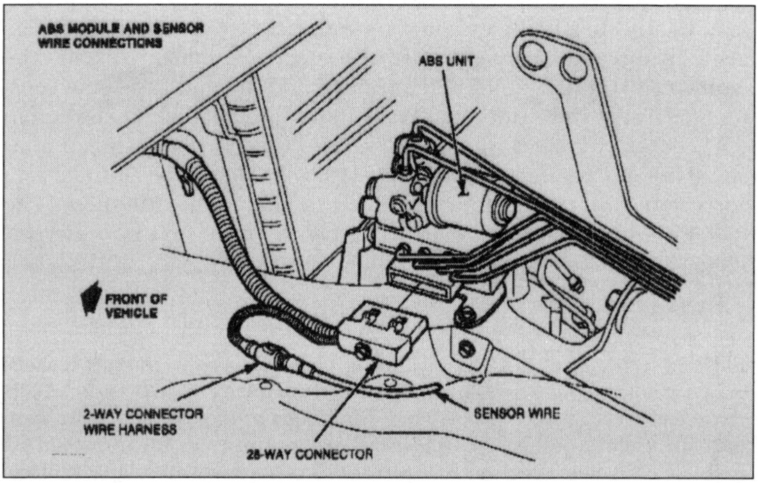

This three-hour job should fall under Ford's base warranty.

1996–97—Spark knock during acceleration can be silenced by replacing the MAF sensor and reprogramming the power control module. **All**

models/years—Three components that frequently benefit from Ford "goodwill" warranty extensions are catalytic converters, fuel pumps, and computer modules. If Ford balks at refunding your money, apply the emissions warranty for a full or partial refund. • Paint delamination, fading, peeling, hazing, and "microchecking" (see page 112 for details on claiming a refund).

Crown Victoria, Grand Marquis Profile

	1991	1992	1993	1994	1995	1996	1997	1998
Cost Price ($)								
Crown S/LTD	19,795	22,095	23,195	22,495	24,695	27,195	29,895	30,995
Grand Marquis GS	21,795	23,995	24,595	23,795	26,695	30,295	32,195	32,895
Used Values ($)								
Crown S/LTD ↑	5,500	6,500	8,500	11,000	13,500	16,000	19,000	22,000
Crown S/LTD ↓	4,500	5,500	7,500	10,000	12,500	15,000	18,000	20,000
Grand Marquis GS ↑	6,500	8,000	9,000	12,500	14,000	17,000	21,000	23,000
Grand Marquis GS ↓	5,500	7,000	8,500	11,500	13,000	16,000	20,000	22,000
Extended Warranty	Y	Y	Y	N	N	N	N	N
Secret Warranty	Y	Y	Y	Y	Y	Y	Y	N
Reliability	③	③	③	④	④	④	⑤	⑤
Air conditioning	❷	❷	❷	❷	③	③	③	④
Automatic transmission	❶	❶	❷	❷	❷	❷	④	④
Body integrity	❶	❶	❶	❶	❷	❷	❷	③
Braking system	❶	❶	❶	❶	❷	❷	❷	③
Electrical system	❷	❷	❷	❷	③	③	③	④
Engines	④	④	④	④	④	④	⑤	⑤
Exhaust/Converter	❶	❶	③	③	③	④	⑤	⑤
Fuel system	❷	❷	③	③	③	③	③	④
Ignition system	❷	❷	❷	③	③	③	③	④
Rust/Paint	❷	❷	❷	❷	③	③	③	④
Steering	❷	❷	③	③	③	③	③	④
Suspension	❷	❷	❷	❷	❷	❷	③	③
Crash Safety	—	⑤	❷	❷	④	⑤	⑤	⑤
Side Impact	—	—	—	—	—	—	④	④

Note: The 1993–94 models' low crash scores represent severe leg trauma.

GENERAL MOTORS

Caprice, Custom Cruiser, Impala SS, Roadmaster, Wagon

Rating: Above Average (1995–96); Average (1994); Not Recommended (1985–93). The 1995–96 versions carry the best crash ratings. This group of cars ceased production in 1995. A good alternate choice would be any rear-drive Buick LeSabre or a Ford Crown Victoria or Grand Marquis. **Maintenance/Repair costs**: Inexpensive and easy to perform, plus repairs can be done by any corner garage. **Parts**: Average parts cost can be cut further by shopping at independent suppliers who are generally well stocked.

Strengths and weaknesses: These cars are large, comfortable, and easy to maintain. The trunk is spacious. Overall handling is acceptable, but expect a queasy ride from the too-soft suspension. Gas mileage is particularly poor. Despite the many generic deficiencies inherent in these rear-drives, they still score higher than GM's front-drives for overall reliability and durability. The Impala SS is basically a Caprice with a 260-hp Corvette engine and high-performance suspension.

On 1988–91 models, engine problems include crankshaft and head gasket failures, cracked cylinder heads, injection pump malfunctions, and oil leaks. Engine knocking is another common problem on early models that's hard to correct inexpensively due to the various possible causes that have to be eliminated. Early V8s, in particular, suffer from premature camshaft wear, and the 350-cubic-inch V8s often fall prey to premature valve guide wear caused by a faulty EGR valve. Cars equipped with the 5.7L diesel V8 should be approached with caution; they aren't very durable and cost an arm and a leg to troubleshoot and repair. The 4-speed automatic transmission was troublesome until 1991, with burnt-out clutches and malfunctioning torque converters being the most common failures.

The 1991–96 models have elicited few engine and transmission complaints, although they have shown the following deficiencies: chronic AC and ignition glitches; prematurely worn brakes (lots of corrosion damage), steering, and suspension components, especially shock absorbers and rear springs; serious electrical problems; and poor-quality body and trim items.

Body assembly is not impressive, but paint quality and durability is fairly good, considering the delamination one usually finds with GM's other models. Wagons often have excessive rust around cargo-area side windows and wheel wells; and hubcaps on later models tend to fly off. During my stint as an MP on Parliament Hill, RCMP Caprices were constantly losing their hubcaps.

Dealer service bulletins: **1993**—Buzzing at idle coming from the 4L60E automatic transmission, along with a maladjusted shift linkage, causing loss of Reverse or lack of power in second gear. • Improperly adjusted automatic transmission shift linkage leading to a burnt-out Low/Reverse clutch and increased rpms with downshifts. • Noisy power-steering units. • Intermittent cruise control operation. • Extreme temperature difference when the AC is put in a bi-level mode, and inoperative AC compressors. **1994**—Excessive oil consumption. • Faulty door-mounted radio speakers. • Poor AC performance. • Inadequate interior air flow. **1995**—1–2 and 2–3 shifting malfunctions (Caprice). • Front brake pulsation. • An exhaust system moan or boom; inadvertent horn honking; a low voltage reading or dim lights at idle. • Automatic door lock may not work (Roadmaster). • A steering system crunch or popping noise. • Wiper blade chatter. • Poor paint application and rust spots. **1996**—Engine noise (install new valve-stem oil seal). • 3–2 part throttle downshift flare and delayed transmission engagement. • Transmission chuggle/surge. • Transmission fluid leak from pump body (replace bushing). • Steering column noise. • Radio frequency interference diagnosis. • AC odours.

Safety summary/Recalls: **1994–96 vehicles with heavy-duty cooling systems**—NHTSA probes are looking into reports that coolant may leak into the rear of the engine compartment, where it may ignite. **1996**—Left rear wheel came off while driving. • Airbags failed to deploy. • Airbag light stays lit for no apparent reason. • ABS brakes lock up. • Transmission torque converter/engine flywheel breakage. • Broken torque converter bolts. • Steering box loosens up despite new bolts. • Steering lockup. • Sometimes vehicle fishtails uncontrollably while at moderate speed. • Front coil spring failure. • Premature tire wear caused by faulty suspension components that can't be fixed by repeated alignments. • Fuel pump failure caused by wiring harness short. **Recalls**: **1985**—Leaking fuel feed and return pipe may cause a fire. • Cars equipped with a 4.3L engine could have a battery-cable short, which creates a fire hazard. **1985–88**—Faulty cruise control could lead to sudden acceleration. **1987**—Cars equipped with 200-4R automatic transmission could start in gear or engage the wrong gear. **1989**—GM will inspect and replace the AC condenser inlet pipe. **1991**—Seatbelts malfunction because of defective shoulder belt guide loop. **1991–92**—Rear seatbelts that are uncomfortable will be changed for free by GM. • Corrosion could prevent the hood from latching properly, making it hard to open or causing it to fly open. **1992**—A rattling front door lock rod can be silenced by installing a corrective kit (#10222731). **1994**—A leaking oil cooler inlet hose is a fire hazard. • Fractured wheel studs may allow the wheel to separate from the car. • Faulty fuel tank strap fasteners. **1994–95**—Accelerator pedal may stick. **1995–96 Roadmaster, Caprice**—Loose wheel lug nuts could cause the wheel to fall off.

Secret Warranties/Service Tips

1982–91—Hydramatic 4L60/700R4 automatic transmission may have no upshift or appear to be stuck in first gear. The probable cause is a worn governor gear. It would be wise to replace the retaining ring as well. **1982–93**—Vehicles equipped with a Hydramatic 4L60 transmission that buzzes when the car is in Reverse or at idle may need a new oil pressure regulator valve. **1985–88**—The 307-cubic-inch V8 is the most reliable and durable engine in this lineup. In 1985–86, it was plagued with camshaft/lifter problems due to the small offset from lifter bore to cam lobe. The problem was rectified in 1987 when GM adopted roller-lifters. The engine's aluminum intake manifold and gasket can be eaten away by extensive corrosion in each of the four corners where they contact a coolant port; serious coolant leaks result. Flush the radiator yearly and use a good-quality antifreeze in order to extend the life of these components. **1987–91**—A bulletin lists the different types of tailpipe smoke signals that alert you to the need for different repairs: cars with V8 engines that emit blue or bluish-white smoke may require a valve seal kit (#12511890); models with port fuel-injection that emit black smoke after long starter cranks may have fuel leaking into the engine from the injectors; models with throttle body injection and port fuel-injection that emit white or bluish smoke and have normal oil consumption likely have poor sealing between the intake manifold joint and the cylinder head. **1989–91**—If the transmission won't go into Reverse or is slow to shift into Reverse, GM suggests that the Reverse input clutch housing be changed. **1991**—A poor-running 5.0L engine may require an EGR valve kit and PCV valve in addition to a new PROM. **1994**—Excessive oil consumption is likely due to delaminated intake manifold gaskets. Install an upgraded intake manifold gasket kit. • GM campaign 94C15 will adjust at no charge a maladjusted automatic transmission shift linkage that could, if left alone, burn out the Low/Reverse clutch. **1994–96**—Excessive engine noise can be silenced by installing an upgraded valve stem oil seal. **1994–96 with 5.7L engine**—A chuggle or surge condition will require a reflash calibration. **1991 Roadmaster**—If the starter makes a grinding noise or won't engage the engine, the fault could be a weak starter spring. • Front doors that won't stay open on an incline or are hard to open need an improved hold-open door spring. **1992**—Water leaks into the trunk from the fixed mast antenna; install an upgraded antenna base (#25610633) and bezel (#25609129). • If the front seat doesn't have enough upward travel, GM will provide a special kit (#12520796) to raise it. • A chuggle or shudder at 44 mph on vehicles equipped with a 5.7L engine may be corrected by installing a new torque converter assembly. **1994**—Excessive oil consumption is likely due to delaminated intake manifold gaskets. Install an upgraded intake manifold gasket kit. • Poor AC performance can be improved by replacing the temperature control cable. • Delayed automatic transmission shift engagement is a common problem addressed in DSB #47-71-20A. **1995–96**—Delayed automatic transmission shift engagement may require the replacement of the pump cover assembly. **All models/years**—A rotten-egg odour coming from the exhaust is usually the result of a malfunctioning catalytic converter covered by the emissions warranty. • Paint delamination, peeling, or fading (see pages 76, 113).

Caprice, Custom Cruiser, Impala SS, Roadmaster, Wagon Profile

	1990	1991	1992	1993	1994	1995	1996
Cost Price ($)							
Caprice	20,128	19,698	20,298	20,806	22,500	25,455	28,345
Caprice wagon	21,347	20,398	21,598	22,138	24,985	29,120	30,980
Cruiser wagon	25,116	25,098	25,350	—	—	—	—
Impala SS	—	—	—	—	—	29,005	30,675
Roadmaster	—	—	25,398	26,298	29,798	32,930	34,230
Wagon	—	24,898	26,150	27,398	31,798	35,015	36,315
Used Values ($)							
Caprice ↑	4,000	6,000	7,000	9,000	12,500	14,000	16,000
Caprice ↓	3,000	5,000	6,000	8,000	10,500	12,500	15,000
Caprice wagon ↑	6,500	7,000	7,500	9,500	13,500	15,500	17,500
Caprice wagon ↓	5,000	6,000	6,500	8,500	12,500	14,000	16,000
Cruiser wagon ↑	6,500	7,000	7,500	9,500	—	—	—
Cruiser wagon ↓	5,000	6,000	6,500	8,500	—	—	—
Impala SS ↑	—	—	—	—	—	17,500	21,000
Impala SS ↓	—	—	—	—	—	16,000	19,500

Roadmaster ↑	—	—	9,000	11,000	14,000	17,500	21,000
Roadmaster ↓	—	—	8,000	9,500	13,000	16,000	19,000
Wagon ↑	—	8,000	9,500	10,500	14,500	18,500	22,000
Wagon ↓	—	7,000	8,500	9,500	13,500	16,500	20,000
Extended Warranty	Y	Y	Y	Y	N	N	N
Secret Warranty	Y	Y	Y	Y	Y	Y	Y
Reliability	❷	❷	❷	❷	③	③	③
Air conditioning	❷	❷	❷	❷	③	③	④
Automatic transmission	❷	③	③	③	③	④	④
Body integrity	❷	❷	❷	❷	❷	③	③
Braking system	❷	❷	❷	❷	❷	❷	③
Electrical system	❷	❷	❷	❷	❷	❷	③
Engines	③	③	③	③	③	④	④
Exhaust/Converter	❷	❷	❷	❷	③	④	④
Fuel system	③	③	③	③	③	③	③
Ignition system	❷	❷	❷	❷	❷	❷	③
Rust/Paint	❷	❷	③	③	③	③	③
Steering	❷	❷	❷	❷	③	④	④
Suspension	❷	❷	❷	③	③	④	④
Crash Safety							
Caprice	—	❷	❷	❶	❶	④	④

Note: The 1995–96 Caprice models scored four stars for driver protection but only two for front passenger protection in government-run crash tests.

Olds 98, Regency, Park Avenue

Rating: Average (1997–98); Below Average (1991–96); Not Recommended (1985–90). Although a new Park Avenue commands a higher price than a new Olds 98, Regency, the value of a used Park Avenue trails slightly. The last model year for the Olds 98, Regency was 1996, although it continued to be sold in the States. **Maintenance/Repair costs**: Higher than average, but repairs aren't dealer dependent. **Parts**: Higher-than-average cost (independent suppliers sell for much less), but not hard to find. Nevertheless, don't even think about buying one of these front-drives without a three- to five-year extended warranty backed by the automaker.

Strengths and weaknesses: These attractive, luxurious cars are billed as six-seaters, but only four passengers can ride in comfort. Although the 1991–96 Park Avenue and Olds 98, Regency were improved over the years, they've compiled the worst repair history among large cars. Main problem areas are the engine, automatic transmission, fuel system, brakes, electrical system (including defective PROM and MEMCAL modules), starter and alternator, and badly assembled, poor-quality body hardware. The 3.0L V6 engine is inadequate for cars this heavy, and the 3.8L has been a big quality disappointment. Stay away from the failure-prone diesel engine.

Under-hood servicing is complicated. Automatic transmission and engine computer malfunctions are common. The fuel-injection system is temperamental. Window mechanisms are poorly designed. The power-steering assembly is failure-prone. There are frequent electrical failures. Front brake pads and rotors require frequent replacement. Shock absorbers leak or go soft very quickly. Extensive surface corrosion has been a problem because of poor and often incomplete paint application at the factory.

Like the LeSabre, the 1997 Park Avenue and Ultra were redesigned to include a reworked powertrain, a stiffer body, improved interior amenities, upgraded four-wheel disc brakes, and an upgraded ventilation system.

Aficionados of full-sized luxury sedans love the flush glass, wraparound windshield and bumpers, and clean body lines that lend the latest makeover an aerodynamic and pleasing appearance. But these cars are more than just pretty packages; they provide lots of room, luxury, style, and, dare I say, performance. Plenty of power is available with the 205-hp 3.8L V6 engine and the Aurora's 240-hp supercharged version of the same powerplant. It does 0–100 km/h in under 9 seconds (impressive, considering the heft of these vehicles), and improves low- and mid-range throttle response. Power is transmitted to the front wheels through an electronically controlled transmission that features "free-wheeling" clutches designed to eliminate abrupt gear changes. Both the Park Avenue and Ultra use a stretched version of the more rigid Buick Riviera and Olds Aurora platform.

The revised 1997 Park Avenue models have had fewer complaints, perhaps because of the short time they've been out and the fact that GM's warranty is still in effect. Nevertheless, the main problem areas continue to be the following: powertrain malfunctions; engine and transmission leaks; a concerto of squeaks, rattles, moans, and whines; AC not performing properly; and numerous body and trims defects. Since the 1998 returned relatively unchanged, don't expect any major improvements quality-wise.

Dealer service bulletins: **1993**—3.8L engines frequently stall when decelerating or lead to extended cranking; also, they often stall after a cold start. • Excessive front-end vibration when passing over bumps. • Whistling when the AC is in recirculate mode. **1994**—Lots of bulletins address automatic transmission and heater/AC problems. **1995**—Noise while driving up medium grades or accelerating. • Stall upon deceleration. • Cold-start rattle with the 4T60E automatic transmission. • Grinding/growling when in Park on an incline. • A coolant leak from the belt tensioner assembly. • Throttle body noise upon acceleration from stop. • Brake vibration and/or pedal pulsation. • Whistle coming from the heater/AC unit. • Low voltage reading or dim lights at idle. • Headlights/parking lights remain on. • Squeak/creak from rear of vehicle. • Poor paint application and rust spots. **1996**—Second-gear starts, poor shifting 1–3. •

AC defrost valve creaks when changing modes. • AC odour at start-up in humid climates. • AC flutter or moan. • Noisy steering column. • Cold-start rattle. • Engine cranks but won't start. • Engine oil leaks. • Excess vibration of electrochromatic rear view mirror. • Exterior light condensation. • Fluttering, popping, ticking, and clunking noises. • Inoperative door locks. • AC thump noise upon start-up. • Intermittent Neutral/loss of Drive at highway speeds. • Instrument panel buzz or rattle when the brakes are applied. • Popping noise from front of vehicle when turning. • Troubleshooting radio frequency interference. • Rattle in rear door when closing. • Transaxle gear whine. • Whistle noise from HVAC (climate control system). • Wind noise around front and rear doors. **1997**—AC thump noise. • Driver's seat rocks. • Noisy front brakes. • Engine oil leaks. • Engine oil level indicates over-full. • Harsh shift from Reverse to Drive. • Excess vibration of electrochromic rear view mirror. • Erratic wiper operation in automatic mode. • Spontaneous horn activation or inoperative horn. • Instrument panel buzz or rattle when the brakes are applied. • Popping noise from front of vehicle when turning. • Growl or vibration when turning or accelerating. • Paint chips on rear compartment lid. • Roof panel running front to back has a wavy/dimple appearance and transmits excessive rain noise. • Steering vibration/moan when parking. • Transaxle gear whine.

Safety summary/Recalls: **Recalls**: **1986**—Faulty headlight switches. • Power-steering hose leaks could cause an under-hood fire. • ABS may suddenly fail due to fluid leaking from system. **1987**—A faulty fusible link could cause an under-hood fire. • Some throttle cables may not return to idle position. **1991**—Parking brakes might not hold. • Console shift lever could disengage. **1992–93**—Transmission-cooler line could separate at low temperatures. **1994–95 Olds 98, Regency**—Headlight switch may not work. **1995 models with Twilight Sentinel**—Current leakage can cause loss of headlights and parking lights, or the lights may suddenly come on while the car is parked. **1996**—Damaged capacitor could cause confusing electronic warnings to be displayed. • Backfire upon start-up can damage the intake manifold and cause hard starting or a fire. **1997 Park Avenue**—Centre seatbelt anchor bolts were improperly installed. • Brake/Traction Control Module could cause ABS to lose its effectiveness and make for longer stopping distances.

Secret Warranties/Service Tips

1988–93—Vehicles equipped with a 3800 engine that stalls upon deceleration or is hard to start may need a new air control motor (IAC). **1991–94**—Harsh automatic transmission upshifts can be corrected by installing an upgraded accumulator valve in the control valve body. • Loss of Drive or erratic shifts may be caused by an intermittent short to ground on the A or B shift solenoid or an electrical short circuit in the transaxle. • Water leaking from the doors into the passenger compartment has a number of causes and remedies (DSB #431003). **1993–94**—Owners who complain of automatic transmission low-speed miss, hesitation, chuggle, or skip may find relief with the improved MEMCAL module that GM developed to remedy the problem. **1994**—Inadequate heating, ventilation, and AC operation is addressed in DSB #431219. **1995–96**—Wind noise around front and rear doors; diagnosis and repair. **1995–97**—Transmission gear whine at 40–70 km/h means the final drive assembly may have to be replaced. **All models/years**—A rotten-egg odour coming from the exhaust is probably caused by a malfunctioning catalytic converter and may be covered under GM's emissions warranty. • The THM 44C-T4 automatic transaxles on front-drive models equipped with V6 engines are particularly failure-prone: their pinched or kinked vacuum lines result in low oil pressure. • Paint delamination, peeling, or fading (see pages 76, 113).

Olds 98, Regency, Park Avenue Profile

	1991	1992	1993	1994	1995	1996	1997	1998
Cost Price ($)								
Olds 98, Regency	29,490	27,795	28,798	29,298	34,498	36,600	—	—
Park Avenue	29,995	30,898	31,798	33,798	37,115	38,150	40,865	41,850
Used Values ($)								
Olds 98, Regency	7,500	9,000	10,500	14,500	17,000	20,000	—	—
Olds 98,Regency ↓	7,000	8,000	9,500	13,000	15,500	18,500	—	—
Park Avenue ↑	8,500	9,500	11,000	14,000	17,000	21,000	24,000	29,500
Park Avenue ↓	7,500	8,000	10,000	13,000	15,000	18,500	22,500	28,000
Extended Warranty	Y	Y	Y	Y	Y	Y	Y	Y
Secret Warranty	Y	Y	Y	Y	Y	Y	Y	Y
Reliability	❷	❷	❷	❷	❷	❷	③	③
Air conditioning	❷	❷	❷	❷	❷	③	③	③
Automatic transmission	❷	❷	❷	❷	❷	③	③	③
Body integrity	❶	❶	❶	❶	❶	❶	❶	③
Braking system	❷	❷	❷	❷	❷	③	③	③
Electrical system	❷	❷	❷	❷	❷	❷	❷	❷
Engines	❷	❷	❷	③	③	③	③	③
Fuel system	❷	❷	❷	❷	❷	❷	③	③
Ignition system	❷	❷	❷	❷	③	③	③	③
Rust/Paint	❶	❶	❶	❶	❶	❶	❷	③
Steering	❶	❶	❶	❶	❷	❷	❷	③
Suspension	❷	❷	❷	❷	❷	③	③	③

Note: These vehicles weren't crash-tested for the 1991–98 model years.

LUXURY CARS

Used luxury cars are attractive buys because they project a flashy image and can be bought for 50 percent of the new price after three years or so. If the selling price is reasonable, independent servicing is available, and used parts can be found, they can be bargains. On the downside, they're complicated to service and the cost of new parts can be horrendous.

And do remember this: spending more on a luxury car doesn't always mean you'll get more safety or reliability. You may simply get more headaches (see "Safety summary/Recalls" for Cadillac, Lincoln, Jaguar, and Saab).

Traditionally, the luxury-car niche has been dominated by American and German automakers. During the past decade, however, buyers have gravitated towards Japanese models. This shift in buyer preference has forced GM, Ford, and Chrysler to downsize and adopt front-drives. It has also forced GM to drop its rear-drive Caprice, Roadmaster, and Fleetwood. Despite these moves to better respond to buyers' preferences, American luxury cars are still seen by most consumers as overweight and unreliable land yachts.

You don't always have to spend big bucks to get true luxury and ironclad reliability. Smart buyers can target fully equipped Toyota Camrys and Cressidas, Honda's Accord, Nissan's Maxima, and Mazda's 929, all of which offer the same equipment, reliability, and performance as Lexus, Infiniti, and Acura models, but for much, much less.

There are few used American luxury cars that can stand up to a six-year-old Japanese or German luxury vehicle. And this fact is reflected in the head-spinningly high depreciation rates seen with most luxury cars put out by the Big Three. Chrysler's rear-drive New Yorker, Ford's Crown Victoria and Grand Marquis, and Lincoln's Town Car come closest to meeting the imports in overall reliability and durability. On the other hand, the discontinued Chrysler Imperial was more show than go, and the front-drive New Yorker and its LHS variants are unremarkable and are plagued by serious powertrain reliability problems, as can be seen by the numerous dealer service bulletins I've intercepted. GM's Cadillacs aren't even in the running; in fact, *AutoWeek* magazine has dubbed the Catera a "lame duck."

Recommended

BMW 5 Series (1992–98)
Lexus ES 250, ES 300, GS 300, LS 400, SC 400 (1990–98)
Mercedes-Benz 300 Series, 400 Series, 500 Series, E-Class (1993–98)
Nissan Maxima (1996–98)
Toyota Avalon (1995–98)
Volvo 850 Series (1995–97)
Volvo 900 Series (1995–97)

Above Average

Acura Legend (1989–95)
Audi 80, 90, A4, A6, A8, S6, 100, 200, 5000, V8, Coupe, Quattro, (1995–98)
BMW 3 Series (1995–98)
GM Aurora (1995–98)
GM Brougham, Fleetwood RWD (1993–96)
Infiniti G20, I30, J30, Q45 (1995–98)
Lincoln Mark VII, Mark VIII (1995–98)
Lincoln Town Car (1995–98)
Mazda 929 (1993–95)
Mazda Millenia (1995–98)
Mercedes-Benz 190 Series C-Class (1995–98)
Mercedes-Benz 300 Series, 400 Series, 500 Series, E-Class (1992)
Nissan Maxima (1989–95)
Toyota Cressida (1985–92)
Volvo 850 Series (1993–94)
Volvo 900 Series (1989–94)

Average

BMW 3 Series (1994)
BMW 5 Series (1985–91)
GM Brougham, Fleetwood RWD (1984–92)
GM Concours, DeVille, Fleetwood FWD (1995–98)
GM Riviera, Toronado, Trofeo (1995–98)
Infiniti G20, I30, J30, Q45 (1991–94)
Lincoln Continental (1996–98)
Lincoln Mark VII, Mark VIII (1994)
Lincoln Town Car (1988–94)
Mazda 929 (1988–92)
Mercedes-Benz 190 Series, C-Class (1988–94)
Nissan Maxima (1986–88)
Saab 900, 9000 (1998)

Below Average

BMW 3 Series (1984–93)
GM Allanté, Catera, Eldorado, Seville (1992–98)
GM Concours, DeVille, Fleetwood FWD (1985–94)
Lincoln Continental (1984–95)
Lincoln Mark VII, Mark VIII (1986–93)
Mercedes-Benz 190 Series, C-Class (1984–87)
Mercedes-Benz 300 Series, 400 Series, 500 Series E-Class (1985–91)
Saab 900, 9000 (1995–97)

Not Recommended

Acura Legend (1986–88)
Audi 80, 90, A4, A6, A8, S6, 100, 200, 5000, V8, Coupe, Quattro (1984–94)
GM Allanté, Catera, Eldorado, Seville (1986–91)
GM Riviera, Toronado, Trofeo (1986–93)
Saab 900, 9000 (1985–94)
Volvo 700 Series (1985–92)

ACURA

Legend

Rating: Above Average (1989–95); Not Recommended (1986–88). Resale value is high on all Legend models, but especially so on the coupe. Shop instead for a cheaper 1989 or later base Legend with the coupe's upgraded features. In fact, the 1990 Legend combines the lowest price with the best crash rating score. **Maintenance/Repair costs**: Higher than average, and most repairs are dealer dependent. To avoid costly engine repairs, check the engine timing belt every 2 years/40,000 km and replace it every 96,000 km. **Parts**: Higher-than-average cost (some independent suppliers sell for much less under the Honda name), but not hard to find, despite the fact that the Legend was dropped in 1995.

Acura Canada says that its service bulletins published in the U.S. don't apply to Canadian Acuras and that they won't follow the labour time set out in their bulletins. I say this is hogwash. Here's what one angry Acura owner wrote:

> Here's a copy of the fax sent to me by Acura explaining their position on the "warranty issue" concerning the speed sensor failure. The main contention is that the bulletin referred to in your [*Lemon-Aid*] publication is for use in the U.S. only. I fail to see the difference between U.S. Acuras and those in Canada and the rep admitted the cars are the same. They say, however, that driving conditions vary from country to country, hence the warranty differential. I doubt if Acura owners in Michigan or Minnesota, with their comparable climate conditions, would agree.
>
> No, I believe it isn't country of origin that determines their use of "secret" or "goodwill" warranties, it's market conditions and competition. I suspect Acura goes overboard to satisfy U.S. customers because it's simply a bigger market with more potential customers.
>
> The difference also shows up in the cost of repair. The Halifax dealer charged half an hour's labour to diagnose the problem (even though I told them what it was). Then the dealer charged one and a half hours to install the new sensor when the U.S. service bulletin says it should take 0.7 (i.e., less than an hour). If Acura is alone or among the very few that take this approach, then readers should be warned....

Strengths and weaknesses: **1986–90**—Early Legends were upscale, enlarged Accords that were unimpressive performers with either 6-cylinder powerplant. Dependability, acceptable road handling, and a spacious interior are the car's main advantages. In spite of occasional clutch malfunctions, the 5-speed manual gearbox is the transmission of choice. The automatic shifts harshly and its lockup torque converter is constantly cutting in and out, reducing both performance and fuel economy. An overly soft suspension gives the car a bouncy ride, and it easily bottoms out when the vehicle is loaded or traversing rough roads. Front-end components get noisier as time passes. Long-term durability is better than average, with the only problems centred on body hardware and accessory equipment malfunctions, notably poor radio performance.

1991–95—The 3.2L V6 is by far a better performer than the 2.5L and 2.7L engines of earlier models, although some drivers complain it runs hot. Ride quality is improved, power steering is more responsive, and rear seating is more spacious. Automatic shifting is still rough (especially during acceleration), fuel economy is disappointing, and the coupe's rear seat room is a joke. Owners report that suspension struts soften quickly, minor electrical problems occur occasionally, AC performance is erratic, radio reception is poor, brake servicing is too frequent, and body hardware is unimpressive. The windshield washer pump frequently malfunctions on the 1991–92 versions.

Dealer service bulletins: **1993**—Faulty trunk-mounted CD changer magazines. **1994**—Lots of body-panel fit problems. • Poor radio reception. • Steering wheel shimmy. • Faulty power-steering speed sensors. • Door window scratches. **1995**—Malfunctioning seatbelt. • Buzzing seatbelt anchor. • Steering wheel shimmy troubleshooting. • Erratic remote-audio-control operation. • Gap in front door sill moulding at B-pillar. • Clicking noise while turning.

Safety summary/Recalls: **NHTSA probes**: **1986–87**—Fuel tank overpressurization may result in fuel spewing out when the cap is removed. **1986–88**—These Legends have one of the highest rates of accidental sudden acceleration reported by owners. **1992–93**—Inadvertent airbag deployment. **Recalls**: **1986–1990**—Takata seatbelt replacement. **1991**—A faulty automatic transmission shift cable should be replaced. **1992**—Vehicles may not have an igniter for the passenger-side airbag.

Secret Warranties/Service Tips

1986–95—A clicking noise heard while the vehicle is turning means you need to replace the constant-velocity joint. **1990**—Water collects in the spare tire well; diagnosis and repair DSB #89-021. • Malfunctioning security system control unit; diagnosis and repair DSB #90-002. **1991**—Poor AM radio reception or interference from the car's electrical equipment may be due to a poor ground connection between the antenna collar and car body. **1991–92**—Static on the AM radio band may also be caused by a faulty Bose amplifier as well as poor antenna grounding due to loose mounting nuts. • If the remote audio volume controls cause static or operate erratically, change the volume control motor inside the audio unit. • Poor AC performance is often caused by a maladjusted heater valve cable. **1991–93**—Engine knocking after a cold-start is likely due to carbon buildup in the piston ring land. **1991–94**—To prevent the lower bumper face from pulling loose, Acura will reinforce the mounting points and reverse the overlap of the lower bumper face and the splash shield. • Power-steering speed sensor failures. **1995**—You can correct the gap in the front door sill moulding at the B-pillar by trimming back the nylon carpet retainer. **All models/years**—Like Hondas, most Acura DSBs allow for special warranty consideration on a "goodwill" basis even after the warranty has expired or the car has changed hands. Referring to this euphemism will increase your chances of getting some kind of refund for repairs that are obviously factory defects. Acura will repair or replace defective steering assemblies, constant velocity joints, and catalytic converters free of charge for up to 5 years/80,000 km on a case-by-case basis. There's no labour charge or deductible. Used vehicles and repairs carried out by independent garages aren't covered by this special program. But, since the converter is an emissions component, it's almost always automatically covered under the emissions warranty. • Steering wheel shimmy can be reduced by rebalancing the wheel/tire/hub/rotor assembly in the front end. • Poor radio reception or interference is addressed in DSB #94-011, issued August 30, 1994. • The automaker will install air intake screens to keep debris away from the blower motor. • Vertical scratches on the door window result from the window making contact with the moulding's metal clips. • Seatbelts that fail to function properly during normal use will be replaced for free under the company's lifetime seatbelt warranty.

Legend Profile

	1988	1989	1990	1991	1992	1993	1994	1995
Cost Price ($)								
L/LS	29,895	31,995	33,295	38,540	42,360	44,655	50,000	51,300
Used Values ($)								
L/LS ↑	6,000	7,500	9,000	11,000	13,000	15,500	22,000	27,000
L/LS ↓	5,000	6,500	8,000	9,500	11,500	14,000	20,000	25,000
Extended Warranty	N	N	N	N	N	N	N	N
Secret Warranty	N	N	N	N	N	N	N	N
Reliability	③	③	③	④	④	④	⑤	⑤
Air conditioning	③	③	③	④	④	④	⑤	⑤
Body integrity	③	③	③	③	③	③	③	③
Braking system	③	③	③	④	④	④	④	⑤
Electrical system	❷	❷	❷	③	③	③	③	④
Engines	❷	❷	❷	③	④	⑤	⑤	⑤
Exhaust/Converter	④	④	⑤	⑤	⑤	⑤	⑤	⑤
Fuel system	④	④	④	④	④	⑤	⑤	⑤
Ignition system	③	③	④	④	④	⑤	⑤	⑤
Manual transmission	③	❷	④	⑤	⑤	⑤	⑤	⑤
- automatic	③	③	③	③	③	④	④	⑤
Rust/Paint	❷	③	③	③	④	④	⑤	⑤
Steering	④	④	④	④	⑤	⑤	⑤	⑤
Suspension	③	③	③	③	④	⑤	⑤	⑤
Crash Safety	⑤	—	⑤	—	—	❷	❷	③

AUDI

80, 90, A4, A6, A8, S6, 100, 200, 5000, V8, Coupe, Quattro

Rating: Above Average (1995–98); Not Recommended (1984–94). **Maintenance/Repair costs**: Higher than average, and almost all repairs have to be done by an Audi dealer. **Parts**: Way-higher-than-average cost, and independent suppliers have a hard time finding parts. Don't even think about buying one of these front-drives without a three- to five-year supplementary warranty backed by Audi.

Strengths and weaknesses: These cars are attractively styled, handle well, are comfortable to drive, and provide a spacious interior. Yet the pre-1993 models have a worse-than-average reliability record and are plagued by mechanical and electrical components that don't stand up to the rigors of driving in Canada. Furthermore, it seems that Volkswagen has written off these failure-prone cars in favour of the better-built and more recent A4 and A6 models. The dealer body isn't strong enough to adequately service all of these vehicles when things go wrong, so owners of older models are generally left to independent garages to serve their needs. Premium fuel is required for vehicles equipped with the 6-cylinder engine.

A4, A6, A8, S4, S6, and Sport 90

This is the pick of the Audi litter starting with the 1995 model year. Although Audi continued its alphabet name game by grouping its 100 series under a new A6 heading, these cars were substantially improved. Packed with standard features, the A6 is a comfortable, spacious, front-drive or all-wheel-drive luxury sedan that comes with dual airbags and ABS. It uses the same V6 powerplant as the A4, its smaller sibling. The S4 is a limited-production, high-performance spin-off that carries a 227-hp turbocharged rendition of the old 5-cylinder powerplant. Restyled for 1995, the A4 and S4 were rebadged as the A6 and S6. The S6 wasn't sold in 1998, however.

These cars are conservatively styled, slow off the mark (in spite of the V6 addition when hooked to an automatic), and plagued by electrical glitches. The 4-speed automatic shifts erratically and the 2.8L V6 engine needs full throttle for adequate performance. Handling is acceptable, but the ride is a bit firm and the car still exhibits considerable body roll, brake dive, and acceleration squat when pushed. Handling is on a par with the BMW 300 series, and acceleration times beat out those of the Mercedes. Quattro's AWD is extended to entry-level models at a time when most automakers are dropping the option on passenger cars.

Overall quality control has really improved during the past several years, with fewer body, brake, and electrical glitches than exhibited by previous models. A perusal of this year's internal service bulletins shows a dramatic improvement in quality control over earlier versions.

Dealer service bulletins: **1995**—Oil leakage from the rear and upper engine. • Coolant circulation pump leakage. • Delayed first- and second-gear shift. • A malfunctioning climate control. • Inaccurate fuel gauge and temperature gauges. • Door lock key binding. • Skipping wiper blades. **1996**—ABS light comes on inadvertently. • Front speaker buzzing. • Cruise control won't maintain set speed. • Delayed 1–2 shift on cold-start warm-up. • Inaccurate Delta Bose radio display. • Friction noise from front and rear door seals. • Fuel gauge doesn't register full. • Carbon buildup in the intake valve and combustion chamber. • Exhaust popping and rasping noise. • Radio volume goes to maximum when adjusted. • Binding rear ashtray lid. • Rear differential noise and vibration while driving. • Loose rear reading light. • Tachometer sticking and erratic AC display. **1997**—False ABS light warning. • Delayed 1–2 shift on cold-start warm-up. • Friction noise from front and rear door seals. • Rasping noise from exhaust.

80, 90, and 100 series

Launched in 1988, these entry-level Audis share the same wheelbase, front-drive, and 4X4 components. Equipped with an efficient but wimpy 4-cylinder (dropped in 1991) or the more powerful 2.3L 5-cylinder engine, 4-wheel disc brakes, and galvanized body panels, these small sedans are leagues ahead of Audi's mid-1980s vehicles. The 1991 models are clearly a better choice; they use an improved 4-speed automatic transmission hooked up to a more powerful engine. 1992 was basically a carry-over year in which unsold 1991 models were recycled. Some of their more common problems include AC, electrical system, and brake malfunctions.

Dealer service bulletins: **1993 Audi 100s**—Misaligned deck lids that are hard to close. • Noisy steering column and squeaking and rattling coming from the front seats and inside the B-pillar. • Excessive wind noise because of poor window sealing. • Smearing, chattering wiper blades. • The 1992–94 models have quirky AC systems that often malfunction and have to be constantly readjusted. The fresh air fan for the same model years will also misbehave, suddenly going into high-speed operation for no reason.

4000

Gasoline engines require constant and expensive adjustments, the cooling and electrical systems are weak, front brakes require frequent and expensive servicing, and the front MacPherson struts don't last long. Additionally, exhaust system parts rust through quickly and the fuel-injection system needs constant adjusting. Door panel trim peels away from the body on 1988–89 models. Later models have fewer complaints, probably due to their comprehensive base warranty.

5000 and 5000 Quattro

Renamed the 100 and 200 in 1989, these sedans have had a terrible repair record. The diesel engine is a nightmare, as are the turbocharged gasoline engines coupled to automatic transmissions. Front-drive mechanicals are unreliable and very complex to troubleshoot. Steering racks aren't durable and the engine cooling and electrical systems require frequent repair. Front brake rotors and calipers wear out very quickly and exhaust system components have a short life span. Other common problems include electrical short circuits in the taillights, AC failures, hot starting difficulties, and leaking transmission seals. The top of the front fenders and the front edge of the hood and rear-wheel housings are susceptible to rust perforations. Furthermore, the persistently poor alignment of lower side moldings is a common body problem. Resale value is much lower than average and owners who've been sucked into buying these vehicles bccause of their bargain-basement price face frequent and outrageously expensive "routine" servicing that's anything but routine.

Safety summary/Recalls: **1993 90, 100**—Under-hood fires reported. **1998–99 A6 sedan and Avant Quattro wagon**—During refueling, gasoline spits back violently from the filler pipe. **Recalls**: **1984–90 80, 90, 4000**—Dealers will fix leaky or cracked brake hoses. **1985–86**—Defective idle stabilizer valve on the 5-cylinder version with automatic transmission may cause the vehicle to surge and buck. **1985–91**—Evaporation of the differential oil could result in bearing and gear failure, which saps power from the front wheels or locks them up. **1986**—Rear or right-front seatbelt locks may malfunction. **1990**—Faulty steering lock bolts may block steering wheel movement. A faulty idle control module may cause erratic or increased idle speed. **1990 Quattro**—Faulty front seatback hinges. **1984–85 100, 200, 5000 Quattro, V8**—A faulty idle control module may cause erratic or increased idle speed. **1984–86**—Audi will install a transmission/brake interlock. **1984–88**—Audi will install a vent line valve or modify the fuel-filler neck to prevent fuel vapor ignition on Turbo models. **1984–89 Quattro**—Hardened fuel-injector seals allow fuel to leak onto the engine. **1985–86**—A maladjusted fuel distributor may cause fuel vapors to collect in the air filter and ignite if the engine backfires. • Defective idle stabilizer valve on the 5-cylinder version with automatic transmission may cause the vehicle to surge and buck. **1985–91**—An extension of the previous recall to prevent bearing and gear failure from sapping power from the front wheels or locking them up. **1986**—Malfunctioning rear or right-front seatbelt locks. **1990–91 V8 with cruise control**—Cruise control may not return to idle. **1992 100 series**—Audi will modify the brake vacuum booster system to improve brake pedal assist. **1993–95 90, 100, and all V6-equipped Cabriolets**—Leaky fuel-injectors will be replaced, free of charge. **1994–96 90, 100, A6, Cabriolet**—Defective ignition switches will cause malfunctioning turn signals, windshield wipers, lights, power windows, and air conditioners. **1995–96 90, A4, A6, and Cabrio**—If the cruise control won't maintain speed, add additional vacuum or change the vacuum servo unit. **1995–97**—Static electricity can set off driver airbag. **1996 A4**—A fuel gauge that won't register full signals a short circuit. • Inoperative horn is due to insufficient electrical ground contact. **1997–98**—Dealer will install a retaining ring to hold the air screen in place.

Secret Warranties/Service Tips

100, 200, 5000 Quattro, V8: **1991–92**—A clacking noise occurring whenever the vehicle passes over a bump is caused by plastic-to-plastic contact between the door lock latch and the door wedge. **1992–94**—DSB #94-09 troubleshoots common AC problems. • DSB #94-05 recommends replacing the fresh air fan with an upgraded part. **1995–97**—Delayed 1–2 shift on cold-start warm-ups is a normal condition resulting from the emissions control settings, according to Audi. **A4: 1996–99**—Audi will install upgraded brakes to "fix" problems related to premature corrosion (*Automotive News*, February 8, 1999). **All models/years**—Defective catalytic converters that cause a rotten-egg smell may be replaced free of charge under the emissions warranty.

80, 90, A4, A6, A8, S6 Profile

	1991	1992	1993	1994	1995	1996	1997	1998
Cost Price ($)								
80	—	27,995	—	—	—	—	—	—
90	32,920	—	28,450	—	—	—	—	—
90S/4X4	37,065	—	38,750	35,250	38,750	—	—	—
A4	—	—	—	—	—	36,250	31,600	32,700
A6	—	—	—	—	47,480	48,904	49,270	48,800
A8	—	—	—	—	—	—	—	90,540
S6	—	—	—	—	59,900	61,400	63,550	—
Used Values ($)								
80 ↑	—	6,000	—	—	—	—	—	—
80 ↓	—	5,500	—	—	—	—	—	—
90 ↑	7,000	—	9,000	—	—	—	—	—
90 ↓	6,000	—	8,000	—	—	—	—	—
90S/4X4 ↑	8,500	—	11,000	17,000	22,000	—	—	—
90S/4X4 ↓	7,500	—	9,500	15,000	21,000	—	—	—
A4 ↑	—	—	—	—	—	23,000	24,000	27,000
A4 ↓	—	—	—	—	—	22,000	22,000	26,000
A6 ↑	—	—	—	—	28,000	32,500	37,500	42,500
A6 ↓	—	—	—	—	27,000	31,000	35,500	39,500
A8 ↑	—	—	—	—	—	—	—	75,000
A8 ↓	—	—	—	—	—	—	—	72,000
S6 ↓	—	—	—	—	36,000	39,000	47,000	—
S6 ↓	—	—	—	—	34,000	37,000	44,000	—
Extended Warranty	Y	Y	Y	Y	Y	Y	Y	Y
Secret Warranty	N	N	N	N	N	N	N	N
Reliability	❶	❶	❷	❷	③	④	④	④

Air conditioning	❶	❶	❶	❶	❷	❷	❷	③
Body integrity	④	④	④	④	④	④	④	④
Braking system	❷	❷	❷	❶	❷	❷	❷	③
Electrical system	❶	❶	❶	❶	❶	❷	❷	③
Engines	❶	❶	❶	❷	③	③	③	④
Exhaust/Converter	❷	❷	❷	❶	④	⑤	⑤	⑤
Fuel system	❶	❷	❷	⑤	④	④	④	④
Ignition system	❷	❷	❷	❶	③	③	③	③
Manual transmission	③	③	③	③	③	④	⑤	⑤
- automatic	❷	❷	④	④	④	⑤	⑤	⑤
Rust/Paint	❷	❶	④	④	④	④	④	④
Steering	❷	❷	③	③	③	③	③	④
Suspension	❷	③	③	③	③	③	③	③
Crash Safety								
A4	—	—	—	—	—	④	④	—
A6	—	—	—	—	—	⑤	⑤	⑤
A8	—	—	—	—	—	—	—	⑤

100, 200, 5000, V8 Profile

	1988	1989	1990	1991	1992	1993	1994
Cost Price ($)							
100	—	39,100	41,200	38,915	—	37,600	48,250
100Q	—	43,250	43,850	40,945	—	—	—
200	—	46,250	48,600	45,365	—	—	—
200TQ	—	49,900	51,740	48,295	—	—	—
5000	38,375	—	—	—	—	—	—
5000Q	46,125	—	—	—	—	—	—
V8	—	—	70,800	66,560	—	69,900	73,050
Used Values ($)							
100 ↑	—	5,000	7,000	9,000	—	11,000	21,000
100 ↓	—	4,500	6,000	7,500	—	10,000	19,500
100Q ↑	—	6,000	7,500	9,500	—	—	—
100Q ↓	—	5,500	6,500	8,500	—	—	—
200 ↑	—	7,000	9,000	10,500	—	—	—
200 ↓	—	6,000	8,500	9,500	—	—	—
200TQ ↑	—	8,000	10,000	12,000	—	—	—
200TQ ↓	—	7,000	9,000	10,500	—	—	—
5000 ↑	4,000	—	—	—	—	—	—
5000 ↓	3,000	—	—	—	—	—	—
5000Q ↑	5,500	—	—	—	—	—	—
5000Q ↓	4,500	—	—	—	—	—	—
V8 ↑	—	—	11,000	13,000	—	19,000	26,000
V8 ↓	—	—	9,500	11,500	—	17,000	24,500
Extended Warranty	Y	Y	Y	Y	Y	Y	Y
Secret Warranty	N	N	N	N	N	N	N
Reliability	❶	❶	❶	❷	—	❷	③
Air conditioning	❷	❷	③	③	—	③	③
Body integrity	④	④	④	④	—	④	⑤
Braking system	❷	❷	❷	❷	—	❷	③
Electrical system	❶	❶	❶	❶	—	❷	❷

Engines	❶	❶	❶	❷	—	④	⑤
Exhaust/Converter	❷	❷	❷	③	—	⑤	⑤
Fuel system	❶	❷	❷	❷	—	③	③
Ignition system	❷	❷	❷	❷	—	③	③
Manual transmission	④	④	④	—	—	⑤	⑤
- automatic	❷	❷	④	④	—	⑤	⑤
Rust/Paint	❷	③	④	④	—	④	⑤
Steering	❷	❷	❷	❷	—	④	④
Suspension	❷	❷	❷	③	—	⑤	⑤
Crash Safety							
100	—	⑤	—	⑤	—	—	—

BMW

3 Series, 5 Series

Rating: **3 series**: Above Average (1995–98); Average (1994); Below Average (1984–93). **5 series**: Recommended (1992–98); Average (1985–91). These cars come with a reputation that far exceeds what they actually deliver. Don't take my word for it; look at the BMW Lemon websites at: *www.bssound.com.au/, bmw/lemon.htm,* and *www.bmwlemon.com/.* **Maintenance/Repair costs**: Higher than average, but many repairs can be done by independents who specialize in BMW repairs. Unfortunately, the specialists are concentrated around large urban areas. **Parts**: Higher-than-average cost, and often back ordered.

3 series

Strengths and weaknesses: **1984–91**—Great 6-cylinder performance with the manual gearbox, and ride and handling are commendable. The 318's small engine is seriously compromised by an automatic transmission; the 325e is more pleasant to drive and delivers lots of low-end torque. The 4-cylinder engines that first appeared on the 1991 models aren't well suited to the demands of the automatic gearbox. These cars are fun to drive as long as you keep in mind that the rear end is unstable on slippery pavement. Rear passenger and cargo room is limited and overall reliability is average. Whenever a problem arises, however, repair costs are particularly high, due to the small number of dealers, the relative scarcity of parts, and the acquiescence of affluent owners.

The electrical system is the source of most complaints. The automatic transmission isn't durable and front brakes require frequent attention. Owners also report chronic surging at idle and a rotten-egg smell from the exhaust (all models/years).

Door seams, rocker panels, rear-wheel openings, and fender seams are particularly rust-prone. Early BMWs were poorly rust-proofed and deteriorated very quickly, especially along the door bottoms and within the front- and rear-wheel wells. Check the muffler bracket for premature wear, and weather seals and door adjustments for leaks.

1992–98—Peppy 4-cylinder acceleration only with high revs and a manual transmission. Keep in mind that city driving requires lots of manual gear shifting characterized by an abrupt clutch. If you must have an automatic, look for a used model with the 6-cylinder engine or see if you can get by with the larger 1.9L 4-cylinder that went into the mid-1996 models. 1997 models came with traction control.

Dealer service bulletins: **1994**—Coolant leakage from the timing case profile gasket. • A binding/sticking ignition leading to starter failures. • Sagging convertible headliners. • Water leaking into the driver-side footwell and the E-box area, flooding the DME control module. • Self-activating emergency flashers and flickering instrument lighting. • A rotten-egg smell caused by faulty catalytic converters. **1995**—Excessive brake squealing. • Insufficient interior air distribution. • Leaking door contact switches; binding air-distribution knobs; and noisy, ineffective wipers. **1996–97**—AC evaporator ices up. • ABS warning light glows for no reason. • CD skips when road surface is rough. • Delayed gear engagement or adapter case leak. • Electrical troubleshooting tips. • Electronic mobilizer malfunctions. • False alarms caused by glass breakage sensor. • General module malfunctions. • Noise from AC expansion valve on 318ti. • Noise from AC compressor area. • Sun visor pops out of clip. • Sunroof fails to close. • Transmission shudders. • Whistle noise when cooling fan runs on high speed. • Wind noise troubleshooting tips.

5 series

No problem with rear seat or cargo room with this Bimmer. Handling and ride are superb, although these weighty upscale models do strain when going over hilly terrain if they have the automatic gearbox. There was no 1996 version.

5-series owners report numerous electrical and fuel glitches, faulty turn signal indicators, starter failures, self-activating emergency flashers, rotten-egg odours from the exhaust, and excessive steering wheel/brake vibration.

Dealer service bulletins: **1995**—AC belt tensioner noise. • Leaking door contact switches. • Faulty trunk lock actuator. • Noisy, ineffective wipers. **1997 528, 540**—Brakes momentarily won't release. • CD skips when road surface is rough. • Electronic mobilizer malfunctions. • Glove-box lock broken. • Radio with DSP switches off intermittently. • Right front door won't lock. • Wind noise troubleshooting tips.

Safety summary/Recalls: **1995 318**—During an accident, the driver-side seatbelt and airbags failed to operate as they should. • Other reports of airbags not deploying. • Sudden steering lockup. • AC expansion valve failures. • Noisy exhaust manifold. • Erratic automatic transmission shifting: hesitation and jerky shifts. • Transmission slips when accelerating, causing vehicle to stall. • Clutch pressure plate failures. • Electrical system malfunctions. **1996**—Airbags failed to deploy. • AC failures. **1997**—Gas and brake pedals are too close together. • Airbag safety light keeps coming on. **1998**—Sudden acceleration. **Recalls**: **1984**—Sudden acceleration on 318i models may be caused by a faulty engine idle control valve. • A binding heater control valve can cause the solenoid to malfunction, resulting in an instrument panel fire. **1984–91**—Leaks caused by a cooling system malfunction will be corrected at no charge. **1985–86**—Supporting plates will be installed to reinforce the steering column. **1985–90**—Fuel pump relay failures that cause no-starts require a new relay, which may be replaced free under the emissions warranty. **1986–87**—Faulty centre-mounted brake light switch needs to be replaced. **1988 325iX**—A bent oil dipstick tube could prevent the throttle from returning to idle when the accelerator is released. **1988–95**—All 3- and 5-series models will have their radiator caps replaced free of charge. Original caps failed to relieve excess pressure causing heater cores or hoses to rupture. **1991–92 318**—Ice buildup in the throttle housing can lead to increased idle speed or impaired deceleration. **1991 318, 325**—Erratic wiper operation caused by an electrical short circuit. **1991 318i, 325iS**—The knee booster will be relocated so that it won't interfere with the steering column's absorption of crash forces. **1991 325i**—Faulty windshield wiper switch ground screw. **1991 325iX**—The transfer case leaking at the weep hole signals the need to revise the depth of the inner seal; repair may be covered under BMW's emissions warranty. **1992 318**—A faulty cooling system may spray hot water into the interior. **1992 318i, 318iS, 325i, 325iS**—Airbag may not deploy. **1992–93**—Fuel lines can harden and "set" over time. **1992–94 325i, 325iS and 1994 325C**—Brake light switch may fail. **1993–94 325i, 325iS**—Defective front transmission cross-member support.

Secret Warranties/Service Tips

3 series: **1986–89**—A cracked intake manifold purge valve is a common problem. **1990–91 models with E34 or M50 engines**—Cylinder head cover leaks require a new, upgraded cover (#11-12-1-722-385 or #11-12-738-171 for later models). **1990–93**—Vehicles equipped with Jurid 506 brake pads may exhibit a steering vibration. **1992**—If the heater blower performs erratically, chances are the blower resistor unit needs replacing. **1994**—A starter motor failure may be caused by a faulty steering lock and electrical switch. • A sagging convertible headliner requires reinforced plastic clips that hold the headliner bracket to the bow. • Water intrusion into the driver's footwell area and the subsequent corrosion of the X13 and X14 connectors can be prevented by sealing off the hood's mounting flange just inboard of the left gas strut. • Water leaks into the E-box can be prevented by installing an improved capacity air intake drain hose in the right side drain, in addition to installing another drain hose. **1996**—Transmission shuddering requires the installation of a modified transmission control module. **1996–97**—Delayed gear engagement or adapter case leak requires the installation of new transmission seal kit (#21-41-422-762). **5 series**: **1989**—A fusible link could fail, shorting out all electrical power. **1990–93**—Change the Jurid 506 brake pads. **1990–94 525i**, **1992–94 525iT**, **and 1994 530i, 530iT**—Brake light switch may fail. **1991**—Inoperative power accessories may require the installation of a Reinshagen general module (version 5.2 or later). **1991–92 525i, 535i**—Airbag may not deploy. **1993–94**—Emergency flashers may self-activate. • Tips on fixing starter motor failures. **7 series**: **1990–92 735i and 735iL**, **1993–94 740i and 740iL**, **and 1990–92 750i**—Brake-light switch may fail. **1991–92 735i, 735iL, 750iL**—Airbag may not deploy. **All models/years**—According to BMW, a rotten-egg smell coming from the exhaust is due to fuel impurities and can't be resolved by changing the catalytic converter. The company says that it won't pay for any converters replaced for this reason.

3 Series Profile

	1991	1992	1993	1994	1995	1996	1997	1998
Cost Price ($)								
318iT	—	—	—	—	24,900	25,900	26,900	27,800
318i 4d	23,271	25,700	28,430	26,900	28,900	30,900	32,300	33,300
Convertible	31,682	33,270	34,760	—	40,900	42,900	43,900	44,900
325i/328i	32,336	38,800	39,750	38,200	41,900	43,900	46,900	47,900
Convertible	40,654	42,365	45,780	52,900	53,900	55,300	57,900	58,900
Used Values ($)								
318iT ↑	—	—	—	—	16,000	18,000	21,000	24,000
318iT↓	—	—	—	—	14,500	16,500	19,000	22,000
318i 4d ↑	8,500	11,000	13,000	15,500	19,000	22,000	25,000	28,000
318i 4d ↓	7,500	9,000	11,500	14,000	17,500	20,000	23,000	26,000
Convertible ↑	12,500	14,000	16,000	—	25,000	30,000	33,000	37,000
Convertible ↓	11,500	13,000	15,000	—	23,000	28,000	31,500	35,000
325i/328i ↑	11,000	13,000	18,000	23,000	28,000	32,000	35,000	41,000
325i/328i ↓	9,500	12,000	16,500	21,000	26,000	30,500	33,000	39,000
Convertible ↑	14,500	16,000	21,000	29,000	36,000	40,000	44,000	47,000
Convertible ↓	14,000	15,000	19,500	27,500	34,000	37,500	42,000	45,000
Extended Warranty	Y	N	N	N	N	N	N	N
Secret Warranty	N	N	N	N	N	N	N	N
Reliability	❷	❷	③	③	③	③	③	④
Air conditioning	❷	❷	❷	❷	③	④	④	④
Body integrity	❷	❷	❷	③	③	③	③	④
Braking system	❷	❷	❷	③	③	③	③	③
Electrical system	❷	❷	❷	❷	❷	❷	❷	③
Engines	❷	❷	❷	③	④	④	④	⑤
Exhaust/Converter	❷	❷	❷	③	③	③	③	④
Fuel system	❷	❷	③	③	③	③	④	④
Ignition system	③	③	③	③	④	④	④	④
Manual transmission	③	③	③	④	⑤	⑤	⑤	⑤
- automatic	③	③	③	③	③	③	④	④
Rust/Paint	❷	❷	③	④	④	⑤	⑤	⑤
Suspension	④	④	④	④	④	⑤	⑤	⑤

Crash Safety								
325i	❷	⑤	❷	—	④	—	—	—
325i Convertible	—	—	—	⑤	—	—	—	—
328i	—	—	—	—	—	④	④	—

5 Series Profile

	1991	1992	1993	1994	1995	1996	1997	1998
Cost Price ($)								
525i/528i	45,200	47,830	51,210	49,750	52,900	—	54,900	56,200
535i	52,150	54,760	58,520	—	—	—	—	—
Used Values ($)								
525i/528i↑	13,000	15,500	21,000	26,000	31,000	—	39,000	44,500
525i/528i ↓	12,000	14,000	19,000	23,500	28,500	—	36,500	41,500
535i ↑	16,000	18,000	23,000	—	—	—	—	—
535i ↓	15,000	17,000	21,000	—	—	—	—	—
Extended Warranty	Y	Y	Y	Y	Y	Y	Y	Y
Secret Warranty	N	N	N	N	N	N	N	N
Reliability	❷	③	③	③	④	⑤	⑤	⑤
Air conditioning	❷	③	④	④	⑤	⑤	⑤	⑤
Body integrity	❷	❷	❷	③	③	③	③	④
Braking system	❷	❷	❷	③	③	③	③	④
Electrical system	❷	❷	❷	❷	③	③	③	③
Engines	❷	③	③	③	④	④	⑤	⑤
Exhaust/Converter	③	③	③	③	③	③	④	⑤
Fuel system	③	③	③	③	④	④	⑤	⑤
Ignition system	③	③	③	④	④	④	⑤	⑤
Manual transmission	③	③	④	⑤	⑤	⑤	⑤	⑤
- automatic	❷	③	③	③	④	⑤	⑤	⑤
Rust/Paint	❷	③	④	④	④	④	⑤	⑤
Suspension	④	④	④	④	⑤	⑤	⑤	⑤

Note: The above models haven't been crash-tested.

GENERAL MOTORS

Allanté, Catera, Eldorado, Seville

Rating: Below Average (1992–98); Not Recommended (1986–91). **Maintenance/Repair costs**: Higher than average, but repairs aren't dealer dependent. **Parts**: Higher-than-average cost (independent suppliers sell for much less), but not hard to find. Don't buy one of the front-drives without a three- to five-year supplementary warranty.

Strengths and weaknesses: The early Cadillacs are luxury embarrassments and later models barely pass muster. Even though most use the same mechanical components with the same deficiencies as the Riviera and Toronado models, they're far more failure-prone due to the complexity of their different luxury features.

Allanté

Introduced in 1987, the Allanté was essentially a "kit car" Cadillac two-seat roadster with body components flown in from Italy and mechanicals assembled in Detroit. This car got the more powerful Northstar V8 in 1993—the same year the car was discontinued. Originally selling for $77,475, a 1987 model would fetch no more than $9,000 today. Does it have any value as a collector's item? Not a chance.

Owners report many reliability problems—especially with brake, electronic control, and powertrain glitches—and woefully poor-quality body assembly. Also, the convertible top is a chore to put up or down. Allanté owners have expressed a few other concerns: chronic and costly ABS brake repairs; emergency brake rattles, particularly if the emergency brake is not used often (the cause is prematurely worn rear brake pads, or defective brake "springs" or clips); turn signals sticking (simply squirt TV Tuner Cleaner from Radio Shack into the assembly); defective intake manifolds; and faulty alarms. For more detailed information from Allanté owners who both love and hate their car, access *www.allante.org* on the Internet.

Catera

Assembled in Germany and based on the Opel Omega, the rear-drive, mid-sized Catera comes with a 200-hp V6 engine, 4-speed automatic transmission, 16-inch alloy wheels, 4-wheel disc brakes, a limited-slip differential, traction control, and standard dual front airbags. The conservatively styled 1997 Catera (the uninspired styling has Lumina written all over it) was designed to compete with the BMW 328i, Lexus ES 300, and Mercedes Benz C280. GM hoped not to drive away its more traditional "empty-nesters" (I guess that's a polite term for "old folks"), while luring more baby boomers to its higher-performance variations.

The last time Cadillac introduced an entry-level model was 1981, when the automaker launched the Cimarron, a fully loaded, Chevrolet Cavalier–derived Cadillac that carried a $4,000 premium over the comparably equipped Cavalier. Back then, most auto critics and consumer advocates considered the Cimarron to be at the rear of the pack as far as performance and quality control were concerned. The Catera, on the other hand, has received good reviews from the European press for its quiet, spacious, and comfortable interior, responsive handling, precise steering, fine-tuned suspension, and almost nonexistent lean or body roll when cornering. Without a doubt, this is one Cadillac that's meant to be driven. On the downside, though, the controls aren't easy to figure out, some gauges are hard to read, and the driver's rear view is hindered by the large rear head restraints and narrow-back windshield. Furthermore, owners report chronic stalling and hard starts, possibly due to a malfunctioning idle control valve, constant warning-light illuminations, poor AM radio reception (requiring an additional amplifier), and loose interior panels. Two other performance problems reported by owners: when you pass over a large expansion joint, the floorpan vibrates annoyingly; drive over a bump when turning and the steering wheel kicks back in your hands.

A few other points you may wish to consider: GM dealers are notoriously bad when it comes to understanding and repairing European-transplanted cars (just ask any Saab owner). As well, low-volume cars generally don't have an adequate supply of replacement parts in the pipeline until they've been on the market for a while. Add in the Catera's European connection and you'd best be ready to endure long service waits and high parts costs for those repairs not covered under warranty.

Finally, the fact that Cateras are European-built doesn't necessarily mean that these Cadillacs will be reliable or durable. GM first learned that lesson with the British-built Vauxhall Firenza it unleashed on an unsuspecting Canadian public in the '70s. A few years later, it settled out of court on several class actions that I piloted, and paid a $20,000 fine to the federal government for misleading advertising. (On a nationally advertised roadtrip across Canada, GM said the cars excelled. Truth is, they were a mess. They required a team of engineers just to get started.)

Eldorado and Seville

From 1992 on, the Eldorado's styling became more distinctive, even though the vehicle shares most powertrain and chassis components with the Seville. Although the base 4.9L V8 provides brisk acceleration, the 32-valve Northstar V8, first found on the 1993 Touring Coupe, gives you almost 100 more horses with great handling and a comfortable ride. Overall, the Touring Coupe or Sport Coupe will give you the best powertrain, handling, and braking features. Of course, you'll have to contend with poor fuel economy, rear visibility that's obstructed by the huge side pillars (a Seville problem, as well), confusing and inconvenient climate controls, and a particularly complex engine compartment.

Sitting on the same platform as the Eldorado, the Seville has European-style allure with a more-rounded body than the Eldorado. Apart from that, since its redesign in 1992, its engine, handling, and braking upgrades have followed in lockstep fashion the Eldorado's improvements.

Whether you buy a used Eldorado or Seville, keep in mind that the improved versions came out with the 1995 models, which carried on unchanged until their redesign for 1998. So, if you must buy one of these models, remember that the only distinguishing feature between them is styling, not performance.

The 4.1L V8 is best avoided. It may be simpler to work on, but it too is fuel thirsty and easily overpowered by the Eldorado's weight.

These cars have generic deficiencies that fall into the same categories: poorly calibrated and failure-prone engines, transmissions, and fuel and ignition systems; a multiplicity of electrical short circuits; and sloppy body assembly using poor-quality components. Specifically, engines and fuel systems often produce intermittent stalling, rough idling, hesitation, and no-starts; the Overdrive automatic is prone to premature failure; oil pumps fail frequently; front brakes and shock absorbers wear out quickly; often, paint is poorly applied, fades, or peels away prematurely; fragile body hardware breaks easily; and there are large gaps between sheet-metal panels and doors that are poorly hung and not entirely square. Other body problems include cracking of front outside door handles, door rattles (Eldorado), poor bumper fit, loose sun visor mounting, rear taillight condensation, fading and discolouring appliqué mouldings (Seville), interior window fogging, "creaking" body mounts, water leaking into trunk from licence-plate holder (Eldorado), noisy roof panels and seatback lumbar motors, and a creaking noise at the front-door upper hinge area.

Dealer service bulletins: **1993 models with 4.9L engines**—May be hard to start, frequently stall, and lose power; and may not maintain the cruise control set speed. • Condensation in the taillights. • Radio static. • Excessive front-end vibrations when passing over bumps. • Faulty engine oil pumps (Allanté). • Transmission gear whine and rear suspension noise. **Eldorado**—Squeaks and squealing after a cold start. • Warped windshield and rear-window moulding. **1994**—Faulty engine oil pumps and noisy fuel pumps. • Engine oil leaks. • 4.6L engines may run roughly, miss, surge, or hesitate. • Engine accessory belt noise. • An inoperative cruise control or brake/transmission interlock. • Erratic shifting. • AC hissing noise and rear suspension noise. • Condensation dripping from the heater duct and poor heat distribution (driver's feet get cold). • Doors won't stay open on slight grades and parking brake binds. **1995**—Engine cranks but won't start; no fuel pressure; loss of oil pressure or lack of power. • Engine oil leak at the rear main seal or T joint. • Starter motor runs continuously. • Metallic knocking sound from the engine. • A grinding or scraping noise in Park or Neutral. • Front brake vibration and/or pedal pulsation. • Annoying AC odour. • A low-voltage reading or dim lights at idle. • Remote keyless entry malfunctions. • Rattle noise from front of vehicle, front-seat clicking noise, clicking noise from under the hood or dash, and rubbing noise when the front wheels are turned all the way. • Door window slaps or rattles when closing door. • Excessive radio static. • Wet or smelly carpet from water leaks. • Poor paint application and rust spots. **1996**—Engine oil leaks from upper to lower crankcase joint. • Loss of oil pressure and/or lack of power. • Oil leakage from the oil pan to lower crankcase attaching bolts. • Second-gear starts, poor 1–3 shifting. • Steering column noise. • Poor AC cooling in traffic and high ambient temperatures. • Insufficient heat distribution or lack of air flow. • Coolant temperature gauge always reads hot. • Erratic or inaccurate instrument panel displays and gauges. • Insufficient remote keyless entry operating range. • Positive battery cable won't tighten properly. • Radio clock stuck on 12:00. • Radio frequency interference diagnosis. • Whistle noise from HVAC. • Exhaust rattles. • Rattling heard when vehicle goes over bumps. • Clicking noise from front seats. • Squawk noise from rear of vehicle in cold temperature. • Slow parking brake release or slow HVAC mode changes. • Water leak from rear-door side-window area. • Wet or smelly carpet from water leaks. • Abnormal condensation in taillights and other lights. • Faulty cup holders. **1997**—Front brake noise. • Remote won't open doors or trunk. • Excessive vibration of electrochromatic mirror. • Fuel tank won't fill to capacity/always reads full. • Horn hard to operate when cold. • Inoperative power door locks. • Metallic buzz or rattle at the instrument panel when the brakes are applied. • No warning light or driver information centre message. • Power window inoperative after express down feature is used. • Torque converter clutch buzz or moan. **1997 Catera**—Brake squealing. • Clunk noise from engine compartment on vehicles with cruise control. • Ignition key binds, drags, or sticks in ignition. • Loose fuel cap. • Low-voltage reading or dim lights at idle. • Noise in radio AM band when rear defogger on. • Oil leak from engine timing cover. • Inoperative power door locks. • Inoperative rear defogger. • Windshield wiper slap/flop noise.

Safety summary/Recalls: **1989–92 Allanté**—The NHTSA is investigating Allantés for brake problems following a petition from a 1989 Allanté owner who collected 29 complaints from his website (Allante Appreciation Group, *www.allante.org*). The government is looking primarily into reports of the brakes failing or malfunctioning. **1994 Seville**—Owners report that the throttle may stick in the open or closed position. Dealers will modify the throttle control cable to secure it in place. Drivers also complain that the shoulder belt chafes at the neck. NHTSA is looking into dash-trim panels that fly apart when the airbag is deployed. **Recalls**: **1986–87**—Faulty headlight circuit. **1989**—ABS may be defective. **1990**—Gear indicator may be poorly aligned. • Faulty rear seatbelt shoulder retractors. **1990–91**—Front outer shoulder belt web may stick in retractor. **1992**—Possible loss of steering control. **1995–96 Eldorado, Seville**—Excessive moisture (due to a window left open when it rains, for example) may cause the airbag to deploy when the ignition is switched on. There have been 70 reported incidents, 39 of them with injuries to drivers, passengers, or both. Before NHTSA forced this recall, GM sent notices to owners advising them to take precautions if their vehicle interiors became wet. (No, I haven't read the letter and I can't imagine what kind of precautions one should take.) A short circuit caused by wet carpets could cause the airbags to deploy suddenly. **1993 models with 4.6L engine**—Fuel leakage in engine compartment. **1993–94 models with 4.6L engine**—Leaking engine oil cooler hose may cause a fire. **1994**—Faulty throttle cable.

Secret Warranties/Service Tips

1990–94—A rear suspension squawk can be eliminated by installing upgraded stabilizer shaft insulators. **1991–95 models with 4.9L engines**—

GM will install a new computer chip that reduces stalling, pursuant to an agreement with the U.S. Justice Department and EPA. **1993**—DSB #476506 gives lots of tips on fixing 4.6L engines that run roughly, miss, surge, or hesitate. **1993–94**—DSB #476003 goes into great detail about troubleshooting the various engine oil leaks afflicting 1993–94 models. • Noisy fuel pumps can be silenced only by installing an upgraded fuel pump under warranty. • Poor heat distribution (driver's feet get cold) can be fixed by replacing the floor outlet assembly. **1994**—Condensation dripping from the heater duct requires the installation of a watertight dam in the HVAC case. • An inoperative cruise control or brake/transmission interlock may signal a maladjusted stop-light switch assembly. • A binding parking brake may need a new park-brake vacuum-release switch. **1996–97**—A torque converter clutch buzz or moan requires the installation of an upgraded case-cover-assembly spacer plate and the upper-control-valve body. **1997**—Excessive front brake noise can be cured by installing upgraded front brake pads. **1997 Catera**—Brake squealing can be silenced by installing redesigned calipers. • Oil leakage from the engine timing cover can be corrected by installing a new oil pump gasket.

Allanté, Catera, Eldorado, Seville Profile

	1991	1992	1993	1994	1995	1996	1997	1998
Cost Price ($)								
Allanté	69,998	74,198	71,998	—	—	—	—	—
Catera	—	—	—	—	—	—	42,690	43,250
Eldorado	39,198	39,840	40,998	46,498	50,430	50,745	52,015	53,000
Seville	42,598	42,640	44,488	49,998	55,315	55,635	57,000	59,900
Used Values ($)								
Allanté ↑	25,000	31,000	38,000	—	—	—	—	—
Allanté ↓	22,500	30,000	35,000	—	—	—	—	—
Catera ↑	—	—	—	—	—	—	26,000	30,000
Catera ↓	—	—	—	—	—	—	24,000	29,000
Eldorado ↑	9,500	11,000	14,000	18,000	23,000	29,000	34,500	40,000
Eldorado ↓	8,500	10,500	12,500	16,000	21,000	26,000	31,500	37,500
Seville ↑	11,000	13,000	15,500	20,000	25,500	31,000	37,500	44,000
Seville ↓	10,000	12,000	14,500	18,000	22,000	29,500	35,000	41,500
Extended Warranty	Y	Y	Y	Y	Y	Y	Y	Y
Secret Warranty	Y	Y	Y	Y	Y	Y	Y	Y
Reliability	❷	❷	❷	❷	❷	❷	③	③
Air conditioning	❷	❷	❷	❷	❷	③	③	④
Automatic transmission	❷	❷	❷	❷	③	③	③	④
Body integrity	❷	❷	❷	❷	❷	❷	❷	❷
Braking system	❷	❷	❷	❷	❷	❷	③	③
Electrical system	❷	❷	❷	❷	❷	❷	❷	③
Engines	❷	❷	❷	❷	③	③	③	③
Exhaust/Converter	❷	❷	❷	③	③	④	④	⑤
Fuel system	❷	❷	❷	③	③	③	④	④
Ignition system	❷	❷	❷	❷	③	③	③	③
Rust/Paint	❷	❷	❷	❷	❷	❷	③	③
Steering	❷	❷	❷	❷	④	④	④	④
Suspension	❷	❷	❷	❷	❷	❷	③	④

Note: Reliability figures apply to the Eldorado and Seville only; Allanté and Catera reliability information is given in the text. None of the above vehicles has been recently crash-tested.

Aurora, Riviera, Toronado, Trofeo

Rating: **Riviera, Toronado, Trofeo**: Average (1995–98); Not Recommended (1986–93). GM skipped the 1994 model year and introduced an all-new 1995 version. **Aurora**: Above Average (1995–98). **Maintenance/Repair costs**: Higher than average, but repairs aren't dealer dependent. **Parts**: Higher-than-average cost (independent suppliers sell for much less), but not hard to find. Nevertheless, don't even think about buying one of the earlier front-drives without a three- to five-year supplementary warranty.

Strengths and weaknesses: Although the redesigned 1988–93 cars got performance, handling, and ride upgrades, they kept the same low level of quality control with multiple design and manufacturing defects, including serious fuel-injection, engine computer, and electrical system problems. One particularly poor design was the complex Graphic Control Center, which used an oversensitive video screen and small push buttons. It's both distracting and expensive to repair. The automatic transmission is notoriously failure-prone, and brakes wear out prematurely and perform poorly. Surface rust and poor paint quality are the most common body complaints on all years. Shock absorbers wear out quickly and the diesel engine seldom runs properly. Mechanical parts are easy to find, but body panels have to be ordered from GM at a premium.

The 1995–98 models offer many more luxury features but continue their checkered repair history. GM improved the quality over the years, but generic deficiencies affecting the automatic transmission, engine, computer modules, suspension, and fit and finish make these cars less than luxury from a quality-control standpoint.

Dealer service bulletins: **1995**—Engine misfires, rough idle, white exhaust smoke, and a continuous spark knock. • Excessive oil consumption. • A cold-start rattle with the 4T60E automatic transmission. • A starter motor that runs continuously. • Reduced heater performance on the driver's side. • Brake vibration and/or pedal pulsation. • Instrument panel trim may separate from the upper-trim pad. • Low voltage reading or dim lights at idle. • A scraping noise and increased effort needed to open the doors. • Inoperative fuel-filler door. • ABS/traction light stays on. • Poor paint application and rust spots. **1996**—Second-gear starts and erratic 1–3 shifts. • AC odour in humid climates. • Steering column noise. • Cold-start rattle noise. • Engine cranks but won't start. • Engine oil leaks. • Excessive vibration of electrochromatic mirror. • Exterior light condensation. • Troubleshooting tips for fluttering, popping, ticking, and clunking noises. • Gap between door opening weather strip and roof line. • Inoperative power door locks. • Intermittent Neutral/loss of Drive at highway speeds. • Instrument panel buzz or rattle when the brakes are applied. • Popping noise, growl, or vibration when making a right turn. • Reduced heater performance on the driver's side. • Reduced retention of door trim lace. • Roof panel running front to back has a wavy/dimpled appearance and transmits excessive rain noise. • Underbody noise from rear of vehicle. • Wet or smelly carpet from water leaks. **1997**—Front brake noise. • Engine cranks but won't start. • Engine oil leaks. • Engine oil level indicates over-full. • Excessive vibration of electrochromatic mirror. • Harsh shift from Reverse to Drive. • Inoperative power door locks. • Instrument panel buzz or rattle when the brakes are applied. • Popping noise, growl, or vibration when making a right turn. • Steering vibration, shudder, or moan when parking. • Roof panel running front to back has a wavy/dimpled appearance and transmits excessive rain noise.

Aurora

This front-drive luxury sedan is aimed at the Acura, Infiniti, and Lexus crowd. It uses the same basic design as the Riviera but, fortunately, doesn't share the same major mechanical features or styling. The Aurora's main advantages are its sporty handling and unusual aero styling. In contrast to the Riviera and Toronado, the Aurora seats five only and uses a 4.0L V8 derived from the Cadillac 4.6L V8 Northstar engine. Acceleration is underwhelming (this is a heavy car) but adequate for highway touring. Road and wind noise is omnipresent and the rear trunk's small opening compromises the large trunk's ability to handle odd-sized objects.

Dealer service bulletins: **1995**—White exhaust smoke, engine oil leak at the rear main seal or T joint, and loss of oil pressure or lack of power. • A popping noise is heard during cranking, or the starter motor continues to run or crank after shutoff. • Reduced heater performance on the driver's side. • Brake vibration and/or pedal pulsation. • A clicking noise emanating from the dash or hood, creak noise at the right side of vehicle, and a thumping from the rear of the vehicle. • Loose, rattling headliner and clicking emanating from the sunroof. • Excessive radio static. • Instrument panel trim that may separate from the upper-trim pad. • Wet, smelly carpet from water leaks. • Wind noise at the front door A-pillar. **1996**—Engine oil leaks from upper to lower crankcase joint. • Oil leakage from the oil pan to lower crankcase attaching bolts. • Loss of oil pressure and/or lack of power. • Steering column noise. • Popping noise from front of vehicle when turning. • Underbody noise from rear of vehicle. • Reduced heater performance. • Rear compartment lid-assist rod separating from lid. • Gap between door opening weather strip and roof line. • Reduced retention of door trim lace. • Condensation on exterior light. • Wet or smelly carpet from water leaks.

Safety summary/Recalls: **Recalls**: **1986–87**—The power-steering pump hose may leak, creating the possibility of an engine fire. **1989**—ABS brakes may fail due to a faulty brake line. **1990**—Faulty transmission cable may indicate that the wrong gear is engaged. **1991–93**—Front shoulder seatbelt may stick in retractor. • Poorly aligned rear shoulder seatbelt retractor assemblies. **1992**—Possible steering loss due to the disengagement of the steering shaft. **1996**—Driver warning alarms and displays may malfunction. • Backfire can break the intake manifold and cause a fire.

Secret Warranties/Service Tips

1989–90—Stalling from a cold start and a chuggle at 70 km/h require a new MEMCAL. **1990–91**—A revised headrest guide loop will be installed on Rivieras if the shoulder seatbelt rests uncomfortably on the neck. **1990–93**—An engine ticking at idle can be traced to rattling piston pins, which must be replaced with upgraded parts. **1991–93**—Loss of Drive or erratic shifts may be caused by an intermittent short circuit to ground on the A or B shift solenoid, or an electrical short circuit in the transaxle. • Harsh automatic transmission upshifts can be corrected by installing an upgraded accumulator valve in the control valve body. • A front-end engine knock troubleshooting chart and extensive diagnostic tips are found in DSB #306001. **1995–96**—Intermittent Neutral/loss of Drive at highway speeds can be fixed by replacing the control valve body assembly. **1990–92 Toronado, Trofeo**—A body-mount creak that occurs whenever the vehicle passes over a bump may be due to one or more of the body mounts being poorly positioned in the frame; correct by installing a new, lower insulator. • Chronic wind noise coming from the front door window can be corrected by reinstalling the run channel retainer. • A shake or vibration in the front end when going over smooth roads may be caused by an internal leak in the engine mount. **All models/years**—A rotten-egg odour coming from the exhaust is likely the result of a malfunctioning catalytic converter covered by GM's emissions warranty. • Paint delamination, peeling, or fading (see pages 76, 113).

Aurora Profile

	1995	1996	1997	1998
Cost Price ($)				
Aurora	43,020	43,695	46,045	47,250
Used Values ($)				
Aurora ↑	21,500	26,000	30,000	35,500
Aurora ↓	19,000	23,500	28,000	33,000
Extended Warranty	Y	Y	Y	Y
Secret Warranty	Y	Y	Y	Y
Reliability	④	④	④	④
Crash Safety	—	③	③	③

Riviera, Toronado, Trofeo Profile

	1991	1992	1993	1995	1996	1997	1998
Cost Price ($)							
Riviera	29,698	29,898	30,790	39,525	40,700	42,415	44,950
Toronado	29,698	28,799	—	—	—	—	—
Trofeo	32,698	31,998	—	—	—	—	—
Used Values ($)							
Riviera ↑	8,000	9,000	11,500	18,000	23,000	26,000	33,000
Riviera ↓	7,000	8,000	10,000	16,500	20,500	25,000	31,000
Toronado ↑	8,500	9,500	—	—	—	—	—
Toronado ↓	8,000	9,000	—	—	—	—	—
Trofeo ↑	8,500	9,500	—	—	—	—	—
Trofeo ↓	8,000	9,000	—	—	—	—	—
Extended Warranty	Y	Y	Y	Y	Y	Y	Y
Secret Warranty	Y	Y	Y	Y	Y	Y	Y
Reliability	❷	③	③	③	③	④	④
Air conditioning	③	③	④	④	④	④	④
Automatic transmission	❷	❷	③	③	③	③	③
Body integrity	❷	❷	❷	❷	③	③	③
Braking system	❷	❷	❷	❷	❷	③	③
Electrical system	❷	❷	❷	❷	❷	❷	③
Engines	③	③	③	④	④	④	④
Exhaust/Converter	④	④	④	⑤	⑤	⑤	⑤
Fuel system	❷	③	③	③	③	③	④
Ignition system	③	③	③	③	③	③	④
Rust/Paint	❷	❷	❷	❷	❷	❷	③
Steering	③	③	③	④	④	④	④
Suspension	❷	③	③	③	③	④	④
Crash safety	—	—	—	—	—	—	—

Brougham, Fleetwood RWD

Rating: Above Average (1993–96); Average (1984–92). A smart car for retirees, it's on par with the Ford Crown Victoria and Mercury Grand Marquis when it comes to comfort and reliability. **Maintenance/Repair costs**: Average, and repairs aren't dealer dependent. **Parts**: Reasonably priced (independent suppliers sell for much less) and not hard to find, despite the fact that these rear-drives were dropped in '96.

Strengths and weaknesses: The quintessential land yacht, these cars emphasize comfort over handling with their powerful engines and large chassis. Nevertheless, with their spacious interior and many convenience features, these large cars are ideal for vacationing and light trailer-pulling.

The most serious problem areas are the fuel-injection system, which frequently malfunctions and costs an arm and a leg to repair;

automatic transmissions that shift erratically; a weak suspension; computer module glitches; poor body assembly; and paint defects. From a reliability/durability standpoint, the rear-drives are much better made than their front-drive counterparts.

GM dealer service bulletins show that these vehicles also have noisy power-steering units and cooling fans, the AC bi-level mode produces extreme temperature differences, the instrument panel squeaks and rattles, there are rear quarter-panel gaps and rusting at the rear side door window molding, and water leaks into the passenger side of the front compartment.

Safety summary/Recalls: **1995**—Dashboard reflects into windshield. • Vehicle accelerated while braking. • Frequent stalling. • Trunk lid opened while driving. • Brakes require extended stopping distances. **1996**—Airbags failed to deploy. • Seatbelts didn't restrain driver and passenger during a collision. • Chronic stalling due to fuel sending unit failure. • Engine head gaskets failures. • Transmission pounds when shifting gears. • Water pump leakage on the serpentine belt may cause steering to lock up. • Power-steering hose and pump failure. • Brakes often lock up when applied. • Excessive brake noise caused by the premature wearout of brake rotor and drum. • AC cooling switch and high pressure hose failures. • Instrument cluster hard to read in daylight. • Power door locks and trunk lock frequently fail to operate properly. • Loose windshield molding. • Defective keyless entry module. **Recalls**: **1984–88**—A sticking throttle could lead to sudden acceleration. **1986**—Vehicles with a 5.0L engine could also have a sticking throttle. **1987**—Cars equipped with an automatic transmission could start in gear or engage the wrong gear. **1993**—Passenger-side airbag may have a defective igniter. **1994**—Oil-cooler inlet hose leaks. • Lug nuts may loosen. • Fuel-tank strap fasteners can detach. **1994–95**—Throttle control spring may stick in low temperatures.

Secret Warranties/Service Tips

1994—Instrument panel displays that are too dim. GM installed an upgraded cluster in mid-1994. • Intermittent loss of power door locks, seats, and mirrors due to a short circuit in the door lock circuit. **1995**—A popping noise during cranking; the engine cranks but won't start; no fuel pressure; or extended crank time after cold soak. • Lack of power. • Engine oil leak at the rear main seal or T joint. • Low oil pressure, loss of oil pressure, or lack of power. • AC odour. • Grinding or scraping noise in Park or Neutral • A cold-start rattle with the 4T60E automatic transmission. • Front-brake vibration and/or pedal pulsation. • A rubbing noise when the front wheels are turned all the way. • Door window scraping noise or sticking, rattle noise from front of vehicle, and excessive radio static. • A clicking noise from under the dash or hood • A front-seat clicking noise. • Erroneous fuel gauge readings, a low voltage reading or dim lights at idle, and frequent blown fuse or battery drain. • Malfunctioning remote keyless entry. • Wet or smelly carpet from water leaks. • Poor paint application and rust spots. **1996**—3–2 part throttle downshift flare. • Engine noise (install new valve stem oil seal). • Transmission chuggle/surge. • Transmission fluid leak from pump body (replace bushing). • Crunch/pop noise in steering system. • AC odours. • Radio frequency interference diagnosis. **1993 Fleetwood**—Noisy transmissions, power-steering units, and cooling fans. • AC bi-level mode produces extreme temperature differences. • Instrument panel squeaks and rattles. • Rusting at the rear side door window molding. • Water leaks into the passenger side of the front compartment. **All models/years**—Defective catalytic converters that cause a rotten-egg smell in the interior may be replaced free of charge under the emissions warranty. • Paint delamination, peeling, or fading (see pages 76, 113).

Brougham, Fleetwood RWD Profile

	1990	1991	1992	1993	1994	1995	1996
Cost Price ($)							
Brougham/Fleetwood (RWD)	39,816	37,298	37,488	39,988	41,798	46,830	46,965
Used Values ($)							
Brougham/Fleetwood (RWD) ↑	7,500	9,000	11,000	14,500	18,500	22,000	28,000
Brougham/Fleetwood (RWD) ↓	6,500	8,000	9,000	13,000	17,000	20,500	26,000
Extended Warranty	Y	Y	Y	Y	Y	Y	Y
Secret Warranty	N	N	N	Y	Y	Y	Y
Reliability	③	③	③	③	③	④	④
Air conditioning	③	③	③	③	③	④	④
Automatic transmission	③	③	③	③	③	④	④
Body integrity	❷	❷	❷	③	③	④	④
Braking system	❷	❷	❷	③	③	③	③
Electrical system	❷	❷	❷	❷	❷	❷	❷
Engines	③	③	③	③	④	④	④
Exhaust/Converter	❷	❷	❷	❷	③	③	④
Fuel system	❷	❷	❷	③	③	③	③
Ignition system	❷	❷	❷	③	③	③	③
Rust/Paint	❷	❷	❷	③	③	③	③
Steering	❷	❷	❷	③	④	④	④
Suspension	❷	❷	❷	③	③	③	③

Note: The Brougham and Fleetwood haven't been crash-tested.

Concours, DeVille, Fleetwood FWD

Rating: Average (1995–98); Below Average (1985–94). **Maintenance/Repair costs**: Higher than average, and most repairs must be done by a dealer. **Parts**: Higher-than-average cost (independent suppliers sell for much less), but not hard to find. All of these front-drives require a three- to five-year supplementary warranty.

Strengths and weaknesses: Although they have better handling and are almost as comfortable as the old series, the early models of these luxury coupes and sedans aren't worth considering because of their dismal reliability and overly complex servicing. Redesigned 1995–98 versions have posted fewer complaints; however, they are still far below the industry norm for quality and reliability. As with the Eldorado and Seville, you get the best array of handling, braking, and performance features with the 1996 and later versions. They do ride more quietly and comfortably, but fuel economy is still poor, the dash controls and gauges are confusing and not easily accessible, and the rear view is obstructed by the high trunk lid and large side pillars.

The 4.3L V6, 4.1L V8, and 4.5L V8 engines and 4-speed automatic transmission suffer from a variety of terminal maladies including oil leaks, premature wear, poor fuel economy, and excessive noise. The electrical system and related components are temperamental. The suspensions go soft quickly, and the front brakes often wear out after only 18 months/20,000 km. Problems with the digital fuel-injection and engine control systems are very difficult to diagnose and repair. Poor body assembly is characterized by premature paint peeling and rusting, excessive wind noise in the interior, and fragile trim items.

Dealer service bulletins: **1994 all models except Fleetwood**—4.6L engines may run roughly, miss, surge, or hesitate. • Engine oil leaks. • Engine accessory belt noise and AC hissing. • Erratic shifting. • Inoperative cruise control or brake/transmission interlock. • Parking brake binding. • Condensation dripping from the heater duct and poor heat distribution (driver's feet get cold). • Doors won't stay open on slight grades. • Rear compartment water leaks. **1995–96**—Engine oil leaks from upper to lower crankcase joint. • Loss of oil pressure and/or lack of power. • Oil leakage from the oil pan to lower crankcase attaching bolts. • Poor AC cooling in traffic and high ambient temperatures. • AC defrost creaking noise. • AC odours. • Steering column noise. • Insufficient heat distribution or lack of air flow. • Erratic or inaccurate instrument panel displays and gauges. • Insufficient remote keyless entry operating range. • Positive battery cable won't tighten properly. • Radio clock stuck on 12:00. • Radio frequency interference diagnosis. • Inaccurate fuel gauge. • Exhaust rattles. • Rattling heard when vehicle goes over bumps. • Clicking noise from front seats. • Squawk noise from rear of vehicle in cold temperature. • Slow parking brake release or slow HVAC mode changes. • Vertical seat height adjustment tips. • Abnormal condensation in taillights and other lights. • Wet or smelly carpet from water leaks. **1997**—Fuel tank won't fill to capacity/always reads full. • Front brake noise. • Remote won't open doors or trunk. • Excessive vibration of electrochromatic mirror. • Front-door front auxiliary weather strip loose. • Uncomfortable front seatback. • Horn hard to operate when cold. • Inoperative power door locks. • Metallic buzz or rattle at the instrument panel when the brakes are applied. • No warning light or driver information centre message. • Power window inoperative after express down feature is used. • Torque converter clutch buzz or moan. **Concours**—Harsh suspension performance. **1993 DeVille**—Buzzing or whining 4L60 automatic transmissions. • Fuel tank popping during start-up. • Noisy fuel pumps and faulty engine oil pumps. • A noisy cooling fan, power-steering moaning, and rattles from the rear strut area. • Rear quarter-panel gaps. • Squeaks and squealing after a cold start, and whistling when the AC is in recirculate mode.

Safety summary/Recalls: **1994–97 DeVille**—Rear turn signal lights may not flash. **1995–96 DeVille and Concours**—A short circuit caused by wet carpets could cause the airbags to suddenly deploy. Excessive moisture in the vehicles—for example, due to a window left open when it rains—may cause the airbag to deploy when the ignition is switched on. **Recalls**: **1986**—Faulty headlight switch. **1986–87**—Anti-lock brake fluid may leak onto the pump motor and cause partial or complete loss of front or rear braking. **1994**—Engine oil cooler inlet hoses may be too close to the steering gear, causing it to leak and creating a fire hazard. **1994–95**—Accelerator pedal may stick. **1996**—Hood may fly up. **1997**—Faulty Brake Traction Control module could increase stopping distance. **1991–93 DeVille with 4.9L engines**—Upper transaxle oil cooler hose could come loose and create a fire hazard. **1991–95 models with 4.9L engines**—GM will install a new computer chip that also reduces stalling, pursuant to an agreement with the U.S. Justice Department and EPA.

Secret Warranties/Service Tips

1991–94—Loss of Drive or erratic shifts may be caused by an intermittent short to ground on the A or B shift solenoid, or an electrical short circuit in the transaxle. **1994**—Condensation dripping from the heater duct requires the installation of a watertight dam in the HVAC case. • An inoperative cruise control or brake/transmission interlock may signal a maladjusted stoplight switch assembly. • DSB #476003 goes into great detail about how to troubleshoot the various engine oil leaks afflicting 1994 models. • Doors that won't stay open on slight grades require upgraded door springs. • Noisy fuel pumps can be silenced only by installing an upgraded fuel pump under warranty. • DSB #476506 gives lots of tips on fixing 4.6L engines that run roughly, miss, surge, or hesitate. • Poor heat distribution (driver's feet get cold) can be fixed by replacing the floor outlet assembly. • Rear compartment water leaks are addressed in DSB #311510. **1994–96 with 5.7L engine**—A chuggle or surge condition will require a reflash calibration. **1994–96**—Excessive engine noise can be silenced by installing an upgraded valve-stem oil seal. **1995**—Intermittent Neutral/loss of Drive at highway speeds can be fixed by replacing the control valve body assembly. **1996–97**—A torque converter clutch buzz or moan requires the installation of an upgraded case-cover-assembly spacer plate and the upper-control-valve body. **All models/years**—Defective catalytic converters that cause a rotten-egg smell in the interior will be replaced free of charge under the emissions warranty. • Paint delamination, peeling, or fading (see pages 76, 113).

Concours, DeVille, Fleetwood FWD Profile

	1991	1992	1993	1994	1995	1996	1997	1998
Cost Price ($)								
Concours	—	—	—	49,498	53,690	54,340	56,985	58,600
DeVille	36,598	37,388	39,685	41,998	46,635	48,125	49,400	50,495
Fleetwood	37,598	37,389	39,988	41,798	46,830	46,965	—	—
Used Values ($)								
Concours ↑	—	—	—	18,500	23,000	29,000	34,000	41,000
Concours ↓	—	—	—	17,500	21,000	27,000	32,000	38,500
DeVille ↑	8,000	9,500	12,500	17,000	21,000	26,000	31,000	37,000
DeVille ↓	7,000	8,500	11,500	16,000	19,000	24,000	29,000	34,500
Fleetwood ↑	10,000	11,500	14,000	18,000	22,000	28,000	—	—
Fleetwood ↓	9,000	10,500	13,000	17,000	21,000	26,000	—	—
Extended Warranty	Y	Y	Y	Y	Y	Y	Y	Y
Secret Warranty	N	N	N	Y	Y	Y	Y	Y
Reliability	❷	❷	❷	❷	❷	③	③	③
Air conditioning	❷	❷	❷	③	③	③	④	④
Automatic transmission	④	④	④	④	④	⑤	⑤	⑤
Body integrity	❷	③	③	❷	❷	❷	③	③
Braking system	❷	❷	❷	❷	❷	❷	❷	③
Electrical system	❷	❷	❷	❷	❷	❷	③	③
Engines	❷	❷	❷	③	③	③	④	④
Exhaust/Converter	③	③	④	④	④	④	⑤	⑤
Fuel system	❷	❷	❷	❷	③	③	③	③
Ignition system	❷	❷	❷	❷	③	③	③	③
Rust/Paint	❷	③	③	③	④	④	④	⑤
Steering	❷	❷	❷	❷	③	④	④	④
Suspension	❷	❷	③	③	③	③	④	④
Crash Safety								
DeVille	⑤	⑤	❶	❶	—	③	④	④
Side Impact								
DeVille	—	—	—	—	—	—	④	④

Note: The 1993–94 DeVilles' low crash scores represent severe leg trauma.

INFINITI

G20, I30, J30, Q45

Rating: Above Average (1995–98); Average (1991–94). The fully equipped Maxima, Accord, Camry, Avalon, Millenia, and 929 are better buys from a price/quality standpoint, but they don't have the same luxury cachet. The 1997–98 Q45s were "decontented" by Nissan, meaning they sold for less because they were made more cheaply, came with fewer standard features, and were equipped with a smaller, less powerful engine. **Maintenance/Repair costs**: Higher than average, and they must be done by either an Infiniti or a Nissan dealer. This is worrisome inasmuch as Nissan dealers have had a string of poor sales years and may be tempted to make up the profit loss through their service bays. **Parts**: Higher-than-average cost, but not hard to find (except for body panels).

Strengths and weaknesses: With its emphasis on sporty handling, the Infiniti series takes the opposite tack from the Lexus, which puts the accent on comfort and luxury. Still, the Infiniti comes fully equipped and offers owners the prestige of driving a comfortable, reliable, and nicely styled luxury car. One serious weakness, however, is the 1994 airbag-equipped J30's poor crash rating—all the more surprising when one considers that the 1992 Nissan Maxima passed the NHTSA 57 km/h crash tests with flying colours.

G20

The least expensive Infiniti, the G20 is a front-drive luxury sports sedan that uses a base 2.0L 140-hp 16-valve, twin-cam, 4-cylinder powerplant to accelerate smoothly, albeit noisily, through all gear ranges. Dual airbags came on line midway through the 1993 model year and ABS is standard. Towing capacity is 454 kg (1,000 lbs). Cruise control is a bit erratic, particularly when traversing hilly terrain. The automatic transmission is silent and engine power is reduced automatically when shifting. Steering is precise and responsive on the highway. However, the rear end tends to swing out sharply following abrupt steering changes. Early Infiniti G20s rode a bit too firmly, which led to the suspension being softened on the 1994 model. Now drivers say that the suspension tends to bounce and jiggle occupants whenever the car goes over uneven pavement or the load is increased.

Overall, however, the Infiniti G20s aren't as refined as their entry-level Lexus counterparts in interior space, drivetrain, or convenience features. Owners have complained that the engine's lack of low-speed torque means that it has to work hard above 4,000 rpm—while protesting noisily—to produce brisk engine response in the higher gear ranges. The automatic transmission shifts roughly, particularly when passing (a problem corrected in the 1994 models);

the power steering needs more assist during parking manoeuvres; and the dealer-installed fog lights cost an exorbitant $500 to replace. Poorly thought-out control layout is best exemplified by the hard-to-reach heat/vent controls, an armrest-mounted trunk and filler release that's inconvenient to operate, and centre-console-mounted power window switches that are difficult to find while driving. Tall drivers will find the leg room insufficient. The trunk is spacious, but its small opening is limited by the angle of the rear window.

There are three helpful bulletins containing troubleshooting tips for AC compressor leaks and noise, Code 45 driveability alerts, and brake shudder and steering wheel shimmy.

I30, J30

Resembling the 929 Serenia, the rear-drive, four-door J30 and its high-performance variant, the I30, are sized and priced midway between the G20 and the top-of-the-line Q45. The J30 uses a modified version of the Nissan 300ZX's 3.0L 210-hp V6 engine. Although the vehicle is replete with important safety features and accelerates and handles well, its engine is noisy, passenger and cargo room have been sacrificed to styling, and fuel economy is underwhelming. The more spacious, better-performing I30 replaced the J30 in 1997.

The J30 comes with a standard airbag (or dual airbags, depending on the model year), ABS, and traction control. It's changed very little over the years, meaning that there's no reason to choose a more recent model over a much cheaper older version. Or consider buying an Acura Legend, Lexus ES 300, or a fully equipped Accord, Maxima, 929, Cressida, or Camry.

Dealer service bulletins list the following two defects affecting the 1995s: a front seat rattling noise and a loose B-pillar lower finisher. For all J30 model years, there are a number of other helpful bulletins containing troubleshooting tips for AC compressor leaks and noise, brake clunking noises and pedal pulsation, hard starts, rough idle, water leaks, Code 45 driveability alerts, and brake shudder and steering wheel shimmy.

Q45

This luxury sedan provides performance while its chief rival, the Lexus ES 400, provides luxury and quiet. Faster and glitzier than other cars in its category, the Q45 uses a 32-valve 278-hp 4.5L V8 tire burner not frequently found on a Japanese luxury compact. It accelerates faster than the Lexus, going 0–100 km/h in 7.1 seconds without a hint of noise or abrupt shifting. Unlike the base engine of the G20, though, the Q45's engine supplies plenty of upper-range torque as well. The suspension was softened in 1994, but the car still rides much more firmly than its Lexus counterpart. The four-wheel steering is precise, but the standard limited-slip differential is no help in preventing the car's rear end from sliding out on slippery roads, due mainly to the original equipment "sport" tires designed mainly for 200-km/h autobahn cruising. There's not much foot room for passengers, and cargo room is disappointing. Fuel economy is nonexistent. ABS is standard, but a passenger-side airbag wasn't available before 1994. A redesigned 1994 version got a restyled front end, a chrome grille, and an updated instrument panel. Three years later, the car was again made over with the addition of a downsized 4.1L V8 set on a smaller platform.

Owners report excessive wind noise around the A-pillars, sunroof wind leaks, tire thumping noise, cellular telephone echoing, faulty CD players, and a popping sound from the radio.

Dealer service bulletins list the following defects affecting the 1994–96 models: AC not blowing cold, front brake pad noise, low or rough idle, doors locking/unlocking themselves, and windshield cracking. You may also be interested in reviewing other helpful bulletins that contain troubleshooting tips for AC compressor leaks and noise, brake clunking noises and pedal pulsation, booming/drone noise and vibration, cold-weather hard starts, rough idle, suspension noise, Code 45 driveability alerts, and brake shudder and steering wheel shimmy.

Bulletins for the 1997 model don't cover much that's new, concerning themselves principally with clunking noises and pedal pulsation, cold-weather starting tips, hard starting, and rough idle and suspension noise diagnostic tips.

Safety summary/Recalls: **1995 G20**—Driver seriously injured by airbag deployment when vehicle was pushed into a wall a low speed. • Faulty cruise control wouldn't disengage, brakes failed, and collision ensued. • Defective fuel-filler tube/fuel vent tube. • Frequent stalling. • Large sun visor blocks driver's view of stoplights. • Passenger-side power windows operate erratically. • Defective speaker amplifier. • Centre console handle broke. **1996**—Violent deployment of airbag during collision resulted in permanent eye damage. • Brakes were applied to disengage cruise control and vehicle suddenly accelerated. • Noisy brakes; dealer cannot fix the problem. • Oil pressure switch failure. **1995 J30**—Airbag failed to deploy. • Airbags deployed at low speed (11 km/h) fender-bender, causing extensive injuries to occupants. • Rear end swings out when accelerating. • Excessive vibrations when accelerating make vehicle difficult to steer. • Driver's seatbelt won't retract. **1995 Q45**—Sudden acceleration when coming to a stop. • Gas and brake pedals are set too close together. • Failure of driver's power seat adjuster motor. • Hood flew up while driving along the highway at about 100 km/h. **1996**—Airbag indicator flashes due to ECM failure. • Premature failure of the shock absorber and power window. **1997**—Owner alleges that vehicle design causes the vehicle to hydroplane where other cars wouldn't. **Recalls**: **1991–92 G20**—Rear seatbelt buckle may only partially engage. **1991–96**—Possible fuel leakage from a corroded fuel-filler tube. **1993–95**—Harness connector protector near seatbelt pre-tensioner can ignite. **1993–95 G20, 1993–94 J30**—Cabin may catch fire in a collision. **1991–92 Q45 with Bose speakers and internal amplifier**—Circuit board may short out, overheat, and burn. **1997–98**—Key can be removed when shift lever is not in Park and engine is shut off; dealers will replace the shift lock control unit.

Secret Warranties/Service Tips

1991 G20—An intermittent slip of the sunroof motor requires the installation of an improved motor assembly. • Noisy 2–3 shifting requires the installation of a countermeasure input shaft and an improved high clutch assembly with less free play. • Reverse gear blocking can be corrected by installing an improved Reverse idle gear set. **1991–92**—If the engine's timing chain rattles, install an upgraded chain tensioner (#13070-53J03). • If the air conditioner emits a stale odour or blows out small white flakes, Infiniti will install an improved evaporator core that should correct the problem. **1991–94**—Condensation may prevent the CD player from reading the CD. The only recourse is to wait until the optic sensor dries (10–20 minutes; longer for trunk-mounted players). **1990 Q45**—An idle vibration felt through the steering, floor, and seat can be corrected by adjustments to the idler. • Air and wind noise coming from the windshield, dash, or A-pillar may be corrected by resealing the problem areas. • The following noises require the following repairs, according to DSB #ITB90-039: valve ticking—replace valves/guides; front engine block knocking—replace tensioners; tapping from valves during warm-up—replace pivot/rocker; tapping from valves at all times—check cam bearings. **1990–91**—Front-suspension clicking may require that the shock absorber upper bushing be regreased. • A driveline vibration or drone at moderate speeds can be eliminated by installing a new balance propshaft assembly. • Takata seatbelt replacement. **1991**—Reports of transmission overheating and failures have forced Infiniti to extend the warranty to 7 years in order to compensate owners whose transmissions have insufficient cooling and filtration. Furthermore, the company will install an external cooler and filter at no charge. **1994–96**—Doors that intermittently lock by themselves require the installation of countermeasure front door lock actuators (DSB #NTB96-027). **All models/years**—Troubleshooting tips to correct hard starts. • Vehicles with sunroofs may have wind noise coming from the sunroof area because of a small pinhole in the body sealer at the rear C-pillar. • Windshield cracking. • Erratic operation of the power antenna requires that the antenna rod be replaced.

G20, I30, J30, Q45 Profile

	1991	1992	1993	1994	1995	1996	1997	1998
Cost Price ($)								
G20	24,300	25,275	23,440	26,440	29,540	31,440	—	—
I30	—	—	—	—	—	40,600	41,000	41,350
J30	—	—	41,500	45,000	48,100	51,600	52,600	—
Q45	50,840	54,000	59,500	72,000	75,450	72,000	65,000	66,500
Used Values ($)								
G20 ↑	8,000	9,000	11,000	13,000	16,000	19,000	—	—
G20 ↓	7,000	8,000	10,000	11,500	15,000	17,500	—	—
I30 ↑	—	—	—	—	—	25,000	29,000	33,000

I30 ↓	—	—	—	—	—	23,500	28,000	31,000
J30 ↑	—	—	15,000	19,000	24,000	30,000	36,000	—
J30 ↓	—	—	13,000	17,500	22,000	28,000	34,000	—
Q45 ↑	12,500	15,000	19,000	27,000	34,000	42,000	47,000	51,000
Q45 ↓	11,000	13,000	17,500	24,000	31,000	39,000	44,000	48,000
Extended Warranty	N	N	N	N	N	N	N	N
Secret Warranty	N	N	N	N	N	N	N	N
Reliability	❷	❷	❷	④	④	⑤	④	④
Air conditioning	❷	③	③	③	④	⑤	④	⑤
Automatic transmission	❷	❷	❷	③	④	④	④	⑤
Body integrity	❷	❷	❷	③	③	③	❷	③
Braking system	❷	❷	❷	❷	③	③	❷	③
Electrical system	❷	❷	❷	③	③	③	❷	③
Engines	❷	③	④	④	⑤	⑤	④	④
Exhaust/Converter	④	④	④	④	⑤	⑤	⑤	⑤
Fuel system	❷	❷	③	④	④	④	④	④
Rust/Paint	③	③	④	⑤	⑤	⑤	⑤	⑤
Steering	❷	③	④	④	④	⑤	③	⑤
Suspension	③	③	③	④	⑤	⑤	③	⑤
Crash Safety								
I30	—	—	—	—	—	—	④	④
J30	—	—	—	❷	④	④	—	—
Side Impact								
I30	—	—	—	—	—	—	④	—

LEXUS

ES 250, ES 300, GS 300, LS 400, SC 400

Rating: Recommended (1990–98). A bit more reliable and better built than the Infiniti, but more costly too. A fully equipped Legend, Accord, Maxima, or Camry will provide airbags, comparable highway performance, and reliability at far less initial cost. But, if you do pay top dollar for a used Lexus, its slow rate of depreciation virtually guarantees that you'll get much of your money back. Keep in mind Lexus airbags pack a mighty wallop and may cause severe injuries; install an off-switch or look for depowered airbags starting with the 1997 models. **Maintenance/Repair costs**: Higher than average, and repairs must be done by either a Lexus or a Toyota dealer. **Parts**: Higher-than-average cost, but not hard to find (except for body panels).

Strengths and weaknesses: Like the Acuras and Infinitis, Lexus models all suffer from some front brake, electrical, body, trim, and accessory deficiencies that are confirmed by confidential dealer service bulletins.

ES 250

The ES 250's base engine gives exceptional, smooth acceleration and gets 19 mpg. The 4-speed Overdrive transmission makes subtle changes by reducing the engine power just before shifting. Trailers of up to 2,000 lbs (900 kg) can be towed by cutting out Overdrive. The ES 250 is one of the best-riding front-drive cars money can buy, and its seating offers plenty of support and a comfortable driving position that can be automatically adjusted for two drivers. Braking and handling, though, aren't very impressive.

ES 300

Resembling an LS 400 dressed in sporty attire, the ES 300 was launched in 1992 to fill the gap between the ES 250 and the LS 400. In fact, the ES 300 has many of the attributes of the LS 400 sedan for much less money. A five-passenger sedan based on the Camry, it comes equipped with a standard 3.0L 24-valve engine that produces 181 horsepower coupled to either a 5-speed manual or a 4-speed electronically controlled transmission. Unlike the Infinitis, the ES 300 accelerates smoothly and quietly, while averaging about 20 mpg in mixed driving. The suspension is soft and steady. Passenger and cargo room are plentiful, with lots of leg and head room (except on sunroof-equipped versions). ABS is standard, but a second airbag is available only on the 1994 model.

Surprisingly, for a vehicle this well made, government-reported safety-related defects are legion. Airbag-induced injuries, sudden acceleration, ABS brake and Goodyear tire failures, interior window fogging, and AC toxic emanations are only a few of the ES 300 complaints that carried over many years and have been reported with other Lexus models.

GS 300

The rear-drive GS 300 is a step up from the front-drive ES 300 and just a rung below Lexus's top-of-the-line LS 400. It carries the same V6 engine as the ES 300, except it has 20 horses more. This produces sparkling performance at higher speeds, though, the car is a bit sluggish from a start. Fuel economy is sacrificed for performance, however, and the base suspension and tires pass noisily over small bumps and ruts. Visibility is also less than impressive with large rear pillars and a narrow rear window restricting the view. There's not much usable trunk space either, and the liftover is unreasonably high.

LS 400

The Lexus flagship, the LS 400 rear-drive outclasses all other luxury sedans in reliability, styling, and function. The base engine is a 242-hp 4.0L V8 that provides smooth, impressive acceleration and superior highway passing ability at all speeds. Its transmission is smooth and efficient. The suspension gives an easy ride without body roll or front-end plow during emergency stops, thereby delivering a major comfort advantage over other luxury compacts. There's an absence of engine and wind noise. ABS and dual airbags are standard.

Owners have complained that the brakes don't inspire confidence, owing to their mushy feel and average performance. Furthermore, there's limited rear foot room under the front seats, and the rear-middle passenger has to sit on the transmission hump. This car is a gas guzzler that thirsts for premium fuel.

SC 300, 400

These two coupes are pratically identical, except for their engines and luxury features. The cheaper SC 300 gives you the same high-performance 6-cylinder engine used by the GS 300 and Toyota Supra, while the SC 400 uses the same 4.0L V8 engine found in the LS 400. You're likely to find fewer luxury features with the SC 300 because they were sold as options. Nevertheless, look for an SC with traction control for additional safety during poor driving conditions. On the downside, V8 fuel consumption is horrendous, rear seating is cramped, and trunk space is unimpressive. Also, invest in a good anti-theft device, or your Lexus relationship will be over almost before it begins.

Dealer service bulletins: **1996**—Ambient temperature displays –220°C. • Front brake groan. • Static noise on weak AM stations. **1995 LS 400**—Engine knocking, front seat cushion noise, front stabilizer bushing noise, strut bar cushion noise, and sun visor rattling. **1995 SC 300 and SC 400**—Rear suspension rattling or clacking, probably caused by a faulty rear spring bumper (replace with an upgraded spring bumper). **1995–96 ES 300**—CD changer won't eject magazine. • Direct clutch improvements. • Front brake groan. • Moon roof panel wind noise. • Static noise on weak AM stations.

Safety summary/Recalls: **1994**—Airbags failed to deploy. • During collision, driver's seat and shoulder belts did not retract, but released. • Driver-side seatbelt buckle released after vehicle hit on the left side. • Accelerator pedal fell off and jammed under the accelerator bar. • Frequent reports that the interior windows fog up, causing reduced visibility. • Frequent reports of Goodyear tire blowouts and ABS brake failures. • Rear brake pad failures. • Transmission pan leaks and failures. • Inadequate middle console illumination. **1995**—Airbags failed to deploy. • Both front seatbelts failed to lock during a collision. • Seatbelt failed to restrain passenger during collision. • Goodyear Eagle tire blowout. • Premature failure of the electronic control unit. • Transmission leaks. • Brake pedal design makes it difficult for someone with a shoe size larger than men's nine to use the brakes. • Excessive noise when braking. • Faulty windshield moldings cause excessive wind noise. **1996**—Fire in the engine compartment. • Frequent reports that the airbag failed to deploy. • Weak seatback collapsed rearward in a collision. • Entire vehicle shimmies. • Sudden steering failure when turning. **1997**—Vehicle left in Park rolled backwards and hit another car. Same thing occurred with a vehicle parked in a garage, but this time cause was isolated to a failure of the shift lock actuator fuse. • Owner says airbags are a hazard for short people. • Also, floor pedals are located too high and are too far apart for short people. • Middle rear shoulder belt locks up, making it very difficult to get occupant out; owner had to cut the belt. • Instrument cluster lights aren't bright enough for night driving. • AC assembly panel failure. **1998**—Vehicle accelerated as brakes were applied. • Premature front brake pad wearout. • Seat adjustment motor failure. **1993 ES 300**—Engine compartment fire. • Airbags fail to deploy. • Airbag deployed and caught on fire. • Airbag deployment burned driver severely. • Airbags deployed when vehicle at a standstill, causing severe burns to driver's nose and hands. • In another incident, car was bumped slightly, airbags deployed 10 seconds later, and driver received severe facial injuries. • ABS system failures. • Goodyear tires experience blowouts and are worn out prematurely at 15,000 km. • Steering tie rod broke while driving, causing loss of steering and an accident. • Transmission doesn't engage right away, but when it does engage, it jumps forward. • When vehicle goes over a bump, the steering wheel jumps out of hand. • Foul odour from AC caused occupants to fall ill with bronchial infection. • Several reports that dashboard emits fumes that fog inside window glass. • Headlights provide poor illumination. • Driver can't read speedometer when sun shines onto dash. • Seatback collapsed backwards when vehicle was rear-ended. **Recalls**: **1990–91 ES 250**—Fuel tank leak. **1990 LS 400**—Cruise control may not return to its former position. • Prolonged illumination of the centre-mounted brake light. **1993–94 GS 300**—Premature ball joint wear could affect steering and handling. An upgraded ball joint socket will be installed. **1995–97 LS 400, 1996–97 SC 400**—A faulty starter-motor magnetic switch may cause a fire or render the starter inoperative.

Secret Warranties/Service Tips

1990 ES 250—To reduce front brake squeaks, Lexus has changed the rotors (#43517-32020). **1990–91**—Front brake squeaking can also be reduced by using revised brake pads (#04491-32390). **1991**—Cellular telephone antenna wind noise can be eliminated by installing a newly designed antenna. • A cruise control that cancels after setting needs a filter circuit added to the cruise control ECU. **1992 ES 300**—Inaccurate fuel gauges require an improved indicator needle. **1992–93**—Sun visor rattles can be fixed by using the Lexus Squeak and Rattle Repair Kit. • Rear brake squeaks can be reduced by using upgraded rear brake pads. • Rear stabilizer bar bushing noise can be eliminated by installing upgraded bushings. **1993**—Problems with hot start or poor engine performance when going downhill require the installation of an upgraded ECM. • Front seat headrest rattles can be corrected by installing an improved headrest support. **1992 SC 300/SC**

400—Popping from the Nakamichi radio has been eliminated with an upgraded model containing improved volume control resistors. **1990 LS 400**—Front brake popping can be corrected by installing a modified pad support plate and applying new adhesive. • Moon roof wind noise may be corrected by realigning the roof panel. • Faulty cruise control assemblies will be replaced with an improved assembly (SSC 901). • Warped centre high-mounted brake-light housings will be replaced free under a goodwill program (SSC 902). **1990–91**—A difficult-to-open rear door may be helped by modifying the rear door check arm. • To prevent transmission clicking when shifting from Neutral to Drive or Reverse, reduce the depth of the flange yoke assembly. • Windshield upper moulding noise requires that the moulding be changed and a thicker adhesive tape be applied. • AC groaning can be eliminated by reducing the expansion valve flow rate and adding an O-ring to the EPR piston.

ES 250, ES 300, GS 300, LS 400, SC 400 Profile

	1991	1992	1993	1994	1995	1996	1997	1998
Cost Price ($)								
ES 250	29,300	—	—	—	—	—	—	—
ES 300	—	34,400	37,300	41,300	44,700	45,600	42,960	—
GS 300	—	—	54,700	57,800	66,100	71,400	71,400	58,900
LS 400	57,000	63,000	67,200	71,100	75,900	78,700	78,700	78,300
SC 400	—	57,000	61,000	64,600	74,600	80,800	82,100	84,000
Used Values ($)								
ES 250 ↑	9,000	—	—	—	—	—	—	—
ES 250 ↓	8,000	—	—	—	—	—	—	—
ES 300 ↑	—	14,000	18,000	23,000	26,000	29,000	31,000	—
ES 300 ↓	—	13,000	17,000	21,000	25,000	27,000	28,000	—
GS 300 ↑	—	—	23,000	28,000	34,000	38,000	44,000	48,000
GS 300 ↓	—	—	22,000	27,000	32,000	36,000	42,000	46,000
LS 400 ↑	18,000	22,000	28,000	35,000	42,000	50,000	58,000	67,000
LS 400 ↓	15,000	19,500	25,000	32,000	39,500	47,000	55,000	64,000
SC 400 ↑	—	20,000	28,000	36,000	43,000	50,000	59,000	69,000
SC 400 ↓	—	18,500	26,000	33,000	40,000	47,000	56,000	66,000
Extended Warranty	N	N	N	N	N	N	N	N
Secret Warranty	N	N	N	N	N	N	N	N
Reliability	③	④	④	⑤	⑤	⑤	⑤	⑤
Air conditioning	❷	❷	③	④	⑤	⑤	⑤	⑤
Automatic transmission	③	④	⑤	⑤	⑤	⑤	⑤	⑤
Body integrity	❷	❷	③	④	⑤	⑤	⑤	⑤
Braking system	❷	❷	❷	❷	❷	③	④	④
Electrical system	❷	❷	③	③	④	④	④	④
Engines	❷	③	③	③	⑤	⑤	⑤	⑤
Exhaust/Converter	④	④	⑤	⑤	⑤	⑤	⑤	⑤
Fuel system	④	④	⑤	⑤	⑤	⑤	⑤	⑤
Ignition system	③	④	⑤	⑤	⑤	⑤	⑤	⑤
Rust/Paint	⑤	⑤	⑤	⑤	⑤	⑤	⑤	⑤
Steering	④	④	⑤	⑤	⑤	⑤	⑤	⑤
Suspension	❷	❷	③	④	⑤	⑤	⑤	⑤
Crash Safety								
ES 250	⑤	—	—	—	—	—	—	—

ES 300	—	—	—	—	—	⑤	—	④
GS 300	—	—	—	⑤	③	—	③	—
Side Impact								
ES 300	—	—	—	—	—	—	—	⑤

LINCOLN

Continental, Mark VII, Mark VIII, Town Car

Rating: **Continental**: Average (1996–98); Below Average (1984–95). **Mark VII, Mark VIII**: Above Average (1995–98); Average (1994); Below Average (1986–93). **Town Car**: Above Average (1995–98); Average (1988–94). Redesigned for 1998. **Maintenance/Repair costs**: Higher than average, and they must be done by a Lincoln dealer. **Parts**: Higher-than-average cost, but not hard to find (except for electronic components and body panels).

Strengths and weaknesses: These large luxury cruisers are proof that quality isn't proportional to the money you spend. Several designer series offer all the luxury options anyone could wish for, but the two ingredients most owners would expect to find—high quality and consistent reliability—are sadly lacking, especially when it comes to the automatic transmission, electrical system, brakes, body hardware, and fit and finish.

These cars aren't lemons (at least the later models aren't), but they don't offer the kind of trouble-free driving one would normally expect in a new vehicle selling for over $40,000. The automatic leveling air-spring suspension system makes for a stiff ride (especially on early models), while still allowing the Continental to "porpoise" due to its heavy front end. The Continental's anemic V6 powertrain is poorly suited to a car of this heft. The engine hesitates in cold weather and the automatic transmission shifts roughly due to malfunctioning computer modules.

Major mechanical defects affect the engine (frequent flywheel replacements), transmission, ABS, electrical, suspension, and steering systems, as well as the electronic modules. The mass of electrical gadgets increases the likelihood of problems as the cars age. For example, automatic headlight doors fail frequently, and the electronic antenna seldom rises to the occasion. The computerized dashboard is particularly failure-prone.

Other reliability complaints concern transmission fluid leakage due to misplaced bolts, rough upshifting caused by a defective valve body, and air conditioning and heating that sometimes work in reverse order (you often get heat when opening the AC, and air conditioning frequently comes on when the heater is engaged).

Continental (front-drive)

When the Continental went front-drive in 1988, it made a bad situation worse. The frequency and cost of repairs increased considerably, and parts became more difficult to find. The automatic transmission continued to malfunction, particularly on 1988–91 models; electrical components became even less reliable; stopping performance was compromised by premature brake wear and rear wheel lockup; and body hardware continued to be second class. The redesigned 1995 Continental featured a new V8 powerplant, more aerodynamic styling, and fibreglass panels. However, powertrain, electrical system, and brake problems remained.

Town Car

The rear-drive Town Car is the pick of the Lincoln litter. Nevertheless, it's still afflicted by the Lincoln's generic problems: transmission, AC, and electrical glitches, and body hardware deficiencies. Nevertheless, thanks to its rear-drive configuration, it's relatively inexpensive to repair and parts aren't hard to find. Try to find a 1991 or better version with the 4.6L engine to optimize performance and economy.

Dealer service bulletins: **Continental**: **1993**—Poorly performing AC systems caused by a slipping clutch at high ambient temperatures, and AC FX-15 compressor growling. • Service Engine light may go on for no apparent reason. • A-pillars may make cracking noises in cold weather. • Power-steering units may be noisy and power door locks may not work. • Frequent under-hood squeaks, chirps, and knocks. • Premature wear on the inner edge of the rear tires, corrected by a new camber kit. **1994**—Faulty 3.8L engine rocker arm assemblies could cause squeaking, chirping, and knocking. • Inadequate AC operation caused by a faulty cold engine lockout switch and hose assembly. • Faulty electric rear window defrosters. **1995**—Engine misfires, no-start, hard start, and spark plug fouling. • Faulty in-tank fuel delivery modules causing the engine to stall and not restart will be replaced for free under Ford Program #95B71. • The Virtual Image Cluster will also be replaced at no charge under Program #95B71. (Ford will pay for a loaner car at $30 (U.S.) per day whenever carrying out these Program repairs.) • Harsh shifts, no 3–4 shift, erratic shifts, delayed transmission engagement and shift errors, low transaxle fluid level improperly setting the DTC and causing serious transmission malfunctions, and clicking noise when shifting into Reverse. • Insufficient AC cooling or excessive clutch end gap. • Brake roughness upon application or a thumping, clacking noise when braking. • Power-steering grunt or groan. • Rocker arm noise. • A ticking noise from the cooling fan or AX4N transaxle. **1996**—Stall or hard start after 1–4 hour soak. • Vehicle drifts or pulls while driving. • Transaxle click noise during Reverse engagement. • Transaxle driveline noises. • Case breaks at rear planet support. • Air suspension leaks down overnight. • Fog/film on windshield/interior windows. • Musty and mildew-type odours. **1997**—Harsh automatic shifting. • Chronic dead battery. • Excessive blower motor noise. • Acceleration or deceleration clunk. • Front suspension clunk. • A-pillar creaking noise. • Front end accessory drive belt (FEAD) may slip during wet conditions, causing a reduction in steering power-assist. • Steering wheel grunt or honk noise. • Loose catalyst or heat shields. • Stall or surging with automatic transmission engagement.

Town Car: **1994**—A rough idle, hesitation, excessive fuel consumption, and a poor heater output that's likely caused by a thermostat sticking in an open position or opening before it should. • Transmission shudder under light to moderate acceleration. • Radio speaker noise caused by fuel pump malfunctions. • Faulty electric rear window defrosters. **1995**—Problems with the 1995 Town Car are identical to those listed for the Crown Victoria, except for one bulletin addressing a rough idle after a cold soak start-up. **1996**—Stall or hard start after 1–4 hour soak. • Squealing noise from the engine compartment. • Transmission valve body cross-leaks. • Chatter during turns. • Creaking from A-pillar area. • Rough idle or stalling at high altitudes. • Inoperative power-assisted deck lid. • Fog/film on windshield/interior windows. • Musty and mildew-type odours. **1997**—Erratic transmission shifting, especially the 1–2 shift. • Spark knock upon acceleration. • Door latch stuck open. • Fog/film on windshield/interior windows. • Loose catalysts or muffler heat shields. • Diagnosing and correcting wind noise around doors. • Shudder or vibration while in third or fourth gear. • Steering wheel noise and vibration. • Suspension leans to right side. • Whining noise from rear axle at highway speeds.

Mark VIII: **1994–97**—An erratic or prolonged 1–2 shift can be cured by replacing the cast aluminum piston with a one-piece stamped steel piston with bonded lip seals and by replacing the top accumulator spring. These upgraded parts will increase the transmission's durability, says Ford. **1995**—Intermittent loss of torque at 3–4 upshift, shifts to Neutral at heavy throttle, delayed transmission engagement and shift errors, and irregular or no torque converter operation. • Brake roughness upon application. • Insufficient AC cooling or excessive clutch end gap. • Faulty fuel gauge. • Hard start in low temperatures. **1996**—Stall or hard start after 1–4 hour soak. • Transmission valve body cross-leaks and shudder/downshift bump. • Growling noise from steering column when vehicle turns. • Squealing noise from the engine compartment. • Fog/film on windshield/interior windows. • Musty and mildew-type odours. **1997**—Erratic transmission shifting, especially the 1–2 shift. • Chirping and squeaking from blower motor. • No start or crank; anti-theft system not responding. • Radio AM band static while driving. • Troubleshooting wind noise around doors.

Safety summary/Recalls: **Continental:** Investigators are looking into reports of fires erupting in the engine compartment or wheel wells. **1995**—While parked, vehicle rolled backwards and suddenly accelerated forward. • Steering locks when driving through deep puddles. • Frequent reports of seatbelts failing to retract during an accident. • Defective inner tie-rod and strut causes severe steering shake. • Brakes failures due to premature wear of rear drums and rotor warpage. • Inadequate defogging leaves two small circles to see through. • Console lights reflect into driver's portion of windshield and rearview mirror. • Rear deck stereo hump obstructs rear visibility. **1996**—Sudden acceleration while driving on the highway. • While parked, vehicle suddenly jumped out of Park and rolled forward and hit another car. • Several reports that the front suspension collapsed. • Front strut failures. • Clock spring broke causing loss of steering. • Steering wheel jams intermittently. • Power steering is inoperative when it rains. • Transmission jumped out of gear while driving. • Power seats malfunction. • Vehicle cannot be properly aligned, causing premature tire wear.

Town Car: **1992–93**—Engine compartment fires near the left front wheel well or brake master cylinder. **1996**—Vehicle fires reportedly ignited in the engine compartment wiring. • Frequent reports of sudden acceleration. • Many reports of sudden acceleration while in Reverse, leading to the replacement of the speed control servo and servo cable. • Frequent complaints of chronic stalling. • ABS brakes often malfunction. • Airbag failed to deploy during collision. • Airbag packing gets hard in cold weather, leaving horn inoperative. • Trunk lid flew open on highway. • Many incidents where seatbelts either are too short for adults, won't retract, or ratchet tighter while being worn. • Two front seatbacks collapsed rearward when vehicle was rear-ended. • Excessive noise when braking. • Wind noise comes in through windshield. **1998**—Fuel may spit out of filler pipe when refueling.

Recalls: **Continental**: **1986**—Throttle may fail to return to idle. • Defective brake master cylinder warning sensor. **1986–87**—Fuel-line coupling may leak. **1987–88**—Body mounts will be inspected and a reinforcement plate installed. • Faulty transmission shift cable. • Power-steering pump pulley could separate from engine. **1988–93**—Advanced front subframe fastener corrosion. **1989**—Faulty left rear seatbelt retractor. **1992–94**—Engine cooling fan motor may overheat and cause a fire. **1994**—Missing or improperly installed brake pedal push rod retainer. **1995–96**—"Autolamp" control module may fail. **1996–97**—Transmission may not engage; PRNDL may give a false reading.

Town Car: **1988–89**—Faulty ignition switches may be a fire hazard. **1988–94**—Subframe engine and transmission mounts could fail due to excessive corrosion, which could cause the subframe to drop and make steering difficult. **1990**—Ford will install new seat tracks and a memory track on the driver's side. **1990–91**—Hood can pop open during highway driving. **1991–92**—Secondary hood latch may not engage. **1991–94**—Faulty cruise control units and speed control cable. **1992–94**—Overheated engine cooling fan may start a fire. **1994**—Rear brake adapter to axle housing flange could loosen or separate. • Missing or improperly installed brake pedal push rod retainer. **1995**—Fuel could leak from tank. • Passenger airbag may not deploy properly. **1996**—Driver's door may not sustain specified load. • Seatbelts with switchable retractors for child restraints may not have the right components. **1997**—Driver's airbag may malfunction.

Secret Warranties/Service Tips

Continental: **1984–94**—A hum from the air suspension system can be corrected by replacing the compressor isolators with upgraded parts. **1988–90**—Hard cold-start, hesitation, and stalling during idle or when decelerating may be corrected by removing excessive sludge deposits or an oil film from the throttle body bore and plate or from the idler air bypass valve. • High effort during wide turns at highway speeds signals the need to install a new power-steering short rack. • The intermittent loss of the AC may be due to a defective suction accumulator (#E6VY-19C836-A). **1988–94**—A squeak or chirp coming from the blower motor can be stopped by installing an upgraded blower motor with improved brush-to-commutator friction. • A cracked forward clutch piston may cause forward/reverse problems; install an improved clutch piston. • A speaker whine or buzz caused by the fuel pump can be stopped by installing an electronic noise RFI filter. **1989**—A rough idle or lean fuel flow may require the installation of deposit-resistant injectors. **1989–93**—A persistent fuel odour in the interior when the AC is running signals the need to install a new auxiliary vapor tube kit and relocate the tube near the rear bumper. **1991**—When Ford modified the automatic transmission for easier shifting, it also made it less reliable. In fact, *Consumer Reports*' annual member survey shows that almost 20 percent of the owners of the 1991 Sable, Taurus, and Continental said that they had experienced a serious transmission failure in the preceding 12 months—four times the average for 1991 vehicles, according to *Consumer Reports*. Ford has extended the warranty on these transmissions to up to six years. • Delayed transaxle 3–2 downshifts may require a new, more durable spring retainer clip (#F1DZ-7F194-A). **1991–94**—A sloshing noise from the fuel tank when accelerating or stopping requires the installation of an upgraded tank. Cost is covered under the emissions warranty. **1992–94**—A corroded solenoid may be the cause of starter failures. **1993–94**—An inoperative AC blower probably needs an improved cold engine lockout switch and hose assembly. • Stalling or hard starts in high ambient temperatures or high altitudes may be due to fuel tank contamination damaging the fuel pump. Ford paid for a fuel tank flush and a new fuel pump/sender and in-line fuel filter until May 31, 1997, under Service Program 94B48. **1994–95**—A thumping or clacking heard from the front brakes signals the need to machine the front disc brake rotors. Ford will pay for this repair under its base warranty. **1995–97**—An acceleration or deceleration clunk is likely caused by the rear lower subframe isolators allowing movement between the mounts and the subframe. • A front suspension clunk may signal premature sway bar wear. **Mark VII, Mark VIII**: **1985–92**—An exhaust buzz or rattle may be caused by a loose heat shield catalyst. **1986–94**—The in-tank fuel pump is the likely cause of radio static. Install an electronic noise RFI filter (#F1PZ-18B925-A). **1993–94**—A squeak or chirp coming from the blower motor can be stopped by installing an upgraded blower motor with improved brush-to-commutator friction. • Automatic transmissions with delayed or no forward engagement, or a higher engine RPM than expected when coming to a stop, are covered in DSB #94-26-9. **Town Car**: **1983–90**—A buzzing or humming from the fuel pump when the engine is shut off, a low battery, and hard or no-starts signal the need to install an upgraded pump relay (#F19Z-9345-A). **1987–90**—Excessive oil consumption is likely caused by leaking gaskets, poor sealing of the lower intake manifold, defective intake and exhaust valve stem seals, or worn piston rings. Install new guide-mounted valve stem seals for a more positive fit and new piston rings with improved oil control. **1989–90**—No Overdrive or an extended 3–4 shift may require the installation of a new Overdrive band, transmission separator plate kit, overhaul kit, and reverse drum. **1990**—A no-start condition may be caused by a defective Neutral start switch. • A pull to the right when braking can be corrected by installing a new brake pad and lining kit (#F1VY-20001-A). **1990–92**—Brake pedal and steering vibration when braking can be reduced by installing improved rotors (#F1VY-1125-A) and linings (#F3AZ-2001-A). **1991–92**—Secondary hood latch may not engage. **1992–94**—Automatic transmissions with delayed or no forward engagement, or a higher engine rpm than expected when coming to a stop, are covered in DSB #94-26-9. **1993–94**—A speaker whine or buzz caused by the fuel pump can be stopped by installing an electronic noise RFI filter. **1997**—Driver-side airbag may lack two mounting bolts. **All models/years**—Three components that frequently benefit from Ford "goodwill" warranty extensions are catalytic converters, fuel pumps, and computer modules. If Ford balks at refunding your money, apply the emissions warranty for a full or partial refund. • Paint delamination, fading, peeling, hazing, and "microchecking" (see page 112 for details on claiming a refund).

Continental Profile

	1991	1992	1993	1994	1995	1996	1997	1998
Cost Price ($)								
Continental	38,495	37,895	38,842	40,295	50,995	51,896	49,995	51,995
Used Values ($)								
Continental ↑	6,500	9,000	11,000	15,000	20,500	24,000	29,000	34,000
Continental ↓	6,000	7,500	9,500	13,000	18,000	22,500	27,000	31,500
Extended Warranty	Y	Y	Y	Y	Y	Y	Y	Y
Secret Warranty	Y	Y	Y	Y	Y	Y	Y	N
Reliability	❶	❷	❷	❷	❷	③	③	④
Air conditioning	❶	❶	❶	❶	❷	③	③	③
Automatic transmission	❶	❷	❷	❷	❷	③	③	③
Body integrity	❷	❷	❷	❷	❷	③	③	③
Braking system	❶	❶	❶	❶	③	③	③	④
Electrical system	❶	❶	❶	❷	❷	❷	❷	③
Engines	❶	❶	❷	❷	③	③	④	④
Exhaust/Converter	③	④	④	⑤	⑤	⑤	⑤	⑤
Fuel system	❷	③	③	③	④	④	④	④
Ignition system	❶	❷	③	③	③	③	④	④
Rust/Paint	③	③	③	④	④	④	④	④

Steering	❷	③	③	④	④	④	④	④
Suspension	❶	❶	❶	❷	❷	❷	③	④
Crash Safety	⑤	⑤	❶	❶	—	—	④	—

Note: The Continental registered low scores for the 1993–94 model years due to severe leg trauma.

Mark VII, Mark VIII Profile

	1991	1992	1993	1994	1995	1996	1997	1998
Cost Price ($)								
Mark VII/VIII	38,895	41,010	43,968	47,995	50,996	51,895	53,695	56,595
Used Values ($)								
Mark VII/VIII ↑	9,500	11,000	14,000	18,000	22,000	27,000	32,000	38,000
Mark VII/VIII ↓	8,500	9,500	13,000	17,000	20,000	25,000	30,000	36,000
Extended Warranty	Y	Y	Y	Y	Y	Y	Y	Y
Secret Warranty	N	Y	Y	Y	Y	Y	Y	N
Reliability	❷	❷	❷	❷	❷	③	③	③
Air conditioning	❷	❷	❷	③	③	③	③	③
Automatic transmission	❷	❷	❷	❷	❷	③	④	⑤
Body integrity	❷	❷	❷	❷	❷	❷	❷	③
Braking system	❷	❷	❷	❷	❷	③	④	④
Electrical system	❷	❷	❷	❷	❷	❷	❷	❷
Engines	④	④	③	③	④	④	④	④
Exhaust/Converter	❷	❷	③	④	④	④	⑤	⑤
Fuel system	③	③	③	③	③	③	④	④
Ignition system	③	③	③	③	③	③	④	③
Rust/Paint	❷	❷	❷	❷	③	③	③	④
Steering	❷	❷	③	④	④	④	⑤	⑤
Suspension	❷	❷	❷	③	③	③	③	④

Note: The Mark series hasn't been crash-tested.

Town Car Profile

	1991	1992	1993	1994	1995	1996	1997	1998
Cost Price ($)								
Town Car	37,195	37,695	38,637	40,495	44,495	44,895	45,895	52,795
Used Values ($)								
Town Car ↑	8,500	10,000	12,000	17,000	21,000	23,000	27,000	36,000
Town Car ↓	7,000	9,000	11,000	16,000	18,000	22,000	26,000	35,000
Extended Warranty	N	N	N	N	N	N	N	N
Secret Warranty	Y	Y	Y	Y	Y	Y	Y	N
Reliability	❷	❷	❷	③	④	④	④	④
Air conditioning	❶	❷	❷	③	③	③	③	④
Automatic transmission	③	❷	❷	❷	③	③	④	④
Body integrity	❶	③	③	③	③	③	③	③
Braking system	❶	❶	❶	❶	❶	❷	③	③
Electrical system	❶	❶	❶	❷	❷	③	③	③

Engines	④	④	④	④	④	⑤	⑤	⑤
Exhaust/Converter	③	③	④	④	④	⑤	⑤	⑤
Fuel system	❷	❷	③	③	③	③	③	④
Ignition system	❶	❷	❷	③	③	③	③	③
Rust/Paint	❷	❷	③	③	③	③	④	⑤
Steering	❷	❷	③	④	④	③	③	④
Suspension	③	③	③	③	③	④	④	④
Crash Safety	⑤	⑤	❷	❷	⑤	④	④	—

Note: The Town Car registered low scores for the 1993–94 model years represent severe leg trauma.

MAZDA

929

Rating: Above Average (1993–95); Average (1988–92). The only real negative is the 929's consistently poor crash test scores. **Maintenance/Repair costs**: Higher than average, and they must be done by a Mazda dealer. **Parts**: Higher-than-average cost, but not hard to find through cheaper independent suppliers, despite '95 being its last model year.

Strengths and weaknesses: The key word for the 929 is understatement: the engine is unobtrusive, the exterior is anonymous, and the interior is far from flashy. In spite of its lack of pizzazz and imprecise power steering, the 929 will accelerate and handle curves as well as the best large European sedans. The 929 has proven to be fairly reliable.

In addition to poor crashworthiness, the car's main drawbacks are its limited interior room and trunk space. The driver's seat doesn't have enough rear travel for tall drivers, and head room is tight. Owners report some problems with premature disc brake wear, electrical glitches, exhaust system rust-out, electronic shock absorber durability (particularly with the 1989–91 models), and fit and finish deficiencies.

Redesigned in 1992, the 929 got a more rounded body and a longer wheelbase that added to interior room while sacrificing trunk space. Despite these improvements, tall occupants will still feel cramped.

Shocks are very expensive to replace. Manual transmission isn't offered, and the automatic's many settings can be confusing. Furthermore, the transmission's lockup feature frequently cuts in and out. The rear end sometimes wants to slide out a bit on slippery surfaces, and the front end bounces around on bumpy roads. The optional automatic adjusting suspension does little to improve the car's ride or handling.

Dealer service bulletins: **1995**—Heater and AC unit noise after long storage. • Rattling sunroof and outer door handles. • A slightly off-centre steering wheel. • Brake pulsation repair. • Inoperative rear door window. • Brake shudder repair. • Scratched or peeling B-pillar trim. • Troubleshooting various causes of engine noise.

Safety summary/Recalls: **Recalls**: **1988–91**—Mazda will replace door handles with ones that won't break as easily. • Takata seatbelt replacement.

Secret Warranties/Service Tips

1988–91—Cold-engine piston slapping requires replacement pistons to fix the problem. • Constant brake pulsation or shudder is likely caused by an uneven rotor surface or excessive rotor run-out. • Hard shifting after cold-weather starts can be corrected by installing upgraded synchronizer rings and clutch hub assemblies. **1988–92**—Valve train noise occurring just after start-up may be caused by air trapped in the hydraulic lash adjuster. Correct by installing redesigned rocker arm shafts that promote better oil flow. **1989**—Mazda has modified the bushing assembly to reduce shudder and make for a more comfortable ride. The new assembly part is #H260-34-230C. **1990**—A power seat that won't adjust up or down smoothly likely has a broken gear in the seat motor. **1990–91**—Difficult starts in hot weather can be corrected by installing an upgraded cold-start thermo-switch (#JE27-18-870). **1992–94**—Freezing door and hatch lock cylinders are addressed in DSB #021/94. **All models/years**—DSB #006/94 covers all the possible causes and remedies for excessive vibrations when braking. • Water intrusion into the lock actuator connectors may cause the unintended operation of the rear defroster.

929 Profile

	1988	1989	1990	1991	1992	1993	1994	1995
Cost Price ($)								
Sedan	27,538	31,888	32,708	31,158	35,238	38,550	40,295	48,320
Used Values ($)								
Sedan ↑	5,000	6,500	8,000	9,000	10,000	13,000	17,500	24,000
Sedan ↓	4,500	5,500	7,000	8,000	9,500	12,500	16,500	23,000
Extended Warranty	Y	Y	Y	Y	N	N	N	N
Secret Warranty	N	N	N	N	N	N	N	N
Reliability	❷	❷	③	③	③	④	⑤	⑤
Air conditioning	❷	❷	❷	❷	③	④	⑤	⑤
Automatic transmission	❷	❷	❷	③	④	④	⑤	⑤
Body integrity	❷	❷	❷	❷	❷	❷	③	③
Braking system	❷	❷	❷	❷	❷	❷	❷	③
Electrical system	❷	❷	❷	❷	❷	③	③	③
Engines	❷	③	③	④	⑤	⑤	⑤	⑤
Exhaust/Converter	❷	❷	❷	❷	③	③	④	⑤
Fuel system	③	③	③	④	④	④	④	④
Ignition system	❷	③	④	⑤	⑤	⑤	⑤	⑤
Rust/Paint	③	③	④	④	⑤	⑤	⑤	⑤
Steering	③	③	③	③	④	⑤	⑤	⑤
Suspension	❷	❷	❷	③	④	⑤	⑤	⑤
Crash Safety	❷	❷	❷	❷	❷	❷	❷	—

Millenia

Rating: Above Average (1995–98). Lots of power and sophisticated mechanicals make this luxury tourer a winner. From a performance standpoint, the Camry V6 with its less complicated powertrain outruns the Millenia. **Maintenance/Repair costs**: Higher than average, and repairs must be done by a Mazda dealer. **Parts**: Higher-than-average cost, but parts are easily found.

Strengths and weaknesses: Smaller than the Mazda 929, the front-drive Millenia carries the same 2.5L 170-hp V6 used by the 626. An optional 2.3L Miller-Cycle "S" 6-cylinder engine, although smaller than the base powerplant, still manages to pump out 210 horsepower. Both engines use a standard 4-speed automatic transmission that shifts a bit harshly when pushed. As with all luxury cars, the Millenia comes with a wide array of standard features that would normally cost thousands of dollars more. Although billed as a five-passenger car, the middle occupant in the rear seat is cramped and has to sit on a hump—a problem that 929 owners are familiar with.

Assembly quality and component quality are fairly high, however there have been reports of powertrain failures, front brake failures, and electrical glitches. Among the powertrain problems, owners cite transmission failures and defective engine head gaskets.

Dealer service bulletins: 1995—Cruise control surging. • Transmission position indicator light failure. • AC evaporator freeze-up. • Battery discharge due to the trunk light staying on. • Snap noise around the A-pillar. • Steering wheel may be off-centre. • Roof insulator peeling off. **1996**—Brake pulsation repair. • Cracked centre sun visor holder. • Creaking or knocking noise from rear of vehicle. • Rattle noise from rear package tray. **1997**—Brake pulsation repair.

Safety summary/Recalls: **1995**—Many complaints that vehicle surges when shifted into Reverse. • Inadvertent airbag deployment; failure to deploy. • Steering wheel back is open, allowing objects to jam the steering mechanism. • Driver's automatic seatbelt came loose when vehicle was rear-ended. • While vehicle was being driven on the highway, engine locked up due to leakage from the oil pan. • ABS brake failures. • Many reports of cracked engine valve cover gaskets, which allowed oil to leak onto the wiring and spark plugs. • Because the engine intake valve is set so low, whenever it comes in contact with water, the car stalls. • Premature AC and CV joint failures. • Sudden electrical system failure. • Defroster doesn't do an adequate job in cold weather; collects water or quits altogether. • When AC is on, the headlights go dim. • Low-beam lights are inadequate for night driving. • Battery connections become loose, making it impossible to unlock or start vehicle. • When driver turned the ignition switch, the battery cable popped off. This caused the automatic door locks to become inoperative, trapping occupants. • Trunk won't open with inside latch release. • The location of the cup holder allows drinks to be spilled into the shifter mechanism, causing the shifter to lock up. • Cup holder doesn't always release. Seatbelt buckle on passenger's side fell completely apart in driver's hand. • Aluminum alloy wheel cracked. **1996**—Airbag failed to deploy. • Transmission failures. • Intermittent starting problems. • Loose dashboard and interior trim. • It's easy to hit your head on the low trunk lid and latch. **1997**—Traction control system failure. **Recalls**: N/A.

Secret Warranties/Service Tips

1995–96—A creaking or knocking noise from the rear of the vehicle is likely caused by loose diagonal braces behind the rear seat.

Millenia Profile

	1995	1996	1997	1998
Cost Price ($)				
Base	38,055	42,035	42,900	38,660
Used Values ($)				
Base ↑	18,000	22,500	26,000	29,000
Base ↓	17,000	21,500	25,000	28,000
Extended Warranty	N	N	N	N
Secret Warranty	N	N	N	N
Reliability	③	④	⑤	⑤
Crash Safety	④	④	④	—

MERCEDES-BENZ

190 Series, C-Class

Rating: Above Average (1995–98); Average (1988–94); Below Average (1984–87). Although these cars are above average in reliability and comfort, every model—other than the 300 series—is overpriced and overrated. The 190 version is not recommended. Consider buying a 1997 C-class only if you feel you need the 2.3L engine's 12 additional horses and improved automatic gearbox. Keep in mind, though, that you'll have to keep the car much longer to amortize its higher cost. **Maintenance/Repair costs**: Higher than average, and most repairs must be done by a Mercedes dealer if you don't live in an area where independent shops have sprung up. **Parts**: Higher-than-average cost. Parts supply and servicing have become problematic now that the 190 series has been off the market for almost six years.

Strengths and weaknesses: Mercedes introduced the 190 "baby" Benz in 1984 in an effort to downsize its entry-level compact and make it more affordable. It never caught on due to its serious drivetrain deficiencies, cramped interior, and rounded styling (a real departure from Mercedes's traditional squared-off look). The best choice from a quality/price standpoint is any post-1989 version equipped with the in-line 6-cylinder powerplant.

The 1994 models were renamed the C-class, and gained interior room and more powerful engines. The standard 2.6L 6-cylinder motor is a real powerhouse in this small car, and its power is used effectively when coupled to the manual 5-speed transmission. The 4-speed automatic is a big disappointment—it requires a lot of throttle effort to downshift and prefers to start out in second gear. The 1994 versions add much-needed horsepower, but lack the manual 5-speed transmission that would set those extra horses free. Rear seat room is limited and lots of road noise intrude into the passenger compartment.

The base 2.3L engine is acceptable around town, but highway cruising requires more grunt to handle the car's heft and accessories. And speaking of grunt, the 2.8L engine ideally suited to these small cars. Although its power is used most effectively when coupled to the new 5-speed manual transmission, its performance with the 5-speed automatic is quite good. On 1996 models, some owners have complained that the 4-speed automatic requires a lot of throttle effort to downshift.

Keep in mind that owner surveys give the entry-level C-class cars a *just*-better-than-average rating, while the 300 and higher series have always scored way above average in owner satisfaction. The 190's reliability is a notch below that of other Mercedes; the 1994 model C-class is the better buy from a quality and reliability standpoint, although 1995 and later models will have fewer bugs. Nevertheless, owners report frequent problems with drivetrain noise and vibration, and slipping or soft shifts. Brakes, AC, and the electrical system, are also failure prone.

Dealer service bulletins: **1993 190, 1994 C220**—Drivetrain noise and vibration. • Harsh shifts or erratic shift quality. • High coolant temperature. • Rough reverse release. • Handling of fuel system complaints. **1995 C-class**—Drivetrain noise and vibration. • High coolant temperature. • Slipping or soft shifts.

Safety summary/Recalls: **1993**—Vehicle stalled while on the highway. • After vehicle was towed home, fire ignited in the engine compartment. • ABS brake failures cause the brakes to lock up. • Premature replacement of the front wheel bearings. • Climate control module failures. • Hard starting. • Driver's master door lock failed to unlock other doors. • Power door lock vacuum control pump failure. • Bent upper control arm/camber link caused right rear wheel to turn out. **1994 C280**—Vehicle suddenly accelerated forward in garage. • Another report that while coming to a stop at an intersection, vehicle will accelerate when foot is on the brake. • Airbag failed to deploy. • Upon braking, ABS pedal went to the floor, resulting in extended stopping distance. • Extreme pressure is needed to push horn button. • Cruise control failure. • Premature rear wheel bearing failure. • Windshield wiper is inadequate; poor visibility. **1995**—Check Engine light remains on for no apparent reason. • Frequent stalling due to defective idle speed control unit. • Missing left rear wheel well inside panel allows tire to spray debris into area. **1996**—Airbags failed to deploy. • Airbag light stays on for no apparent reason. • After parking the vehicle and turning off the ignition, the vehicle lurches forward or rocks backwards. • In another incident, vehicle was put in Park on an incline and keys were removed; vehicle rolled backwards down the hill. • Brakes are noisy when applied and don't brake well. • While car is being driven, engine light comes on for no apparent reason. • Complete electrical system failure. • Windshield washer fluid sensor failure. **1997**—While in Reverse, vehicle suddenly accelerated in reverse. **Recalls**: **1994–95**—Secondary hood latch may not work properly.

Secret Warranties/Service Tips

1989–92—Excessive oil consumption may be corrected by replacing the valve stem seals with upgraded Viton seals. **1990–91**—A jerking that occurs when driving downhill with the cruise control engaged can be corrected by installing a relay to disable the deceleration fuel shut-off switch. **All models/years**—Excessive engine valve train noise may be caused by a stretched timing chain. After 48,000 km, the camshaft and timing chain drive should be checked carefully, especially if excessive noise is heard.

190 Series, C-Class Profile

	1991	1992	1993	1994	1995	1996	1997	1998
Cost Price ($)								
190E 2.3	36,700	38,150	38,050	—	—	—	—	—
190E 2.6	43,500	45,250	45,050	—	—	—	—	—
C220	—	—	—	34,350	34,995	35,995	—	—
C230	—	—	—	—	—	—	36,950	37,550
C280	—	—	—	47,650	48,750	49,995	50,995	49,950
Used Values ($)								
190E 2.3 ↑	11,500	12,000	15,000	—	—	—	—	—
190E 2.3 ↓	10,000	10,500	13,500	—	—	—	—	—
190E 2.6 ↑	14,000	16,000	19,000	—	—	—	—	—
190E 2.6 ↓	13,000	14,500	17,500	—	—	—	—	—
C220 ↑	—	—	—	17,000	21,000	25,000	—	—
C220 ↓	—	—	—	15,000	18,000	23,100	—	—
C230 ↑	—	—	—	—	—	—	29,000	33,000
C230 ↓	—	—	—	—	—	—	27,500	32,000
C280 ↑	—	—		24,000	28,000	33,000	39,000	44,000
C280 ↓	—	—	—	22,500	27,000	31,000	37,000	42,500

Extended Warranty	Y	Y	Y	Y	Y	Y	Y	Y
Secret Warranty	N	N	N	N	N	N	N	N
Reliability	❷	❷	❷	③	③	④	④	⑤
Air conditioning	❷	❷	❷	③	③	④	④	⑤
Automatic transmission	④	④	⑤	⑤	⑤	⑤	⑤	⑤
Body integrity	❷	❷	❷	❷	③	③	③	③
Braking system	❷	❷	❷	❷	③	④	⑤	⑤
Electrical system	❷	❷	❷	❷	❷	③	③	③
Engines	❷	③	③	④	④	⑤	⑤	⑤
Exhaust/Converter	④	④	④	⑤	⑤	⑤	⑤	⑤
Fuel system	❷	③	③	③	④	⑤	⑤	⑤
Ignition system	④	④	④	④	④	⑤	⑤	⑤
Rust/Paint	④	④	④	④	④	④	⑤	⑤
Steering	③	③	④	④	⑤	⑤	⑤	⑤
Suspension	③	③	④	④	⑤	⑤	⑤	⑤
Crash Safety								
190	⑤	—	—	—	—	—	—	—
C220/230	—	—	—	❶	④	④	④	④
Side Impact								
C230	—	—	—	—	—	—	—	③

300 Series, 400 Series, 500 Series, E-Class

Rating: Recommended (1993–98); Above Average (1992); Below Average (1985–91). The 1994 and later models are referred to as E-class with the entry-level model a 300 diesel. **Maintenance/Repair costs**: Higher than average, and repairs must be done by a Mercedes dealer. **Parts**: Higher-than-average cost and limited availability.

Strengths and weaknesses: These cars are ideal mid-sized family sedans. They're reliable, depreciate slowly, and provide all the interior space that the pre-1994 190 series and C-class leave out. Their only shortcomings are a high resale value that discourages bargain hunters and a weak dealer network that limits parts distribution and drives up parts costs. The 300 series offers a traction control system that prevents wheel spin upon acceleration—somewhat like ABS in reverse.

Another interesting feature is a 24-valve 220-hp high-performance version of the in-line 6-cylinder engine that powers the 300 series. All this has its price, though. If, ironically, you'd like to drive one of these cars but are of an economical frame of mind, choose the 260E—it offers everything the 300 does, but for much less. The 300CE is a coupe version, appealing to a sportier crowd, while the 300TE is the station wagon variant.

Dealer service bulletins: Bulletins show that the 1995 300 series may have transmission gasket leaks, in addition to drivetrain noise and vibration, and slipping or soft shifts. The 1996–97 models have no bulletins listing factory-related problems or troubleshooting tips.

Safety summary/Recalls: **1994 300 series**—Sudden acceleration caused accident and injuries. • Defective ABS brakes; brakes locked up and caused driver to lose control of vehicle. • Emergency brake cable failed twice. • Vehicle rolls backwards when stopped on a hill. • Driver's seatbelt failed. • Accelerator pedal is too hard to press, causes fatigue in driver's leg. • Frequent power door lock and automatic antenna failures. • Front panels were cracked and loose. • Rear axle and fuel pump failures. **1995**—Sudden acceleration caused an accident. • Brake rotor and caliper failure. • Headlights cast dark shadows. • Windshield wipers fail to keep glass clean. • Automatic transmission slips or shifts erratically, as if hunting for the right gear. **1996**—Check Engine light flashes on and off at will. • AC fan starts and stops intermittently. • Door rattles. • Failure prone headlight bulbs. **Recalls**: **1992–95 300 series and 300E-class**—Front passenger's footrest could abrade the wiring harness underneath. The ensuing short circuit could stall the engine or deploy the airbag.

Secret Warranties/Service Tips

1987–90—A cruise control that surges or maintains a speed that's 4–6 km above the set speed may have a faulty amplifier or reference resistor gasket. **1987–92**—Excessive brake vibrations can be reduced by installing upgraded Jurid 226 front brake pads. **1988–90**—A gurgling heater core noise can be silenced by Mercedes's "gurgling kit." **1990–91**—Excessive exhaust noise between the exhaust manifold flange and rear muffler may be caused by a leak at the O_2 sensor or the clamped joints. If this isn't the cause, change the catalytic converter.

300 Series, 400 Series, 500 Series, E-Class Profile

	1991	1992	1993	1994	1995	1996	1997	1998
Cost Price ($)								
300E	64,700	67,275	68,050	55,995	—	—	—	—
300CE	76,800	79,800	80,800	—	—	—	—	—
300D	58,551	58,550	—	—	—	—	—	—
E300D	—	—	—	55,995	55,995	58,500	59,950	59,950
E320 4d	—	—	—	58,895	60,950	64,750	65,900	66,450
E420/430	—	—	—	71,000	72,950	101,900	73,300	73,950
E500/500S	—	—	—	108,00	124,900	132,500	132,950	117,900
Used Values ($)								
300E ↑	19,000	23,000	27,000	29,000	—	—	—	—
300E ↓	16,000	21,000	25,000	26,000	—	—	—	—
300CE ↑	22,000	27,000	32,000	—	—	—	—	—
300CE ↓	20,000	24,000	29,000	—	—	—	—	—
300D ↑	—	19,000	23,000	—	—	—	—	—
300D ↓	—	17,000	21,000	—	—	—	—	—
E300D ↑	—	—	—	31,000	33,000	38,000	43,000	49,000
E300D ↓			—	29,000	31,000	36,000	41,000	47,000
E320 4d ↑	—	—	—	31,000	36,000	42,000	47,000	55,000
E320 4d ↓	—	—	—	28,500	34,000	39,000	45,000	53,000
E420/430 ↑	—	—	—	34,000	40,000	65,000	58,000	66,000
E420/430 ↓	—	—	—	32,000	38,000	63,000	55,000	64,000

E500/500S ↑	—	—	—	53,000	65,000	77,000	90,000	105,000
E500/500S ↓	—	—	—	51,500	62,000	74,000	85,000	100,000
Extended Warranty	**N**	**N**	**N**	**N**	**N**	**N**	**N**	**N**
Secret Warranty	**N**	**N**	**N**	**N**	**N**	**N**	**N**	**N**
Reliability	③	③	③	③	④	④	⑤	⑤
Air conditioning	❷	③	③	③	③	④	⑤	⑤
Automatic transmission	③	③	③	④	④	⑤	⑤	⑤
Body integrity	③	④	⑤	⑤	⑤	⑤	⑤	⑤
Braking system	❷	③	③	③	③	③	③	③
Electrical system	❷	❷	❷	❷	❷	❷	③	③
Engines	❷	❷	❷	❷	③	④	⑤	⑤
Exhaust/Converter	③	④	④	④	⑤	⑤	⑤	⑤
Fuel system	③	③	④	④	⑤	⑤	⑤	⑤
Ignition system	③	③	④	⑤	⑤	⑤	⑤	⑤
Rust/Paint	④	④	④	④	④	④	⑤	⑤
Steering	③	③	③	③	④	④	④	⑤
Suspension	③	③	③	③	⑤	⑤	⑤	⑤

Note: The above model years haven't been crash-tested by NHTSA.

NISSAN

Maxima

Rating: Recommended (1996–98); Above Average (1989–95); Average (1986–88). The redesigned 1995–98 version offers a peppier engine, more-rounded styling, and a bit longer wheelbase. **Maintenance/Repair costs**: Higher than average, but repairs can be done practically anywhere. **Parts**: Higher-than-average cost, but easy to find.

Strengths and weaknesses: These front-wheel drive sedans are very well equipped and nicely finished, but cramped for their size. Although the trunk is spacious, only five passengers can travel in a pinch (in the literal sense). The 6-cylinder engine, borrowed from the 300ZX, offers sparkling performance; the fuel injectors, however, are problematic. Early Maximas are less expensive to buy, but more costly to maintain—for example, the exhaust manifold, a component that commonly fails, will set you back $300–$500 to replace. Owners report that the '95 Maxima's suspension was cheapened to the detriment of both the ride and handling.

Minor electrical and front suspension problems afflict early Maximas. Brakes and engine timing belts need frequent attention in all years. Newer models have a weak automatic transmission and the ignition system can malfunction. There have also been reports of "cooked" transmissions. This is due to a poorly designed transmission cooler. Mechanics say that this breakdown can be avoided by installing an externally mounted transmission cooler with a filter and replacing the transmission filter cooler at every oil change.

Owners report that the V6-equipped Maxima is sometimes hard to start in cold weather due to the engine's tendency to flood easily. The cruise control unit is another problematic component. When it's engaged at moderate speeds, it hesitates or "drifts" to a lower speed, acting as if the fuel line were clogged. It operates correctly only at much higher speeds than needed. Incidentally, owners say that a new fuel filter will *not* correct the problem. Additionally, though warped manifolds were once routinely replaced under a "goodwill" warranty, Nissan now makes the customer pay. The warpage causes a manifold bolt to break off, thereby causing a huge exhaust leak. Most fuel-injector malfunctions are caused by carbon clogging up the injectors; there are additives you can try that might reduce this buildup. There have also been internal problems with the coil windings on the fuel-injectors. Your best bet is to replace the entire set.

Nissan has had problems with weak window regulators for some time. If the window is frozen, don't open it. The rubber weather stripping around the window is also a problem. It cuts easily and causes the window to go off track, which in turn causes stress on the weak regulators. Driver-side window breakage is common and can cost up to $300 to repair. Costly aluminum wheels corrode quickly and are easily damaged by road hazards. There have been a few reports of surface rust and paint problems. Pre-1990 Maximas suffer from rust perforation on the sunroof, door bottoms, rear wheel wells, front edge of the hood, and bumper supports. The underbody should also be checked carefully for corrosion damage. Premature wearout of the muffler is a frequent problem; it's often covered by Nissan's "goodwill" warranty, wherein the company and dealer will contribute 50 percent of the replacement cost.

1995–98 models

These redesigned Maximas have a longer wheelbase (adding to interior room), a new 3.0L engine, and more rounded styling. They compete well with fully equipped Camrys, entry-level Infinitis, and Lexus models. Nevertheless, tall passengers will find the interior a bit cramped, and the automatic transmission is often slow to downshift and isn't always smooth. Quality control and overall reliability are apparently much better with these more recent iterations.

Dealer service bulletins: 1995—Hard cranking, no or low line pressure rise (control valve), and poor driveability (Code 45). • Clutch slippage. • AC compressor leaks or is noisy. • Troubleshooting excessive front and rear brake noise. • Rear brake, spare tire cover, and exhaust heat shield rattling. • Faulty centre console lid latch. • Wind noise around the front door. • Certain windshield cracks are covered under warranty. **1996**—Brake squeak or squeal. • Centre console lid latch replacement. • Hard starts or no-starts. • Timing chain noise.

Safety summary/Recalls: **1996**—Fire ignited from shorted wires under the passenger-side seat. • Airbag failed to deploy. • ABS failures. • Vehicle constantly pulls to the right. • Engine warning light flashes for no apparent reason. • Sudden loss of power resulting in inoperative brakes and steering. • Chronic stalling. • Power steering failure. • Erratic transmission performance. • Hard to shift transmission out of Park. • Key can be taken out of ignition while vehicle isn't in Park. • Power door locks failed. • Defective AC expansion valve. • Several reports of headlight explosions. **1997**—Door latch won't engage in cold weather. • Front wheel suddenly locked up while driving. • Power steering leaks fluid. • Trunk lid opened while driving. • Pedal went to the floor when brakes were applied, resulting in extended stopping distance. **1998**—Vehicle intermittently accelerates while braking. • Defective AC compressor. • Hazy, milky pattern on glass exterior causes poor visibility. • Frequent windshield wiper failures. **Recalls**: **1986**—Windshield may detach in a collision. **1992–93**—Dealers will install a new airbag sensor so that the airbags won't inadvertently deploy whenever the car passes over a speed bump. **1993–94**—Loose wheel nuts on aluminum wheels could allow wheels to fall away.

Secret Warranties/Service Tips

1985–91—Excessive brake noise can be corrected with upgraded front and rear pads, caliper pins, and baffle plates. **1989**—Insufficient heating may be caused by air bubbles trapped in the heater core. • Ice-induced heater motor failure requires a new blower (#27200-85E02). **1989–91**—Starting difficulties can often be traced to a connector that is not fully seated in the ECU. **1989–92**—Nissan has developed a variety of brake pads to respond to a number of customer complaints regarding excessive noise when braking. • Clutch shudder can be eliminated by installing an upgraded pressure plate release lever. **1995–96**—Timing chain rattling noise can be silenced by replacing the timing chain tensioner and slack guide. • Brake squeak or squeal can be corrected by installing front and rear brake kits. **All models/years**—Defective catalytic converters that cause a rotten-egg smell may be replaced free of charge under Nissan's emissions warranty. • Bulletin P195-006 looks at the many causes and remedies for excessive brake noise.

Maxima Profile

	1991	1992	1993	1994	1995	1996	1997	1998
Cost Price ($)								
Base	22,790	23,190	24,690	25,690	25,990	27,998	27,998	27,998
Used Values ($)								
Base ↑	8,000	9,000	11,000	14,000	16,000	19,000	21,000	23,000
Base ↓	7,000	8,500	10,000	13,000	15,000	17,000	20,000	22,000

Extended Warranty	N	N	N	N	N	N	N	N
Secret Warranty	N	N	N	N	N	N	N	N
Reliability	③	④	④	④	④	④	⑤	⑤
Air conditioning	③	④	④	⑤	⑤	⑤	⑤	⑤
Body integrity	❷	❷	❷	③	④	④	④	④
Braking system	❷	❷	❷	❷	③	③	④	④
Electrical system	❷	❷	❷	❷	③	③	④	④
Engines	③	④	④	④	⑤	⑤	⑤	⑤
Exhaust/Converter	❷	③	③	③	③	⑤	⑤	⑤
Fuel system	③	③	③	③	③	④	⑤	⑤
Ignition system	③	③	③	③	③	④	⑤	⑤
Manual transmission	❷	❷	③	③	③	⑤	⑤	⑤
- automatic	③	③	③	④	④	⑤	⑤	⑤
Rust/Paint	③	③	③	④	④	⑤	⑤	⑤
Steering	③	③	④	④	④	⑤	⑤	⑤
Suspension	③	③	④	④	④	⑤	⑤	⑤
Crash Safety	⑤	⑤	❷	❷	④	④	④	④
Side Impact	—	—	—	—	—	—	—	④

Note: The low crash safety ratings represent severe leg trauma.

SAAB

900, 9000

Rating: Average (1998); Below Average (1995–97); Not Recommended (1985–94). Interestingly, the upscale 9000 series isn't as crashworthy as the cheaper 900 versions, nor is it more reliable, exhibiting similar generic deficiencies as its entry-level brother. **Maintenance/Repair costs**: Higher than average, and repairs must be done by a GM/Saab dealer. **Parts**: Higher-than-average cost and limited availability.

Strengths and weaknesses: These Swedish-built luxury cars don't offer the refinement and ride comfort of most other cars in their class; the redesigned 1999 models will, however. They do combine excellent handling and great interior ergonomics (one of the few imports with the EPA's "large car" label), but without all the bells and whistles found in domestic luxury breeds. Convenience items like a fuse box in the glove box, a toolbox in the hatchback, and easy-to-replace bulbs, etc., add to your comfort.

Unfortunately, Saabs don't live up to the Swedish reputation for exceptional reliability and are quirky in design. Servicing is inadequate and will probably get much worse in the future. Parts are already costly and hard to find outside major urban areas. Take the tires, for example. Those for the 1992 9000 aren't a common size (205/50ZR16), and therefore command a not-so-common price—about $250 per tire depending on the manufacturer.

Generally, the 900 and 9000 series have similar deficiencies affecting the engine cooling—biodegradable water pumps and electrical systems, brakes, automatic transmission (clutch O-rings), and body hardware. The 9000 is assembled with greater care, but owner reports show only a marginal improvement in overall reliability and durability.

Short circuits are legion and run the gamut from minor annoyances to fire hazards (see "Safety summary/Recalls"). Electrical glitches in the Traction Control System's relay module give a false reading that the tires are spinning, which shuts the engine down.

Turbos produce much stronger acceleration and better handling than other 9000s without compromising their overall reliability. Nevertheless, they should be approached with caution because owner abuse or poor maintenance can quickly lead to turbocharger deterioration. Air conditioners and exhaust system parts have a short life span, and leaky seals and gaskets are common. Rust perforations tend to develop along door bottoms and the rocker panels. The underbody, especially the floor, should be inspected for corrosion damage on older models.

Dealer service bulletins: **900**: **1995**—Battery drain due to climate control system. • Excessive exhaust vibration. • Loose centre cover on rear belt member. • Rear floor duct noise. • Creaking noise in right hand A-pillar. • Evaporative loss system leaks. • Excessive clutch pedal freeplay. • Hard brake pedal when starting a cold engine. • Binding ignition switch contacts. • Inoperative blower motor. • Leaking sunroof or wet headliner. • Loosening seat upholstery. • Noise from rear speaker, secondary air injection pump, vacuum pump, dash, or front seat. • Pedal raising kit. • AM band radio interference. • Airbag warning light comes on. • Sulfur smell from exhaust. • Vibration caused by tires or rims. **1996**—Rear floor duct noise. • Excessive clutch pedal freeplay. • Inoperative blower motor. • Leaking sunroof or wet headliner. • Loosening seat upholstery. • Front seat or dash noise. • Pedal raising kit. • AM band radio interference. • Clutch plate rattling. • Sulfur smell from exhaust. **9000**: **1995**—Poor EDU illumination. • Loose dome light. • Evaporative loss system leaks. • Excessive fuel line pressure. • Failing fan control. • Secondary air injection pump noise. • Noise when shifting from second to third gear (automatic). • Rain or washer fluid running over rear window. • Dashboard squeaks and rattles. • Smell from climate control system. • Inoperative speedometer. • Airbag warning light comes on for no apparent reason. • Sulfur smell from exhaust. • Tire or rims causing excessive vibration. • Whistling noise from windshield. **1996**—Failing fan control. • Interior lights won't go out. • Noise when shifting from second to third gear (automatic). • Rattling noise from clutch plate. • Climate system smells. • Whistling noise from windshield. • Inoperative speedometer.

Safety summary/Recalls: 1996 900—Under-hood fire ignited as vehicle was idling with transmission lever in the Park position. • Several reports that when shifting from Reverse to Drive, vehicle suddenly accelerated forward without braking or steering control. • Airbags failed to deploy and the warning light comes on for no apparent reason • Fuel filter failures caused fuel leak. • Car drifts to the right while driving. • Defective exhaust check valve lifters. • Transmission failures and excessive vibrations. • Repeated gear shifting problems cause the shifter bushings and motor mount housing to be prematurely worn. • Sundry electrical system shorts and failures affecting the stereo and CD player, anti-theft system, and battery. • Frequent trunk latch release, door handle, and passenger seat adjuster failures. • Glass headlight cover is often broken. • Left sideview mirror can't be adjusted properly to see blind spots to left rear of vehicle. **1986–91 9000**—Reports of passenger compartment fires. The American government is investigating reports that Saab officials destroyed key documents that described the problem. A class action suit has been filed in Los Angeles, and a second suit has been filed in Atlanta, Georgia. **Recalls**: **1989 900**—Recall No. 274 provides for the free replacement of the fuel filter. • Front seats may not lock into position properly. **1994 hatchback 900**—Fatigue cracks could allow seat to suddenly fold backwards. **1985 Turbo 900**—Flexible fuel hose may leak. **1986**—Steering shaft could pull out of joint. **1986–87**—Fuel hose may rupture. **1987–88**—Wiring harness may chafe, creating a fire hazard. **1988**—Front lower control arm may fail. **1989**—Leaking fuel filter may create a fire hazard. • Heater fan resistor may overheat. **1993**—Front brakes are very vulnerable to salt and slush, which compromises braking. **1994–95**—Transmission may be in Neutral when shifter is in Reverse. • Car could roll away if it's parked with the parking brake disengaged. **1996**—Seatbelt anchorage may not hold. **1994–95 5d and Coupe with manual transmissions**—A shift linkage defect may cause an unexpected movement if transmission is in gear. **1995 with 2.3L engines**—Faulty electronic system. **1995 Saab 900 convertibles**—Possible loss of steering control due to a misaligned steering shaft. **1986 9000**—Wiring harness may short-circuit, creating a fire hazard. **1986–91**—Fire may be caused by a short circuit in the driver console. **1988**—Faulty cruise control may lead to unintended acceleration. • Recall No. 272 provides for the free replacement of the lower control arm attachment bolts. **1988–89**—Fluid leakage may compromise braking effectiveness. **1988–90**—Fire may erupt in the backup light circuit. **1989**—Leaking fuel filter may create a fire hazard. **1991**—Car may be started in gear. **1992–93**—Dealers will install a new fuel filler and fuel-filler vent hose to prevent fuel leakage. • Vehicles equipped with an engine oil cooler may catch fire in a collision. **1992–94**—ABS corrosion can lead to loss of full braking power. **1993–94**—Brake lights may operate erratically.

Secret Warranties/Service Tips

1988–91 900—Cold weather starting problems may be fixed with service kit #8819070. **1990–91**—A new valve and new O-rings will correct starting problems caused by a sticking fuel pump check valve. **1993–94**—Binding ignition switch contacts can lead to electrical failures. **1994**—A-pillar wind noise is addressed in DSB #08194-0486. • DSB #88/94-0480 lists the causes and remedies of AC malfunctions. **1992–94 9000**—A noisy climate control unit may have excess pressure building up at the fresh air intake. **1993–94**—A stuck shift lever may be caused by a blown #3 fuse. A faulty sun visor/vanity mirror causes the short circuit.

900, 9000 Profile

	1991	1992	1993	1994	1995	1996	1997	1998
Cost Price ($)								
900	24,965	24,965	26,715	27,000	26,995	28,500	29,901	30,700
9000	33,680	31,795	33,875	35,000	35,995	36,400	37,100	50,300
Used Values ($)								
900 ↑	6,000	5,500	7,500	11,000	13,000	16,000	20,500	25,000
900 ↓	4,500	5,000	6,500	10,000	11,500	14,500	19,000	23,500
9000 ↑	7,000	7,500	10,000	13,000	17,500	22,000	25,000	40,000
9000 ↓	6,500	7,000	8,500	11,000	16,000	21,000	23,000	38,000
Extended Warranty	Y	Y	Y	Y	Y	Y	Y	Y
Secret Warranty	N	N	N	N	N	N	N	N
Reliability	❷	❷	❷	❷	❷	❷	③	③
Air conditioning	❷	❷	❷	❷	③	③	③	③
Body integrity	❷	❷	❷	❷	③	③	④	④
Braking system	❷	❷	❷	❷	❷	❷	❷	③
Electrical system	❶	❶	❶	❷	❷	❷	❷	③
Engines	❷	❷	❷	③	③	③	④	④
Exhaust/Converter	❷	❷	❷	❷	③	③	④	④
Fuel system	③	③	③	③	③	③	③	③
Ignition system	❷	❷	③	③	③	③	③	③
Manual transmission	③	③	③	④	④	⑤	⑤	④
- automatic	❷	❷	❷	❷	❷	③	③	④
Rust/Paint	③	③	③	③	③	③	④	⑤
Steering	③	③	④	⑤	⑤	⑤	⑤	⑤
Suspension	❷	❷	③	③	③	③	④	④
Crash Safety								
900	—	—	—	—	④	④	④	④
9000	—	—	❷	—	—	—	—	—

TOYOTA

Avalon

Rating: Recommended (1995–98). A Camry knock-off; if you want a more driver-involved experience in a Toyota, consider a Lexus ES 300 or GS 300. **Maintenance/Repair costs**: Higher than average. Repairs must be done by a Toyota dealer. **Parts**: Higher-than-average cost and limited availability.

Strengths and weaknesses: This near-luxury four-door offers more value, interior space, and reliability than do other cars in its class that cost thousands of dollars more. A front-engine, front-drive mid-sized sedan based on a stretched Camry platform, the Avalon is bigger than the rear-drive Cressida it replaced and similar in size to the Ford Taurus. Sure, there's a fair amount of Camry in the Avalon, but it's quicker on its feet than the Camry, better attuned to abrupt manoeuvres, and two inches longer. In fact, there's more rear-seat leg room than you'll find in either the Taurus or the new Chevrolet Lumina. It's close to the Dodge Intrepid in this respect.

Quality control is above reproach. Owners have some performance gripes, however. They include numerous electrical system glitches, premature front brake repairs, power steering that's a bit too light, excessive body lean, and under-steer when cornering. Body construction and assembly are solid, although the first two model years exhibited more deficiencies than the last few years.

Dealer service bulletins: **1995–96**—Air conditioning odours. • AM band radio static on vehicles with an automatic antenna. • Front brake groan and rear brake moan. • Rear seat popping noise. • Wind noise from front door A-pillar area.

Safety summary/Recalls: **1995**—Researchers are looking into 20 incidents of turn signal failures after the hazard warning lights have been activated. **1997**—Brakes may fail due to freezing vacuum hose.

Secret Warranties/Service Tips

1995–96—Use an upgraded rear brake pad material to eliminate rear brake moan. • To reduce wind noise from the front door A-pillar area, consult DSB #B0010-97. **1996**—Use an upgraded brake pad material to eliminate brake groan noise. • Tips on reducing engine noise, front door wind noise, front suspension, and rear popping noise. • Upgraded hazard switch. **1997**—AC odour troubleshooting. • Fixing front suspension crunch.

Avalon Profile

	1995	1996	1997	1998
Cost Price ($)				
XL	31,058	33,368	33,718	34,688
XLS	34,458	35,778	36,188	37,868
Used Values ($)				
XL ↑	19,000	22,000	25,000	28,000
XL ↓	17,000	20,500	23,500	27,000
XLS ↑	20,500	24,000	27,000	30,500
XLS ↓	19,500	22,000	25,500	29,000
Extended Warranty	N	N	N	N
Secret Warranty	N	N	N	N
Reliability	⑤	⑤	⑤	⑤
Crash Safety	—	④	④	④
Side Impact	—	—	—	③

Cressida

Rating: Above Average (1985–92). The Cressida was dropped in 1992 and then replaced with the Avalon for the 1995 model year. **Maintenance/Repair costs**: Higher than average, but repairs can be done by an independent garage. **Parts**: Higher-than-average cost and limited availability.

Strengths and weaknesses: The Cressida ages well and offers an excellent combination of dependable, no-surprise, rear-drive performance, comfort, and luxury. There is little to find at fault when it comes to overall reliability, and the engine is a model of smooth power. Its only shortcomings: a bit less interior and trunk space than one would find with the Nissan Maxima or Acura Legend, inconvenient and confusing dash controls, and poor fuel economy.

The later models are more crashworthy, reliable, and trouble-free than earlier versions, but they're also much more expensive. Two complaints, however, continue to surface throughout the years: premature front brake wear and excessive brake pulsation/vibration. Toyota has issued a drawerful of bulletins to eliminate the problems, but they seem to reappear each year in some form or another. AC glitches and electrical short circuits are also commonplace. Exhaust system parts rust quickly.

Safety summary/Recalls: N/A.

Secret Warranties/Service Tips

1989—Speaker static when the power mirror is activated requires a noise filter in the power-mirror circuit. **All models/years**—Older Toyotas with stalling problems should have the engine checked for excessive carbon buildup on the valves before any more extensive repairs are authorized. • The brake pulsation/vibration problem is fully outlined and corrective measures are detailed in DSB #BR94-002, issued February 7, 1994.

Cressida Profile

	1985	1986	1987	1988	1989	1990	1991	1992
Cost Price ($)								
Base	19,998	23,228	25,968	27,538	31,888	32,708	31,158	35,238
Used Values ($)								
Base ↑	3,000	4,000	4,500	5,500	7,000	8,500	10,000	12,000
Base ↓	2,500	3,500	4,000	5,000	6,000	7,500	9,000	11,000
Extended Warranty	N	N	N	N	N	N	N	N
Secret Warranty	N	N	N	N	N	N	N	N
Reliability	❷	❷	❷	❷	③	③	③	③
Air conditioning	❶	❶	❶	❷	❷	❷	③	③
Body integrity	④	④	④	④	④	④	④	④
Braking system	❷	❷	❷	❷	❷	③	③	③
Electrical system	❷	❷	③	③	③	③	③	④
Engines	④	④	④	④	④	④	④	④
Exhaust/Converter	❷	❷	❷	❷	❷	❷	❷	③
Fuel system	④	④	④	④	④	④	④	④
Ignition system	③	③	③	③	③	③	③	④
Manual transmission	③	③	③	③	—	—	—	—
- automatic	④	④	④	④	④	④	④	④
Rust/Paint	④	④	④	④	④	④	④	④
Steering	③	③	③	③	③	④	④	④
Suspension	❶	❶	❶	❷	③	③	③	④
Crash Safety	❶	—	—	—	❶	—	⑤	⑤

VOLVO

700 Series

Rating: Not Recommended (1985–92). Actually, the 700 series has fewer defects than other cars, however, the components that do fail are hard to diagnose and also cost a great deal to correct. As these cars age, these problems become more severe and dealers become more reluctant to invest their time in fixing them. **Maintenance/Repair costs**: Higher than average. Repairs must be done by a Volvo dealer. **Parts**: Higher-than-average cost and limited availability.

Strengths and weaknesses: The 700 series is more spacious, luxurious, and crashworthy than the entry-level 240. Its standard engine and transmission perform well, but aren't as refined as the 850. Furthermore, the 700 series suffers from the same generic brake, electrical, engine cooling, air conditioning, and body problems as does its cheaper cousin. Brakes tend to squeak or grind, wear rapidly, and require expensive service. Exhaust systems usually need replacing after a few years.

Owners complain of hard cold starts on 1987–92 models. Power windows fail to operate as a result of either dirt getting into the mechanism or the wiring short-circuiting. Air conditioning units that emit a musty odour or fail to work properly when the car is idling are a common problem. Body and interior trim pieces are fragile—dashboard cracks often appear after the third year. The windshield wiper motor malfunctions after two years. There have been some complaints of sunroof rattles, premature surface rust (the paint chips easily), and rusted-out exhaust systems.

Safety summary/Recalls: **1986–91**—Front seat/centre console fires. **Recalls**: **1985**—Water pump pulley may be defective. **1985–87**—Possible engine wiring harness short circuit. **1985–88**—On vehicles with B230F engines, the driveshaft could separate from the transmission. **1988**—Headlight switch may short-circuit. **1989–90**—Fuel may seep from fuel tank. Cars are eligible for new fuel tanks. **1992**—Faulty front seatbelts could detach from anchors. • The seatbelt webbing guide may break. **1986 wagons**—Locked tailgate can be opened from inside the vehicle.

Secret Warranties/Service Tips

1988–90—Brake pulsation, a common problem, is addressed in DSB #51/111. • If you're having problems starting your Volvo on cold mornings, see DSB #23/135 and DSB #23/21A. **1989–90**—Service Campaign No. 54 calls for the free installation of a cable harness. **1989–92**—To improve cold starting, Volvo will install an improved fuel-injection control module gratis if the emissions warranty applies. Under another program, Volvo will replace the MFI E PROM to improve cold starting. **1990–91**—AM band radio interference will be stopped with kit #3533250-1. **1992**—Volvo Special Service Campaign No. 59 provides for the free replacement of AC pressure switches and harness, the Regina fuel control units, and Rex ignition control units. These repairs are to be carried out regardless of vehicle mileage or the number of previous owners. • A decrease in idling speed when the AC engages can be corrected by installing a capacitor kit. **All models/years**—Check the valve cover nuts at every servicing interval to prevent oil leakage.

700 Series Profile

	1986	1987	1988	1989	1990	1991	1992
Cost Price ($)							
740 GL	—	—	—	25,995	26,420	23,175	—
Used Values ($)							
740 GL ↑	—	—	—	4,500	5,500	6,500	—
740 GL ↓	—	—	—	4,000	5,000	6,000	—
Extended Warranty	Y	Y	Y	Y	Y	Y	Y
Secret Warranty	N	N	N	Y	Y	Y	Y

Reliability	③	③	③	③	③	③	④
Air conditioning	❶	❶	❶	❷	❷	③	④
Body integrity	❷	❷	❷	❷	❷	❷	③
Braking system	❶	❶	❶	❶	❶	❶	❷
Electrical system	❶	❶	❶	❶	❶	❶	❷
Exhaust/Converter	❶	❶	❶	❷	③	④	④
Fuel system	③	③	③	③	③	③	④
Ignition system	③	③	④	③	③	③	③
Manual transmission	③	③	③	③	③	③	④
- automatic	❷	❷	④	③	③	③	④
Rust/Paint	③	③	③	③	③	③	④
Steering	③	③	③	③	③	③	③
Suspension	③	③	③	③	③	③	④
Crash Safety	—	—	❶	—	—	⑤	⑤

850 Series

Rating: Recommended (1995–97); Above Average (1993–94). Surprisingly, for a car company that emphasizes its commitment to safe cars, the 850 series has quite a few safety-related defects reported by owners, including engine fires, loss of steering, sudden acceleration, transmission failures and tire blowouts. Don't waste your money on a 1997 850: the 1996 models are virtually identical to the more expensive 1997 versions and are a real bargain if the selling price has been reduced sufficiently. **Maintenance/Repair costs**: Higher than average. Repairs must be done by a Volvo dealer. The 1998 model 850s were redesignated the S70/V70. **Parts**: Costs are reasonable and parts are not hard to find.

Strengths and weaknesses: Bland, but practical to the extreme, with plenty of power, good handling, and lots of capacity. For 1997, the 850 GLT got a bit more lower-end torque, while the turbo version was upgraded with electrically adjusted front passenger seats and an in-dash CD player. The base 850 sedan uses a 2.4L 24-valve 168-hp 5-cylinder engine hooked to a front-drive powertrain. An all-wheel drive version is available in only Canada and Europe. Wagons use the same base powerplant, hooked to a 5-speed manual or optional 4-speed electronic automatic. GLTs have a torquier, turbo variant of the same powerplant that boosts horsepower to 190.

The "sports" sedan T5 is a rounder, sportier-looking Volvo that delivers honest, predictable performance but comes up a bit short on the "sport" side. Volvo's base turbo boosts horsepower to 222, but its new T-5R variant uses an upgraded turbocharger that boosts power to 240 horses—for up to seven seconds.

Passenger space, seating comfort, and trunk and cargo space are unmatched by the competition. Braking on dry and wet pavement is also exemplary. The ride of both the sedan and the wagon deteriorates progressively as the road gets rougher and passengers are added. Turbo versions are particularly stiff, and passengers are constantly bumped and thumped.

The 850 hasn't escaped the traditional AC, electrical system, and brake problems that afflict its predecessors. Additionally, owners have complained that the early models have uncomfortable seatbelts, insufficient rear travel for the front seats, and some body hardware deficiencies, resulting in excess noise invading the interior.

Dealer service bulletins: **1993–94**—Faulty engine accessory mounting brackets. • Electrical short circuits. • Water contamination of the accessory drive belt. • Poor radio reception on the FM band. **1995**—Faulty cruise control. • Climate control system odour. • Steering knock at full lock, rattle from tailgate lock, sunroof rattling, doorstop and interior fittings noise, and auxiliary belt noise. • Poor performance of the engine belt tensioner, and prematurely worn crankcase ventilation hoses. **1996**—AM band radio interference. • Cruise control won't engage. • Damage to tailgate by tailgate handle. • Generator noise in the loudspeakers. • High oil consumption. • Starting difficulties. • Improvements to ventilation and defrosting. • New shims to reduce rear resonance vibrations. • Remedies for noise from evaporative canister hose. • Repair tips for loose door stops in A and B posts. • Repair tips for roof panel unevenness. • Loose windshield upper trim strip.

Safety summary/Recalls: **1996**—A handful of reports that while vehicle was being driven, it suddenly lost all power and the engine compartment caught fire. • Inadvertent airbag deployment injured driver. • ABS brake failures. • Vehicle suddenly downshifts while cruising on the highway. • Transmission slipped out of Park, rolled down incline, and hit a house. • During rainy periods, the steering wheel locks up or the dashboard suddenly lights up. • Repeated tire blowouts. • Premature failure of the engine cooling fan and the evaporator pump fan relay for the exhaust manifold. Owners also cite cruise control, power steering pump, and front seatbelt failures. • Electrical glitches affect the speedometer, radio cassette player, and battery. • Broken driver side door hinge. **1997**—Pirelli (205/45-17) tires blew out; the design and size for this vehicle is inappropriate. • Vehicle drifts left when being driven, even after the steering mechanism was replaced. • Gas odour in the interior after driving a short distance. • Driver's seatbelt retractor locks up. **Recalls**: **1993–96**—Block heater could loosen and overheat, seriously damaging the engine. **1994**—Frozen throttle linkage could result in erratic engine operation. **1995**—The threaded insert that attaches the seatbelt catch to the front seat was incorrectly manufactured. • Some jacks may fail. **1996–97**—Throttle may not return to idle when foot is taken off the accelerator pedal.

Secret Warranties/Service Tips

1993—A jerking sensation while accelerating may be caused by electrical interference between the rpm sensor wiring and the secondary ignition system.

• Accessory drive belt noise due to water infiltrating into the system can be corrected by installing a special right front fender liner extension manufactured by Volvo to fix the problem. • Headlight wiper and washer motors may cause radio interference on the FM band. Eliminate this noise by installing suppressed wiper motors and a suppressor between the washer pump and the existing wiring. • Under Service Campaign No. 62, Volvo dealers will install at no charge an improved engine accessory mounting bracket. **1993–94**—Steering column spring noise can be silenced by using upgraded bolts to secure the upper bracket to the airbag retaining plate. **1993–96**—Tips on repairing roof panel unevenness. **1995**—If the cruise control won't engage, check the vacuum supply and vacuum supply pipe first. **1997**—Correcting high oil consumption. New shims will minimize low-speed braking vibrations. • Tips on correcting hard starting. • AC odour troubleshooting tips. • Measures for improved ventilation and defrosting. • Uneven panel above roof arches. • Engine may run too lean. **All models/years**—Check the valve cover nuts at every servicing interval to prevent oil leakage.

850 Series Profile

	1993	1994	1995	1996	1997
Cost Price ($)					
Base	29,950	29,095	29,995	31,995	32,995
Turbo	—	37,495	40,640	41,695	43,995
TLA/AWD	—	—	—	48,695	48,495
Used Values ($)					
Base ↑	16,000	16,000	17,500	19,500	24,000
Base ↓	15,000	14,500	16,000	18,000	22,500
Turbo ↑	—	20,000	24,000	27,000	29,000
Turbo ↓	—	19,000	23,000	26,000	28,000
TLA/AWD ↑	—	—	—	31,000	34,000
TLA/AWD ↓	—	—	—	30,500	33,500
Extended Warranty	Y	Y	Y	Y	Y
Secret Warranty	Y	N	N	N	N
Reliability	③	④	⑤	⑤	⑤
Air conditioning	❷	❷	③	③	④
Automatic transmission	❷	③	④	⑤	⑤
Body integrity	❷	❷	③	④	④
Braking system	③	③	③	③	③
Electrical system	❷	❷	❷	❷	③
Engines	⑤	⑤	⑤	⑤	⑤
Exhaust/Converter	⑤	⑤	⑤	⑤	⑤
Fuel system	④	④	④	⑤	⑤
Ignition system	③	③	④	⑤	⑤
Rust/Paint	③	④	⑤	⑤	⑤
Steering	⑤	⑤	⑤	⑤	⑤
Suspension	③	④	⑤	⑤	⑤
Crash Safety	❶	⑤	⑤	—	—

Note: The low crash ratings represent severe leg trauma.

900 Series

Rating: Recommended (1995–97); Above Average (1989–94). The 1998 model 900s were redesignated the S90/V90 and have apparently inherited similar brake and electrical deficiencies. **Maintenance/Repair costs**: Higher than average, and repairs must be done by a Volvo dealer. **Parts**: Higher-than-average cost and limited availability.

Strengths and weaknesses: Essentially repackaged 760s, these flagship rear-drive sedans and wagons have a much better reliability record than do the 240 and 700 series, and are on par with the 850 over the last five model years. Both the 940 and 960 offer exceptional roominess and comfort, and are capable of carrying six people with ease. The wagon provides lots of cargo space and manages to do it in great style. Some owner gripes: the base 114-hp 2.3L engine is overpowered by the car's weight; excessive fuel consumption with the turbo option; and excessive road and wind noise at highway speeds.

Dealer service bulletins: **1993–94**—Hard starting and stalling caused by low fuel volatility. • Excessive vibrations when idling on vehicles equipped with an automatic transmission. • Whistling from the bulkhead and wiper-well cover panel. **1995**—Evaporator odour treatment. • Front brake squealing. • Poor AC performance.

Safety summary/Recalls: These cars come with reinforced sides, which Volvo claims exceed federal regulations. **1986–91**—Front seat/centre console fires. **1991**—Sudden airbag deployment. **1994–95**—Airbags deploy for no reason. **Recalls**: **1989–90**—Fuel may seep from fuel tank. **1991 Turbo**—Throttle may jam. • Child car seat may not conform to federal safety standards. **1992–93**—Front seatbelts may detach from anchorage. • The seatbelt webbing guide may break. **1995**—Driver-side airbag may not deploy properly. **1996–97**—Throttle may not return to idle when foot is taken off the accelerator pedal. **1993 Turbo 944/945**—Erratic throttle operation.

Secret Warranties/Service Tips

1989–93—To improve cold starting, Volvo will install an improved fuel-injection control module free of charge if the emissions warranty applies. Under another program, Volvo will replace the MFI E PROM to improve cold starting. **1992**—Volvo Special Service Campaign No. 59 provides for the free replacement of AC pressure switches and harness, the Regina fuel control units, and Rex ignition control units. These repairs are to be carried out regardless of the vehicle mileage or the number of previous owners. **1992–94**—A decrease in idling speed when the AC engages can be corrected by installing a capacitor kit. **All models/years**—Check the valve cover nuts at every servicing interval to prevent oil leakage.

900 Series Profile

	1991	1992	1993	1994	1995	1996	1997
Cost Price ($)							
940 GLE	27,885	33,275	29,995	26,995	28,460	—	—
Wagon	28,565	33,900	30,995	27,995	29,490	—	—
960	—	40,986	43,045	35,495	41,300	46,400	47,400
Used Values ($)							
940 GLE ↑	9,000	10,000	11,500	12,000	16,000	—	—
940 GLE ↓	8,000	8,500	10,000	10,500	14,000	—	—
Wagon ↑	11,000	13,000	12,000	13,000	17,000	—	—
Wagon ↓	9,500	12,000	10,500	11,500	15,000	—	—
960 ↑	—	14,500	16,500	18,500	22,000	30,500	34,000
960 ↓	—	14,000	15,000	17,000	22,500	29,000	33,000
Extended Warranty	Y	Y	Y	Y	Y	N	N
Secret Warranty	Y	Y	Y	N	N	N	N
Reliability	③	③	④	⑤	⑤	⑤	⑤
Air conditioning	❷	❷	❷	③	④	⑤	⑤
Automatic transmission	❷	③	③	④	⑤	⑤	⑤
Body integrity	❷	❷	❷	❷	③	③	③
Braking system	❷	❷	❷	❷	❷	③	③
Electrical system	❷	❷	❷	❷	❷	③	③
Engines	④	④	④	④	⑤	⑤	⑤
Exhaust/Converter	⑤	⑤	⑤	⑤	⑤	⑤	⑤
Fuel system	④	④	⑤	⑤	⑤	⑤	⑤
Ignition system	③	④	④	⑤	⑤	⑤	⑤
Rust/Paint	⑤	⑤	⑤	⑤	⑤	⑤	⑤
Steering	⑤	⑤	⑤	⑤	⑤	⑤	⑤
Suspension	③	③	④	⑤	⑤	⑤	⑤
Crash Safety	—	—	—	—	—	④	—

SPORTS CARS

The average sports car model, like the Mazda Miata, should be able to go from 0 to 60 mph in under 10 seconds and top 80 mph at the end of a quarter mile. Meanwhile, luxury sports sedans, like the Infiniti Q45 and Lexus LS 400, have produced exceptional acceleration times of 60 mph in less than 8 seconds and have exceeded 90 mph after a quarter mile. But most sports cars, or "high-performance vehicles" as they're euphemistically named, don't offer the comfort or reliability of an Infiniti or Lexus. Instead, they sacrifice reliability, fuel economy, interior space, and comfortable suspension for speed, superior road handling, and attractive styling. They also need a whole slew of expensive high-performance packages, because many entry-level sports cars aren't very sporty in their basic form.

By carefully browsing through classified ads and dealer car lots, you should find many fully loaded offerings at a fraction of their original cost. Remember that models that have been taken off the market, like the Toyota Supra, Nissan 300ZX, and Chevrolet Corvette ZR1, aren't likely to become collectors' cars with soaring resale values. In fact, discontinued Japanese sports cars like the Nissan 1600 haven't done nearly as well as some of the British roadsters taken off the market at about the same time.

Recommended

Chrysler Stealth, R/T (1993–95)
Ford Mustang, Cobra (1996–98)
Ford Probe (1995–97)
GM Camaro, Firebird, IROC-Z Trans Am, Z28 (1996–98)
Honda Prelude (1993–98)
Mazda MX-3, Precidia (1994–96)
Mazda MX-5, Miata (1996–98)
Mazda RX-7 (1991–95)
Nissan 240SX (1995–98)
Toyota Celica (1995–98)
Toyota MR2 (1990–93)

Above Average

Chrysler Avenger, Sebring (1998)
Chrysler Laser, Talon, 4X4 (1997–98)
Chrysler Stealth, R/T (1992)
GM Corvette (1996–98)
Honda Prelude (1985–92)
Hyundai Scoupe (1995)
Mazda MX-3, Precidia (1992–93)
Mazda MX-5, Miata (1990–95)
Mazda RX-7 (1985–90)
Nissan 200SX (1995–98)
Nissan 240SX (1989–94)
Nissan 300ZX (1994–96)
Toyota Celica (1986–94)
Toyota MR2 (1988–92)
Toyota Supra (1995–97)

Average

Chrysler Avenger, Sebring (1995–97)
Chrysler Stealth, R/T (1991)
GM Corvette (1994–95)
Nissan 300ZX (1993)
Toyota MR2 (1986–87)

Ford Probe (1993–94)
GM Camaro, Firebird, IROC-Z Trans Am, Z28 (1994–95)
Toyota Supra (1994)

Below Average

Chrysler Daytona, Laser (1990–93)
Ford Mustang, Cobra (1980–95)
Ford Probe (1989–92)
GM Camaro, Firebird, IROC-Z Trans Am, Z28 (1992–93)
Hyundai Scoupe (1991–94)
Nissan 300ZX (1988–92)
Toyota Supra (1986–93)

Not Recommended

Chrysler Daytona, Laser (1984–89)
Chrysler Laser, Talon, 4X4 (1990–96)
GM Camaro, Firebird IROC-Z, Trans Am, Z28 (1982–91)
GM Corvette (1980–93)

CHRYSLER

Avenger, Sebring

Rating: Above Average (1998); Average (1995–97). The Avenger and its more luxuriously appointed Sebring twin have had fewer factory-related defects than other new Chrysler designs, probably because they're built by Mitsubishi at its Illinois plant. **Maintenance/Repair costs**: Higher than average. Repairs must be done by a Chrysler dealer. **Parts**: Higher-than-average cost and limited availability.

Strengths and weaknesses: Like the Eagle Talon and Mitsubishi Eclipse, the Avenger and Sebring are both surprisingly agile—which isn't actually that surprising considering they share many of the same components. The convertible, made in Mexico, is six inches longer than the Sebring coupe and is powered by a standard 2.0L twin cam, while the upscale JXi gets a performance injection with the 2.5L 6-cylinder powerplant.

Acceleration is fairly good with the base engine, but the optional V6 powerplant is the engine of choice to overcome the power-hungry automatic transmission. Engine, tire, and road noise at higher speeds can be disconcerting. Rear seat access can be a pain, literally, and rear seating may be uncomfortable for long trips.

Mitsubishi quality control has slipped badly as of late, with owners singling out the brakes, electrical system, ignition, and body construction as the areas most needing attention.

Dealer service bulletins: **1995**—Poor engine performance, hard starts, and stalling caused by a faulty powertrain control module. •

Sports Cars

Tendency to drift to the left requires upgraded compression lower arm assemblies. • Loss of air conditioning (corrected by installing a revised AC pressure transducer). • Headliner sagging. • Difficulty in closing the sunroof owing to a faulty lever mechanism. • Whistle from front windshield. **1996**—Troubleshooting excessive brake noise. • Door buzz and rattle. • Hard start, misfire, or rough idle. • Interior window film buildup. • No-start condition. • Poor driveability. • Engine compartment popping or knocking. • Power seat switch can stick. • Ratcheting sound when coming to a stop. • Rear shock noise. • Sag or hesitation on acceleration. • Shudders during upshift. • Sunroof inoperative or opens by itself. • Sunroof ratcheting sound or jerky operation; binding or coming off its track. • Speed control undershoot or overshoot. • Engine compartment ticking noise. • Transaxle shudder. • Vehicle drift or lead. Whistle from front windshield. **1997–98**—Delayed transaxle engagement. • Excessive cold-crank time, start die-out, or weak run-up. • Knocking noise in engine compartment/possible oil pump damage. • Light knocking noise from the rear shock area. • Squeaking or rubbing noise from the rear shock area. • Transaxle desensitization to intermittent faults. • Whistle from front windshield.

Safety summary/Recalls: **1995**—Front ball joints fail, suspension collapses. **1996–97**—Sebring convertible top flies off. • Small horn buttons on the steering wheel spokes may be hard to reach in an emergency. **Recalls**: **1996**—Faulty power mirror switch. • Power brake booster hose mislocated. **1997 Avenger, Sebring**—Front passenger head restraint support bracket may break. Dealers must replace the entire seatback assembly. **1996–98 Sebring convertible**—Faulty ignition switches, console shifter, and cables may cause the vehicle to roll away or render the ignition-park interlock system inoperative.

Secret Warranties/Service Tips

1995—One of the causes of premature brake wear, shudder, and noise is a maladjusted brake light switch. **1995–96**—Engine compartment ticking can be silenced by replacing the duty cycle purge solenoid with a quieter solenoid assembly. **1995–97**—Engine compartment popping or knocking may require an upgraded EGR valve. • Tips for reducing transaxle shudder are found in DSB #21-05-97. **1995–98**—Wind noise coming from the front windshield area is caused by wind lifting the windshield molding at the glass. Add sealant the full length of the windshield (see following illustration).

NO.: 23-49-98
GROUP: Body
DATE: Oct. 23, 1998
SUBJECT:
Whistle at "A"
Pillar Area
MODELS:
1995–**1998** (FJ) Avenger/Sebring
SYMPTOM/CONDITION:

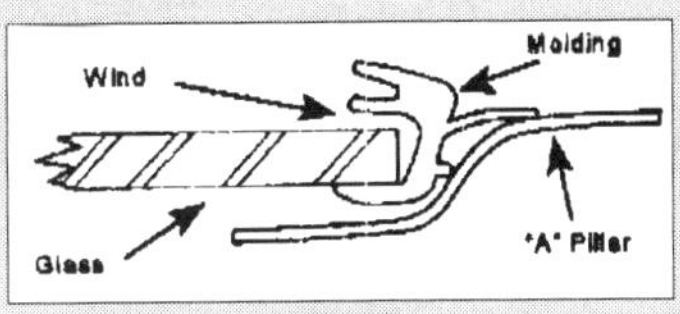

FIGURE 1

Some vehicles may exhibit a whistle (windnoise) coming from the "A" pillar area. This noise is more noticeable at highway speeds. The noise can be caused by wind lifting the windshield molding at the glass (Figure 1).
DIAGNOSIS:
Verify the noise is coming from the "A" pillar area by test driving the vehicle. Using tape, seal the **right or left** "A" pillar glass area the full length of the "A" pillar. If this noise is eliminated, perform the Repair Procedure.
PARTS REQUIRED
1 04318025 Silicone Rubber Adhesive Sealant
REPAIR PROCEDURE:
This bulletin involves filling the gap between the "A" pillar and molding the full length of the "A" pillar area.
1. Clean the "A" pillar and molding area of any foreign material with isopropyl alcohol.

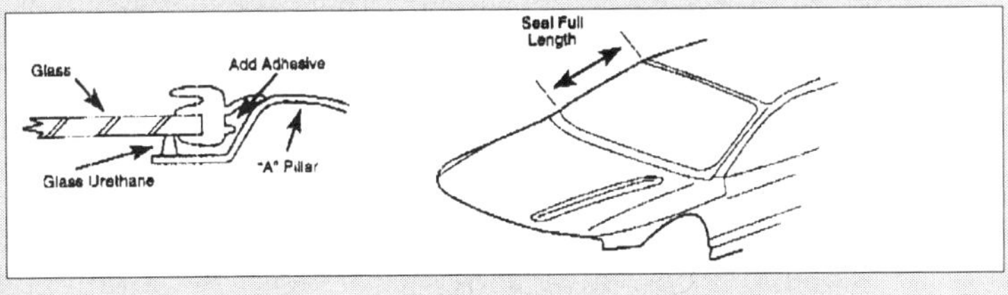

Chrysler estimates that this repair will take about 20 minutes.

Avenger, Sebring Profile

	1995	1996	1997	1998
Cost Price ($)				
Avenger	17,175	18,954	18,780	19,280
Avenger V6	21,275	22,544	23,820	24,320
Sebring	18,155	19,514	21,420	21,500
Sebring V6	24,025	25,538	24,135	24,360
Used Values ($)				
Avenger ↑	11,000	12,500	14,000	16,000
Avenger ↓	10,500	12,000	13,500	15,500
Avenger V6 ↑	13,000	14,000	15,000	17,500
Avenger V6 ↓	12,500	13,000	14,000	17,000
Sebring ↑	11,500	13,000	15,000	17,000
Sebring ↓	11,000	12,500	14,000	16,500
Sebring V6 ↑	14,000	15,500	16,000	19,000
Sebring V6 ↓	13,500	14,000	15,500	18,000

Extended Warranty	Y	Y	Y	Y
Secret Warranty	N	N	N	N
Reliability	③	③	④	④
Crash Safety				
Avenger	—	⑤	⑤	—
Sebring	—	—	⑤	—

Daytona, Laser

Rating: Below Average (1990–93); Not Recommended (1984–89). Sports car thrills combined with wallet-busting bills. **Maintenance/Repair costs**: Higher than average, but repairs can be done by any garage. **Parts**: Higher-than-average cost and limited availability.

Strengths and weaknesses: The high-performance Daytonas—Turbo Z, Pacifica, and Shelby Z—can run with the best of them for a little while, and then the service bills start piling up. Without the failure-prone turbocharged engine and sport suspension, these coupes provide mediocre handling and acceleration.

All engines are troublesome. If they're not maintained meticulously from the very beginning, the turbocharger is likely to fail around the 80,000 km mark and cause serious damage to the engine and your wallet. Fuel system problems are common on all versions, requiring the frequent replacement of electronic computer modules under the 5-year/80,000 km emissions warranty. The manual transmission has a sloppy shift linkage and a heavy clutch that doesn't stand up to hard use. Models loaded with electrical accessories have a higher failure rate than stripped-down versions. Electronic instrument panels and other electrical items are temperamental. The body is particularly poorly assembled, and water/wind leaks are common.

Safety summary/Recalls: **Recalls**: **1985**—Fatigue cracks could allow the seat to suddenly move backwards. **1987–89**—Exhaust system heat may melt carpet. Install free heat shield kit #4549356. **1989–90**—Leaking cylinder heads and gaskets will be replaced free of charge under Recall No. 467. **1990**—Airbag may be defective on cars with a gray interior. **1991**—Front disc brake caliper guide pin bolts may be too loose. **1992**—Coupling bolts on steering column shaft may be faulty. **1992–93**—Dealers will install additional bolts to better secure the dash panel. Inadequate spot welds attaching the front rails to the dash panel could cause structural damage, including door-opening interference and sheet metal cracking. **1985–87 Turbo**—Fire hazard caused by fuel supply hose leak.

Secret Warranties/Service Tips

1989—Leaking fuel-injectors are a common problem. • Corrosion of the oxygen sensor connector is a common problem. **1989–90**—Defective valve stem seals are the likely cause of high oil consumption with 2.2L and 2.5L engines (DSB #HL-49-89C). **1990–92**—Erratic idle speeds occurring after deceleration from a steady cruising speed can be corrected by replacing the idle air control motor with a revised motor. **1991–92**—Engines with a rough idle and stalling following a cold start may require a new single board engine controller (SBEC). **1992**—If the heater and ventilation system change to the defrost mode during acceleration, trailer towing, or hill climbing, the installation of a revised vacuum check valve should cure the problem. • Long crank times, a rough idle, and hesitation may be corrected by replacing the intake manifold assembly. • A leak in the oil filter area may be corrected by installing a special oil filter bracket gasket (#MD198554). **1992–93**—Some 41TE transaxles may produce a buzzing noise when shifted into Reverse. Replace the valve body assembly or valve body separator plate. **All models/years**—A rotten-egg odour coming from the exhaust is probably caused by a defective catalytic converter.

Daytona, Laser Profile

	1986	1987	1988	1989	1990	1991	1992	1993
Cost Price ($)								
Daytona	—	11,642	10,995	11,475	12,425	11,900	12,370	12,855
Turbo	13,940	—	—	—	—	—	—	—
Pacifica	—	15,930	16,385	—	—	—	—	—
Shelby Z	—	15,409	15,935	16,788	18,575	—	—	—
ES	—	—	—	13,155	14,250	13,675	14,200	14,745
IROC	—	—	—	—	—	15,585	16,150	16,735
Laser	—	—	—	—	14,300	13,000	13,735	14,145
Used Values ($)								
Daytona ↑	—	800	1,500	2,000	2,500	3,000	3,500	4,000
Daytona ↓	—	500	1,200	1,500	2,000	2,500	3,000	3,500
Turbo ↑	1,500	—	—	—	—	—	—	—
Turbo ↓	1,000	—	—	—	—	—	—	—
Pacifica ↑	—	1,000	1,500	—	—	—	—	—
Pacifica ↓	—	800	1,200	—	—	—	—	—
Shelby Z ↑	—	1,500	2,500	3,500	4,300	—	—	—
Shelby Z ↓	—	1,200	2,000	3,000	4,000	—	—	—
ES ↑	—	—	—	2,500	3,000	3,500	4,000	5,000
ES ↓	—	—	—	2,000	2,500	3,000	3,500	4,500
IROC ↑	—	—	—	—	—	4,500	5,500	7,000
IROC ↓	—	—	—	—	—	4,000	5,000	6,500
Laser ↑	—	—	—	—	3,000	4,000	5,000	6,000
Laser ↓	—	—	—	—	2,500	3,500	4,500	5,500
Extended Warranty	Y	Y	Y	Y	Y	Y	Y	Y
Secret Warranty	N	N	N	N	N	N	N	N
Reliability	❷	❷	❷	❷	③	③	③	③
Crash Safety								
Daytona	—	—	—	⑤	—	⑤	⑤	⑤

Laser, Talon, 4X4

Rating: Above Average (1997–98); Not Recommended (1990–96). The 1995 version offers fresh styling, dual airbags, and a more powerful engine; reliability, however, didn't improve until the 1997 model arrived. These cars have also elicited an unusually high number of safety-related complaints. **Maintenance/Repair costs**: Higher than average, but repairs can be done practically anywhere. Make sure the engine timing chain is inspected regularly—I've received many reports of owners having to pay huge repair bills for new engines. **Parts**: Limited availability. Dealers have had some trouble adequately servicing these high-tech vehicles, and CAA says that parts are a bit more expensive than other cars in this class.

Strengths and weaknesses: These Mitsubishi-made sporty cars combine high performance with a low price. The base 1.8L engine is adequate, and the suspension is comfortable although a bit soft. The optional 16-valve turbocharged 2.0L comes with a firmer suspension and gives more horsepower for the dollar than most other front-drive sports coupes, without much turbo lag. The 5-speed manual is the gearbox of choice. Torque steer makes the car appear to try to twist out of your hands when all 195 turbocharged horses are unleashed. The 4-speed automatic transmission cuts into the Laser's highway performance. Overall handling is impressive, with the 4X4 system giving sure-footed foul-weather stability. Keep in mind that the all-wheel drive (AWD) model has a smaller trunk area than the front-drive.

The 1997 Talon was dramatically restyled with wide air openings under the front bumper, new body cladding, and a redesigned rear end which includes a spoiler housing the centre-mounted stop light. A new, decontented (a fancy word that means it lacks many of the bells and whistles of the ESi, TSi, and TSi AWD) base model was also added. The TSi got 17-inch wheels and better high-performance tires.

The ESi comes with a 140-hp 2.0L engine (borrowed from the Neon Sport), while the TSi is powered by a 210-hp 2.0L Mitsubishi-bred engine. Two additional powerplants are available: a 2.0L double overhead cam with 135 hp, and a turbocharged version of the same engine rated at a sizzling 195 hp. A manual 5-speed is standard, and an optional 4-speed automatic is available on all models except the Turbo RS. AWD is offered on both 1996 car models, however only the 1997 TSi offers AWD.

All high-performance models cost thousands of dollars less than their Japanese competitors, without compromising performance. The 16-valve Talon, 4X4 is at the top of the trim list and provides five more horses than the turbocharged TSi.

Mitsubishi products usually have an above-average reliability record, as several decades of Colts have shown. Nevertheless,

1990–96 model Lasers and Talons have aged badly. Owners report chronic electrical shorts, transmission and engine failures, premature brake wear and excessive noise, and poor fit and finish. Other problems reported: Unstable idling and reduced rpm when the AC engages (requires an upgraded powertrain control module); cold-weather hard starts and stalling; cold-weather transmission shift delays (2–3 and 3–4) that take up to two minutes; transmission defaults into second gear (limp-in mode); vehicle tends to drift or lead to the right, requiring upgraded compression lower arm assemblies; speed control undershoot or overshoot; false theft alarm; centre exhaust pipe heat shield buzz; door buzz and rattle; maladjusted door glass, causing water leaks and wind noise; noisy clutch pedal; interior window film buildup; headliner sagging; power seat switch can stick; stress marks on the quarter trim panel; buzz or rattle from the rear quarter trim; inoperative, noisy, and jerky sunroof operation; faulty lever latch pin; and sunroof may open by itself.

Dealer service bulletins: **1994**—Turbocharger bolts may be poorly torqued (Chrysler will retorque them for free—DSB #JE-41-89). • Excessive engine noise caused by carbon buildup on the top of the pistons. • Excessive driveline vibration. • Rear brake squeak, and standard brakes often lock up or require long stopping distances. **1995**—Poor idling and reduced rpm when the AC engages (requires an upgraded powertrain control module). • Cold-weather hard starts and stalling, and cold-weather transmission shift delays (2–3 and 3–4) that take up to two minutes. • Vehicle tends to drift or lead to the right, requiring upgraded compression lower arm assemblies. • Centre exhaust pipe heat shield buzz and noisy clutch pedal. • Headliner sagging. • Stress marks on the quarter trim panel. • Faulty sunroof lever latch pin. **1996**—No-start due to faulty Neutral safety switch. • Hard start, misfire, or rough idle. • Erratic idle. • Single cylinder misfire. • Driveability improvements. • Shuddering during upshifts or when torque converter is engaged. • Reduced limp-in default sensitivity. • Difficulty going into second gear or Reverse after a cold start. • Speed control overshoots or undershoots. • Vehicle drifts or leads at high speeds. • Excessive brake noise. • Intermittent theft alarm activation. • Power seat switch can stick. • Ratcheting sound when coming to a stop. • Sunroof is inoperative or opens by itself. • Sunroof operation produces a ratcheting sound or is jerky. • Sunroof or sunshade binds or comes off its track. • Door buzz and rattle. • Rear quarter trim buzz or rattle. • Engine compartment ticking noise. • Interior window film buildup. **1997**—Excessive brake noise. • Inoperative CD player. • Poor driveability. • Transmission shudder. • Loose side body mouldings.

Safety summary/Recalls: Standard brakes often lock up or require long stopping distances. • Three deficiencies that compromise comfort and safety on many of these cars are the absence of airbags and

ABS on the Laser; a small, shallow trunk with a high sill; and head restraints that block rear visibility. **1995**—Parked vehicle caught on fire in the engine compartment. • Fuel tank leaks around the sending unit. • Airbags failed to deploy. • Sudden acceleration when engine suddenly dropped down while vehicle was going around a curve at slow speed; caused an injury and one death. • Many reports of sudden acceleration due to a stuck throttle control cable. • Gas pedal stuck several times, causing transmission failure due to car stopped in gear. • When going into gear, the rear wheel separated from the vehicle. • The left ball joint failed while driving, causing the front tire to come off. • Transfer case will suddenly lock up (three reports). • Rear wheels and rear brakes suddenly lock up while driving. • Transmission will downshift on its own, without shifter lever being moved. • Power-steering failure due to a bent crank pulley, allowing the belt to fall off. • Many reports that the Check Engine light comes on for no apparent reason. • Many other reports concerning crank sensor, timing belt, and engine cylinder head failures. • Frequent brake failures. • Excessive braking noise, and premature replacement of the front brake rotors and pads required. • Parking brake failed to hold vehicle on a slight incline. • Frequent reports of transfer case fluid leakage and transmission slipping. • Windows leak and are often inoperative (fuse failure?). • Electric sunroof rattles and won't close properly. • Inoperative door locks and door latch failures. • When brakes are applied, driver's seat moves forward. • Frequent battery replacement. • AC compressor failures. • Front passenger-side seatbelt ratchets too tight, or the seatbelt retractor fails to retract. • Horn will self-activate when vehicle is parked for several hours; at other times, it won't work at all. • Gas tank failed shortly after recall repair. • Fuel gauge is often inaccurate or inoperative. **1996**—Fuel tank leaks around the sending unit. • Smell of fuel vapours in the passenger compartment. • While being driven, the vehicle dropped down on the frame due to axle shaft failure. • Front right side tire fell off while driving. • Both sides of the airbag module came apart. • ABS brake warning light comes on for no apparent reason. Dealer says cause is most probably corroded wires and ABS pump. • Engine head gasket had to be replaced. • In another report, owner says carbon monoxide entered the passenger compartment just after the head gasket was replaced. • Many reports that the transfer case leaks fluid. • Sudden electrical system failure caused rear brake lights to burn out. • Entire lighting system failed three times due to water leaking into vehicle and creating an electric short. • Design of rear seats prohibits the use of a child safety seat. Windows are all constantly fogged up. • Frequent radio malfunctions. **Recalls**: **1990**—Sunroof glass may detach from roof. • Poor windshield retention during a collision. **1990 Laser**—Headlight wiring harness may short, causing the lights to fail. **1990 Talon 4X4**—Early production oxygen sensors that can't withstand the turbo engine's high temperatures will be replaced for free under an emissions recall. **1990–91**—Takata seatbelt replacement. **1995–96**—Incorrectly installed fuel gauge/pump gaskets may cause a fire. Requires a new fuel tank. **1997 Talon, 4X4** —Front passenger head restraint support bracket may break. Dealers must replace the entire seatback assembly.

Secret Warranties/Service Tips

1990–91—A loose rear-quarter trim panel may require new clips. **1990–94**—Driveline vibrations on smooth roads can be eliminated by installing upgraded engine and transmission mounting brackets. **1991–92**—Tappet/lash adjuster noise is a common problem that's covered by DSB #09-53-91. **1992–94**—Rear brake squeaks can be silenced with a Mitsubishi shim kit. **1997**—Tips available on repairing loose side body mouldings. **1990 Laser**—To correct transmission case gasket leaks, use gasket kit #MD730803. **1990 Talon**—Turbocharger bolts may be poorly torqued. • Warping of the headliner moulding can be fixed by installing an improved moulding. • If the headlights won't retract when the switch is turned off, replace the passing control relay, heat shrink tube, and tie straps. • Hard shifting/gear clash can be prevented by installing a modified 1–2 synchronizer sleeve, spring, and 3–4 spring. **All models/years**—A rotten-egg odour coming from the exhaust may be the result of a malfunctioning catalytic converter, which may be covered by the emissions warranty.

Laser, Talon, 4X4 Profile

	1991	1992	1993	1994	1995	1996	1997	1998
Cost Price ($)								
Laser	13,000	13,735	14,145	14,145	—	—	—	—
Turbo RS	14,900	15,820	16,310	16,310	—	—	—	—
Talon	15,505	16,205	14,475	16,000	19,425	20,695	19,885	20,290
Talon TSI	18,100	19,365	19,365	19,975	28,360	29,985	30,225	30,225
Used Values ($)								
Laser ↑	4,000	4,500	6,000	7,500	—	—	—	—
Laser ↓	3,500	4,000	5,500	6,500	—	—	—	—
Turbo RS ↑	5,000	5,500	8,000	9,000	—	—	—	—
Turbo RS ↓	4,500	5,000	7,500	8,500	—	—	—	—
Talon ↑	4,500	5,000	5,500	7,000	10,500	12,500	13,500	17,000
Talon ↓	4,000	4,500	5,000	6,500	9,500	11,500	12,500	16,000
Talon TSI ↑	6,000	7,500	9,000	10,500	16,000	18,500	22,500	25,000
Talon TSI ↓	5,000	7,000	8,500	10,000	15,000	17,500	21,500	24,000
Extended Warranty	**Y**	**Y**	**Y**	**Y**	**Y**	**Y**	**N**	**N**
Secret Warranty	**N**	**N**	**N**	**N**	**N**	**N**	**N**	**N**
Reliability	❷	❷	❷	❷	❷	❷	④	⑤
Air conditioning	③	③	④	⑤	⑤	⑤	④	④
Body integrity	❷	❷	❷	❷	❷	❷	③	③
Braking system	❷	❷	❷	❷	❷	❷	③	④
Electrical system	❷	❷	❷	❷	❷	❷	❷	③
Engines	❶	❷	❷	❷	❷	❷	③	④
Exhaust/Converter	③	③	③	④	⑤	⑤	⑤	⑤
Fuel system	❶	❷	❷	③	③	③	④	④

Ignition system	③	❷	❷	③	③	④	⑤	⑤
Manual transmission	③	③	③	④	④	④	⑤	⑤
- automatic	❷	❷	❷	❷	❷	❷	③	③
Rust/Paint	❷	③	③	③	③	③	③	④
Steering	③	③	③	④	④	④	④	⑤
Suspension	③	③	③	③	④	④	④	④
Crash Safety	—	—	—	—	—	—	④	—
Side Impact	—	—	—	—	—	—	—	❶

Stealth, R/T

Rating: Recommended (1993–95); Above Average (1992); Average (1991). The Stealth is a serious, reasonably priced sports car that's as much go as show. Although 1995 was the Stealth's last model year in Canada, a 1996 model was carried over in the United States. It's still sold in the States as the Mitsubishi 3000GT. **Maintenance/Repair costs**: Average. Repairs must be done by a Chrysler or Mitsubishi dealer. **Parts**: Higher-than-average cost and limited availability. The small engine compartment means that some of the simplest jobs will require special tools and take an inordinate amount of time to complete.

Strengths and weaknesses: An impressive highway performer with a good reliability record. The base V6 engine accelerates well and provides more than enough power for all driving conditions. The awesome twin turbo power on the R/T Turbo rivals the engine performance of sports cars selling for far more. Power steering is crisp and predictable and provides just the right amount of road feel. Standard disc brakes work very well, but the ABS is particularly impressive in stopping the car in a short distance without any loss of steering stability or fading after repeated application. Where the Stealth disappoints is in its limited driver head room (a Ford Probe problem, too), confusing and hard-to-find interior instrumentation and controls, hard-to-service engine compartment, and mediocre quality control when it comes to body hardware, fit, and finish. The Stealth isn't a car for short drivers, since they may have trouble seeing over the hood or reaching the clutch pedal. When the seat is moved forward, short drivers may be unable to see the overhead lights. Rear seating is very cramped—the norm with most sports cars. The dash control that adjusts the exhaust system sound is more gimmick than innovation. The Turbo's excessive weight taxes fuel economy.

Dealer service bulletins and a small number of owner complaints show that the first-year (1991) Stealth had an unusually large number of factory defects that were corrected in the second year. For example, owners report frequent gearbox and fit problems, including drivetrain noise, grinding noise when shifting, gear clash, hard shifting into all gears, difficulty shifting into Reverse, door glass rattling, excessive wind noise along door glass, water leaks, and sluggish window operation. Other problems reported by owners of later models and confirmed by dealer service bulletins: cruise control failure due to improper wiring connection, noisy engine lash adjuster, power transfer unit and viscous coupling failure, 1–2 gear clash, faulty exhaust manifold nuts, noisy steering column, rear suspension tapping noise, and faulty turn signals.

Safety summary/Recalls: **1992–93**—Sunroof glass shattering. **Recalls**: **1991**—Takata seatbelt replacement. **1991 R/T Turbo**—Oil leaking from the AWD transfer case may cause bearing damage/failure. **1991–94**—Front-wheel brake hoses may crack.

Secret Warranties/Service Tips

1991–94—Rear cargo cover rattling or failure to stay in the holder requires a new clip. **1992**—Vehicles that won't start may have corroded wiring in the A-67 12-connector. **1994**—Door glass weather stripping may pull out, requiring the installation of upgraded weather stripping. **All models/years**—A rotten-egg odour coming from the exhaust is probably the result of a malfunctioning catalytic converter, which may be covered by the emissions warranty.

Stealth, R/T Profile

	1991	1992	1993	1994	1995
Cost Price ($)					
Base	20,458	21,908	—	—	—
ES	22,204	25,435	26,695	—	—
R/T	32,080	33,176	36,115	30,965	35,345
R/T Turbo	42,235	38,820	42,235	48,015	49,405
Used Values ($)					
Base ↑	7,000	8,500	—	—	—
Base ↓	6,000	8,000	—	—	—
ES ↑	8,500	10,000	11,000	—	—
ES ↓	7,500	9,000	10,500	—	—
R/T ↑	10,500	12,000	16,000	18,500	22,000
R/T ↓	9,500	11,000	14,500	17,000	21,000
R/T Turbo ↑	14,000	15,000	19,000	24,500	29,000
R/T Turbo ↓	13,000	14,000	17,500	22,500	27,500
Extended Warranty	N	N	N	N	N
Secret Warranty	Y	Y	Y	Y	Y
Reliability	③	③	④	⑤	⑤
Air conditioning	③	⑤	⑤	⑤	⑤
Body integrity	❷	❷	❷	❷	④
Braking system	③	③	❷	③	④
Electrical system	❷	❷	❷	❷	③
Engines	⑤	⑤	④	④	⑤
Exhaust/Converter	⑤	⑤	⑤	⑤	⑤
Fuel system	⑤	⑤	⑤	⑤	⑤

Ignition system	❷	③	④	⑤	⑤
Manual transmission	③	③	③	③	⑤
- automatic	③	③	③	③	⑤
Rust/Paint	③	③	③	③	③
Steering	⑤	⑤	⑤	⑤	⑤
Suspension	③	③	③	④	⑤
Crash Safety	—	—	❶	—	—

FORD

Mustang, Cobra

Rating: Recommended (1996–98); Below Average (1980–95). Unfortunately, Mustangs don't perform well in rough weather and they have had a frighteningly high number of safety-related mechanical failures. GM's Camaro and Firebird are the Mustang's traditional competition as far as performance is concerned. Ford has the price advantage, with a base Mustang costing a bit less than the cheapest Camaro, but it lags from a performance standpoint—10 horses with the V6 and 60 horses with the V8. The GM models also offer more sure-footed acceleration, crisper handling, standard ABS, a 6-speed transmission, and more comfortable rear seats. Don't waste your money on a 1997 Mustang if a 1996 version, in good condition, is available. The 1996 models were substantially improved and can be a real bargain if the price is right. All 4-cylinder versions should be shunned. **Maintenance/Repair costs**: Average, particularly because repairs can be done anywhere. **Parts**: Average cost, and parts are often sold for much less through independent suppliers.

Strengths and weaknesses: As stated earlier, rear-wheel drive Mustangs remain popular because they offer sporty styling and high-performance thrills, usually for less money than GM's Camaro and Firebird, the Mustang's main domestic rivals. The 4-cylinder engine, used through 1993, isn't just failure-prone—it also doesn't carry half the horses as the 5.0L V8 and has no redeeming qualities. A limited-edition Cobra, sporting a 245-hp V8 debuted in 1993, but few were sold in Canada. Unfortunately, the 1993 Mustang fell behind the GM competition when the Camaro and Firebird were radically restyled that year, gaining additional safety features, a more rigid, dent-resistant body, better body fit and finish, and a more powerful base engine.

1994–98 models

Ford fought back with its own redesign of its 1994 model, replacing the 4-banger with a V6, adding four-wheel disc brakes, making the chassis more rigid (especially the convertible version), and dropping the hatchback. Mustangs now carry a base 3.8L V6 and an optional 4.6L V8. In addition, the high-performance, limited-edition Cobra variation delivers 90 more horses than what the stock 4.6L V8 offers. The single and twin cam V8 options make the Mustang a powerful—if a bit unsophisticated—street machine. V6 models are an acceptable compromise, even though the engines fail to deliver the gobs of power most performance enthusiasts expect from a Mustang and, up to the 1995 model year, they have a propensity for blowing up their head gaskets.

Base models come equipped with a host of luxury and convenience items, which can be a real bargain once the base price has sufficiently depreciated, say, after the first three or four years. Off-lease models are particularly good buys these days.

This is definitely not a family car. For example, a light rear end makes the car dangerously unstable on wet roads or when cornering at high speeds. But for those who want a sturdy and stylish second car, or who don't need room in the back or standard ABS, the 1996–98 Mustang is a pretty good sports car buy. And if GM carries out its threat to drop its rear-drive Camaro and Firebird after the year 2000, the Mustang will be the main alternative for rear-drive sports car enthusiasts.

Like Camaros, Mustangs have never been very reliable cars. The first Mustang, launched in 1964 and now worth more than $25,000 (U.S.), had serious rusting, electrical, and suspension problems. And guess what? Thirty-five years later, Mustangs still have electrical systems and electronic modules that are constantly breaking down, transmissions that jump from Park to Reverse, and a base suspension and front brakes that wear out in the blink of an eye.

The less said about the infamous 2.3L 4-cylinder engine, the better. The V6 is also failure-prone (head gaskets, again), leaving the V8 engine with a definite performance and reliability edge. Turbocharged models aren't recommended because of their frequent and expensive mechanical breakdowns. If you want high-performance action, you'll have to pay a premium—and be prepared for some monstrous repair bills and white-knuckle acceleration on wet roadways. Sport trim models feature an upgraded suspension and wheel package that improves handling considerably.

Keep in mind that the electronic modules that govern engine and transmission performance are often on the fritz, producing chronic hard starts, stalling, and overall poor city and highway performance. Furthermore, the 3.8L 6-cylinder engine has begun to tally up a record number of head gasket failures around the 150,000 km mark. Other problem areas: the front brakes, fuel pumps and front suspension remain consistent weak spots, and MacPherson struts and various steering components are likely to wear out before their time. The parking brake cable also seizes easily. The EEC IV engine computer can be temperamental, and electrical problems are common. Assembly quality is still not on par with Japanese vehicles.

Dealer service bulletins: **1995**—Misaligned or worn accessory drive belt. • Delayed transmission engagement and shift errors, intermittent loss of torque at 3–4 upshift, and shifts to Neutral during heavy throttle. • Insufficient AC cooling or excessive clutch end gap. • Rear brake squeak, brake roughness upon application, and a thumping/clacking sound when braking. • Water intrusion of the MLP/TR sensor. • Faulty rear view mirror, loose rocker panel mouldings, a temperature control knob that doesn't reach full cool position, and a malfunctioning temperature gauge. • Fuel pump buzz/whine heard through the radio speaker in addition to poor radio reception. **1996–97**—Stalling or hard starts. • Noisy 4.6L engine. • Transmission valve body cross leaks. • Erratic or prolonged 1–2 shift. • Squeal or hoot noise from engine compartment. Accessory belt damaged or comes off. • Coolant leaks from upper radiator hose. • Chatter during turns. • Front suspension rattle. • Fuel pump buzz/whine is still heard through the radio speaker. • Musty and mildew-type odours. • Blower motor noisy or inoperative in cold weather. • Convertible top leaks. • Sticking or clicking odometer/trip meter. • Overheating; AC shuts off. • Hard starting, long crank, stalling. • Loose catalyst or muffler heat shields. • Fog/film on windshield/interior glass. **1998**—Loose rocker panel moldings. • Excessive clutch vibrations. • Buzz or rattle from the exhaust system. • Musty odours from the AC. • Radio speaker whine or buzzing.

Safety summary/Recalls: Regularly equipped Mustangs, like most rear-drive Fords, don't handle sharp curves very well. The rear end swings out suddenly, and the car tends to spin uncontrollably. Furthermore, the car loses traction easily on wet roads and braking is barely adequate. **1995**—Fuel line failure caused several fires. • Front fan belt caught on fire with no prior warning. • Several reports of sudden acceleration due to a stuck throttle. • Airbag failed to deploy. • Airbag deployment caused severe injuries. • Inadvertent airbag deployment. • Several reports that when driver applied the brakes, the airbag deployed. • ABS brake failures, warped rotors, and frequent pad replacement. • Engine head gasket failures. • Power-steering fails, steering wheel is off-centre, or steering wheel locks up. • Chronic stalling thought to be caused by defective fuel pump. • Transmission failure while driving caused accident. • Torque converter failure causes transmission to slip. • Many reports of vehicle jumping from Park to Reverse and rolling away. • Rear axle broke during normal driving conditions. • Front-end alignment doesn't hold, causing premature wear of steering and tie-rods. • On one occasion, tie-rods broke and caused vehicle to go out of control, resulting in several fatalities. • Front struts suddenly collapsed. • Weak rear struts allow vehicle to "bottom out." • Front stabilizer bar rusted and cracked. • Driver's door hinge broke. • Fuel gauge inoperative or gives an inaccurate reading. • Gas tank too small and requires frequent fillups. • Trunk leaks cause premature rusting. • Fog lights fill with water and blow their bulbs. • Frequent AC failures. • Seatbelt failed to restrain driver. **Recalls**: **1984–85**—Front seatbelt buckles may break and separate from the belt webbing. **1985**—Defective ignition module. • Faulty plastic sleeve in front seatbelt tongue assembly. • Power brake booster could come apart. **1986–87**—Fuel line coupling may leak fuel. **1987**—Faulty fuel-injection tube assemblies may cause fuel leakage and create a fire hazard. **1988–93**—Faulty ignition module may pose a fire hazard. **1991**—Vehicle could roll away with the shift lever in Park position. **1994–96**—Hood may be defective. **1995**—Defective tie-rod ends could cause an excessive shake or shimmy, resulting in an accident. **1998**—Cruise control may stick. • Steering may fail. • Fuel leak may cause a fire. **1995 GT**—Front seat cushion supports could abrade wiring harness, posing a fire hazard.

Secret Warranties/Service Tips

1980–88—Inoperative power door locks need an upgraded retainer clip assembly (#E8AZ-5421952-A). **1982–86**—Wind noise and water leaks with T-roofs and convertibles are usually caused by misaligned doors. Install door alignment kit #E72Z-6123042-A or -B. **1982–90**—An unusual engine metal-to-metal noise may be caused by the flexing of the torque converter. Install six new flywheel bolts with reduced head height to provide additional clearance. **1985–97**—A buzz or rattle from the exhaust system may be caused by a loose heat shield catalyst. **1986–94**—The in-tank fuel pump is the likely cause of all that radio static you hear. Stop the noise by installing an electronic noise RFI filter (#F1PZ-18B925-A). **1987–90**—Excessive oil consumption is likely caused by leaking gaskets, poor sealing of the lower intake manifold, defective intake and exhaust valve stem seals, or worn piston rings. Install new guide-mounted valve stem seals for a more positive fit and new piston rings with improved oil control. **1988–92**—Cold hesitation when accelerating, rough idle, long crank times, and stalling may all signal the need to clean out excessive intake valve deposits. These problems also may result from the use of fuels that have low volatility, such as high octane, premium blends. **1990**—Excessive transmission noise, delayed shifts, or no engagements may be due to thrust washer metal particles that have plugged the filter or burnt out the clutch plates. **1990–93**—Noise heard from the power-steering pump may be caused by air in the system. **1991**—A rough idle or lean fuel flow may require the installation of deposit-resistant injectors. **1994**—Automatic transmissions with delayed or no forward engagement, or a higher engine rpm than expected when coming to a stop, are covered in DSB #94-26-9. • A cracked cowl top vent grille should be replaced with an upgraded version. • A driveline boom can be silenced by replacing the rear upper control arms. • A noisy fuel pump needs to be replaced by an improved "guided check valve" fuel pump. • Models with laser red paint may have serious paint decay problems requiring a repainting of the entire body. • A ticking or tapping sound coming from the engine at idle can be silenced by installing an improved fuel hose/damper assembly. **1994–95**—A thumping or clacking heard from the front brakes signals the need to machine the front disc brake rotors. Ford may pay for this repair under its base warranty. • Loss of torque during or just after 3–4 shift may be caused by a hydraulic condition in the transmission or an intermittent signal from one of the powertrain system sensors.

1994–97—An erratic or prolonged 1–2 shift can be cured by replacing the cast aluminum piston with a one-piece stamped steel piston with bonded lip seals and by replacing the top accumulator spring. These upgraded parts will increase the transmission's durability, says Ford. **1994–98**—Loose rocker panel moldings will be fixed under the bumper-to-bumper warranty. **1996**—Stalling or hard starts may be due to the idle air control valve sticking. **All models/years**—Three components that frequently benefit from Ford "goodwill" warranty extensions are catalytic converters, fuel pumps, and computer modules. If Ford balks at refunding your money, apply the emissions warranty for a full or partial refund. • Paint delamination, fading, peeling, hazing, and "microchecking" (see page 112 for details on claiming a refund).

Mustang, Cobra Profile

	1991	1992	1993	1994	1995	1996	1997	1998
Cost Price ($)								
LX/Coupe	11,295	12,095	10,995	15,595	17,595	18,595	19,795	22,595
Convertible	19,195	19,995	20,495	24,995	26,095	25,895	26,795	29,295
Cobra GT	16,895	17,995	18,995	—	—	—	—	—
Used Values ($)								
LX/Coupe ↑	3,500	4,000	5,500	9,000	11,000	13,000	14,000	16,000
LX/Coupe ↓	2,500	3,500	5,000	8,500	10,500	12,000	13,000	15,000
Convertible ↑	6,000	7,000	8,500	12,000	15,000	17,000	18,000	21,000
Convertible ↓	5,000	6,500	8,000	11,500	14,000	16,000	17,000	20,500
Cobra GT ↑	6,000	7,000	9,500	—	—	—	—	—
Cobra GT ↓	5,500	6,500	8,000	—	—	—	—	—
Extended Warranty	Y	Y	Y	N	N	N	N	N
Secret Warranty	Y	Y	Y	Y	Y	Y	Y	N
Reliability	❷	❶	❷	❷	❷	③	③	④
Air conditioning	❶	❶	❶	❶	❶	❷	④	⑤
Body integrity	❶	❶	❷	❷	❷	❷	③	③
Braking system	❶	❶	❷	❷	❷	❷	③	③
Electrical system	❶	❶	❶	❶	❷	❷	③	④
Engines	③	❷	❶	❶	❶	③	④	④
Exhaust/Converter	❷	③	③	③	③	④	⑤	⑤
Fuel system	❷	❷	❷	③	③	③	③	④
Ignition system	③	③	③	③	④	④	④	④
Manual transmission	③	③	③	③	④	⑤	⑤	⑤
- automatic	❷	❷	❷	❷	③	③	③	④
Rust/Paint	❶	❶	❶	❶	❶	③	③	④
Steering	❷	③	③	③	③	④	⑤	⑤
Suspension	❷	③	③	③	③	④	⑤	⑤
Crash Safety								
Base	—	—	—	❶	④	④	④	⑤
Convertible	⑤	⑤	❶	—	—	⑤	⑤	—
Side Impact								
Base	—	—	—	—	—	—	—	③

Probe

Rating: Recommended (1995–97); Average (1993–94); Below Average (1989–92). **Maintenance/Repair costs**: Average, and repairs can be done by independent garages or Mazda dealers. Nevertheless, the under-hood layout is crowded, making for high routine maintenance costs. **Parts**: Despite the fact that 1997 is the Probe's last model year (Mazda's MX-6 bit the dust as well), parts should remain plentiful and reasonably priced.

Strengths and weaknesses: The four-seater Probe sporty coupe was launched in May 1988 as the front-wheel drive replacement for the aging Mustang. Don't get the impression that the Mazda MX-6 and Probe are twins because they share most mechanical features. In fact, they differ markedly in handling and appearance. Ford engineers took more control of the chassis tuning and suspension geometry to give the car a smoother and firmer sporty demeanour, and its stylists chopped and pulled the body to give it a more aerodynamic, aggressive personality.

Available only as a two-door hatchback, there are two models, each with its own powerplant: a 2.0L 4-cylinder and a 2.5L V6 engine. These engines give the car much-needed power and smoothness not found in the anemic and brutish powertrains used in the past.

Mazda's mechanicals are above reproach, but despite gobs of torque, the GT doesn't give the muscle-car performance found in the less refined 5.0L Mustang GT. Performance is sapped considerably by the automatic transmission. Early models are beset by severe "torque steer," a tendency for the chassis to twist when the vehicle accelerates. Nevertheless, overall handling is precise and predictable on all models without sacrificing ride quality, which is a bit on the hard side.

Refinements of the 1993 and later versions include two new Mazda-designed engines that give the car a small horsepower boost, more interior room, all-disc brakes on the GT, and less torque steer.

The Probe's overall mechanical reliability is fairly good, but like its Mustang cousin, body assembly is the pits (no surprise here—Mazda did the mechanicals; Ford, the body work.) Paint quality and rust protection are mediocre at best. The turbocharged engine has been relatively trouble-free, but owners have complained of frequent stalling and stumbling with the base 2.2L powerplant. Many experience excessive ABS noise and vibrations when braking, and the front brakes tend to wear out very quickly. AC components have a short lifespan of three to five years and are outrageously expensive to troubleshoot and repair.

The car is essentially a 2+2 with the rear reserved for children or cargo. The interior is short on head room for tall drivers (especially on vehicles equipped with a sunroof), but cargo room is increased with the folding rear seatbacks. Multiple squeaks and rattles, wind and water leaks, and cheap interior appointments are the most common body complaints. The digital read-outs are distracting and often incorrect.

Dealer service bulletins: **1993**—Poorly performing 4EAT automatic transmissions produce harsh 3–2 downshifts. • The Service Engine light may go on for no apparent reason. **1994–1995**—Fluid leak from the transaxle pump seal. • Brake roughness upon application. • Inaccurate fuel gauge. • Erratic headlight door operation. • Off-centre steering wheel. **1996–97**—Engine misses, stalls, runs rough, or has reduced power. • Excessive engine noise. • Upper engine noise service tip. • Hesitation, low power, rough idle. • 3–4 shift hunt. • CD4E transmission shifts harshly, seeps fluid from the vent, and produces a buzz or a clicking vibration. It may also make a whistling noise in Park. • Fluid leaks at axle shaft. • Inaccurate fuel gauge reading when the tank is full. • Loose catalyst or muffler heat shields. • Fog/film on windshield/interior glass. • Door glass makes a pop or snap noise when rolled up or down. • Musty and mildew-type odours.

Safety summary/Recalls: The driver's motorized shoulder belt is literally a pain in the neck. It rides high on the neck, fails to retract properly, tangles easily, and often hangs too loose. It also requires that the driver attach the lapbelt separately. Beware of coil spring corrosion causing the spring to suddenly break. Ford will pay for the repair on a case-by-case basis. **Recalls**: **1985**—A defective ignition may make for hard starting. • Defective plastic sleeve in front seatbelt tongue assembly. • Power brake booster could come apart (except in police cars and taxicabs). **1986–87**—Fuel line coupling could leak fuel. **1990–92**—Motorized shoulder belts may not work. **1991**—Vehicle could roll away when in Park. **1993**—A faulty rear hatch strut may cause the hatch to drop without warning. Dealer will replace rivets that have undersized heads on the strut pivot pins. **1994–95**—Passenger side airbag may not inflate properly. **1996**—Incorrect warning regarding rear-facing child safety seats. **1997**—Timing belt tensioner spring may break and jam the belt. **1991 GL**—There's a free fix for welds that anchor the front shoulder belt retractors.

Secret Warranties/Service Tips

1989—2.2L engines with hairline cracks in the cylinder head must have the head replaced. • Hard cold-starts or chronic stalling may require a cold stall pressure regulator kit (#900809A). • No-starts/low battery may mean that you need a new alternator pulley and belt kit (#E92Z-10344-D). **1989–91**—Constant fogging and moisture condensation on the interior windows and windshield signals the need to adjust the recirculation/fresh control cable. **1989–92**—An exhaust buzz or rattle may be fixed by installing new clamps to secure the heat shield. **1989–94**—A no-start condition, or inoperative heater or lights, may be caused by water and corrosion in the wiring connector, or a short-to-ground at splice 102 (circuit 9). **1990–91**—Rear disc brake squeal can be corrected by installing revised brake pads (#E92Z-2200-A). • Taillight condensation can be prevented by installing a new outer lens kit. • Engine knocking at idle may require the installation of a thicker thrust plate. **1990–94**—A speaker whine or buzz caused by the fuel pump can be stopped by installing an electronic noise RFI filter. **1993–94**—A clunk or knock from the steering assembly when turning the steering wheel is likely due to an insufficient yoke plug (pinion) preload. • Wind/water leaks require the readjustment of the front door glass, as outlined in DSB #994-5-4. • Inoperative power door locks may have a corroded wiring harness connection. • Frozen door locks, a common problem, are addressed in DSB #94-8-6. • A ticking noise coming from the 2.0L engine's hydraulic lash adjusters can be stopped by a longer oil pump control plunger that prevents air from passing through to the oil pump. **1994**—A rough idle affecting 2.0L engines could be caused by spark leakage from a damaged number 1 or number 2 spark plug wire. **1994–97**—Transaxle fluid seepage can be corrected by servicing with a remote vent kit or by replacing the main control cover. **1990 GT**—Service Program No. 96 provides for the rerouting of the wiring harness and hose clamp to prevent transmission failure. **All models/years**—Three components that frequently benefit from Ford "goodwill" warranty extensions are catalytic converters, fuel pumps, and computer modules. If Ford balks at refunding your money, apply the emissions warranty for a full or partial refund. • Paint delamination, fading, peeling, hazing, and "microchecking" (see page 112 for details on claiming a refund).

Probe Profile

	1990	1991	1992	1993	1994	1995	1996	1997
Cost Price ($)								
GL	14,972	13,795	14,795	15,165	16,495	17,895	18,395	19,095
GT	21,760	19,895	17,795	18,240	20,195	21,595	22,195	22,995
Used Values ($)								
GL ↑	3,500	4,000	5,000	7,500	9,000	11,000	13,000	14,500
GL ↓	3,000	3,500	4,500	6,500	8,500	10,500	12,000	13,500
GT ↑	5,000	5,500	6,000	8,000	12,000	14,000	16,000	17,500
GT ↓	4,500	5,000	5,500	7,500	11,500	13,000	15,000	16,500
Extended Warranty	Y	Y	Y	N	N	N	N	N
Secret Warranty	Y	Y	Y	Y	Y	Y	Y	Y
Reliability	③	③	③	③	④	④	⑤	⑤
Air conditioning	❷	❷	❷	❷	❷	③	③	③
Body integrity	❶	❶	❶	❶	❶	❶	❷	❷
Braking system	❶	❶	❶	❶	❶	❷	❷	❷
Electrical system	❶	❶	❷	❷	❷	❷	❷	③
Engines	③	③	③	④	⑤	⑤	⑤	⑤
Exhaust/Converter	❷	❷	❷	③	⑤	⑤	⑤	⑤
Fuel system	③	③	③	③	④	⑤	⑤	⑤
Ignition system	❷	③	③	⑤	⑤	⑤	⑤	⑤
Manual transmission	④	④	④	④	⑤	⑤	⑤	⑤
- automatic	③	③	③	④	④	⑤	④	⑤
Rust/Paint	❶	❶	❶	③	③	③	③	③

Steering	③	③	③	③	④	④	④	⑤
Suspension	③	③	③	③	④	④	④	⑤
Crash Safety	—	⑤	⑤	❶	❶	⑤	⑤	⑤

Note: The low crash ratings represent severe leg trauma. Although the 1990 Probe wasn't crash-tested, the 1989 crash test results showed minimal injury would be sustained by the driver and front-seat passenger.

GENERAL MOTORS

Camaro, Firebird, IROC-Z, Trans Am, Z28

Rating: Recommended (1996–98); Average (1994–95); Below Average (1992–93); Not Recommended (1982–91). As with the Mustang, the Camaro and Firebird have elicited many safety-related complaints, including airbag deployment injuries, sudden acceleration, brake failures, and steering loss. The 1996 Camaro and Firebird are essentially the same as the more expensive 1997 versions. A V8-equipped Camaro or convertible is the best choice for retained value a few years down the road. But you can do quite well with a used base coupe equipped with the performance handling package and 235 tires. GM has announced it may drop the Camaro and Firebird after the year 2000. If it does, this move is unlikely to affect the cars' resale values or parts supply. **Maintenance/Repair costs**: Average, and repairs can be done by any independent garage. **Parts**: Reasonably priced and easy to find.

Strengths and weaknesses: When compared with the Mustang, Camaros and Firebirds are better-performing rear-drive muscle cars that produce excellent crash protection scores and high resale values. They also take the lead over the Mustang with their standard ABS and slightly better reliability record. However, brute power is combined with almost-as-brutal repair charges. The Camaro's and Firebird's overall performance varies a great deal depending on the engine, transmission, and suspension combination in each particular car. Base models equipped with the V6 powerplant accelerate reasonably well, but high performance enthusiasts will find them slow for sporty cars. Handling is compromised by poor wet road traction, minimal comfort, and a suspension that's too soft for high-speed cornering and too bone-jarring for smooth cruising. The Z28, IROC-Z, and Trans Am provide smart acceleration and handling, but at the expense of fuel economy.

1982–93

Much like Ford's embarrassing 4-banger, the puny and failure-prone 2.5L 4-cylinder powerplant was the standard engine up to 1986—part of the legacy of an earlier fuel crisis and the subsequent downsizing binge. The turbocharged V8 offered on some Trans Am models should be viewed with caution because of its many durability problems.

Body hardware is fragile, poor paint quality and application is common and leads to premature rusting, and squeaks and rattles are legion. Body integrity is especially poor on cars equipped with a T-roof. Areas particularly vulnerable to rusting are the windshield and rear wheel openings, door bottoms, and rear quarter panels. The assorted add-on plastic body parts found on sporty versions promote corrosion by trapping moisture along with road salt and grime. Also note that Camaro flat seats don't offer as much support as the better-contoured Firebird seats.

These cars are also plagued by chronic fuel-system problems, especially on the Cross-Fire and multi-port fuel-injection controls. Automatic transmissions, especially the 4-speed, aren't durable. The standard 5-speed manual gearbox has a stiff shifter and a heavy clutch. Clutches fail frequently and don't stand up to hard use. The 2.8L V6, used through 1989, suffers from leaky gaskets and seals and premature camshaft wear. The larger 3.1L 6-cylinder has fewer problems. Malfunctioning dash gauges and electrical problems are common. Exhaust parts rust quickly. Dual outlet exhaust systems on V8 engines are expensive to replace. Front suspension components and shock absorbers wear out very quickly.

1994–98

Of this grouping, you should stick with the 1996–98 models for the best quality and price. Nevertheless, all of these cars are much better overall performers, with a more powerful, reliable base engine; increased body rigidity that hushes some of the squeaks and rattles so prevalent with previous years; a redesigned, easier-to-read dash; a bit more rear head room; and additional standard safety features. A more powerful 3.8L V6 arrived in spring 1995 and became standard for 1996.

On 1997 models, GM offered a 30th birthday styling package for the Camaro and some interior upgrades, V6 engine dampening for smoother running at high speeds, optional Ram Air induction, and racier-looking ground-effects body trim for the Firebird.

These sporty convertibles and coupes are almost identical in their pricing and in the features they offer (the Firebird has pop-up headlights, a more pointed front end, a narrower middle, and a rear spoiler). As noted above, both cars got a complete make-over in 1995, making them more powerful and aerodynamic with less spine-jarring performance. They were given new plastic skins, dual airbags, standard anti-lock brakes, new suspension and steering systems, a new 5-speed manual transmission, a reworked interior, and minor cosmetic improvements.

As one moves up the scale, overall performance improves considerably. The V8 engine gives these cars lots of sparkle and tire-spinning torque, but there's a fuel penalty to pay. A 4-speed automatic transmission is standard on the 5.7L-equipped Z28; other versions come with a standard 5-speed manual gearbox or an optional 6-speed. Many of these cars are likely to have been ordered with lots of extra performance and luxury options, including a T-roof package guaranteed to include a full assortment of creaks and groans.

Both the Camarao and the Firebird may be equipped with an impressively effective pass key theft-deterrent system similar to the one used successfully in the Corvette. A resistor pellet in the ignition disables the starter and fuel system when the key code doesn't match the ignition lock.

Not everything is perfect, however. Owners report that the base engine is noisy, though the 3.8L doesn't have the head gasket failures seen with Ford's 3.8L powerplant. Fuel economy is practically nonexistent, the air conditioner malfunctions (but not to the same extent as Ford-produced units), front brakes and MacPherson struts wear out quickly, servicing the fuel-injection system is an exercise in frustration, and body problems are worse than with the Mustang and just won't go away.

These cars are still afflicted by door rattles, misaligned doors and hatch, a sticking hatch power release, and poor fit and finish. Owners also complain that the steering wheel is positioned too close to the driver's chest, the low seats create a feeling of claustrophobia, visibility is limited by wide side pillars, and trunk space is sparse with a high liftover.

Dealer service bulletins: **1994**—Poor radio reception. • A rough-running AC compressor on V8-equipped vehicles requires an upgraded compressor pulley. **1995**—Transmission fluid leak from the pump body. • Excessive rear axle noise. • Steering pull, usually to the right. • Front brake vibration and/or pedal pulsation and unevenly worn rear brake pads. • A low-voltage reading or dim lights at idle. • Rear compartment lid water leaks. • Door glass rattles. • Poor paint application and rust spots. **1996**—Engine rpm flare during gear shifts. • Engine oil leak at rear of engine near flywheel cover, and spark knock at idle. • 3–2 part throttle downshift flare and delayed transmission engagement. • Transmission fluid leak from pump body (replace bushing). • Steering column noise. • Steering lead or pull, usually to the right. • Steering column click or snap noise. • Uneven brake pad wear. • Convertible top leaks water. • Coolant leak at engine coolant heater assembly. • AC odours. • Radio frequency interference diagnosis. • Condensation on exterior light. • Water leaks at rear compartment lid when vehicle is parked on an incline. • Water leak diagnostic guide. • Cannot remove ignition key. • Anti-theft alarm may sound if vehicle gets wet. • Inoperative power door locks. • Rattle or buzz from parking brake handle release button. **1997**—Oil leaks between the intake manifold and engine block. • Troubleshooting a malfunctioning convertible top. • Erratic headlight or turn signal operation. •

Inoperative power door locks. • Low voltage reading or dim lights at idle. • Premature burnout of rear taillight/stop light bulbs. • Rattle or buzz from parking brake handle release button.

Safety summary/Recalls: **1995**—Four reports that car caught fire on the left rear side near the fuel tank. • Other fires ignited near the radio and in the engine compartment. • Many reports claiming sudden acceleration. • Airbag failed to deploy. • Inadvertent airbag deployment. • Severe injuries caused by airbag deployment. • Airbag deployed, projecting dash panel into the passenger's face. • Airbag deployed and caught fire. • Frequent ABS brake failures; pedal goes to the floor. • ABS particularly ineffective in rainy weather. • Rear brakes rusted out. • Left rear caliper grabs when making turns. • On one occasion, the rear brake caliper bolt came off, causing the brakes to fail. • Emergency brake fails frequently, allowing car to roll away even though shift lever is placed in Park. • Faulty cruise control. • Fuel line retaining clamp could fail, causing the plastic fuel lines to come into contact with the exhaust manifold cover. • The tie-rod broke while driving, forcing car off the road. • In wet conditions, or when passing over a puddle, the power steering pump fails, resulting in steering lockup. • Broken gas tank weld. • Stabilizer bar failure. • Care must be taken when shifting the floor shifter to avoid cutting hand on the dash. • Frequent transmission failures while driving. • Transmission bolts came off. • Trunk latch failure allows trunk to open while driving. • Two reports that double-locking T-top flew off from passenger side while driving with both locks engaged. • In another incident, the sunroof flew off. • Convertible top latch failure. • Door windows leak. • Inadequate defroster; causes windows to fog. • Headlight dimmer switch failures. • Driver's seatbelt failed during accident. **Recalls**: **1985**—Shoulder seatbelt may not retract. **1985–86**—For vehicles with rear disc brakes and manual transmissions, the parking brake adjuster may not hold the vehicle when parked on an incline. **1985–90**—Defective seatbelt buckle assemblies may not latch. **1986**—Faulty push-pull headlight switch. **1988**—GM will inspect and repair the power-steering pump support brace and mounting bracket. **1988–89**—On vehicles equipped with a 2.8L V6 engine, fuel feed hoses may leak. **1990**—Fuel return hoses may leak on vehicles equipped with a V8 engine. **1991**—Poor windshield retention in an accident. Defective seatbelt latchplates will be fixed for free. • Defective front seats. **1991–92**—Fuel-filler neck may leak. **1992**—The automatic transmission shift control cable may separate and hamper shifting. **1994**—Fuel line on V8-equipped cars may leak fuel into the engine compartment. **1995**—Faulty lower steering shaft coupling could lead to loss of steering. **1997**—Seatbelt retractor may fail. **1989 convertible**—Rear seatbelts may be too long.

Secret Warranties/Service Tips

1982–93—Vehicles equipped with a Hydramatic 4L60 transmission that buzzes when the car is in Reverse or idle may need a new oil pressure regulator valve. **1985–91**—Hydramatic 4L60/700R4 automatic transmission may have no upshift or appear to be stuck in first gear. The probable cause is a worn governor gear. It would be wise to also replace the retaining ring. **1989**—If the rear brakes moan when they're applied slightly while backing up or turning, a redesigned caliper mounting plate may be needed (DSB #89-162-5). • Knocking from a cold 5.0L engine means that the PROM module has to be replaced (DSB #89-284-6E). • No-start or stalling may be caused by a disconnected fuel pump coupler. **1989–90**—Campaign 90-C-11 provides for free convertible top latch handles. **1990**—Cold-start stalling with the 3.1L engine may be corrected by replacing the MEMCAL. **1993–94**—DSB #431028 covers all aspects of securing loose door outer panels. • A loose, rattling instrument panel or upper trim panel requires new dual lock riveted fasteners. • Rear brake squeal can be silenced by installing upgraded disc brake pads. **1993–96**—Uneven rear brake pad wear or premature wear can be corrected by replacing the caliper anchor bracket, guide pins, and the brake pads with upgraded parts. **1993–97**—Oil leaks between the intake manifold and engine block are most often caused by insufficient RTV bonding between the intake manifold and cylinder block. **1994**—Excessive oil consumption is likely due to delaminated intake manifold gaskets. Install an upgraded intake manifold gasket kit. • Delayed automatic transmission shift engagement is a common problem addressed in DSB #47-71-20A. • Install a new "flash" PROM to cure engine surging or hesitation and stalling upon acceleration for cars with automatic transmissions. **1997**—Tips on correcting Park brake handle noise. • New switches will fix inoperative door locks. • Fixing automatic transmission slippage. • Engine oil leak diagnosis. • Theft alarm sounds when vehicle gets wet. • Diagnosing engine miss and poor driveability. **2.5L engines**—Spark knock can be fixed with the free installation of a new PROM module (#12269198) if the emissions warranty applies. • Frequent stalling may require a new MAP sensor (DSB #90-142-8A). **1990–91**—Stalling on deceleration or at stops requires a new MEMCAL that refines idle speed control and throttle follower operation. **1991**—Poor starting may be caused by the spring in early starter drives compressing too easily. Install an upgraded starter motor drive assembly (#10473700). **1992**—Oil leaks from the rear of a 5.0L or 5.7L engine may be caused by insufficient sealing around the camshaft plug. **1995–96**—Delayed automatic transmission shift engagement may require the replacement of the pump cover assembly. **3.8L V6**—These engines have a history of low oil pressure caused by a failure-prone oil pump. A temporary remedy is to avoid low-viscosity oils and use 10W-40 in the winter and 20W-50 for summer driving. **All models/years**—A rotten-egg odour coming from the exhaust is probably the result of a malfunctioning catalytic converter, which may be covered by the emissions warranty. • Paint delamination, peeling, or fading (see pages 76, 113).

Camaro, Firebird, IROC-Z, Trans Am, Z28 Profile

	1991	1992	1993	1994	1995	1996	1997	1998
Cost Price ($)								
Camaro/RS	15,298	14,298	15,998	16,498	18,995	20,195	22,075	22,790

Firebird	15,698	14,798	15,168	17,198	19,795	20,955	23,120	24,580
Z28	19,098	18,998	19,898	20,498	23,650	25,530	27,270	27,840
IROC Convertible	24,998	—	—	—	—	—	—	—
Trans Am	19,898	20,998	23,265	24,500	27,390	28,755	30,780	34,080
Used Values ($)								
Camaro/RS ↑	4,500	5,000	6,500	8,500	11,000	13,500	15,000	17,000
Camaro/RS ↓	4,000	4,500	5,500	8,000	10,500	12,500	14,500	16,500
Firebird ↑	4,000	5,000	6,000	9,500	11,000	14,000	15,500	18,000
Firebird ↓	3,500	4,500	5,000	8,500	10,500	13,000	14,500	17,000
Z28 ↑	6,000	6,500	9,000	12,500	15,000	17,000	19,000	21,500
Z28 ↓	5,500	6,000	8,500	11,500	14,500	16,000	18,500	20,500
IROC Convertible ↑	8,500	—	—	—	—	—	—	—
IROC Convertible ↓	8,000	—	—	—	—	—	—	—
Trans Am ↑	6,000	7,000	10,500	13,500	16,000	18,500	20,000	22,500
Trans Am ↓	5,500	6,500	9,500	12,000	14,500	17,000	18,500	21,500
Extended Warranty	Y	Y	Y	N	N	N	N	N
Secret Warranty	Y	Y	Y	Y	Y	Y	Y	Y
Reliability	❶	❷	❷	③	③	④	④	⑤
Air conditioning	❷	③	③	③	④	⑤	⑤	⑤
Body integrity	❶	❶	❶	❶	❷	❷	❷	③
Braking system	❷	❷	❷	❷	❷	③	③	④
Electrical system	❶	❶	❶	❶	❶	❶	❷	❷
Engines	❷	④	④	③	③	③	③	❷
Exhaust/Converter	❶	❶	③	③	③	④	④	⑤
Fuel system	❶	❷	❷	③	③	④	④	⑤
Ignition system	❷	❷	❷	❷	③	③	④	④
Manual transmission	③	③	③	④	④	⑤	⑤	⑤
- automatic	③	③	③	④	④	⑤	⑤	❷
Rust/Paint	❶	❶	③	③	③	④	⑤	⑤
Steering	③	③	③	③	③	③	④	⑤
Suspension	④	③	④	④	④	④	④	⑤
Crash Safety								
Camaro	⑤	⑤	—	⑤	⑤	⑤	⑤	④
Side Impact								
Camaro	—	—	—	—	—	—	③	③

Note: Although the 1990 Camaro wasn't crash tested in a frontal collision, the 1987 crash test results showed minimal injury would be sustained by the driver and front-seat passenger.

Corvette

Rating: Above Average (1996–98); Average (1994–95); Not Recommended (1980–93). The cheaper 1996 Corvette won't have the cachet or the mechanical and body refinements of the redesigned 1997 version. If you choose the 1997 model, try to get a second-series car that was made after June 1997. Keep in mind that premium fuel and astronomical insurance rates will further drive up your operating costs. **Maintenance/Repair costs**: Higher than average, although most repairs can be done by any independent garage. **Parts**: Pricey, but easy to find. Surprisingly, it is often easier to find parts for older Corvettes, through collectors' clubs, than to find many of the high-tech components used today.

Strengths and weaknesses: Corvettes made in the late '60s and early '70s are acceptable buys, due mainly to their value as collector cars and their uncomplicated repairs. The Corvette's overall reliability has declined over the years as its price and complexity have increased. This is due in large part to GM's updating its antiquated design with high-tech, complicated add-ons, rather than coming up with something original. Consequently, the car has been gutted and then retuned using failure-prone electronic circuitry. Miles of emissions plumbing have also been added to make it a fuel-efficient, user-friendly, high-performance vehicle. Unfortunately, the Corvette has missed these goals by a large margin.

The electronically controlled suspension systems have been glitch-plagued over the past several years. Servicing the different sophisticated fuel-injection systems isn't easy—even (especially) for GM mechanics. The noisy 5.7L engine frequently hesitates and stalls, there's lots of transmission buzz and whine, the rear tires produce excessive noise, and wind whistles through the A- and C-pillars. These, and the all-too-familiar fibreglass body squeaks, continue to be unwanted standard features throughout all model years. The electronic dash never works quite right (speedometer lag, for example).

On the other hand, Corvette ownership of more recent models does have its positive side. For example, the ABS vented disc brakes, available since 1986, are easy to modulate and fade-free. The standard European-made Bilstein FX-3 Selective Ride Control suspension can be preset for touring, sport, or performance. Under speed, an electronic module automatically varies the suspension setting, finally curing these cars of their earlier endemic over-steering, wheel spinning, breakaway rear ends, and other nasty surprises.

If you're planning to buy a used Corvette, be wary of any model made 1979–93—and make sure that you get a GM-backed supplementary warranty with later models! The frequency of repairs and the high repair costs make maintenance outrageously expensive on all but the more recent versions. Although older cars are likely to be junked or already restored, here are some of the things that can put a large dent in your wallet if they haven't been fixed already:

1977–83—Major mechanical failings affect the air conditioning, transmission, clutch, shift linkage, camshaft lifters and rear half-shaft soft yokes, carburetor, steering, rear brakes, and starter and electrical system, including the lights. As far as body assembly goes, the major deficiencies are poor panel fits, faulty and fragile interior/exterior parts and trim items, quirky instruments, cheap upholstery, and defective window lifts. Owners also complain of poor workmanship/shoddy assembly causing a cacophony of squeaks and rattles, poor dealer servicing, unavailable and expensive parts, and excessive labour charges.

1984–90—Incredibly difficult to service. One *Lemon-Aid* reader had a faulty engine bearing at 16,000 km and spent $2,500 to remove the engine. It takes half a day to change the spark plugs on the passenger side. Likely mechanical problem areas are the emission-control system (injectors, computer-controlled sensors, fuel injection, and engine gaskets), air conditioning, and ignition/distributor. Owners also experience engine and drivetrain failures, Bosche radio malfunctions, and the need to make frequent wheel alignments. Fragile body hardware, poor fit and finish, wind/road noise, and water intrusion into the interior are still major weaknesses. Owners complain of faulty controls and window lifts, and defective paint (base coat comes through the finish), interior/exterior parts and trim, glass and weatherstripping, instruments, lights, door locks, upholstery, and carpeting.

1991–96—Though the restyled 1991 Corvette remained the same mechanically, all models got the convex tail and square taillights previously used only on the upscale ZR-1 coupe. A new LTI engine with 55 more horses came on the scene with the 1992 'Vette, and the following year the ZR-1 got a 405-hp variant of the same powerplant, shortly before the model was replaced in the spring of 1995 by the Grand Sport. A more substantial redesign was carried out for the 1997 model year.

Owners rave about the redesigned models' improved performance, better handling, and additional safety features, but continue to find fault with the stiff ride, poor fuel economy, and excessive interior noise. From a reliability standpoint, 1993 and later models are much improved, with problems affecting mostly the electronic and electrical system, body hardware, fit and finish, suspension, and air conditioning.

Dealer service bulletins: **1995**—A high effort to shift into gear in cold, wet weather; and high shift effort into Reverse. • A 3–2 downshift flare and erratic downshifting with the automatic transmission. • Starter clicks but won't start engine. • Front brake pulsation. • A low-voltage reading or dim lights at idle. • Right-hand wiper blade chatter • Excessive radio static. • Door glass scratches. • Poor paint application. **1996**—Oil leaks between the intake manifold and engine block. • 3–2 part throttle downshift flare and delayed transmission engagement. • Transmission fluid leak from pump body (replace bushing). • Steering column noise. • Radio frequency interference diagnosis. • Low-voltage reading or dim lights at idle. • Door trim armrest lid hard to open. • Water leak diagnostic guide. • Condensation on exterior light. **1997**—Air temperature from

HVAC outlets doesn't change. • Erratic fuel gauge readings. • Low-voltage reading or dim lights at idle. • Hatch won't pop up when activated in cold weather. • Water drips into rear compartment.

Safety summary/Recalls: **Recalls**: **1988–89**—The rear wheel tie-rod end could fracture. GM will replace any defective tie-rod assembly free of charge. **1990**—Inoperative parking brakes will get a new brake lever assembly. • The fuel feed and return line connectors may leak fuel. **1992–93**—Power-steering hose leakage may cause an engine compartment fire.

Secret Warranties/Service Tips

1982–91—Hydramatic 4L60/700R4 automatic transmission may have no upshift or appear to be stuck in first gear. The probable cause is a worn governor gear. It would be wise to replace the retaining ring as well. **1982–93**—Vehicles equipped with a Hydramatic 4L60 transmission that "buzzes" when the car is in Reverse or idle may need a new oil pressure regulator valve. **1987–89**—Revised cylinder head gaskets provide better sealing (DSB #89-283-6A). **1990–91**—Engine wiring short circuits may be caused by an abraded electrical harness at the mounting clamp just below the oil pressure sensor. **1992**—5.7L engines may develop leaks at the oil filter area. GM Campaign 92COS has covered the repair cost in the past. • 5.7L engines may backfire excessively when shifting from first to second gear with the throttle wide open. Correct by changing the MEMCAL module. • Oil leaks from the rear of a 5.7L engine may be caused by insufficient sealing around the camshaft plug. Replace the plug and reseal. **1992–96**—Oil leaks between the intake manifold and engine block are most often caused by insufficient RTV bonding between the intake manifold and cylinder block. **1994**—Excessive oil consumption is likely due to delaminated intake manifold gaskets. Install an upgraded intake manifold gasket kit. • Delayed automatic transmission shift engagement is a common problem, and is addressed in DSB #47-71-20A. **1995–96**—Delayed automatic transmission shift engagement may require the replacement of the pump cover assembly. **1997**—Correcting water leaks into the rear compartment. • Inadequate heating, defrosting.• Correcting drivebelt noise. • **All models/years**—A rotten-egg odour coming from the exhaust is probably caused by a defective catalytic converter, which may be covered by the emissions warranty. • Clearcoat paint degradation, whitening, and chalking, long a problem with GM's other cars, is also a serious problem with the fibreglass-bodied Corvette, says DSB #331708. It too is covered by a secret warranty for up to six years.

Corvette Profile

	1991	1992	1993	1994	1995	1996	1997	1998
Cost Price ($)								
Base	42,798	40,398	41,398	43,398	47,580	48,080	48,895	50,430
Convertible	48,398	48,098	48,298	50,498	55,020	56,335	—	58,430
ZR-1	—	76,998	78,898	80,798	86,765	—	—	—
Used Values ($)								
Base ↑	15,000	17,500	20,000	24,000	28,000	32,000	37,000	46,000
Base ↓	13,500	16,000	17,500	22,000	26,000	30,500	35,000	44,000
Convertible ↑	18,500	21,000	23,000	27,000	31,000	35,000	—	51,000
Convertible ↓	16,500	18,500	21,000	25,000	29,000	34,000	—	50,000

ZR-1 ↑	—	26,000	29,000	36,000	44,000	—	—	—
ZR-1 ↓	—	24,000	27,000	33,000	41,000	—	—	—
Extended Warranty	Y	Y	Y	Y	Y	Y	Y	Y
Secret Warranty	N	N	N	Y	Y	Y	Y	N
Reliability	❷	❷	❷	③	③	④	④	④
Air conditioning	❷	❷	❷	③	③	③	③	④
Body integrity	❷	❷	❷	❷	❷	❷	❷	③
Braking system	❷	❷	❷	❷	❷	❷	③	③
Electrical system	❷	❷	❷	❷	❷	❷	③	④
Engines	❷	❷	③	③	③	③	③	④
Exhaust/Converter	❷	❷	❷	③	③	④	④	⑤
Fuel system	❷	❷	❷	❷	❷	③	③	④
Ignition system	❷	❷	❷	❷	❷	③	④	⑤
Manual transmission	❷	❷	③	③	③	④	④	⑤
- automatic	❷	❷	❷	❷	③	③	③	③
Rust/Paint	❷	❷	❷	❷	③	③	③	④
Steering	❷	❷	❷	③	③	④	④	④
Suspension	❷	❷	❷	❷	❷	③	④	④

HONDA

Prelude

Rating: Recommended (1993–98); Above Average (1985–92). **Maintenance/Repair costs**: Average. Repairs aren't dealer dependent. To avoid costly engine repairs, check the engine timing belt every 2 years/40,000 km and replace it every 96,000 km ($300). **Parts**: Higher-than-average cost, but independent suppliers sell parts for much less.

Strengths and weaknesses: Unimpressive as a high-performance sports car, the Prelude instead delivers a stylish exterior, legendary reliability, and excellent resale value.

The 1978–87 first-generation Preludes were described as luxury, sporty cars, but didn't offer much of either. They should be inspected carefully for engine problems and severe underbody corrosion, particularly near the fuel tank. Noisy front brakes, premature disc warpage, high oil consumption, and worn engine crankshaft/camshaft lobes are the main problem areas with these models. They're also prone to extensive rusting around wheel openings, door bottoms, the trunk lid, fenders, rear taillights, bumper supports, chassis members, suspension components, and the fuel tank.

The 1988–91 models offer more and smoother engine power, excellent handling, and improved reliability. There are some generic complaints that continue to crop up, including rapid front brake wear, scored and warped front brake rotors, automatic transmission failure, defective constant velocity joints, premature exhaust system rust-out, and a warping hood.

The 1992–96 models are shorter, wider, and heavier. They're not very fast, except for the VTEC-equipped Si. The four-wheel steering found on the 1992 4WS version is more gimmick than anything else. It was dropped after 1994. The automatic transmission is smoother, although it still saps some of the Prelude's power. Both the rear seating and tiny trunk are inadequate for most people. You can expect fewer but all-too-familiar glitches, including minor electrical problems, body and accessory defects, brake squealing, and prematurely warped front brake rotors. Most independent mechanics are ill-equipped to service these cars because Preludes have become increasingly complicated to repair.

The year for big Prelude changes was 1997. The car was restyled, repowered, and given handling upgrades that make it a better-performing, more-comfortable-riding sports coupe: an additional five horses for the base 2.2L VTEC engine, a new Automatic Torque Transfer System (ATTS), an upgraded suspension, standard ABS, air conditioning, 16-inch wheels, and a CD player with six speakers. The Sequential SportShift automatic transmission (a variation of the one used in the NSX) equips the base Prelude. Overall, the car is roomier (the extended wheelbase gives added stability and provides more room in the rear seating area), has a more solid body structure, and includes a totally redesigned, user-friendly dash with analogue gauges.

Owners report that air-conditioner condensers frequently fail after a few years and often need cleaning to eliminate disagreeable odours. Other problems include minor electrical glitches, brake squealing, and prematurely warped front brake rotors. A host of new technical features adds to the Prelude's complexity and guarantees that you'll never stray far from the dealer's service bay. In fact, most corner mechanics are poorly equipped to service these cars and the Automatic Torque Transfer System (ATTS) won't make their job any easier. According to CAA, parts are a bit more expensive than for other cars in this class.

Dealer service bulletins: **1993**—Poor radio reception on the AM band. • Excessive steering wheel vibrations. • CD magazine changer in trunk won't eject. • Noisy window regulators. **1994**—Buzzing in the driver's door. • Cup holders that don't stay closed. • Fuel filler doors that won't stay open. • Noise from the driver's seat track. • Steering wheel shimmy continues to be a problem. **1995**—Creaking clutch pedal. • A jingling in the left side of the dashboard. • A slow-to-retract seatbelt. • A warped console. **1996**—Headliner rattles. • The seatbelt is still slow to retract.

Safety summary/Recalls: **Recalls**: **1978–85**—Severe chassis corrosion around fuel tank and body seams prompted Honda to provide free chassis repairs and replace suspension/steering components and fuel tanks under a "silent" recall campaign. **1983–87**—Road salt could cause the fuel filler and/or breather pipe to rust through, resulting in leaks. A recall campaign has been organized to fix this defect. **1986–91**—Takata seatbelt replacement. **1988**—Corroded coil spring breakage could cause loss of control. • Power-steering hose leaks could cause a fire.

Secret Warranties/Service Tips

1988–93—Poor AM reception or popping from the speakers is likely due to a poor ground connection between the antenna collar and car body. Correct the reception by improving the ground connection and tightening the antenna-assembly mounting nuts in the proper sequence. • Creaking from the window regulator can be corrected by installing an upgraded regulator spiral spring. **1990**—Defective AC compressor pickup sensors are a common problem. **1992**—A rattle from the front of the car that occurs when driving over rough surfaces but goes away when brakes are lightly applied can be corrected by installing new front brake pad retainers. **1994–96**—Silence rear headliner rattling by applying EPT sealer 5T to the rear headliner where it contacts the wiring harness. **All models/years**—Steering wheel shimmy can be reduced by rebalancing the wheel/tire/hub/rotor assembly in the front end. • Seatbelts that fail to function properly during normal use will be replaced for free under Honda's lifetime seatbelt warranty. • Honda will also repair or replace defective steering assemblies, constant velocity joints, and catalytic converters free of charge up to 5 years/80,000 km on a case-by-case basis. There's no labour charge or deductible. Used vehicles and repairs carried out by independent garages aren't covered by this special program. The converter is almost always covered under the emissions warranty. • Most Honda DSBs allow for special warranty consideration on a "goodwill" basis for most problems even after the warranty has expired or the car has changed hands.

Prelude Profile

	1991	1992	1993	1994	1995	1996	1997	1998
Cost Price ($)								
Base/SR	17,575	16,550	26,195	27,300	26,995	27,395	27,300	27,600
2.0L Si	22,190	19,650	—	—	—	—	—	—
VTEC	—	—	27,295	28,295	29,695	29,995	—	—
Used Values ($)								
Base/SR ↑	6,000	7,000	10,500	12,500	16,500	18,500	21,000	24,000
Base/SR ↓	5,000	6,500	9,500	11,500	16,000	17,500	20,500	23,000
2.0L Si ↑	7,000	8,500	—	—	—	—	—	—
2.0L Si ↓	6,500	7,500	—	—	—	—	—	—
VTEC ↑	—	—	13,500	15,500	18,000	20,000	—	—
VTEC ↓	—	—	13,000	14,500	17,000	19,000	—	—
Extended Warranty	N	N	N	N	N	N	N	N
Secret Warranty	N	N	N	N	N	N	N	N
Reliability	⑤	⑤	⑤	⑤	⑤	⑤	⑤	⑤

Air conditioning	③	③	③	④	⑤	⑤	⑤	⑤
Body integrity	③	③	③	③	③	③	③	④
Braking system	❷	③	③	③	③	③	④	⑤
Electrical system	❷	❷	③	③	④	④	④	⑤
Engines	③	③	④	⑤	⑤	⑤	⑤	⑤
Exhaust/Converter	❷	③	⑤	⑤	⑤	⑤	⑤	⑤
Fuel system	④	④	④	④	⑤	⑤	⑤	⑤
Ignition system	④	④	⑤	⑤	⑤	⑤	⑤	⑤
Manual transmission	⑤	⑤	⑤	⑤	⑤	⑤	⑤	⑤
- automatic	③	③	④	⑤	⑤	⑤	⑤	⑤
Rust/Paint	③	❷	③	③	④	⑤	⑤	⑤
Steering	❷	❷	③	③	③	③	④	④
Suspension	③	③	④	⑤	⑤	⑤	⑤	⑤
Crash Safety	❷	⑤	⑤	—	—	—	—	—

HYUNDAI

Scoupe

Rating: Above Average (1995); Below Average (1991–94). An Excel cross-dressing as a sports car. **Maintenance/Repair costs**: Higher than average, although most repairs can be done by any independent garage. **Parts**: Higher-than-average cost, with limited availability, due partly to the discontinuation of the model after '95.

Strengths and weaknesses: This is essentially a cute coupe with an engine more suited to high gas mileage than hard driving. Except for its lighter weight, less luggage space, and shorter overall length, the Scoupe is really an Excel clone, and there are better-performing and more-reliable sports coupes available. Plenty of interior room up front, but the roof is too angular; if you're short and have to pull the seat up, your head almost touches the roof. The rear interior room is inadequate. The Scoupe is fuel efficient and has a good heating and ventilation system.

This is far from a performance car, mainly due to the Excel's underpinnings and drivetrain. The 4-speed automatic transmission robs the base engine of much-needed horsepower, making for poor acceleration on inclines and constant shifting between 2,500 rpms and 3,500 rpms when pushed. Suspension may be too firm for some. There's plenty of body roll in turns, and mediocre steering. Brakes are hard to modulate.

Customer service has improved, if only because dealers have more factory support. Body construction is sloppy, giving rise to wind/water leaks, rattles, and breakage. Lots of engine and road noise intrude into the interior. Owners also report frequent electrical system shorts and brake and wheel-bearing problems.

Dealer service bulletins: **1993**—Engine has difficulty reaching recommended operating temperature (MPI fault code #21); low fuel pressure; and possible fuel leak at fuel gauge connection valve. • Harsh shifting with the automatic transmission when accelerating or coming to a stop. • Excessive disc brake noise. **1994**—Starting difficulties. • Transmission glitches. • AC failures. • Noisy shock absorbers and radio malfunctions. **1995**—Automatic transmission won't engage Overdrive. • Exhaust system releases a rotten-egg odour. • Rear suspension squeaks. • Trunk water leaks. • Troubleshooting tips for locating and plugging other interior water leaks.

Safety summary/Recalls: **1994**—Vehicle suddenly accelerated and would not shut off when key was taken from the ignition. • Airbag failed to deploy. • Complete brake failure. • Loss of power steering when turning. • Idle speed control module failure caused erratic idling along with acceleration and deceleration problems. • While car is being driven at high speeds, the transmission slips into Neutral without warning. • Driver's seat doesn't lock into place. • Seatback collapsed when vehicle was rear-ended. • Cylinder head failures. • Check Engine light comes on for no reason. • Leaky shock absorbers. • Wiper blades don't clean windshield properly. **1995**—Corrosion caused the front inner tie-rod to break, allowing the wheel to fall inward.

Secret Warranties/Service Tips

1991—Stalling when shifting into gear immediately after starting a cold engine may be corrected by installing a Cold Start Enrichment Kit (#39901-24Q00D). **1991–92**—A difficult-to-engage Reverse gear needs an upgraded part. • A harsh shift when coming to a stop or upon acceleration may be due to a maladjusted accelerator switch TCU. **1992–94**—Hyundai has a field fix for manual transaxle gear clash/grind (DSB #9440-004). **1993–94**—Hyundai has also developed an improved spark plug (painted yellow on the tip) for better cold-weather starting. **1994**—Hyundai may replace the AC discharge hose for free under a special service campaign (DSB #94-01-010). **All models/years**—An automatic transmission that won't engage Overdrive may need the oil temperature sensor replaced or may have faulty wiring or a defective Transmission Control Module.

Scoupe Profile

	1991	1992	1993	1994	1995
Cost Price ($)					
Base	9,645	10,125	10,995	11,595	12,095
LS/GLS	11,045	11,595	12,495	12,995	13,495
Used Values ($)					
Base ↑	3,500	4,000	4,500	5,500	7,000
Base ↓	2,500	3,500	4,000	5,000	6,000
LS/GLS ↑	4,000	5,000	5,500	6,500	7,500
LS/GLS ↓	3,500	4,500	5,000	6,000	6,500

Extended Warranty	Y	Y	Y	Y	Y
Secret Warranty	N	N	N	N	N
Reliability	❷	❷	❷	❷	③
Air conditioning	❷	❷	④	④	④
Automatic transmission	❷	❷	❷	❷	③
Body integrity	❷	❷	❷	❷	③
Braking system	❷	❷	❷	❷	❷
Electrical system	❷	❷	❷	❷	❷
Engines	④	④	④	④	⑤
Exhaust/Converter	⑤	⑤	⑤	⑤	⑤
Fuel system	③	③	③	③	④
Ignition system	❷	❷	❷	③	④
Rust/Paint	❷	❷	❷	③	③
Steering	③	③	③	③	④
Suspension	③	③	③	④	④
Crash Safety	⑤	⑤	❶	❶	④

MAZDA

MX-3, Precidia

Rating: Recommended (1994–96); Above Average (1992–93). **Maintenance/Repair costs**: Higher than average, and most repairs have to be done by a Mazda dealer. **Parts**: Higher-than-average cost, with limited availability.

Strengths and weaknesses: The base 1.6L engine supplies plenty of power for most driving situations. When equipped with the optional 1.8L V6 powerplant (the smallest V6 on the market at the time) and high-performance options, the MX-3/Precidia transforms itself into a 130-hp pocket rocket. In fact, the MX-3/Precidia GS sports coupe easily outperforms the 4-cylinder Honda del Sol, Toyota Paseo, and Geo Storm for comfort and high performance thrills. It does fall a bit short of the Saturn SC due to its limited low-end torque, and fuel economy is disappointing. Reverse gear is sometimes hard to engage.

Brake and wheel-bearing problems are commonplace. Most of the MX-3/Precidia's parts are used on other Mazda cars, so their overall reliability should be outstanding.

Body assembly is average and rattles/squeaks are common. Owners have complained of paint defects and sheet metal that's too thin above the door handles. Dents in the metal appear where you would ordinarily place your thumb when closing the door. Some owners report wind and water leaks around the doors and windows. Moderate engine noise increases dramatically above 110 km/h.

Small door openings are a hardship for tall occupants when

entering and exiting. Steering wheel rubs against thighs even when the seat is pushed as far back as it will go. Rear head room and leg room are limited. A high beltline and cowl add to a claustrophobic feeling. The trunk has a high liftover.

Dealer service bulletins: **1995**—Complaints about shift feel. • Brake pulsation troubleshooting tips. • Heater and AC unit noise after long storage. • Breakage of luggage compartment hinge holder. • A slightly off-centre steering wheel.

Safety summary/Recalls: The car's tendency to under-steer can be unnerving when taking corners at moderate speed. **Recalls**: **1990–91**—Malfunctioning automatic shoulder belts will be replaced, gratis.

Secret Warranties/Service Tips

1992—A front-end snapping noise during tight turns at walking speed is probably caused by excessive clearance between the wheel bearing and steering knuckle. Install modified wheel bearings. • Idle fluctuation when applying an electrical load during idle, or after deceleration from a high rpm, can be corrected by installing a new Electronic Control Unit. • Poor AC performance may be due to insufficient airflow across the AC condenser. • A noisy rear hatch can be silenced by changing the rear hatch glass, hinge, or moulding. • Squeaking noise from the rear pillars is due to interference between the rear inner pillar and its reinforcement. • A sulfur odour coming from the exhaust can be eliminated by installing an upgraded catalytic converter. **1992–94**—Freezing door and hatch lock cylinders are addressed in DSB #021/94. • Clutch squealing can be silenced by installing an upgraded part with a thicker clutch cushioning plate. **1994**—Timing belt noise can be silenced by replacing the tensioner pulley with an upgraded part. **All models/years**—DSB #006/94 gives all of the possible causes and remedies for brake vibrations.

MX-3, Precidia Profile

	1992	1993	1994	1995	1996
Cost Price ($)					
Base	13,725	13,995	15,175	16,070	16,580
GS	16,725	17,065	17,495	18,505	19,240
Used Values ($)					
Base ↑	5,000	6,500	7,500	8,500	10,000
Base ↓	4,500	6,000	7,000	8,800	9,500
GS ↑	6,000	7,500	9,500	10,500	12,000
GS ↓	5,000	7,000	9,000	10,000	11,500
Extended Warranty	N	N	N	N	N
Secret Warranty	N	N	N	N	N
Reliability	③	③	④	⑤	⑤
Air conditioning	❷	③	③	④	④
Automatic transmission	③	④	⑤	⑤	⑤

Body integrity	❷	❷	④	④	④
Braking system	❷	❷	❷	❷	③
Electrical system	❷	❷	❷	❷	③
Engines	⑤	⑤	⑤	⑤	⑤
Exhaust/Converter	③	④	④	⑤	⑤
Fuel system	⑤	⑤	⑤	⑤	⑤
Ignition system	⑤	⑤	⑤	⑤	⑤
Rust/Paint	③	③	④	⑤	⑤
Steering	⑤	⑤	⑤	⑤	⑤
Suspension	③	③	⑤	⑤	⑤
Crash Safety	⑤	❷	—	—	—

Note: The low frontal crash ratings represent severe leg trauma.

MX-5, Miata

Rating: Recommended (1996–98); Above Average (1990–95). The 1998 version is merely a carryovered 1997. **Maintenance/Repair costs**: Higher than average, and dealer dependent. **Parts**: Higher-than-average cost, with limited availability (save money with an aftermarket gel battery).

Strengths and weaknesses: The base 1.6L engine delivers adequate power and accelerates smoothly with a top speed of 115 mph. Acceleration from 0 to 60 mph is in the high 8-second range. The 5-speed manual transmission shifts easily and has well-spaced gears. The vehicle's lightness and 50/50 weight distribution make this an easy car to toss around corners, but it's quite jittery on uneven roads.

Owners' top gripes target the same characteristics that make other sports car enthusiasts swoon: inadequate cargo space, cramped interior for large adults, excessive interior noise, and limited low-end torque, which makes for frequent shifting.

Owners also say it's important to change the engine timing chain every 100,000 km. Other reported problems: 1990–91 model crankshaft failures, leaky rear end seals and valve cover gaskets, torn drive boots, electrical system glitches, brake pulsation, valvetrain clatter on start-up (changing oil may help), prematurely worn-out shock absorbers and catalytic converter, a leaking or squeaky clutch, and minor body and trim deficiencies.

Dealer service bulletins: **1995**—Brake pulsation troubleshooting tips. • Clutch release bearing squeal. • Steering wheel may be off-centre. • Clogged side sill drain holes. • Heater unit noise after long storage. **1996**—Brake pulsation repair. • Side sill paint damage from scuff plate. **1997**—Brake pulsation repair.

Safety summary/Recalls: Eighty percent of Miatas are estimated to have some collision damage. Check your choice before signing the contract. **1995–96**—Inadvertent airbag deployment. • Engine and horn failure. • Roof leaks. • Accelerator pedal cut a slit in the carpet allowing the gas pedal to jam. **Recalls**: **1990–93**—The optional hardtop's hoist accessory kit may have plastic buckles that break and allow the hardtop to suddenly fall. **1991**—Faulty ABS.

Secret Warranties/Service Tips

1990—A hard-to-close trunk lid requires an upgraded rubber cushion (#B48156786). • Poor AC performance is likely caused by the misalignment of the AC harness. • A rattling noise coming from the exhaust manifold may require the replacement of the insulator bracket. • Water may damage door speakers unless a speaker cover assembly (#B4Y5 7696X) is installed. • A musty odour coming from the AC system can be cured by installing an upgraded resin-coated evaporator core (#NA0J 61II0A). **1990–91**—Hard shifting into second gear before the vehicle has warmed up can be corrected by installing an upgraded second gear synchronizer ring and clutch hub sleeve (#JM1NA351-M-232720). **1990–95**—Dirt and debris can clog up side sill drain holes, allowing water to collect and corrosion to occur. Drill larger drain holes. **1992–94**—If the window won't open fully, install a new cable fastener. **1994**—Timing belt noise can be silenced by replacing the tensioner pulley with an upgraded part. **All models/years**—DSB #006/94 gives all of the possible causes and remedies for brake vibration.

MX-5, Miata Profile

	1991	1992	1993	1994	1995	1996	1997	1998
Cost Price ($)								
Base	17,255	17,945	18,895	20,165	21,820	24,210	24,695	24,695
Used Values ($)								
Base ↑	7,500	9,000	11,000	13,500	15,000	17,000	18,500	21,000
Base ↓	6,500	8,000	10,000	12,500	14,000	15,500	17,500	19,500
Extended Warranty	N	N	N	N	N	N	N	N
Secret Warranty	N	N	N	N	N	N	N	N
Reliability	④	④	④	④	④	⑤	⑤	⑤
Air conditioning	④	④	④	⑤	⑤	⑤	⑤	⑤
Body integrity	❷	❷	③	③	③	④	④	④
Braking system	❷	❷	③	③	④	④	⑤	⑤
Electrical system	❷	❷	❷	③	③	④	④	④
Engines	❷	❷	③	④	④	⑤	⑤	⑤
Exhaust/Converter	④	④	④	④	④	④	⑤	⑤
Fuel system	③	④	④	⑤	⑤	⑤	⑤	⑤
Ignition system	❷	❷	③	③	③	④	④	④
Manual transmission	⑤	⑤	⑤	⑤	⑤	⑤	⑤	⑤
Rust/Paint	❷	❷	③	③	③	④	④	⑤
Steering	⑤	⑤	⑤	⑤	⑤	⑤	⑤	⑤
Suspension	⑤	⑤	⑤	⑤	⑤	⑤	⑤	⑤
Crash Safety	⑤	⑤	❷	—	—	④	④	—

RX-7

Rating: Recommended (1991–95); Above Average (1985–90). 1995 was the RX-7's last model year. **Maintenance/Repair costs**: Higher than average, and most repairs have to be done by a Mazda dealer. **Parts**: Higher-than-average cost, with limited availability. Parts shortage may become chronic now that the RX-7 has been discontinued.

Strengths and weaknesses: The RX-7 is an impressive performer that goes from 0 to 100 km in 8 seconds and covers the same distance in 6.2 seconds when equipped with a turbocharger. The ride can be painful on bad roads, though, due primarily to the car's stiff suspension. The GSL and Turbo models are very well equipped and luxuriously finished. Except for some oil-burning problems, apex seal failures, and leaking engine O-rings, the RX-7 has served to dispel any doubts concerning the durability of rotary engines.

Nevertheless, careful maintenance is in order, since contaminated oil or overheating will easily damage the rotary engine. Clutches wear quickly if used hard. Disc brakes need frequent attention paid to the calipers and rotors. The MacPherson struts get soft more quickly than average. AC malfunctions and fuel, exhaust system, and electrical glitches are also common. Be wary of leaky sunroofs. Radiators have a short life span. Rocker panels and body seams are prone to more serious rusting. The underbody on older cars should be inspected carefully for corrosion damage. Fuel economy has never been this car's strong suit.

Age has improved the RX-7's reliability. In fact, dealer service bulletins list only two possible problems with the 1995 RX-7: AC unit noise after long storage and a slightly off-centre steering wheel.

Safety summary/Recalls: **1995**—Wide tires compromise the car's stopping ability in the rain. One driver's RX-7 slipped off the highway and hit a tree. **Recalls**: **1979–83** —Idler arms may corrode, freeze, or break after prolonged exposure to road salt. **1986–87**—Road salt may also cause excessive front brake disc pad liner corrosion. **1989–91**—Shoulder belts may fail. **1991–95**—Fuel hoses may crack and leak fuel. • Poor braking after an overnight park is caused by engine oil mist clogging the brake vacuum check valve. **1993–94**—Engine heat could make fuel hoses deteriorate and leak prematurely, posing a fire hazard.

Secret Warranties/Service Tips

1979–91—Hard shifting into second gear before the vehicle has warmed up can be corrected by installing an upgraded second gear synchronizer ring and clutch hub sleeve (#JM1NA351-M-232720). **1986–88**—A hard-to-start cold engine can be fixed by switching to air gap spark plugs and disconnecting the sub-zero starting assist device. • Misfire and hesitation over 6,000 rpm can be corrected by installing an upgraded air flow meter. • A rough idle may be caused by a short circuit in the water temperature switch wiring. • Erratic power window operation may require a new regulator guide assembly, motor bracket, and sprocket. **1987–88**—Turbo models with a rough idle or cold-starting problems may need a new air control valve (#N332 13990). **1993–94**—To eliminate slipping clutch problems, Mazda may install, at no charge, an improved clutch disc. **All models/years**—DSB #006/94 gives all of the possible causes and remedies for brake vibration.

RX-7 Profile

	1988	1989	1990	1991	1992	1993	1994	1995
Cost Price ($)								
RX-7	19,570	21,840	22,300	24,650	—	—	—	—
Convertible	34,520	40,690	41,670	38,950	—	—	—	—
Turbo	33,020	36,440	37,330	35,890	—	41,820	44,175	48,795
Used Values ($)								
RX-7 ↑	4,000	4,500	5,000	6,500	—	—	—	—
RX-7 ↓	3,500	4,000	4,500	6,000	—	—	—	—
Convertible ↑	8,000	8,500	9,500	12,000	—	—	—	—
Convertible ↓	7,500	8,000	9,000	11,000	—	—	—	—
Turbo ↑	6,000	7,000	8,000	9,500	—	18,000	23,000	28,000
Turbo ↓	5,500	6,500	7,500	8,500	—	17,000	22,000	27,000
Extended Warranty	Y	Y	Y	Y	Y	Y	Y	Y
Secret Warranty	N	N	N	N	N	Y	Y	N
Reliability	③	③	④	⑤	⑤	⑤	⑤	⑤
Air conditioning	❷	❷	③	③	③	③	③	③
Body integrity	③	③	③	③	③	③	④	⑤
Braking system	❷	❷	③	③	③	③	③	③
Electrical system	❷	❷	❷	❷	③	③	③	③
Engines	⑤	⑤	⑤	⑤	⑤	⑤	⑤	⑤
Exhaust/Converter	③	③	③	④	⑤	⑤	⑤	⑤
Fuel system	③	⑤	⑤	⑤	⑤	⑤	⑤	⑤
Ignition system	③	③	③	④	⑤	⑤	⑤	⑤
Manual transmission	③	③	③	④	④	④	④	⑤
- automatic	❷	❷	❷	③	③	③	④	⑤
Rust/Paint	❷	❷	③	④	⑤	⑤	⑤	⑤
Steering	④	④	④	⑤	⑤	⑤	⑤	⑤
Suspension	❷	③	④	⑤	⑤	⑤	⑤	⑤
Crash Safety	❷	—	—	—	—	—	—	—

NISSAN

200SX

Rating: Above Average, if you're not a real high-performance enthusiast (1995–98). The '98 SE-R offered standard AC and ABS, in addition to its peppy 140-hp 2.0L engine. **Maintenance/Repair costs**: Average, and most repairs can be done by any garage. **Parts**: Reasonably priced, but hard to find.

Strengths and weaknesses: Nissan calls this front-drive compact their sporty coupe class leader, but the 200SX is merely a Chrysler Neon fighter that has a sports car flair without the substance. The interior is unnecessarily cramped, and you get only average power and handling with the base powerplant. Later models are much sportier and have more passenger room. Prices are reasonable, though, and the repair history for these cars is average.

All engines have performed reasonably well, although the SE-R's 2.0L has been the most suitable for hard-driving thrills. The clutch has been the source of some complaints, along with occasional electrical malfunctions. Rapid front brake wear and rotor damage are common problems. Body construction is much better than average.

Dealer service bulletins: **1995**—AC system shuts off. • AC compressor leakage. • Brake noise. • Cold-weather starting tips. • C-pillar finisher lifting. • Engine cranks but won't start. • Front window misalignment. • Fuel gauge not indicating full tank. • Horn activates randomly. • Squeak and rattle troubleshooting tips. • Wind noise field fix procedure. • Windshield cracking may be covered under the warranty. **1996**—Engine cranks but won't start or is hard to start. • Squeak and rattle repair.

Safety summary/Recalls: **1996**—Inadvertent airbag deployment. • Dual airbags failed to deploy. • Airbag indicator and engine lights come on intermittently for no reason. • Shoulder seatbelt retractors lock up whenever occupant makes the slightest movement. • Clutch makes grinding, chattering noise when fully depressed. • When the weather turns cold, the emergency brake freezes and won't release. • Faulty sunroof. **Recalls**: **1995**—On vehicles with ABS, the hydraulic actuator may malfunction, increasing the stopping distance.

Secret Warranties/Service Tips

All models/years—DSB #006/94 gives all of the possible causes and remedies for brake vibration. • 1994 and earlier models equipped with RE4RO1A transmissions may experience repeated planetary failure due to poor cooler flow in the fin-type cooler. Apparently, cooler line flushing machines can't flush clear this type of cooler. In the past, the entire radiator would need to be replaced to upgrade the cooling system. Nissan now has a spiral cooler replacement kit (#21606-15V25) that can be used instead.

200SX Profile

	1995	1996	1997	1998
Cost Price ($)				
Base	14,590	15,598	16,128	16,500
SE-R	18,191	20,698	21,228	22,998
Used Values ($)				
Base ↑	9,000	10,500	12,000	13,500
Base ↓	8,000	9,500	11,000	12,500
SE-R ↑	11,000	13,000	15,000	17,500

SE–R ↓	9,000	11,500	13,500	16,000
Extended Warranty	N	N	N	N
Secret Warranty	N	N	N	N
Reliability	⑤	⑤	⑤	⑤
Crash Safety	—	—	⑤	—

240SX

Rating: Recommended (1995–98); Above Average (1989–94). Only the convertible version was carried over in 1994. **Maintenance/ Repair costs**: Average. Most repairs can be done by any garage. **Parts**: Reasonably priced and easy to find.

Strengths and weaknesses: Although this rear-drive sport coupe carried a 2.4L 4-cylinder engine to differentiate it from its 2.0L weaker cousin, the 200SX, it still falls far short of producing high-performance thrills. Its 140-hp base engine provides more than enough torque to handle most driving needs—just don't expect fast acceleration times or a comfortable ride. Handling is impressive, though, thanks to the car's independent suspension. The redesigned 1995 coupe features a longer wheelbase, 15 additional horses, and optional ABS/traction control. It still provides a harsh ride and poor traction on slippery roadways. Head room remains limited, cargo space is practically nil, and the trunk is barely large enough to carry your lunch—as long as you eat light and can fit it through the small opening.

This car doesn't have any major serious shortcomings apart from a cramped interior and excessive engine noise. The few deficiencies reported concern electrical malfunctions, early AC burnout, premature clutch wear, noisy brakes that wear out quickly, exhaust system rust-out, and fit and finish deficiencies.

Dealer service bulletins: **1995**—Engine overheating and poor driveability (Code 45). • AC compressor leaks or is noisy. • Excessive brake noise. • Particular windshield cracks are covered under warranty. **1996**—Engine cranks, but won't start.

Safety summary/Recalls: **1996**—ABS brake failure; pedal went to the floor and brakes failed to stop the vehicle. **Recalls**: **1989–91**—Takata seatbelt replacement. **1995**—Faulty brake warning light.

Secret Warranties/Service Tips

1989—Doors that won't lock from the outside may need a door rod clip modification. • A special service campaign covers the free replacement of the engine timing belt and tensioner. It's worth a fight for a partial refund. **1989–91**—Starting difficulties can often be traced to a connector not fully seated in the ECU. **1990**—Excessive squeaking from the front brakes can be fixed by increasing the front brake pad chamfer. **All models/years**—DSB #006/94 gives all of the possible causes and remedies for brake vibration.

240SX Profile

	1991	1992	1993	1995	1996	1997	1998
Cost Price ($)							
Base	17,890	22,090	22,890	26,990	25,598	26,398	26,698
Used Values ($)							
Base ↑	5,500	6,500	10,000	15,000	16,000	17,000	18,500
Base ↓	5,000	6,000	9,000	14,000	15,000	16,000	17,500
Extended Warranty	N	N	N	N	N	N	N
Secret Warranty	N	N	N	N	N	N	N
Reliability	③	③	③	④	④	⑤	⑤
Air conditioning	❷	③	④	④	④	⑤	⑤
Automatic transmission	③	③	③	⑤	⑤	⑤	⑤
Body integrity	❷	❷	③	③	③	④	④
Braking system	❷	❷	❷	❷	③	③	④
Electrical system	❷	❷	❷	❷	③	④	④
Engines	④	④	④	④	⑤	⑤	⑤
Exhaust/Converter	❷	❷	③	③	③	④	⑤
Fuel system	❷	③	③	④	④	⑤	⑤
Ignition system	③	③	③	④	④	⑤	⑤
Rust/Paint	③	③	③	④	④	⑤	⑤
Steering	❷	❷	④	④	④	⑤	⑤
Suspension	④	④	④	④	⑤	⑤	⑤
Crash Safety	⑤	⑤	❷	③	③	③	—

300ZX

Rating: Above Average (1994–96); Average (1993); Below Average (1988–92). Nissan's answer to the Corvette, this car has everything: high-performance capability, a heavy chassis, complicated electronics, and average depreciation. Turbocharged 1990 and later models are much faster than previous versions and better overall buys. The last model year was 1996. **Maintenance/Repair costs**: Way higher than average, and only a Nissan dealer can repair these cars. Just replacing the air cleaner and spark plugs is a task. Turbocharger repairs can put a big dent in your wallet, as well. **Parts**: Expensive and hard to find.

Strengths and weaknesses: Now Nissan's talking high performance with this weighty rear-drive that offers a high degree of luxury equipment along with a potent 300-hp engine. Traction is poor on slippery surfaces, however, and the rear suspension hits hard when going over speed bumps.

The complexity of all the bells and whistles on the 300ZX translates into a lot more problems than you'd experience with either a Mustang or a Camaro—two cars that have their own reliability problems, but are far easier and less costly to repair. The best example of this is the electrical system, long a source of recurring, hard-to-diagnose shorts. Fuel-injectors are a constant problem and lead to poor engine performance. The manual transmission has been failure-prone, clutches don't last very long, front and rear brakes are noisy and wear out quickly, and the aluminum wheels are easily damaged by corrosion and road hazards. The exhaust system is practically biodegradable. The glitzy digital dash with three odometers and weird spongy/stiff variable shock absorbers are more gimmicky than practical. Body assembly is mediocre.

Dealer service bulletins: **1995**—Poor driveability (Code 45). • The AC compressor leaks or is noisy. • Excessive brake noise. • Particular windshield cracks are covered under warranty. **1996**—Engine cranks but won't start.

Safety summary/Recalls: **1996**—Vehicle suddenly accelerated uncontrollably. **Recalls**: **1979–87**—Nissan will install, at no charge, a shift interlock to prevent sudden acceleration.

Secret Warranties/Service Tips

1990—Clunking when braking in Reverse can be eliminated by installing a retainer on each front brake pad. **1990–91**—Starting difficulties can often be traced to a connector not fully seated in the Electronic Control Unit. **1991**—An automatic transmission that shifts poorly or slips out of gear can be fixed by retorquing the band servo retainer bolts. **1992**—An oil leak from the drive pinion oil seal can be fixed by installing an upgraded seal. **All models/years**—Premature rear brake pad wear may be caused by poorly adjusted brakes or rusty rear rotors. Rear brake pad kit (#D4060-01P90) will help to correct the problem. DSB #006/94 gives all of the possible causes and remedies for brake vibration. • 1994 and earlier models equipped with RE4RO1A transmissions may experience repeated planetary failure due to poor cooler flow in the fin-type cooler. Apparently, cooler line flushing machines can't flush clear this type of cooler. In the past, the entire radiator would need to be replaced to upgrade the cooling system. Nissan now has a spiral cooler replacement kit (#21606-15V25) that can be used instead.

300ZX Profile

	1989	1990	1991	1992	1993	1994	1995	1996
Cost Price ($)								
Base	25,100	38,690	37,600	40,990	41,990	46,890	48,890	53,698
Used Values ($)								
Base ↑	8,000	10,000	12,000	14,000	17,000	21,000	26,000	30,000
Base ↓	7,000	8,500	11,000	12,500	15,500	19,500	24,500	29,000
Extended Warranty	Y	Y	Y	Y	Y	Y	Y	Y
Secret Warranty	N	N	N	N	N	N	N	N
Reliability	❷	❷	③	④	④	④	④	④
Air conditioning	❶	❶	③	③	④	④	⑤	⑤

Body integrity	❶	❶	❶	❶	❶	❶	❷	③
Braking system	❶	❶	❶	❶	❶	❶	❷	❷
Electrical system	❶	❶	❶	❶	❷	❷	③	③
Engines	③	③	③	③	③	④	④	⑤
Exhaust/Converter	❶	③	③	③	④	④	④	⑤
Fuel system	❷	❷	③	③	③	③	④	④
Ignition system	❷	❷	③	③	③	③	④	④
Manual transmission	③	④	④	④	④	⑤	⑤	⑤
- automatic	❷	❷	③	③	③	④	④	④
Rust/Paint	❷	❷	③	③	③	④	⑤	⑤
Steering	③	③	③	③	④	④	④	⑤
Suspension	③	③	③	③	④	④	④	⑤
Crash Safety	—	⑤	⑤	❶	—	—	—	—

TOYOTA

Celica

Rating: Recommended (1995–98); Above Average (1986–94). The 1996 and 1997 models are practically identical; choose the cheapest version. Keep in mind that the GT and GTS have a firmer suspension, better-equipped interior, ABS, and a more sporting feel than do other versions. All handle competently and provide the kind of sporting performance expected from a car of this class. The extra performance in the higher-line versions does come at a price, but this isn't a problem given the high resale value and excellent reliability for which Celicas are known. **Maintenance/Repair costs**: Average, and most repairs can be done by any garage. **Parts**: Reasonably priced and easy to find.

Strengths and weaknesses: The pre-1986 Celicas weren't very sporty. Their excessive weight and soft suspension compromised handling and added a high fuel penalty. With the 1986 make-over, Celicas gained more power and much better handling—especially in the GT and GTS versions—but they're still more show than go, with limited rear passenger room.

Redesigned 1994 models are full of both show and go, with more aerodynamic styling, an enhanced 1.8L that gives more pickup than the ST's 1.6L, and better fuel economy. Among the upgraded models available, smart buyers should choose a used 1994 ST for its more-reasonable price, smooth performance, quiet running, and high fuel economy.

All Celicas offer exceptional reliability and durability. Servicing and repair are straightforward, and parts are easily found. The front-wheel drive series performs very well and hasn't presented any major problems to owners. Prices are high for Celicas in good condition, but some bargains are available with the base ST model.

Owner gripes target the excessive engine noise, limited rear seat room, and inadequate cargo space. Pre-1994 models get the most complaints regarding brakes, electrical problems, AC malfunctions, and premature exhaust wearout. The 1994 models may have a manual transmission that slips out of second gear and hard starts caused by a faulty air flow meter (#22250-74200). Areas vulnerable to early rusting include rear wheel openings, suspension components, the area surrounding the fuel-filler cap, door bottoms, and trunk or hatchback lids.

Safety summary/Recalls: Even if your vehicle has 4X4 capability, it's imperative that snow tires be fitted in order to avoid dangerous control problems on snow and ice. **1990**—Campaign L03 provides for the free replacement of the instrument panel light control switch. • Faulty airbag inflator. **1996**—Sudden brake failure. • Broken brake caliper bolt. • Clogged up idle control valve makes for hard starts. **Recalls**: **1988–89 All-Trac Turbos**—The radiator and coolant will be replaced for free by Toyota.

Secret Warranties/Service Tips

1986–89—Reduce excessive engine ping and jerking by installing an upgraded TCCS Electronic Control Unit under the emissions warranty. **1986–91**—Rattling headrests is a common problem addressed in DSB #B091-010. **All models/years**—Older Toyotas with stalling problems should have the engine checked for excessive carbon buildup on the valves before any more extensive repairs are authorized. • Owner feedback and dealer service managers (who wish to remain anonymous) confirm the existence of Toyota's secret warranty that will pay for replacing front disc brake components that wear out before 2 years/40,000 km. • The decade-old problem of brake pulsation/vibration is fully outlined and corrective measures are detailed in DSB #BR94-002, issued February 7, 1994. • To reduce front brake squeaks on ABS-equipped vehicles, ask the dealer to install new, upgraded rotors (#43517-32020).

Celica Profile

	1991	1992	1993	1994	1995	1996	1997	1998
Cost Price ($)								
Base	16,418	17,949	19,078	21,438	23,878	27,968	28,528	34,138
Used Values ($)								
Base ↑	6,000	7,000	8,000	12,500	15,000	19,500	22,500	27,000
Base ↓	5,500	6,500	7,500	11,500	14,000	18,500	21,500	26,000
Extended Warranty	N	N	N	N	N	N	N	N
Secret Warranty	N	N	N	N	Y	Y	Y	Y
Reliability	③	③	④	④	④	⑤	⑤	⑤
Air conditioning	❶	❶	❷	③	③	③	③	④
Body integrity	❷	❷	❷	❷	③	③	⑤	⑤
Braking system	❶	❶	❷	❷	❷	❷	③	③
Electrical system	❶	❶	❶	③	③	④	⑤	⑤

Engines	③	③	③	③	③	③	④	⑤
Exhaust/Converter	❷	❷	❷	❷	③	③	⑤	⑤
Fuel system	④	④	④	④	④	④	④	④
Ignition system	④	④	④	④	④	④	⑤	⑤
Manual transmission	③	④	④	④	④	④	⑤	⑤
- automatic	③	③	③	④	④	④	④	⑤
Rust/Paint	❷	④	④	④	④	④	④	⑤
Steering	④	④	④	④	④	④	⑤	⑤
Suspension	④	④	④	④	④	④	⑤	⑤
Crash Safety	⑤	—	❶	—	—	—	—	—

Note: The low crash ratings represent severe leg trauma.

MR2

Rating: Recommended (1990–93); Above Average (1988–92); Average (1986–87). The best choice is the 1993 model, which had handling and engine performance improved. The MR2 was discontinued in Canada right after the 1993 model year; in the States, it continued for two additional model years. **Maintenance/Repair costs**: Average. However, most repairs have to be done by a Toyota dealer, and engine repairs can get quite expensive due to the limited access to the engine compartment. **Parts**: Reasonably priced, but not easily found.

Strengths and weaknesses: The MR2 does everything a sports car should, without fuss or surprises. Except for vague steering and some front-end instability, handling is practically flawless. The standard 16-valve 4-cylinder motor is smooth and adequate. The supercharged and turbocharged engines, though overpriced, produce more power than even a sports car driver would want to use. They also have a bit of turbo lag and require deft handling. The engine's placement behind the driver's seat also produces excessive noise and vibration, which make long trips very uncomfortable.

In order of frequency, the most common complaints on all MR2s are as follows: the brakes, electrical glitches, and body hardware (fit and finish) deficiencies.

Safety summary/Recalls: **1993**—Seatbelt and airbag failed during accident, resulting in multiple injuries. • Windshield and windows are coated with a foglike film, reducing visibility.

Secret Warranties/Service Tips

1991—Front brake squeal can be reduced by using new anti-squeal springs (#47743-32030). • Sound system interference can be eliminated by changing the alternator. **All models/years**—Toyota has a secret warranty that will pay for replacing front disc brake components that wear out before 2 years/40,000 km. • The brake pulsation/vibration problem is fully outlined and corrective measures are detailed in DSB #BR94-002, issued February 7, 1994.

MR2 Profile

	1987	1988	1989	1990	1991	1992	1993
Cost Price ($)							
Base	17,188	18,748	22,998	23,917	21,918	25,848	26,628
Used Values ($)							
Base ↑	3,500	4,500	5,000	6,000	7,500	8,500	11,000
Base ↓	3,000	4,000	4,500	5,500	6,500	8,000	10,500
Extended Warranty	Y	Y	Y	Y	Y	Y	Y
Secret Warranty	Y	Y	Y	Y	Y	Y	Y
Reliability	③	③	③	③	④	④	④
Air conditioning	❷	❷	③	④	⑤	⑤	⑤
Body integrity	❷	❷	③	③	③	③	③
Braking system	❷	❷	❷	❷	③	③	③
Electrical system	❷	❷	❷	❷	③	③	③
Engines	⑤	⑤	⑤	⑤	⑤	⑤	⑤
Exhaust/Converter	❷	❷	③	③	③	④	⑤
Fuel system	④	④	④	④	④	④	④
Ignition system	③	③	③	③	③	④	④
Manual transmission	③	③	③	③	③	⑤	⑤
- automatic	③	③	④	④	④	④	④
Rust/Paint	③	③	⑤	⑤	⑤	⑤	⑤
Steering	③	③	③	③	③	⑤	⑤
Suspension	③	③	③	③	③	③	③

Note: The MR2 hasn't been crash-tested by NHTSA.

Supra

Rating: Above Average (1995–97); Average (1994); Below Average (1986–93). Supras are average sports cars that are overpriced and overweight. **Maintenance/Repair costs**: Way higher than average. Only a Toyota dealer can repair these cars. **Parts**: Expensive and frequently back-ordered. 1997 was the last model year, although some '97 Supras were recycled as '98s.

Strengths and weaknesses: This is a nicely styled sports car that has caught the Corvette/Nissan 300ZX malady: cumulative add-ons that drive up the car's price and weight, and drive down its performance and reliability. The 6-cylinder engines are smooth and powerful. Handling is sure and precise—better than the Celica because of the independent rear suspension. The Supra has limited rear seating, fuel mileage is marginal around town, and insurance premiums are likely to be much higher than average. Owners report major engine problems, frequent rear differential replacements, electrical short circuits, AC malfunctions, and premature brake, suspension, and exhaust system wear. The engine is an oil-burner at times, and cornering is often accompanied by a rear-end growl.

Seatbelt guides and the power antenna are failure-prone. Body deficiencies are common.

Safety summary/Recalls: **1994**—Sudden acceleration resulting in collision with a tree. • Airbag failed to deploy. • Electrical system retards the timing of the engine, causing extreme lack of power in warmer weather. • Passenger's window intermittently fails to shut. • Oil-plug drain gasket failure. • Transmission, brake, and AC failures. • Several Michelin tire blowouts. **Recalls**: None that are recent.

Secret Warranties/Service Tips

1985–88—Starter/ring gear clash can be corrected with an upgraded starter assembly (#28100-62011 or -62021). **1986–88**—Excessive brake squeak and groaning can be reduced by installing revised brake pads (#04491-14240). **1989–90**—Cold driveability and startability is improved with a modified TCCS Electronic Control Unit. **1990–91**—A new front brake pad material is now used (#04491-14280) to improve brake pad durability. **All models/years**—Older Toyotas with stalling problems should have the engine checked for excessive carbon buildup on the valves before any more extensive repairs are authorized. • Toyota will pay for replacing front disc brake components that wear out before 2 years/40,000 km. • The brake pulsation/vibration problem is fully outlined and corrective measures are detailed in DSB #BR94-002, issued February 7, 1994.

Supra Profile

	1990	1991	1992	1994	1995	1996	1997
Cost Price ($)							
Base	32,358	34,828	39,108	59,800	70,958	78,738	80,758
Used Values ($)							
Base ↑	9,500	12,000	16,000	38,500	47,000	56,000	65,000
Base ↓	8,000	10,500	14,000	36,00	45,000	54,000	59,000
Extended Warranty	N	N	N	N	N	N	N
Secret Warranty	Y	Y	Y	Y	Y	Y	Y
Reliability	❷	③	③	③	③	④	④
Air conditioning	❷	❷	③	③	③	④	④
Body integrity	❷	❷	❷	❷	③	③	③
Braking system	❷	❷	❷	❷	❷	❷	③
Electrical system	❷	❷	❷	❷	❷	③	③
Engines	③	③	③	④	④	④	⑤
Exhaust/Converter	❷	❷	❷	③	③	③	④
Fuel system	❷	❷	❷	③	③	④	④
Ignition system	③	③	④	④	④	④	⑤
Manual transmission	③	③	③	③	③	④	⑤
- automatic	③	③	③	③	③	④	⑤
Rust/Paint	❷	❷	③	④	④	④	⑤
Steering	③	③	③	③	③	④	⑤
Suspension	③	③	③	③	③	③	③

Note: The Supra hasn't been crash-tested by NHTSA.

MINIVANS

Chrysler launched the minivan concept with its 1984 Caravan and Voyager. Although poorly assembled and riddled with deficiencies, the tall, boxy vehicle was an instant success because it combined fuel efficiency, carlike manoeuvrability, and increased cargo/passenger space in a smartly styled "garageable" van. On the other hand, Volkswagen's minivan—morphed into the EuroVan in the early '90s—had been on sale for over 35 years and never did very well. Used mostly as a camper, it suffered from decades-old styling, poor dealer support, insufficient heating, glacial acceleration, and a reputation for being reliably unreliable.

Fortunately, while Chrysler and VW have apparently stood still from a quality improvement standpoint, the competition has caught up and, in the case of Honda and Toyota, surpassed the two originators of the minivan concept. In fact, vehicles produced during the past few years use more powerful engines; offer a greater variety of powertrains; and have better crash ratings, more safety features (airbags, etc.), more responsive road handling, better quality control, more competitive prices, and greater parts availability.

If you must buy a minivan, remember that, much like sport-utilities, they usually fall into two categories: up-sized cars and down-sized trucks. The up-sized cars are "people movers." They're mostly FWD, handle like a car, and get great fuel economy. GM's Lumina, Silhouette, Trans Sport, and Venture are the best examples of this kind of minivan. In fact, their road performance and reliability surpass both Chrysler and Ford's front-drive minivans. GM's Astro/Safari and the Ford Aerostar, on the other hand, are down-sized trucks. Using rear-wheel drive, 6-cylinder engines, and heavier mechanical components, these minivans handle cargo as well as passengers. Towing capacity varies between 3,000 and 5,000 lb. On the negative side, fuel economy is no match for the FWD minivans, and highway handling on rear-drives is also more trucklike. Also, post-1995 rear-drive GM and Ford minivans are more reliable performers than the Ford Windstar or Chrysler offerings.

Minivan ownership costs are quite reasonable, according to the American management consulting firm Runzheimer International, which has concluded that it costs less to operate a minivan than it does many compact cars. Still, most minivans are overpriced, and motorists needing a vehicle with large cargo and passenger carrying capacity should consider a cheaper GM Vandura or Chevy Van, even if it means sacrificing some fuel economy and convenience features (they can be added by most conversion shops at competitive prices). You just can't beat the excellent forward vision and easy-to-customize interiors that these large vans provide.

Recommended

Ford Mercury Villager, Nissan Quest (1997–98)

Above Average

Ford Mercury Villager, Nissan Quest (1995–96)
Honda Odyssey (1995–98)
Nissan Axxess (1993–95)
Toyota Previa (1991–97)
Toyota Sienna (1998)

Average

GM APV, Lumina, Silhouette, Trans Sport, Venture (1995–98)
Ford Aerostar (1995–97)
GM Astro, Safari (1996–98)
Mazda MPV (1996–98)
Nissan Axxess (1991–92)

Below Average

Chrysler Caravan, Ram Van, Town & Country, Voyager (1997–98)
Ford Windstar (1996–98)
Mazda MPV (1988–95)

Not Recommended

Chrysler Caravan, Ram Van, Town & Country, Voyager (1984–96)
Ford Aerostar (1986–94)
Ford Mercury Villager, Nissan Quest (1993–94)
Ford Windstar (1995)
GM APV, Lumina, Silhouette, Trans Sport (1990–94)
GM Astro, Safari (1985–95)
Nissan Axxess (1987–90)
Toyota LE (1984–90)

CHRYSLER

Caravan, Ram Van, Town & Country, Voyager

Rating: Below Average (1997–98); Not Recommended (1984–96). Don't consider any model year without being armed with the longest powertrain warranty you can afford. Other minivans you may wish to consider: the GM front-drive minivans, Mercury Villager, Nissan Quest, or Toyota Sienna. Major generic defects affecting the 1984–96 models make these minivans very risky buys. In spite of the poor reliability of the early versions, Chrysler minivans remained popular because Chrysler's comprehensive 7-year warranty paid for their shortcomings. But that warranty's gone and the defects remain. Interestingly, the defects affecting Chrysler's '96 and later minivans are similar to those failures we have seen for over a decade. All the more reason to be skeptical of Chrysler's claims that quality control was improved with its last redesign. **Maintenance/Repair costs**: Higher than average, but any garage can repair these minivans. **Parts**: Expensive (especially for paint, AC, transmission, and ABS components, which are covered under a number of "goodwill" warranty programs and several recall campaigns).

There's presently an abundance of used Chrysler minivans on the market; however, very few have any of their original warranty coverage left. Count on spending at least another $1,000 for a supplementary warranty to protect yourself from Chrysler's costly generic defects. Just ABS and transmission repairs alone could cost you $5,000. One radio listener recently asked me if my advice to get the extended warranty on the Caravan or Ford's Windstar should be seen as a "red flag" as to the reliability of these vehicles. Yes, that's exactly what it is.

Strengths and weaknesses: Chrysler's minivans dominate the new- and used-minivan market because they offer pleasing styling and lots of convenience features. They also ride and handle better than most truck-based minivans (although there's lots of room for improvement) and can carry up to seven passengers in comfort. Cargo hauling capability is more than adequate, with a 3,000 lb. maximum towing range.

On the downside, these minivans pose maximum safety risks due to their poor crashworthiness and chronic mechanical failures. Worse still, owners report of experiencing a host of bizarre "happenings" with their minivans, suggesting that these vehicles require the services of an exorcist rather than a mechanic (seatbelts that may strangle children, airbags that deploy when the ignition is turned on or the floor gets wet, and stalling when within radar range of airports or military installations.

Powertrain problems

These minivans are way underpowered with the base 4-cylinder engine. It has a history of head gasket failures, but it outshines the larger Mitsubishi 2.6L, which is guaranteed to self-destruct just as the warranty expires. The timing belt, piston rings, and valves are particular weak points on the 2.6L powerplant. The two-piece camshaft oil seals are also prone to sudden leaks. The Chrysler-built 2.5L engine is fairly dependable but sluggish, and the 3.3L V6 is the most reliable. The Mitsubishi 3.0L V6 is much more reliable than its smaller 2.6L version, but it lacks power on long climbs, has multiple fuel-injection and oil leak problems, and produces a loud piston "slapping" noise during cold starts. Canada's climate presents additional problems for Chrysler minivan owners: one 1989 Voyager owner from Red Deer, Alberta, says that dealers in his area claim that his 3.0L engine seized because of poor cold-weather oil lubrication. If this is true, it spells trouble for the many owners who have to drive in cold-weather conditions and are now nearing the end of Chrysler's original warranty coverage.

Automatic transmissions built over the past decade are a nightmare (see "Safety summary/Recalls"): imagine having to count to three in traffic before Drive or Reverse will engage, or "limping" home on the highway in second gear at 50 km/h. The disastrous A604 4-speed automatic was particularly troublesome on 1988–91 models. Then it was renamed—and continued to pile up complaints. Chrysler, stung by critics' charges that these transmissions were lemons, pledged that a free oil cooler bypass valve would be installed on 1989–90 minivans to prevent transmission damage in cold temperatures. This program is outlined in Customer Satisfaction Notification #281T.

Automatic transmissions on the post-1991 versions have been known to leak, gear down to a "limp-in" mode, shift noisily, and hunt for the proper gear—the likely cause of some of the poor gas mileage claims. These factory-related defects on recent minivans and other models are listed in DSB #18-24-95.

NO: 18-24-95
DATE: June 23, 1995
SUBJECT: Improved Transmission Shift Quality
MODELS:

1989 – 1995	(AA) Acclaim/Sprint/LeBaron Sedan
1989 – 1993	(AC) Dynasty/New Yorker/New Yorker Salon
1990 – 1993	(AG) Daytona
1990 – 1995	(AJ) LeBaron Coupe/LeBaron Convertible
1993 –1994	(AP) Sundance/Shadow/Shadow Convertible
1990 – 1991	(AQ) Chrysler TC
1989 – 1995	(AS) Caravan/Voyager/Town & Country
1990 – 1993	(AY) Imperial/New Yorker Fifth Avenue
**1995	(FJ) Sebring/Avenger/Talon**
1995	(JA) Cirrus/Stratus
1993 – 1995	(LH) Concorde/Intrepid/Vision/LHS/New Yorker

NOTE:
THIS BULLETIN APPLIES TO VEHICLES EQUIPPED WITH THE 41TE OR 42LE TRANSAXLE.
SYMPTOM/CONDITION:
Vehicles that operate at speeds where EMCC usage is engaged (vehicle speeds between 34 and 41 MPH), may experience early deterioration of the transmission fluid (15,000 to 30,000 miles), exhibit a pronounced shudder during EMCC operation, harsh upshifts/downshifts, and/or harsh torque converter clutch engagements. Performing REPAIR PROCEDURE # 2, which includes updates to the Transmission Control Module (TCM) calibration and eliminates EMCC, will resolve these symptom/conditions. The TCM calibration used in the 1995 model year 41TE arid 42LE TCM is being made available for all vehicles dating back to the 1989 model year. The shift quality improvements and default issues that will be corrected by the new TCM calibration are:

1. COASTDOWN TIP-IN BUMP: Vehicle is decelerated almost to a stop (less than 8 MPH), then the driver tips back into the throttle to accelerate a noticeable bump may be felt.
2. COASTDOWN SHIFT HARSHNESS: Harsh coastdown shifts on some 4–3, 3–2 and 2–1 downshifts.
3. 1995 LH WITH 42LE TRANSAXLE — SLUGGISHNESS/LACK OF RESPONSE: On some early 1995 LH vehicles built prior to Oct. 24, 1994, a perceived lack of power or transmission responsiveness may be encountered under normal operating conditions. The transmission may not release the converter clutch as desired with increased throttle. This occurs in 4th gear from 35 MPH to 50 MPH.
4. 1989–1994 WITH 41TE & 42LE TRANSAXLES: Harsh shifts and/or vehicle shudder during 3–2 or 2–1 kickdowns at speeds less than 25 MPH.
5. 1993 WITH 41TE TRANSAXLE: Harsh 3–4 upshifts may occur, especially at highway speeds, while using the speed control.
6. 1989–1994 WITH 41TE TRANSAXLE — HARSH/DELAYED GARAGE SHIFTS: Delay is less than 2 seconds and the shift is harsh after the brief delay. NOTE: Delays greater than 2 seconds are caused by transmission hardware malfunction, i.e., valve body, pump, failed lip seals or malfunctioning PRNDL or neutral start switch.
7. 1989–1994 WITH 41TE & 42LE TRANSAXLES—POOR SHIFT QUALITY AFTER A BATTERY DISCONNECT: All transmission learned values are reset to the factory default values if battery power is lost to the TCM. The new 1995 calibration will now retain all learned values in memory after battery disconnect. However, if a transmission is rebuilt or a new transmission or TCM is installed. the Quick Learn procedure must be performed to calibrate Clutch Volume Indexes (CVI)on 1993 and later vehicles. (1992 and prior vehicle cannot be Quick Learned).
8. EARLY 1993 WITH 41TE & 42LE TRANSAXLE — INTERMITTENT SPEED CONTROL DROP OUT: The new service calibration change corrects this condition (this condition was also covered in Technical Service Bulletin 08-09-93 dated Mar. 12, 1993).
9. 1989–1993 WITH 41TE & 42LE TRANSAXLES — New fault code 35 (failure to achieve pump prime) has been added for improved diagnostic capability and fault codes 21, 22 and 24 are de-sensitized to reduce erroneous limp-in conditions.

POLICY: Reimbursable within the provisions of the warranty.

DATE: Feb. 17, 1995
SUBJECT: Delayed Transaxle
MODELS:

1993 – 1995	(AA)	Spirit/Acclaim/LeBaron Sedan
1993	(AC)	Dynasty/New Yorker/New Yorker Salon
1993	(AG)	Daytona
1993 – 1995	(AJ)	LeBaron Coupe/LeBaron Convertible
1993 – 1994	(AP)	Shadow/Shadow Convertible/Sundance
1993 – 1995	(AS)	Town & Country/Caravan/Voyager
1993	(AY)	Imperial/New Yorker Fifth Avenue
1993 – 1995	(ES)	Chrysler Voyager (European Market)
1993 – 1995	(LH)	Concorde/Intrepid/LHS/New Yorker/Vision

SYMPTOM/CONDITION: Intermittent delayed transmission engagement at vehicle start up (garage shift) in excess of 3 seconds. The transaxle may be cold or hot.
Other possible conditions that nay cause delayed transaxle engagement are a sticking or frozen PRNDL switch, or a transaxle front pump with excessive gear clearances.
PARTS REQUIRED:

AR 4659085	Valve Body, 41TE
AR 4504048	Package, Oil Filter 41TE
AR 4659084	Valve Body, 42LE
AR 4796730	Package, Oil Filter 42LE
AR 4467721	Fluid, Mopar Automatic Transmission — Type 7176

REPAIR PROCEDURE: This bulletin involves replacing the transaxle valve body and oil filter.

These bulletins clearly show that Chrysler is well aware of its automatic transmissions' propensity for shuddering, delayed engagement, and eventual self-destruction.

Other mechanical weaknesses in early models include the premature wearing out of front suspension components, wheel bearings, front brake discs, brake master cylinder, water pump, air conditioning unit, engine cooling system, and manual transmission clutch. Fuel-injectors on all engines have been troublesome, and engine supports may be missing or not connected.

Overall fit and finish has gotten worse, not better, over the years (see "Secret Warranties/Service Tips"). Body hardware and interior trim are fragile and tend to break, warp, or fall off (door handles are a good example). After about a year's use, the Caravan and its various spin-offs become veritable rattle boxes, with poorly anchored bench seats being a major player. Finish problems can be summed up in three words: paint, paint, paint. The paint tends to discolour or delaminate after the second year. Chrysler knows about this problem and often tries to get the owner to pay half the cost of a repainting job (about $1,500 on a $3,000 job) before agreeing to the total cost if the owner stands fast, threatens small claims court action, or belongs to a consumer protection group like Vancouver-based CLOG (Chrysler Lemon Owners Group). Moreover, minivan owners fed up with Chrysler's refusal to repaint their vehicles have filed a class action in Washington State, seeking damages for owners of 1986–97 cars, sport-utilities, minivans, and trucks (see Part Two for the full text of the lawsuit).

I understand that similar class actions will be filed in the near future in Canada relative to Chrysler paint delamination and automatic transmission defects. Stay tuned to the *www.lemonaidcars.com* website for the details as they become available.

Mechanical weaknesses on later models include the premature wearing out of the engine tensioner pulley, motor mounts, starter motor, front brake discs and pads (the brake pad material crumbles in your hands), brake master cylinder, suspension components, exhaust system components, ball joints, wheel bearings, water pumps, fuel pumps and pump wiring harnesses, radiators, heater cores, and AC units. Fuel-injectors on all engines have been troublesome, the windshield wiper washers freeze up in cold temperatures, sliding doors malfunction, engine supports may be missing or not connected, and there are frequent power-steering pump leaks. Two further problems are batteries that last 9–12 months and factory-installed tires that fail prematurely (at 30,000–35,000 km) and are hard to find—especially in the LT rating.

Cold-weather problems abound. One Hamilton, Ontario, owner says that her 1992 Voyager's rear heater coolant tubes were so badly corroded they had to be replaced after two years at a cost of $160. Radiators and AC lines also quickly succumb to Canada's climate.

Chrysler owners fight back

Chrysler's repeated attempts to blame drivers for its ABS, automatic transmission, and paint delamination problems continue to arouse consumer anger, despite the fact the company has settled many of its complaints through CLOG. The CBC program *Marketplace* has made several references to these deficiencies; the Internet is replete with postings relating to failures all over the world (Chrysler problems: *www.wam.umd.edu/~gluckman/Chrysler/* or *z.simplenet.com/cc/fix.html*; Jeep problems: *www.goofball.com/keane/badpaint/*); and Chrysler is currently the defendant in half a dozen class actions in the States, including the aforementioned Washington State action.

Furthermore, about 600 Chrysler owners have helped form Chrysler Lemon Owners Groups (CLOG) in British Columbia and New Brunswick. These groups have submitted members' names to the automaker and have succeeded in getting sizeable refunds for brake, transmission, and paint repairs. If you have had any of these problems and want "goodwill" repairs or a refund for repairs already carried out, go through Chrysler's regular customer relations hotline, as outlined below. If this doesn't work within a reasonable period of time, phone, fax, or email Mr. Lou Spadotto, National Service Manager, Chrylser Parts, Service, and Engineering, at: Tel: 519-973-2300; Fax: 519-561-7005; Email: Rar17@Chrysler.com. He works closely with a Review Committee that will review all claims, including those that were previously rejected.

Independent garage repairs

Spadotto and his colleague Larry Latta, V.P., Sales and Marketing, have committed themselves to treating each claim under review in an expeditious, thorough, and courteous manner. In the year since I published last year's *Lemon-Aid* guide, I can say that these gentlemen haven't reneged on their commitment. We do have our differences, however.

A major one is the way we both look at repair claims arising from work done by independent agencies. I'm going to spend some time here on the subject because I'm also getting feedback from Ford owners that they are being routinely denied transmission and engine head gasket refunds because they went to an independent garage. Apparently, even those repairs done by independents to correct well-documented factory defects stand a greater chance of being denied than those claims involving a dealer.

Both Chrysler and Ford maintain that independent agencies make it harder to prove the repairs were really necessary (I'd argue the contrary) and that to pay such claims undercuts the dealers' customer loyalty. My response is the following: We are dealing with well-known product defects, and the independent garage is available to show the repair was necessary to correct a factory-related defect and attest that nothing was done by the owner to either create or exacerbate the problem. I also believe that the independent garage factor should be considered as only one element in assessing the claim. Again, the owner may have had no choice but to go to an independent garage, for such reasons as follows:

- It was an emergency: no dealer was nearby or the dealer was closed.
- The car was off warranty, and the owner had no idea warranty "goodwill" was available. And dealer repairs would have cost a lot more.
- The dealer had been uncooperative in the past.

Spadotto and Latta have agreed that these other elements are important considerations and add they won't routinely deny these kinds of claims in a knee-jerk fashion without looking at the above factors.

But I have another point to add about independent repairs, and this time I'll get just a little bit legalistic. In addition to claiming a warranty, or a "goodwill" refund from Chrysler, use provincial consumer protection laws and jurisprudence to your advantage. Point out that:

- A product must be reasonably durable (use my "Estimated Part Durability" chart in Part Two).
- A defect brought to the dealer's attention or manifesting itself during the warranty period that's not corrected is still warrantable after the warranty expires (see *Lowe v. Fairview Chrysler-Dodge Limited and Chrylser Canada Limited* in Part Two, "Repairs: Faulty diagnosis"; and *Shields v. General Motors of Canada* in Part Two, "Getting a Settlement").
- A defect that was once subject to a special "goodwill" program establishes a benchmark for owners of similar vehicles not covered by that special program. For example, you have already read that Chrysler has a 7-year "goodwill" warranty for AC evaporators on its 1993–97 Jeep Grand Cherokees and LH cars (Concorde, Intrepid, LHS, New Yorker, and Vision), and your Caravan, Cherokee, or truck has the same problem. It stands to reason that you can argue under provincial consumer protection statutes that Chrysler's

own program maintains that an evaporator *should last* a minimum of seven years. You also wouldn't be remiss in adding that your vehicle cost a lot more or was exceptionally well-maintained (see *Bentley v. Dave Wheaton Pontiac Buick GMC Ltd and General Motors of Canada* in Part Two, "Getting a Settlement").

- Finally, you shouldn't be penalized for going to an independent if the dealer didn't disclose the "goodwill" warranty. To rule otherwise would be against public interest and would encourage dealers and automakers to keep these programs secret for as long as possible.

All set to make a claim? Lou Spadotto advises owners to take the following steps before calling or emailing his office:

- Discuss the matter with a Chrysler dealer Service Manager; then with the General Manager, if necessary.
- Give the dealership a reasonable length of time or number of opportunities to satisfy you. The dealer may contact Chrysler on your behalf.
- If the problem still has not been resolved, contact Chrysler at: Chrysler Canada Customer Service, Chrysler Center (why not the Canadian spelling, "Centre"?), P. O. Box 1621, Windsor, Ontario, N9A 4H6. You may also call toll-free: 1-800-465-2001.

When you phone Chrysler, have the following information handy:

- Your name, address, and phone number.
- The name of your dealer.
- Your Vehicle Identification Number (VIN) found on the dash or registration.
- The vehicle's current mileage.

Dealer service bulletins: **1996**—3.3L and 3.8L lower engine oil leaks. • Coolant seepage from rear heater hose connections. • Cold-start stumble. • Rough idle, hesitation, or sags after fuel tank is filled (see "Secret Warranties/Service Tips"). • Intermittent driveability problem near radar (again, explained more fully in "Secret Warranties/Service Tips"). • Intermittent powertrain shudder. • Transmission limp-in caused by a faulty speed sensor. • Reduced limp-in default sensitivity. • Excessive transmission downshifting/upshifting in cruise control. • Dealers will install an upgraded overdrive clutch hub. • Difficulty going into second gear or Reverse after a cold start. • Upshift shuddering. • ABS activates below 10 mph.• Front wipers activate while driving or will not turn off. • Vehicle drifts or leads at high speeds. • Underbody squeaks, buzzes, and rattles. • Exhaust drone at 2,500–2,800 rpm. • B-post area rattling. • Blower motor whine and AC-related moan/whine. • Power steering produces a clunk or popping noise at highway speeds. • Steering noise during parking lot manoeuvres. • Front door squeak/creak noise. • Ratcheting sound when coming to a stop. • Rear wheel rattle or click noise. • Rattling rear bench seat. • Rear brake noise (see "Secret Warranties/Service Tips"). • Integrated child safety seat seatbelt retractor may restrict seatbelt travel. Dealer will replace assembly at no charge under Customer Satisfaction Note

#650. • Child seat shoulder harness won't pull out. • Discoloured cowl grille or outside rear view mirrors. • Poor cowl cover fit. • Dust intrusion into rear of vehicle. • False info on fuel tank capacity and inaccurate fuel gauge. • Hard to unlatch rear bench seats. • HVAC control knobs or buttons may stick. • Interior window film buildup. • Intermittent operation of sliding door locks. • Liftgate to rear fascia gap too small. • Paint chips at the upper front corner of the sliding doors. • Poor AM radio reception and RAS radio cassette malfunctions. • Faulty power vent windows. • Loose quad seats. • Roof panel is wavy or has depressions. • Inoperative sliding door and liftgate power lock motor; the sliding door may be difficult to open from the outside. • Suction-cup marks on door glass. • Unexplained theft alarm activation or dead battery. • Water leaks onto carpet from HVAC housing. • White stress marks on interior trim panel. • Flying hubcaps (see "Secret Warranties/Service Tips"). **1997**—The defects affecting the newest minivans are similar to those failures we've seen for over a decade (see following DSB). All the more reason to be skeptical of Chrysler's claims that its quality control has improved since the 1996 model was redesigned.

DSB SUMMARY	1997 CHRYSLER CARAVAN
1. MAY-97	41TE/42LE TRANSMISSION SERVICE INFORMATION
2. OCT-96	A/C ERROR STARTER, EATX, CRUISE CONTROL...
3. JUL-97	A/C EVAPORATOR ODORS
4. JUN-97	A/C SUCTION AND/OR DISCHARGE LINE SERVICE
5. FEB-97	AIR BAG WARNING LABELS
6. OCT-96	BRAKE NOISE
7. FEB-97	CHALKY RESIDUE ON BLACK PLASTIC BODY COMPONENTS
8. JAN-97	COMPASS (CMTC) AND/OR INSTRUMENT CLUSTER MALFUNCTION
9. MAR-97	COOLANT SEEPAGE AT UNDERBODY AUXILIARY REAR HEATER LINE
10. APR-97	CUSTOMER NOTIFICATION #712—PASSENGER AIR BAG SAFETY
11. JUL-97	CUSTOMER SATISFACTION NOTICE #721—CHILD SAFETY SEAT
12. SEP-96	DINGS FROM INSIDE-OUT AT SLIDING DOOR(S) LATCH AREA
13. MAY-97	ENGINE COOLANT USAGE
14. JAN-97	ERRONEOUS M L ON WITH HEX CODE $29, $2A, $2B
15. MAY-97	FRONT HVAC BLOWER MOTOR RESISTOR CONNECTOR SERVICE
16. JAN-97	FUEL TANK SERVICE
17. MAY-97	HIGH EFFORT REQUIRED TO UNLATCH REAR BENCH SEATS
18. DEC-97	HONK NOISE DURING LOW SPEED MANEUVERS
19. SEP-97	INOPERATIVE CD PLAYER/SALES CODE RAZ RADIO
20. OCT-97	INTERMITTENT MOMENTARY LOSS OF POWER ASSIST
21. NOV-96	INTERMITTENT/INOPERATIVE RADIATOR FANS
22. MAY-97	LEAK DETECTION PUMP (LDP) DEALER TEST MODE
23. SEP-97	NHTSA AUTHORIZED AIRBAG DEACTIVATION / MEDICAL NECESSITY
24. NOV-97	POOR DRIVEABILITY WITH HIGH DI FUEL
25. FEB-97	REAR BRAKE CYCLIC RUBBING NOISE
26. FEB-97	REAR BRAKE MOAN/HOWL NOISE
27. NOV-97	REMOTE KEYLESS TRANSMITTER BATTERY FAILURE
28. NOV-96	REPAIR OF FALLOUT DAMAGED PAINT
29. APR-97	SAFETY RECALL #714/MASTER CYLINDER PRIMARY PISTON
30. MAY-97	SAFETY RECALL #724 — STAMPED STEEL ROAD WHEELS
31. NOV-97	SERVICE MANUAL REVISION
32. JUN-97	SERVICE MANUAL, REVISIONS
33. MAR-97	SLIDING DOOR — POOR FIT AT REAR

34. OCT-97	THUMPING NOISE AT REAR OF VEHICLE DURING COLD OPERATIONS
35. JUN-97	TIRE & WHEEL RUNOUT
36. MAR-97	TRANSAXLE SHUDDER DURING EMCC SHIFT
37. JUN-97	TRANSMISSION FLUID AUXILIARY COOLER
38. JAN-97	UNDER BODY CREAK OR KNOCK SOUND
39. SEP-97	WATER LEAKS ONTO FLOOR FROM HVAC HOUSING
40. JUN-97	WIPERS DON'T PARK OR WIPE IN INTERMITTENT MODE

Déjà vu, all over again!

Safety summary/Recalls: Chrysler continues to downplay the seriousness of its minivan safety defects, whether in the case of ABS failures, inadvertent airbag deployments, or faulty rear latches. Nevertheless, the company has recently lost several important lawsuits concerning airbag dangers (a $69 million U.S. class action) and faulty minivan rear latches (see Part Two).

Following a sustained Internet campaign against Chrysler—led by irate car and minivan owners and buttressed by thousands of complaints registered by safety investigators on both sides of the border—the automaker has announced a safety recall of 1991–93 minivans and passenger cars in order to correct life-threatening ABS malfunctions and extend the base warranty to 10 years/160,000 km on ABS components (not calipers, pads and shoe linings or other maintenance items). Two other ABS components, piston seals (excessive wear) and the pump motor (deterioration), will be repaired free of charge at any time during the life of the vehicle. Drivers of these defective minivans and passenger cars—1991–93 Fifth Avenue, Imperial, New Yorker, Salon, Dynasty, Monaco, and Premier—that experience any kind of ABS malfunction should go to the nearest Chrysler dealer and demand the ABS be checked at Chrysler's expense. **1991–93**—Seatbelts may become unhooked from the floor anchor. **1994**—Airbags deploy when the vehicle is started. **1994–95**—Airbags and wiper motor share the same fuse; if the wiper motor fails, the airbags are deactivated. **1995**—Engine fires. • Thieves love the door lock design. • Broken spare tire suspension cable allows tire to fall away while driving. • Right side rear door suddenly flies open when vehicle passes over a small bump. • Seatbelt buckles jam or suddenly release. • Child shoulder harness clip easily pulls out. • Roof drip rails allow water to leak inside. • Open glove box back section allows for papers to be sucked into the AC blower. • Inoperative horn. **1995–96**—Brake failures, lockup, excessive noise, and premature wear. • Airbag fails to deploy or accidentally deploys. • Injury from airbag. • Sudden acceleration, stalling. • No steering. • Seatbacks fall backwards. • Fuel tank is easily damaged. • Park won't hold vehicle. • Transmission fails, suddenly drops into low gear, won't go into Reverse, delays engagement, or jumps out of gear. • Rear windows fall out or shatter. • Power window and door lock fail. • Sliding door jams, trapping occupants. • Weak headlights. **1996**—Driver's side airbag deploys when the ignition is turned on. • Child safety seat harness over-retracts, trapping children or catching their hair. • Fuel tank leaks from tank top and fuel rail. • Incorrect fuel gauge. • Fuel sloshes around (no baffle), deforms tank, and makes noise. • Fuel leaks from vapour recovery canister. • Broken steering belt tensioner causes the sudden loss of power steering and power brakes. • Wipers self-activate. • Sliding door falls off. • Faulty power door locks. • Cracked axle/drive shaft. • Cruise control drops speed and then surges to former setting. • With AC engaged, vehicle stalls, then surges forward. • Power-steering failure, excessive noise. • Vehicle parked with gear in Park and with emergency brake applied rolled into a lake. **All models/years:** Owners report that cruise control units are often dysfunctional, accelerating or decelerating the vehicle without any warning, and that sudden stalling and transmission failures also create life-threatening situations. One Toronto owner of a 1993 SE equipped with a 3.3L engine calls the transmission malfunction a safety hazard:

> I have experienced a transmission control module failure where the vehicle immediately dropped into second gear. This could have been tragic if it had occurred on Highway 401 in heavy traffic.

Recalls: **1984–95**—All minivans are subject to a voluntary service campaign that will fix a rear liftgate latch that may fail in a collision. **1985**—Dealers will install a protective cover over the brake proportioning valve. • The fuel supply tube leaks on vehicles equipped with a 2.2L engine. **1986–88 passenger models**—First rear seats may detach in an accident. **1988**—Possible fuel tank leakage. **1988–89**—Notification #281T, applicable to vehicles with a trailer towing package, provides for a free oil cooler bypass valve to prevent transmission failure in cold weather. **1989–90**—Engine valve cover gasket may leak oil, creating a fire hazard. • Notification #466 provides for a free engine valve spring. • Safety recall #314T provides for the free installation of a reinforcing plate on the front seatbelt strap. **1990**—An incorrectly mounted proportioning valve may increase the chance of skidding. • Notification #281T extends the free oil cooler bypass valve program to cars and 1990 minivans. **1991**—Faulty turn-signal flasher. **1991 ABS**—Hydraulic fluid leakage. **1991–92**—Steering wheel cracks may cause wheel to loosen. **1991–93**—Recalled to fix two kinds of seatbelt problems: faulty buckle cover may prevent seatbelt from being fully latched; and seatbelt anchor hook could become detached from the anchor. **1991–93**—Transport Canada Defect Investigator Nigel Mortimer says that the Chrysler ABS recall calls for the installation of a new pump and seal kit, if needed. The kit comes with a lifetime warranty, while all other ABS components will henceforth be covered for 10 years/160,000 km. 1989–91 Jeep Cherokees and Wagoneers are also included in this safety recall and service campaign. **1992**—Replace liftgate supports, which may break from fatigue. • Safety recall

#326T requires the replacement of all brake pedals that have been found to lack sufficient strength. • Faulty steering column shaft coupling bolts. • Improperly bent fuel tank flanges could cause a fire. **1993–94**—Rear liftgate struts and bolts will be replaced to prevent liftgate from falling down. **1995–98**—Airbags may deploy for no reason. **1996**—Faulty bench seat attaching bolts. • Filler tube rollover valve. • Installation of a fuel-filler ground strap. • Improved retractors for child safety seats (service action) and upgraded seat module bolts (recall). • Defective engine cylinder head plug could cause a fire. **1997**—Goodyear Conquest tires will be replaced. • Dealers will clean the child safety seat latch and add a belt extender for anchoring the belts. • Brake master cylinder seals may be defective, allowing fluid to be drawn into the power-assist reservoir. • Wheels may have been damaged during mounting. **1998**—Install upgraded child safety seat belts.

Secret Warranties/Service Tips

1987–92—The heater and air conditioning system may suddenly change to the defrost mode during a low vacuum condition, which can occur during trailer towing, hill climbing, and acceleration. Install a revised vacuum check valve to correct this problem. **1987–94**—3.0L engines that burn oil or produce a smoky exhaust at idle can be fixed by installing snap rings on the exhaust valve guides and replacing all of the valve guide stems or the cylinder head. **1988–90**—Intermittent rough running at idle signals a need for a new EGR. **1988–94**—A sticking AC heater blend door can be corrected by spraying an anti-rust penetrant into the assembly. **1989–90**—Oil leaks from the cylinder head cover with 2.5L engines are caused by poor sealing. The original cylinder head cover must be replaced with one that uses silicone sealant (RTV) instead of a gasket (DSB # 09-17-89). • Defective valve stem seals are the likely cause of high oil consumption with 2.5L engines (DSB #HL-49-89C). • A604 automatic transmission clutch slippage is a common problem addressed in DSB #21-09-9. • A surge/buck at 35–55 mph with an automatic transmission can be corrected by installing driveability kit #4419447. **1989–95**—An excellent summary of Chrysler's transmission glitches and corrections covering these seven model years can be found in DSB #18-24-95. **1989–96**—Acceleration shudder that may be accompanied by a whine is likely the result of leakage in the transmission front pump, caused by a worn pump bushing. **1990–92**—Erratic idle speeds occurring after deceleration from a steady cruising speed can be corrected by replacing the idle air control motor with a revised motor. **1990–94**—Harsh automatic shifts can be tamed by installing the following revised parts: kickdown, accumulator, reverse servo cushion springs, and the accumulator piston. • Cold-start piston knocking noise can be eliminated by replacing the piston and connecting rod assembly (see following page).

NO: 09-03-94
GROUP: Engine
DATE: Apr. 1, 1994
SUBJECT: Engine Idle Piston Noise
MODELS:

1990-1993	(AC)	Dynasty/New Yorker/Salon
1990-1994	(AS)	Caravan/Voyager/Town & Country
1990-1993	(AV)	Imperial/New Yorker Fifth Avenue
1993-1994	(LH)	Concorde/Intrepid/Vision

NOTE: THIS BULLETIN APPLIES TO VEHICLES EQUIPPED WITH THE 3.3L AND 3.8L ENGINE
SYMPTOM/CONDITION: Cold start up piston knocking noise at idle that diminishes 3 to 5 minutes after startup. The condition is more prominent in low ambient temperatures and after the vehicle has set overnight. Similar symptoms are described in Technical Service Bulletin 09-21-93 dated Dec. 31, 1993, "Carbon Buildup on Top of Piston". Refer to this Bulletin for additional details.
DIAGNOSIS:
Using the Mopar Diagnostic System or the Scan Tool (DRB II) with appropriate diagnostic procedure manual, verify that all engine systems are functioning as designed. If Diagnostic Trouble Codes (DTC's) are present, record them on the repair order for future reference, and repair as necessary. If no codes are present and all systems are functioning correctly, and carbon buildup on the top of the piston is not evident, proceed with the following repair.
PARTS REQUIRED:

AR	4621978	Piston & Connecting Rod Assy (use w3.3L engine)
AR	4621981	Piston & Connecting Rod Assy (use w3.8L engine)

REPAIR PROCEDURE: This repair involves installing a new coated piston and connecting rod assembly.

This engine repair should be covered by the base warranty or emissions warranty. If neither one applies, demand a pro rata refund.

• Erratic fuel gauge operation can be fixed by installing a revised sender assembly or fuel pump/sender assembly. **1991**—Loss of fuel pressure causing fuel-pump noise, erratic transmission shifting, engine power loss, or engine die-out may be due to a defective fuel pump. • An erratic idle with 2.5L engines can be cured by using an improved SMEC/SBEC engine controller. • Faulty power door locks may have a short circuit, need a new fuse, or require a new door latch with power door lock assembly. **1991–92**—If the engine knocks when at full operating temperature and during light to medium acceleration, it may mean that the single board engine controller (SBEC-Powertrain Control Module) needs replacing. • Engines with a rough idle and stalling following a cold start also may require a new SBEC. • The airbag warning light may continuously illuminate when the vehicle's ignition is in the ON position. This malfunction may be due to corrosion caused by water in the airbag's six-way connector. **1991–93**—Engines that stall following a cold start may need an upgraded Park/Neutral/start switch. **1991–94**—The serpentine belt may come off the pulley after driving through snow. Install an upgraded shield, screw, and retainers. • Noisy fuel pumps need to be replaced with an upgraded pump, wiring harness, fuel tank isolators, and fuel tank straps. • Noise when shifting into Reverse or when turning is addressed in DSB #09-14-94. **1991–95**—Poor AC performance while the AC blower continues to operate is likely due to a frozen evaporator. • DSB #24-05-94 looks at all the causes of, and remedies for, poor heater performance. • If the vehicle tends to drift left, cross-switch the tire and wheel assemblies, readjust the alignment, or reposition the front crossmember. **1992**—The brake pedal may not return to its fully released position, causing the brake lights to remain illuminated. Install a pedal return kit (#4723625). • Front door forward hem separation (the door seems to sag) can be corrected by welding the inner door panel to the outer door panel along the front door forward hem. • Long crank times, a rough idle, and hesitation may be corrected by replacing the intake manifold assembly. • A vehicle that's hard to start may have a corroded ECT/sensor connector. • An oil leak in the oil filter area may be corrected by installing a special oil filter bracket gasket

(#MD198554). • If the heater and ventilation system change to the defrost mode during acceleration, trailer towing, or hill climbing, the installation of a revised vacuum check valve should cure the problem. • Intermittent failure of the power door locks, chimes, wipers, gauges, and other electrical devices can be corrected by replacing defective relays with revised relays (#4713737). **1992–93**—Some 41TE transaxles may produce a buzzing noise when shifted into Reverse. This problem can be corrected by replacing the valve body assembly or valve body separator plate. • A deceleration shudder can be eliminated by replacing the powertrain control module with an upgraded version. • Rough idling after a cold start with 2.5L engines can be corrected by installing an upgraded powertrain control module (PCM). **1992–94**—AC duct odours are addressed in DSB #24-21-93. • Poor heater performance may be the result of a maladjusted clip on the blend air door cable. **1993**—A fuel pump check valve failure can cause start-up die-out, reduced power, or erratic shifting. **1993–94**—Improved automatic shifting can be had by installing an upgraded transmission control module. • AC evaporator whistling requires the installation of upgraded AC expansion valves and gaskets. • An AC moan may be silenced by installing an AC clutch plate with a damper ring. **1993–95**—Delayed automatic transmission engagement may be due to low fluid, a stuck or frozen PRNDL switch, or a transaxle front pump with excessive ground clearance. • Harsh low-speed automatic transmission shifting, accompanied by a fluctuating digital speedometer reading, may be corrected by covering the wiring harness with aluminum wire to prevent the spark plug wires from sending false signals into the outport speed sensor wiring that connects to the TCM. • Constant upshifting/downshifting on vehicles equipped with cruise control has a variety of causes as set out in DSB #08-15-95. • An exhaust that smells like rotten eggs can be corrected by installing an upgraded catalytic converter.

No: 11-03-96
GROUP: Exhaust
DATE: June 17, 1996
SUBJECT: Excessive Sulphur Odor
MODELS:
1992-1995 (AA) Acclaim/Spirit/LeBaron Sedan
1992-1993 (AC) Dynasty/New Yorker/ New Yorker Salon
1992-1993 (AG) Daytona
1992-1995 (AJ) LeBaron Coupe/LeBaron Convertible
1992-1994 (AP) Sundance/Shadow/Shadow Convertible
1993-1995 (AS) Caravan/Voyage/Town & Country
NOTE: THIS BULLETIN APPLIES TO VEHICLES EQUIPPED WITH 2.2L OR 2.5L ENGINES ONLY.
SYMPTOM/CONDITION: New catalytic converters are now available to address customer complaints of excessive hydrogen sulfide odor (rotten egg smell) coming from the tailpipe of their vehicles at idle. These catalysts were specially manufactured with a revised internal coating which minimizes sulphur odor.

1	4882530	Catalytic Converter	(AS 2.5L ATX/MTX FED/CAL)
1	4882531	Catalytic Converter	(AA,AG,AP 2.21/2.51 MTX FED/CAL) 1992-1994
1	4882532	Catalytic Converter	(AA,AC,AG,AJ,AP 2.21/2.51 ATX FED) 1992-1995
1	4882533	Catalytic Converter	(AA,AC,AG,AJ,AP 2.21/2.51 ATX CAL) 1992-1995
			(AA, 2.51 ATX FED) 1994)
			(AA, 2.5L FFV) 1992
1	4882534	Catalytic Converter	(AA, 2.5L FFV) 1993-1994

PARTS REQUIRED:
REPAIR PROCEDURE: This bulletin involves the replacement of the catalytic converter with a revised part. Replace the catalytic converter following the repair procedures given in group 11 of the appropriate service manual.
POLICY: Reimbursable within the provisions of the warranty.
TIME ALLOWANCE:
Labor Operation No: 11-50-01-91 0.7 Hrs.
FAILURE CODE: P8-New Part

If your minivan stinks, blame Chrysler, not your passengers. Repair costs should be covered by the emissions warranty.

1994—Harsh, erratic, or delayed transmission shifts can be corrected by replacing the throttle position sensor (TPS) with a revised part. • A creaking left B-pillar can be silenced by repositioning the metal portion of the left B-pillar baffle. **1994–95**—Intake valve deposits are frequently the cause of poor driveability complaints. • Intermittent no-cranks can be corrected by modifying the battery to starter cable terminal insulator at the starter connection. • A front suspension rapping noise heard when going over bumps can be corrected by providing additional clearance between the front coil springs and strut towers. **1996**—Poor engine performance near military installations or airports is caused by radar interference. Correct by installing a "hardened" crankshaft position sensor and/or reprogramming (flashing) the PCM with new software calibrations. • Rear brake noise that occurs at any time can be silenced by replacing the rear brake shoes and rear wheel cylinders. Another possibility is the addition of rear brake shoe springs. • Rough idle, hesitation, or sags after the fuel tank is filled can only be corrected by the installation of a new fuel tank, according to DSB #18-28-95. The repair is covered under warranty and should take about an hour. The dealer should also give you your gas back. • Steering noise during parking lot manoeuvres may be fixed by installing a new power-steering gear and left side attaching bolt. • As if flying toasters weren't enough, Chrysler minivan wheel covers tend to take flight as well (and I thought that was only a Chevy Caprice problem). Chrysler will install upgraded covers under warranty. **1998**—You may silence a chronic squeaking noise coming from underneath the vehicle by installing a new strut pivot bearing (see following bulletin).

NO: 02-10-98
GROUP: Suspension
DATE: Sept. 25, 1998
SUBJECT:
Squeaking Noise From Strut Bearing(s)
MODELS:
1998–1999 (NS) Town & Country/Caravan/Voyager
1998–1999 (GS) Chrysler Voyager (International Market)
NOTE:
THIS BULLETIN APPLIES TO VEHICLES BUILT THROUGH OCT. 1, 1998.
SYMPTOM/CONDITION:
A squeaking/chirping noise is heard (inside or outside the vehicle) from the strut tower area(s) when turning the steering wheel or when the vehicle is driven over any irregularities in the road surface.
DIAGNOSIS:
With the vehicle on a level surface, hold a front suspension coil spring with your hands near 180° apart, and try to rotate the spring right and left. If a squeaking/chirping noise is heard from the top of the strut tower, perform the Repair Procedure. Repeat for the opposite side strut tower.
Parts Information:
AR (2) 04684418 Bearing, Strut Upper Pivot
POLICY: Reimbursable within the provisions of the warranty.

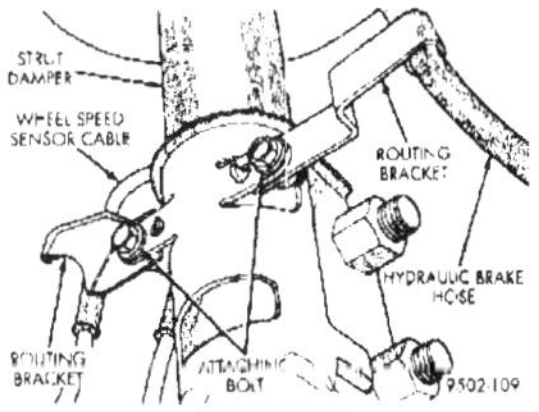

FIGURE 1

This repair should take about an hour for each side and is covered by Chrysler's base warranty.

All models/years—If pressed, Chrysler will pay the full cost of correcting paint defects during the first six years of ownership (see Part Two). • A rotten-egg odour coming from the exhaust may be the result of a malfunctioning catalytic converter, probably covered under the emissions warranty. • Front brakes tend to wear out quickly on front-drive minivans. Owners say that Chrysler has paid half the cost of brake repairs for up to 2 years/40,000 km.

Caravan, Ram Van, Town & Country, Voyager Profile

	1991	1992	1993	1994	1995	1996	1997	1998
Cost Price ($)								
Ram	14,710	15,125	15,925	16,800	20,525	—	—	—
4X4	17,725	18,890	—	—	—	—	—	—
Caravan	15,200	15,810	15,935	16,860	18,840	18,840	19,885	20,255
4X4	19,515	20,750	—	—	—	—	—	—
Grand Caravan	18,170	19,105	19,685	20,425	22,120	20,320	21,465	23,160
4X4	20,050	21,300	21,860	23,680	27,450	—	—	—
Town & Country	30,230	31,470	32,490	34,874	35,980	38,280	40,350	41,040
Used Values ($)								
Ram ↑	3,500	4,000	5,000	6,000	7,000	—	—	—
Ram ↓	3,000	3,500	4,500	5,500	6,500	—	—	—
4X4 ↑	4,500	5,000	—	—	—	—	—	
4X4 ↓	4,000	4,500	—	—	—	—	—	
Caravan ↑	5,500	7,000	8,000	11,000	10,500	12,000	14,000	16,500
Caravan ↓	5,000	6,500	7,500	10,000	11,500	11,500	13,500	16,000
4X4 ↑	7,000	8,000	9,000	—	—	—	—	—
4X4 ↓	6,500	7,500	8,000	—	—	—	—	—
Grand Caravan ↑	6,500	7,500	8,500	10,000	12,000	14,000	16,000	18,000
Grand Caravan ↓	6,000	7,000	8,000	9,500	11,000	13,000	15,000	17,500
4X4 ↑	7,000	8,000	9,000	10,500	13,500	—	—	—
4X4 ↓	6,500	7,500	8,500	10,000	12,500	—	—	—
Town & Country ↑	9,500	10,500	12,000	15,000	18,000	23,000	28,000	33,000
Town & Country ↓	9,000	9,500	11,000	14,000	17,000	22,000	27,000	31,500

Note: Voyager prices and ratings are almost identical to the Caravan and Grand Caravan.

Extended Warranty	Y	Y	Y	Y	Y	Y	Y	Y
Secret Warranty	Y	Y	Y	Y	Y	Y	Y	Y
Reliability	❶	❶	❷	❷	❷	❷	❷	③
Air conditioning	❶	❶	❷	❷	❷	❷	③	③
Automatic transmission	❶	❶	❶	❶	❶	❶	❶	❶
Body integrity	❶	❶	❶	❶	❶	❶	❷	❷
Braking system	❶	❶	❶	❶	❶	❶	❶	❷
Electrical system	❶	❶	❶	❶	❶	❶	❷	③
Engines	❶	❶	❷	❷	❷	③	③	③
Exhaust/Converter	❷	❷	❷	❷	③	③	③	③
Fuel system	❶	❶	❶	❷	③	③	③	④
Ignition system	❷	❷	❷	❷	❷	❷	③	④
Rust/Paint	❶	❶	❶	❶	❶	❶	❶	❶
Steering	③	③	③	③	③	④	④	④
Suspension	③	③	③	③	③	③	③	③

Crash Safety								
Caravan	—	⑤	❶	❶	④	—	④	③
Grand Caravan	—	—	—	—	—	③	③	③
Town & Country	—	—	—	—	—	—	④	③
Town & Country LX	—	—	—	—	—	—	③	③

Note: The low crash ratings represent severe leg trauma.

FORD

Aerostar

Rating: Average (1995–97); Not Recommended (1986–94). More brawny and more reliable than Chrysler's minivans. The Aerostar's last model year was 1997. **Maintenance/Repair costs**: Average, and any garage can repair an Aerostar. **Parts**: Plentiful, but costly (especially AC and electrical components).

Strengths and weaknesses: The 1986–94 models are at the bottom of the evolutionary scale as far as quality control is concerned. However, the last three model years have shown lots of improvement. Repair costs are reasonable (except for high AC costs), but mainly because the Aerostar's myriad mechanical and body defects are covered by Ford warranty extensions (see "Secret Warranties/Service Tips"). The Aerostar's modern, swoopy shape belies its limited performance capabilities: the 3.0L and 4.0L engines are unreliable through the 1991 model year, and the 3.0L is a sluggish performer. Older 2.3L and 2.8L engines can barely pull their own weight. The 4-speed automatic transmission often has a hard time deciding which gear to choose, and the power steering transmits almost no road feel to the driver. The ride is bouncy, handling is sloppy, and braking performance is poor. Reliability problems on all models make early Aerostars risky buys, especially if the previous owner has been less than fastidious in maintenance and repairs.

Valve cover and rear main oil seal leaks are frequent, and leaks from the front axle vent tube often require the replacement of the front axle assembly. Even oil pans, which you wouldn't normally associate with leaks, tend to leak as a result of premature corrosion. Fuel-injectors are either faulty or plugged. Other problems include expensive automatic transmission failures; electronic and electrical system glitches; power steering, suspension, and brake defects; and premature and chronic air conditioner breakdowns involving AC condensers and compressors that tend to fail after the first three years of use. (Ford has consistently produced failure-prone AC systems in the Aerostar van since its birth in 1986; according to *Consumer Reports*, their failure rate was almost three times that of other vehicles in its class.) Auto air conditioning experts refer to the problem as the "Black Death"—in reference to the sludge these

Ford ACs produce that leads to massive internal component failures. One AC specialist describes the problem in these terms:

> The biggest problem is the original FX-15 compressor that came with the truck and the rebuilt replacement, which will likely be another FX-15. These compressors are bad news. When they go bad they fall to pieces internally and contaminate your system. The "gook" settles in your condenser and is hard to get rid of. We sometimes end up having to flush as many as three times, operating the system for 30 to 45 minutes between flushes and changing the orifice tube each time. This allows the heat and pressure to break loose the "gook" in the condenser. When we replace an FX-15 compressor, we will not use a remanufactured FX-15...they simply do not hold up. Motorcraft's FS-10/FX-15 compressors are specially sensitive to high pressures. Faulty fan clutches are one of the reasons why so many late-model Ford ('89 and newer) compressors fail. This also applies to FWD cars with weak or inoperative radiator cooling fans. FS-10 compressor design is excellent. However, something could be done about their fan clutch durability. We have noted an above normal incidence in Aerostars ('91 and up), Explorers ('91 and up), Ford Taurus, Sable (up to 1994), T-Bird/Cougar ('89 and up), and F-series ('92 and up).

A grinding/growling coming from the rear signals that the in-tank electric fuel pump is defective. Many mechanics find that the electronic engine controls are difficult to diagnose if problems arise. Routine repairs are very awkward because most components are buried under the windshield and dashboard. Windshield wipers are badly designed for winter driving. They freeze at the bottom of the windshield and wear out the wiper motor. Body hardware and integrity have earned low marks as well.

Safety summary/Recalls: **1986–90—**A draft of a January 20, 1993, Ford internal document lists "known incidents which are attributed to the ignition switch." It includes the 1986 Aerostar and Econoline full-sized vans. NHTSA documents show that the Aerostar has almost the highest rate of ignition switch fires of any Ford truck, second only to the 1988 Bronco. State Farm Insurance has filed an action against Ford to cover its payouts to policyholders whose Aerostars caught fire from an overheated ignition switch. • The hefty B-pillar (where the edge of the door meets the body behind the front seat) obstructs peripheral vision. • Some Aerostars don't have front seat head restraints, unless the vehicles are equipped with the optional captain's chairs, and even then the restraints sit far too low for the average person. • Owners complain that the brake master cylinder and daytime running lights module may suddenly fail, and that 1990 Aerostars' rear brakes sometimes lock up in wet weather. **1988–90—**Cracked fuel tanks. **1993–94—**NHTSA is looking at electrical malfunctions that may cause overheating and/or fires at the fuel sender assembly on fuel tanks. **Recalls:**

1986—The rear suspension separating from the axle could cause steering loss. • Recall #90S04 provides for a free fuel tank replacement. • Lumbar seat wiring could short-circuit, creating a fire hazard. **1986–87**—Possible fuel line leaks. • Captain's chair may cut into seatbelt webbing. **1986–88**—Trailer towing package taillight wires may short-circuit, creating a fire hazard. **1987–89**—The rear liftgate may fail due to defective ball studs attaching the lift cylinders to the body. **1988–91**—The ignition switch could experience an internal short circuit, creating the potential for overheating, smoke, and possibly fire in the steering column area of the vehicle. **1989–90**—Quad Bucket seat assemblies may fail. **1990**—Ford will replace, gratis, faulty brake master cylinders. **1990–91**—Models with the A4LD automatic transmission may slip out of Park and roll as if in Neutral. **1992–94**—Vehicles may have a defective rear anti-lock brake control module. **1992–97**—Ford will fix wiring shorts that could cause a fire. **1995**—Spare tire may fracture the brake line. • Defective brake rotors. • Fuel pump and wiring electrical shorts. **1996**—Driver's door may not sustain specified load in the secondary latched position. • Defective brake rotors. • Defective electrical relay-ignition switch. **1992–97 Aerostar (AWD)**—The transmission or transfer case may crack or break. Dealers will install a new rear transfer case extension and a new aluminum rear driveshaft. **1997**—Aftermarket brake rotors may crack. Call 1-800-264-3414 for Aimco manufacturer's free replacement.

Secret Warranties/Service Tips

1986–91—3.0L engine knocking at idle may require the installation of a new, thicker thrust plate to reduce camshaft end play. • Hard cold starts, hesitation, and stalling may be caused by sludge in the throttle body and/or an idle by-pass valve. Install an idle air by-pass service kit (#F2DZ-9F939-A). **1986–94**—A speaker whine or buzz caused by the fuel pump can be stopped by installing an electronic noise RFI filter. • Rust perforations may occur in the front and/or rear lower rocker panel area(s) because the panels weren't coated with primer. Ford will replace the panel at no charge for up to 6 years/160,000 km (see following).

95-6-12
03/27/05
ROCKER PANEL - FRONT AND/OR REAR - RUST PERFORATION - SERVICE TIPS
LIGHT TRUCK:
1986–94 AEROSTAR
ISSUE:
Rust perforation may occur in the front and/or rear lower rocker panel area(s) on some vehicles. This may be due to the lower rocker panels having no electrodeposition primer (E-Coat) on the inside surface, for vehicles built from 1985–94.
ACTION:
Weld on lower rocker panel covers and blend to body with body filler. Paint rocker panel(s) to match the vehicle. Refer to the following procedures for service details.

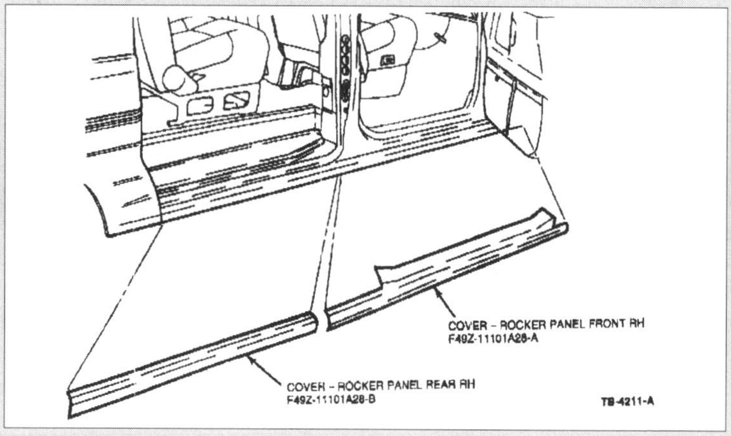

1987–92—Brakes that stick or bind may have corroded brake caliper slide pins. Install corrosion-resistant pins (#E8TZ-2C150-B). **1988**—Delayed Reverse engagement can be corrected by installing a new separator plate. **1989–92**—If there's a popping or clunking and floorboard vibration when braking, it's likely caused by the front brake caliper suddenly springing away from the caliper abutment and returning. Adjusting the front brake pads and knuckle clearance will correct this problem. **1989–96**—A chatter noise during sharp turns is likely due to an insufficient friction modifier, or over-shimming of the clutch packs within Traction-Lock differentials. **1990**—Delayed upshift or no upshift can be corrected by installing a new #4 thrust washer. • AC compressor shaft seal leaks can be best fixed by installing a new seal (#E9SZ-19D665-A). **1990–92**—A 4.0L engine oil leak may occur around the rocker gasket because of variations in the gasket quality. Install two new rocker cover gaskets (Carrier type), along with conical spring screws. • Automatic transmission fluid leakage may be caused by a faulty or loose transfer case rear output seal. **1990–96**—Noisy power steering is likely caused by a pressure spike in the serpentine tube of the power-steering cooler. **1993–95**—A squeak or chirp coming from the blower motor can be stopped by installing an upgraded blower motor with improved brush-to-commutator friction. • An automatic transmission whine heard upon light acceleration can be silenced by replacing the front and rear planetary assembly and the front and rear sun gear. • Overheating or binding rear brakes can be corrected by backing off the self-adjusters. **1995–96**—Hard starts and stalling can be corrected by installing an upgraded idle air control valve under the emissions warranty. **1995–97**—Coolant leakage from the engine block heater requires the replacement of the block heater with an upgraded one that attaches better. **1997**—Delayed transmission engagement or constant shifting may be caused by an improper torque converter apply-and-release schedule (reprogram the powertrain control module). **All models/years**—Press Ford for "goodwill" warranty coverage if your AC is afflicted by the "Black Death" within 5 years/100,000 km. • Use the emissions warranty to cure rotten-egg odours caused by a defective catalytic converter. • Two components that benefit from Ford's "goodwill" warranty extensions are fuel pumps and the computer modules which govern engine, fuel injection, and transmission functions. If Ford balks at refunding your money for a faulty computer module or fuel pump, apply the emissions warranty for a full or partial refund. • Paint delamination, fading, peeling, hazing, and "microchecking" (see page 112 for details on claiming a refund).

Aerostar Profile

	1990	1991	1992	1993	1994	1995	1996	1997
Cost Price ($)								
Cargo	15,219	14,273	16,095	16,899	17,195	18,395	18,495	19,695
4X4	18,141	17,071	18,895	19,840	19,995	20,845	—	—
XL	—	—	16,395	17,215	18,095	20,195	19,995	20,195
Used Values ($)								
Cargo ↑	2,000	2,500	3,000	3,500	5,000	5,500	7,500	10,000
Cargo ↓	1,500	2,000	2,500	3,000	4,500	5,000	7,000	9,000
4X4 ↑	2,500	3,000	4,000	4,500	5,500	6,500	—	—
4X4 ↓	2,000	2,500	3,000	4,000	4,500	6,000	—	—
XL ↑	—	—	5,500	6,000	7,500	9,000	11,500	14,000
XL ↓	—	—	5,000	5,500	6,500	8,000	10,500	13,000
Extended Warranty	Y	Y	Y	Y	Y	Y	N	N
Secret Warranty	Y	Y	Y	Y	Y	Y	Y	Y
Reliability	❶	❷	❷	❷	③	③	④	④
Air conditioning	❶	❷	❷	❷	❷	③	④	④
Body integrity	❷	❷	❷	❷	❷	❷	❷	❷
Braking system	❶	❶	❶	❶	❶	❷	❷	❷
Electrical system	❶	❶	❶	❷	❷	③	③	③
Engines	❷	❷	❷	④	④	⑤	⑤	⑤
Exhaust/Converter	❶	❶	❶	❷	③	④	④	⑤
Fuel system	❶	③	③	③	③	④	④	④
Ignition system	❷	❷	③	③	④	④	⑤	⑤
Manual transmission	③	③	③	③	③	③	④	⑤
- automatic	❶	❷	③	③	③	③	④	④
Rust/Paint	❶	❶	❶	❶	❷	❷	③	③
Steering	❷	❷	③	③	③	④	④	④
Suspension	❷	❷	❷	③	③	③	③	④
Crash Safety	—	—	⑤	❶	❶	④	④	④

Note: The low crash ratings represent severe leg trauma.

Mercury Villager, Nissan Quest

Rating: Recommended (1997–98); Above Average (1995–96); Not Recommended (1993–94). More carlike than most of their competitors. The Quest is practically identical to the Villager, except for some slight styling differences and more standard equipment. By choosing a cargo van over a wagon, you can save between $1,500 and $2,000. **Maintcnancc/Rcpair costs**: Higher than average. Any garage can repair these minivans. **Parts**: Both Mercury and Nissan dealers carry parts, and auto club surveys show that parts are less expensive than those of most other minivans in this class. The exception to this rule: broken engine exhaust manifold studs (a frequent problem), AC, and electrical components. There are plenty of two- and three-year-old models on the market that have just come off lease; the Nissan may cost a bit more, however, because it depreciates more slowly than the Villager. Other minivans you may wish to consider: the GM front-drives, Nissan Axxess, or Toyota Previa and Sienna.

Strengths and weaknesses: The Villager's and the Quest's strongest features are their impressive highway performance (as long as you're not carrying a full load) and easy, no-surprise, carlike handling. Additional assets are a 4-speed automatic transmission that's particularly smooth and quiet, and mechanical components that have been tested for years on the Maxima and other Ford vehicles. The ride on both smooth and uneven highways is comfortable, overall highway stability is above reproach, and braking performance is quite good, aided by standard ABS, which improves directional control by eliminating wheel lockup.

These fuel-thirsty minivans are heavier than the Aerostar by about 700 pounds, and the 3.0L engine has only 6–16 additional horses to carry the extra weight. GM's 2.8L engines produce more torque than what the Villager and Quest can deliver. Precise steering makes the Villager feel more responsive at highway speeds and in emergency manoeuvres than it really is. The control layout can be a bit confusing.

Interior space is impressive. The Villager is nearly a foot longer and two inches wider and higher than Chrysler's short-wheelbase minivans. There's better seating for three adults in the rear than with the Caravan, middle seatbacks fold flat, and the rear seats have tracks that allow them to slide forward all the way to the front or fold flat and convert to a serving area for tailgate parties.

There have been many reports of engine exhaust manifold failures (through 1994), electrical problems, brake failures, premature wear of the front disc brakes, and chronic stalling, possibly due to faulty fuel pumps or a shorted electrical system. One owner of a 1993 Villager GS tells of chronic vibrations at 60 mph that can't be dampened even after frequent tire replacements and suspension retuning. Body integrity is sub-par, with doors opening and closing on their own and poor fit and finish, allowing lots of windnoise to enter the interior. (Yes, Ford got the body work, while Nissan handled the mechanicals. You can tell, eh?) There have also been some reports of panel and paint defects and premature rusting on the inside sliding door track.

Dealer service bulletins: 1994—Transaxle whine in Overdrive at 30–75 km/h. • Insufficient AC cooling or excessive clutch end gap. • Chirp/squeak from the blower motor, and blower motor running after shutoff. • Coolant leaks at the radiator hose (replaced for free at 3 years/60,000 km under Ford's 94B42 Program). • Inoperative heater backlight. • Rattling or chucking noise from the sliding door, and sub-woofer rattling noise in the radio speakers. **1995**—Knocking noise in lower engine. • Coolant leak from the engine below the camshaft. • AC compressor failure. • Insufficient AC cooling or excessive clutch end gap. • Brakes binding and/or overheating. • Front brake squeal or groan. • Fuel pump buzz/whine through radio speaker. • Rear bench seat squeaks or rattles when the vehicle is underway. • Sliding door squeaks, rattles, or is hard to close. • Rear suspension makes a crunch/grunt noise. • Body side window won't stay closed. • Inoperative rear wiper motor. • Fog/film on windshield/interior glass. • Carpet lifts at right sliding seat track. • Sulphur smell from the exhaust system. • Musty and mildew-type odours. **1996**—Knocking noise in lower engine continues for another model year. • AC compressor failure is still a problem. • Intermittent operation of remote keyless entry system. • Sliding door squeaks, rattles, or requires great effort to close. • Rear bench seat squeaks or rattles when the vehicle is moving. • Rear suspension makes a crunch/grunt noise. • Airbag light flashes constantly. • Body side window won't stay closed. • Fog/film on windshield/interior glass. • Fuel pump buzz/whine through radio speaker. • Inadvertent horn operation. • Carpet lifts at right sliding seat track. • Musty and mildew-type odours. **1997**—Hard starts or no-starts in cold temperatures or high altitudes. • Shift lever hard to move from Park. • Front door windows bind. • Broken third-row right-hand seatbelt guide. • Fuel odour in passenger compartment when refueling. • Intermittent operation of remote keyless entry system. • Airbag light flashes constantly. • Chirping/squeaking noise from blower motor at low speeds. • Hissing from B-pillar area. • Noise during light brake application and release. • Rear suspension noise. • Tips on preventing brake vibration. • Fog/film on windshield/interior glass.

Safety summary/Recalls: **1994–96**—NHTSA is looking into fuel leaks into the engine compartment and gas fumes that infiltrate the interior. • Vehicle suddenly accelerated forward. • Inadvertent airbag deployment. •Sudden stalling due to faulty fuel pump. •Both steering and brakes failed when turning into an intersection. • Steering wheel locked up while making a right turn. • Chronic ABS brake failures; brake pads and rotors need replacing every 5,000 km; and drums often need turning. • Loose exhaust system heat shield. • Alarm system will lock and unlock the doors on its own. • Remote control door locks often won't respond, trapping occupants inside. • Faulty rear door latch and lock allows the door to come open while driving. • The sliding door opens for no apparent reason. •Power windows often jam. •Seatbelts don't retract properly and ratchet too tightly. • Shoulder belt got caught around child's neck and had to be cut away. •Frequent rear windshield wiper failures. • Solar window at certain angles distorts vision by reflecting images. **Recalls**: **1993**—A faulty brake master cylinder or leaking fuel-filler hoses will be replaced for free. • Defective automatic seatbelt anchor bolts. • Leaves or other debris can accumulate in the fresh-air intake of the heating and AC system, creating a possible fire hazard. **1994–95**—A faulty electrical socket may cause the rear light to fail to operate. • Defective third-row sliding bench seat. **1996**—Power windows aren't in compliance with safety regulations. **1997**—Fuel line hoses could crack or split. **1997–98 Villager**—Faulty batteries may rupture or cause a fire.

Secret Warranties/Service Tips

1993—Harsh automatic transmission upshifts may be caused by metal contamination in the solenoid assembly. • Excessive exhaust manifold noise is likely due to a broken stud. Install upgraded studs. • An inoperative air conditioning blower motor can be fixed by installing a new blower motor resistor. • Stalling whenever the vehicle is shifted from Park to Drive to Reverse may mean that the torque converter is stuck in the "lockup" mode. Install an upgraded valve body assembly (#F3XY-7A100-D) and a transaxle oil pan gasket to fix the problem. • A ticking or clicking from the suspension strut area signals the need for an upgraded front strut spacer (#F3XY-3A120-A). • Fogging or frosting of the side windows may be caused by a misconnected C261 electrical connector. • If the liftgate light/door ajar light flickers or the vehicle won't start, the rear liftgate latch may need to be changed. • A rear suspension clunk or thump when passing over bumps is likely caused by improperly calibrated rear shock absorbers. • Install an upgraded speed control module (#F3XY-9FS12-B) if the vehicle loses 3–5 mph on hills or grades with the present module. • A clicking or ticking noise when accelerating can be corrected by installing a new EGR tube with a redesigned fitting. • A shudder or vibration that occurs during the 1–2 shift can be corrected by replacing the automatic transaxle valve body with an upgraded valve body and transaxle oil pan gasket. **1993–94**—Ford will cover defective engine exhaust manifold studs on a case-by-case basis up to 5 years/100,000 km (see following bulletin).

Classification: EM94-002
Section: Engine Mechanical
Reference: TECHNICAL BULLETIN NTB94-041
Models: See below
Date: April 14, 1994
BROKEN EXHAUST MANIFOLD STUD
APPLIED MODELS:
- Pathfinder (WD21)
- Maxima (J30 and U11)
- Truck (D21)
- 300ZX (Z31)
- 200ZX (S12)

SERVICE INFORMATION: The exhaust manifold stud has been changed for improved durability. Use the following procedure to repair vehicles that exibit excessive exhaust noise due to exhaust manifold stud breakage.
REPAIR PROCEDURE:

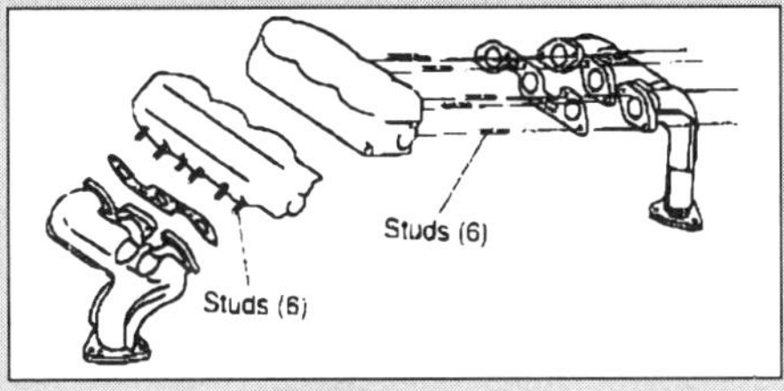

Figure 1

1. Remove the exhaust manifold on the side of the engine with broken stud(s). (Figure 1)
2. Replace the broken stud(s) and the front and rear studs DWI using the improved part, P/N 14065-V5003
3. Replace the exhaust manifold gasket and the exhaust manifold with new parts.

WARRANTY INFORMATION:
Please reference Section "AT-Manifolds (V6 Engine)" of the current "Nissan Flat Rate Manual" and select the appropriate right/left manifold gasket replacement OpCode with the indicated combination OpCode for the replacement of the right/left exhaust manifold stud. Use PNC 14006 and the indicated FRT's for the applicable vehicle.

Nissan will argue they've never heard of this stud problem and then relent and offer some compensation. Ford is easier to deal with.

• A squeak or chirp coming from the blower motor can be stopped by installing an upgraded blower motor with improved brush-to-commutator friction. • Sliding door noise may need a new service spring in the upper hinge and a readjustment of the dovetail. • A speaker whine or buzz caused by the fuel pump can be stopped by installing an electronic noise RFI filter. • Body squeaks and rattles are thoroughly discussed in DSB #04-86-94. **1993–95**—The rear wiper motor may quit or stop intermittently because water has gotten into the motor printed circuit board. • Replace front brake pads that cause excessive brake squeaking or groaning with upgraded pads. **1993–96**—A crunch/grunt noise from the rear suspension may be caused by rear shackle bushings that need lubrication. **1993–97**—Front door windows that bind may have the glass rubber improperly installed into the door sheet metal channel. **1994–98**—If there's a strong fuel odour in the passenger compartment when refueling, it's likely there's a missing sealer between the fuel-filler opening upper flange and the fuel-filler base assembly. **1995–96**—Ford has replaced, at no charge, 3.0L V6 engine blocks that produce excessive engine knock following a cold start. **1996–98**—Power door locks that intermittently self-activate (help, let me out!) are a common occurrence that's covered in DSB #98-22-5 (following).

Article No.:
98-22-5
11/09/98
POWER DOOR LOCK - INTERMITTENT
SELF-ACTIVATION
LIGHT TRUCK:
1996–98 VILLAGER
This TSB article is being republished in its entirety to revise the Power Door Lock Signal and Wiring Check Procedure and provide an Actuator Assembly part number.
ISSUE
Intermittent self-activation of the power door locks may occur on some vehicles. This may be caused by any of the following:

- All 1996–98 Model Year Vehicles:
  - Adjustment of the front door lock rod linkage
  - Water/moisture collecting in the left front door window switch potentially causing an internal short or ground
  - Power door lock wiring
- All 1996-98 Model Year Vehicles Equipped With Anti-Theft:
  - Water/moisture collecting in the front door key lock cylinders potentially causing a grounding condition in the circuit
  - Water/moisture collecting in driver's door key cylinder switch-to-harness connector
- 1996 Vehicles Only:
  - Smart Entry Control (SEC) timer module

ACTION
Replace front window switch, module, adjust front door lock rod linkage, and grease switch and harness connections on anti-theft vehicles.
WARRANTY STATUS: Eligible Under the Provisions of Bumper to Bumper Warranty Coverage.

Talk about being a captive customer…!

1997—Hard starts or no-starts in cold temperatures or at high altitudes can be corrected by replacing the power control module (PCM) under the emissions warranty. **All models/years**—A rotten-egg odour coming from the exhaust is probably caused by a malfunctioning catalytic converter, which is covered by Ford's original warranty and the 5-year/80,000 km emissions warranty. • Two components that benefit from Ford "goodwill" warranty extensions are fuel pumps and computer modules that govern engine, fuel-injection, and transmission functions. • Paint delamination, fading, peeling, hazing, and "microchecking" (see page 112 for details on claiming a refund).

Mercury Villager, Nissan Quest Profile

	1993	1994	1995	1996	1997	1998
Cost Price ($)						
Villager GS	19,695	21,195	22,595	23,695	24,295	24,595
Villager LS	24,195	26,195	27,095	28,095	29,195	29,495
Quest XE	20,231	23,490	24,598	25,598	25,598	25,598
Quest GXE	25,311	27,290	29,598	30,598	30,898	30,898
Used Values ($)						
Villager GS ↑	8,500	10,500	13,000	15,000	18,500	21,000
Villager GS ↓	7,000	10,000	12,000	13,500	16,500	20,000
Villager LS ↑	9,000	12,500	15,000	18,000	21,000	24,000
Villager LS ↓	8,000	11,500	13,500	17,000	19,500	23,000
Quest XE ↑	9,000	10,500	13,000	17,300	18,000	21,500
Quest XE ↓	7,600	10,000	12,000	15,400	17,000	20,000
Quest GXE ↑	10,000	12,500	15,000	21,400	21,000	24,000
Quest GXE ↓	8,000	11,500	14,000	19,400	20,000	23,000

Extended Warranty	N	N	N	N	N	N
Secret Warranty	Y	Y	Y	Y	Y	N
Reliability	③	③	③	④	④	⑤
Air conditioning	③	③	③	③	⑤	⑤
Automatic transmission	④	⑤	⑤	⑤	⑤	⑤
Body integrity	❶	❷	❷	❷	③	④
Braking system	❷	❷	❷	❷	❷	③
Electrical system	❶	❷	❷	❷	❷	③
Engines	❷	③	③	⑤	⑤	⑤
Exhaust/Converter	❶	❷	③	④	⑤	⑤
Fuel system	④	④	⑤	⑤	⑤	⑤
Ignition system	④	④	⑤	⑤	⑤	⑤
Rust/Paint	❷	❷	③	④	⑤	⑤
Steering	④	⑤	⑤	⑤	⑤	⑤
Suspension	③	③	④	④	⑤	⑤
Crash Safety	❶	❶	④	④	④	—

Windstar

Rating: Below Average (1996–98); Not Recommended (1995). Don't go anywhere near this minivan without an extended powertrain warranty. Other more-reliable minivans you may wish to consider: a late-model GM front-drive, a Mercury Villager, Nissan Axxess or Quest, or a Toyota Previa or Sienna. Sure, the Windstar combines an impressive five-star safety rating, plenty of raw power, an exceptional ride, and impressive cargo capacity. But the self-destructing 3.8L engine and erratic automatic transmission are black holes that will suck your wallet into their vortex (see "Dead Ford" and "Ford Windstar Engine Head Gasket Automatic Transmission" website links in Appendix I). And, as a counterpoint to Ford's well-earned Windstar crashworthiness boasting, take a look at the following summary ("Safety summary/Recalls") of safety-related complaints recorded by the U.S. Department of Transportation. Scary, *n'est-ce pas*? **Maintenance/Repair costs**: Average while under warranty; higher than average thereafter. **Parts**: Reasonably priced parts aren't as hard to find as they once were, mainly due to the entry of independent suppliers.

Strengths and weaknesses: Launched as a 1995 model in March 1994, the Windstar is a front-drive minivan that looks a bit like a stretched Mercury Villager. It's longer, larger, and lower than most other minivans. It's also one of the few minivans not built on a truck platform (it uses the Taurus platform instead), and as a result, it has some of the carlike handling distinctions of Chrysler's minivans. It's offered in two body styles—a seven-passenger people-hauler and the less-expensive, basic cargo van.

There are a number of powertrain deficiencies that cut dramatically into the Windstar's performance and overall reliability. The automatic transmission, for example, is anything but smooth and has a scarlet history for running away when unattended, or simply packing it in early due to a failure-prone forward clutch piston, among other defective components. At the best of times, it pauses before downshifting or shifts roughly into a higher gear.

The 3.0L engine is overwhelmed by the Windstar's heft and struggles to keep up, but opting for the 3.8L V6 may get you into worse trouble. Owners of recently minted Windstars report that the 3.8L engine head gaskets need replacing shortly after the 60,000 km mark. And if all these problems get you hot under the collar, too bad, you'll also have to contend with an air conditioning system that's generically dysfunctional and is reportedly also biodegradable.

Dealer service bulletins: **1995**—Insufficient AC cooling or excessive clutch end gap. • Condensation leak at the auxiliary AC unit requires the installation of upgraded hose clamps. • Malfunctioning Forward/Reverse engagement, intermittent loss of torque at 3–4 upshift, and low transaxle fluid improperly setting the DTC causes major transmission malfunctions. • Solenoid corrosion causes a no-crank condition. • Chirp or squeak from the blower motor at low speeds. • Radio static, minor dash rattles, and false door ajar warnings. **1996**—Stall or hard start after 1–4 hour soak. • Case breakage at rear planet support. • Transaxle driveline noises. • Steering drift or pull diagnostic tips. • Clunk in floor pan when accelerating/decelerating. • Clunk when moving gearshift out of Park. • Faulty power door locks. • Fuel gauge won't read over three-quarters. • Fog/film on windshield/interior glass. • Objectionable carpet odour. • AC musty odour. • Windshield wiper chatter/streaking. **1997**—Harsh automatic transmission at 1–2 shift. • Intermittent Neutral condition when coming to a stop. • Parking brake may not release. • Tips on preventing brake vibration. • Water dripping onto floor from AC drain tube. • Inoperative rear auxiliary AC blower. • AC compressor moan. • Chirp or squeak from the blower motor at low speeds. • Clunking noise when moving gear shift lever out of Park. • Squeaking, creaking noise from upper A-pillar on turns. • Static noise heard at the low end of AM radio band. • Right rear quarter panel black soot deposits. • Fog/film on windshield/interior glass.

Safety summary/Recalls: Base models don't have head restraints for all seats, and the digital dash can be confusing. **1996**—Sudden acceleration and chronic stalling. • Severe injuries caused by airbag deployment. • Control arm and inner tie-rod failures cause the wheel to fall off. • Chronic ABS brake and transmission failures. • Almost a dozen reports that the vehicle jumps out of Park and rolls away when on an incline or slips into Reverse with the engine idling. • Faulty fuel pumps. • Built-in child safety seat is easy to get out of, yet securing seatbelts are too tight; child almost strangled. • Engine head gasket failures. • Faulty rear liftgate latches. **1997**—Sudden acceleration. • ABS and transmission failures. • Steering wheel lockup. **1998**—Sudden acceleration and chronic stalling. • Loose or missing front brake bolts could cause the wheels to lock up or a loss of vehicle control. • Frequent ABS brake and transmission failures. • Transmission and axle separation. • Vehicle jumps out of Park and rolls away when on an incline or slips into Reverse with the engine idling. • Trunk lid can fall on one's head. • Faulty fuel pump, sensor, and gauge. **Recalls**: **1995**—Hood may be defective. • A loose connection within the electrical power distribution box could ignite the electrical wiring under the hood. • A pinched instrument panel wire harness could cause a short circuit and start an electrical fire. • The passenger airbag may not deploy properly. • An alternator wiring short circuit could cause a fire. **1996**—Hood may be defective. • Transmission may not engage; PRNDL may give a false reading. **1997–98**—Servo cover can leak, creating a fire hazard. **1998**—Defective steering assembly.

Secret Warranties/Service Tips

1995—Blower motor squeaking or chirping can be silenced by installing an upgraded blower motor. **1995–96**—Intermittent no-starts may be caused by microscopic cracks on the fuel pump relay cover located inside the Constant Control Relay Module (CCRM); install a new CCRM. **1995–98**—The front-end accessory drive belt (FEAD) slips during wet conditions, causing a reduction in steering power-assist. Ford suggests the belt be replaced. • Water leakage onto carpet or headliner in rear cargo area is a factory-related defect covered in DSB #98-5-5, following:

Article No.
98-5-5
03/16/98
WATER LEAK - WET CARPET AND/OR HEADLINER IN CARGO AREA
LIGHT TRUCK:
1995-98 WINDSTAR
ISSUE: The carpet and/or headliner may be wet in the cargo area. This may be caused by missed sealer and/or mispositioned wire loom grommet at the liftgate.
ACTION: Reseal and/or reposition the grommet to reduce the possibility of water leaks.

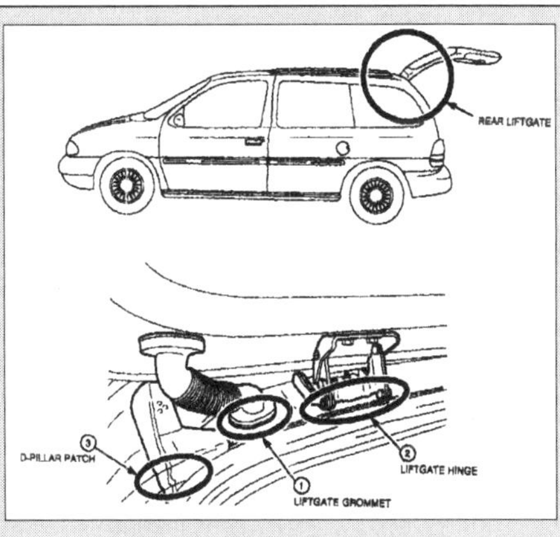

Ford says the resealing of the rear liftgate area or roof rack holes is covered by its bumper-to-bumper warranty.

1996–98—Harsh automatic shifting from 1 to 2 may be caused by a malfunctioning electronic pressure control or the main control valves sticking in the valve body. • An intermittent Neutral condition when coming to a stop signals the need to replace the forward clutch piston and the forward clutch cylinder. • Black soot deposits on the right rear quarter panel can be avoided by installing an exhaust tailpipe extension under Ford's base warranty.

Windstar Profile

	1995	1996	1997	1998
Cost Price ($)				
GL	23,095	22,495	23,495	24,495
LX	27,195	27,495	28,995	28,995
Used Values ($)				
GL ↑	13,500	16,000	18,000	20,000
GL ↓	13,000	15,500	17,500	19,500
LX ↑	15,500	18,000	20,500	22,500
LX ↓	14,500	17,500	19,500	21,500
Extended Warranty	Y	Y	Y	Y
Secret Warranty	Y	Y	Y	N
Reliability	❷	③	③	④
Crash Safety	⑤	⑤	⑤	⑤

GENERAL MOTORS

Lumina, Lumina APV, Silhouette, Trans Sport, Venture

Rating: Average (1995–98); Not Recommended (1990–94). The Lumina APV's last model year was 1996; the Trans Sport was carried over to the 1998 model year. The Venture debuted as a 1997 model. **Maintenance/Repair costs**: Higher than average, but any garage can repair these minivans. **Parts**: Plentiful, but costly. Plastic body panels may soon be in short supply. There are plenty of reasonably priced two- and three-year-old models that have just come off lease on the market. Other minivans you may wish to consider: the Mercury Villager, Nissan Axxess and Quest, or Toyota's Previa and Sienna.

Strengths and weaknesses: Although the 1994s and later models have less of a Dustbuster look, GM's plastic-bodied, front-drive minivan resembles more a swoopy station wagon than the traditional minivan (like the popular, boxy Chrysler Caravan). These vehicles use the Chevrolet Lumina platform, and therefore have more carlike handling than GM's Astro and Safari. Seating is limited to five adults in the standard models (two up front and three on a removable bench seat), but this can be increased to seven if you order optional modular seats. Seats can be folded down flat, creating additional storage space. Incidentally, be careful not to drop your keys between the windshield and the dash because you'll need a fishing rod to get them back.

The chassis and mechanical components come from GM's failure-prone W-bodies (Lumina, Regal, Cutlass Supreme, and Grand Prix), which explains why these minivans have so many of the same factory-related defects as their smaller cousins—notably, electronic module (PROM) and starter failures, short circuits, automatic transmission breakdowns, abysmal fit and finish, chronic sliding door malfunctions, and faulty rear seat latches. Other problems include a 4-speed automatic transmission that isn't as durable as the less fuel-efficient 3-speed automatic; a poorly mounted sliding door; side-door glass that pops open; squeaks, rattles, and clunks in the instrument panel cluster area and suspension; and a wind buffeting noise around the front doors. By the way, don't trust the towing limit listed in GM's owner's manual. Automakers publish tow ratings that are on the optimistic side—and sometimes they even lie. Also, don't be surprised to find that the base 3.1L engine doesn't handle a full load of passenger and cargo, especially when mated to the 3-speed automatic transmission. The ideal powertrain combo would be the 4-speed automatic coupled to the optional "3800" V6 (first used on the 1996 versions).

The large dent- and rust-resistant plastic panels are robot-bonded to the frame with unique new adhesives, and they absorb engine and road noise very well, in addition to having an impressive record for durability. Some body shops complain, however, that the innovative panels are in short supply and that damaged panels can't be recycled. This drives up the cost of repairs and tempts insurance adjusters to simply write off repairable vehicles.

Dealer service bulletins: **1996**—Intermittent Neutral/loss of Drive at highway speeds. • Second-gear starts; poor 1–3 shifting. • Insufficient AC performance, and AC odour at start-up in humid climates. • Steering column click, snapping noise. • Rattles from second- and third-row seat assemblies and in rear compartment area of the roof console. • Radio frequency interference diagnosis. • Erratic fuel gauge needle. • Low-voltage reading or dim lights at idle. • Condensation on exterior lights. • Water leak at the top of the sliding door. **1997**—Engine coolant weeping from coolant recovery reservoir. • Intermittent Neutral/loss of Drive at highway speeds. • Brakes are inoperative, drag, heat up, or wear out early. • Front brake squealing or grinding and rear brake clicking or squealing. • Air temperature from climate control system doesn't change. • Cold-start rattling. • Engine noise in passenger compartment. • Rattle at rear of vehicle or upon closing liftgate. • Rear suspension thud/clunk. • Inoperative power door locks. • Low-voltage reading or dim lights at idle. • Inoperative power window auto-down feature. • Movement of captain's chairs during cornering and turns. • Troubleshooting tips for sliding door malfunctions.

Safety summary/Recalls: Some front door-mounted seatbelts cross uncomfortably at the neck, and there's a nasty blind spot on the driver's side that requires a small stick-on convex mirror to correct. • NHTSA has closed an investigation of transaxle hose separation in exchange for GM's promise to conduct a regional recall. **Recalls**: **1990**—Defective modular rear seat latches. **1993–94**—The rear seatbelt may not retract on vehicles with power sliding doors. **1995**—The throttle cable support bracket needs to be replaced in order to enable the engine to decelerate. • Steering could fail. • Brake pedal arm could fracture. **1998**—Vehicle may roll away while in Park. Install safety guards on seat latches. **1990–91 Trans Sport**—The upper glove box door may not stay closed in an accident.

Secret Warranties/Service Tips

1990–91—Hydramatic 4L60/700R4 automatic transmission may have no upshift or appear to be stuck in first gear. The probable cause is a worn governor gear. It would be wise to replace the retaining ring as well. **1990–92**—Vehicles with a 3T40 automatic transmission may slip in Low or Reverse gear. Correct by replacing the Low/Reverse clutch components, including the Low/Reverse release spring, clutch spring retainer, and piston snap ring. • A shudder or vibration at low speeds in vehicles equipped with a 3.1L engine may be caused by a faulty PROM electronic module. • A power-steering shudder can be corrected by replacing the power-steering return hose/pipe assembly with a revised assembly and internal tuning cable (#26030907). **1990–94**—An engine ticking at idle can be traced to rattling piston pins, which must be replaced with upgraded parts. **1991–92**—A delayed shift between Drive and Reverse is likely caused by a rolled or cut input clutch piston outer seal. **1992**—Power-steering shudder can be reduced by installing an upgraded power-steering outlet hose/pipe assembly (#260337593). **1992–93**—No Reverse gear or slipping in Reverse can be corrected by installing an upgraded Low/Reverse clutch return spring and spiral retaining ring. **1992–94**—Loss of Drive or erratic shifts may be caused by an intermittent short to ground on the A or B shift solenoid, or an electrical short circuit in the transaxle. • A front-end clunking noise when driving over rough roads may require the repositioning of the diagonal radiator support braces. • A front-end engine knock troubleshooting chart is found in DSB #306001. **1993–94**—Owners who complain of automatic transmission low-speed miss, hesitation, chuggle, or skip may find relief with an improved MEMCAL module that GM developed to remedy the problem. **1993–95**—An engine coolant leak from the throttle body assembly may require an upgraded service seal kit. **1994**—Front door glass window scraping/chattering can be eliminated by relocating the front lower guide attachment. • Lazy front seatbelt retractors will be replaced free of charge. • A liftgate that fails to lock may have a loose or missing lock cylinder lock-out pin. • Manual sliding side doors that stick shut require a lock replacement. **1995**—A thud/clunk noise occurs when the fuel tank is more than three-quarters full. GM will replace the fuel tank and sender assembly. **1995–96**—Intermittent Neutral/loss of Drive at highway speeds can be fixed by replacing the control valve body assembly. **1997**—Rear brake clicking or squealing may be caused by a maladjusted park brake cable. • Brakes that don't work, drag, heat up, or wear out early may have a variety of causes, all outlined in

DSB #73-50-27. • A rear suspension thud or clunk may be silenced by installing upgraded rear springs. **1991 with 3.1L engine**—Engine hesitation or stalling may require a new service calibration PROM that revises tip-in fuelling. If the problem is reduced power, it may signal the need to replace the electronic spark control (ESC). **All models/years**—Defective catalytic converters that cause a rotten-egg smell in the interior may be replaced free of charge under the emissions warranty.

Lumina, Lumina APV, Silhouette, Trans Sport, Venture Profile

	1991	1992	1993	1994	1995	1996	1997	1998
Cost Price ($)								
Cargo	15,750	16,698	17,398	17,998	20,110	20,110	—	—
Passenger	17,198	18,098	18,798	19,598	22,730	22,730	—	—
Trans Sport/SE	18,498	18,950	19,698	20,298	23,475	23,475	23,690	24,650
Venture	—	—	—	—	—	—	23,185	24,145
Used Values ($)								
Cargo ↑	3,500	4,000	4,500	6,000	7,000	9,000	—	—
Cargo ↓	3,000	3,500	4,000	5,500	6,500	8,500	—	—
Passenger ↑	5,000	6,000	7,000	9,000	10,100	12,000	—	—
Passenger ↓	4,500	5,500	6,500	8,500	9,500	11,500	—	—
Trans Sport/SE ↑	4,500	5,500	7,000	8,500	10,500	12,500	14,500	17,000
Trans Sport/SE ↓	4,000	5,000	6,500	8,000	10,000	11,500	14,000	16,000
Venture ↑	—	—	—	—	—	—	14,000	17,000
Venture ↓	—	—	—	—	—	—	13,500	16,000
Extended Warranty	Y	Y	Y	Y	Y	Y	Y	Y
Secret Warranty	Y	Y	Y	Y	Y	Y	Y	Y
Reliability	❶	❷	❷	❷	③	③	④	④
Air conditioning	③	③	③	④	④	④	④	⑤
Automatic transmission	❷	❷	❷	③	③	③	③	③
Body integrity	❶	❶	❶	❷	❷	❷	❷	③
Braking system	❶	❷	❷	❷	❷	❷	③	③
Electrical system	❶	❶	❶	❷	❷	❷	③	③
Engines	③	④	④	④	④	④	④	⑤
Exhaust/Converter	③	④	④	④	④	④	④	⑤
Fuel system	④	④	④	④	⑤	⑤	⑤	⑤
Ignition system	❶	❷	❷	③	④	⑤	⑤	⑤
Rust/Paint	❶	❶	❶	❷	❷	❷	❷	③
Steering	④	④	④	④	④	④	④	⑤
Suspension	③	③	③	④	④	④	④	④
Crash Safety	—	⑤	❷	❶	④	④	④	④

Note: The low crash ratings represent severe leg trauma.

Astro, Safari

Rating: Average (1996–98); Not Recommended (1985–95). These vehicles are more minitruck than minivan. Believe it or not, these run-of-the-mill minivans are beginning to look quite good when compared to the powertrain-challenged Chrysler and Ford minivans. **Maintenance/Repair costs**: Average. Any garage can repair these rear-drive minivans. **Parts**: Plentiful and reasonably priced. As with the Aerostar, the classified ads are jam-packed with sellers wanting to unload their Astros and Safaris simply because their vehicles have got high mileage or the owners are small businesses whose needs have moved up to a larger van. Whatever the reason, you can find some real bargains if you're patient. Other choices you may wish to consider: the Ford Aerostar or a full-sized GM or Ford van.

Strengths and weaknesses: Introduced during the 1985 model year, these minivans are basically trucks dressed in minivan garb. Nevertheless, these spacious vans can be fitted to tow up to 6,000 lb. and carry eight people. The base 4.3L V6 gives acceptable acceleration, but the High Output variant of the same engine (first available in the 1991 model) is a far better choice, particularly when it's mated to a manual gearbox. The full-time AWD versions aren't very refined, have a high failure rate, and are expensive to diagnose and repair. The 1995 models got a minor facelift, a bit more horsepower, and extended bodies.

Early versions suffer from failure-prone automatic transmissions, a poor braking system, and fragile steering components. The early base V6 provides ample power, but also produces lots of noise, consumes excessive amounts of fuel, and tends to have leaking head gaskets and failure-prone oxygen sensors. These computer-related problems often rob the engine of sufficient power to keep up in traffic. Even though the 5-speed manual transmission shifts fairly easily, the automatic takes forever to downshift on the highway. Handling isn't particularly agile on these minivans, and the power steering doesn't provide the driver with enough road feel. Unloaded, the Astro provides very poor traction, the ride isn't comfortable on poor road surfaces, and interior noise is rampant. Many drivers find the driving position awkward (no left leg room) and the heating/defrosting system inadequate. Many engine components are hidden under the dashboard, making repair or maintenance awkward.

Even on more recent models, highway performance and overall reliability remain mediocre. The 4-speed automatic transmissions are clunky but much more reliable than the Ford or Chrysler gearboxes. Excessive rear-end clunking and clanging occurs whenever the transmission shifts or downshifts while under load. One worker at GM's Oshawa, Ontario, plant, a 1993 Safari owner, wrote former GM Canada President George Peapples

> Do you as CEO really know what's going on or does middle management filter out the bad news? GM's policy on warranty service is atrocious, your service representative was rude and your parts are substandard...Your bulletin 93-4A-101, "Discouraging dealers from attempting to repair driveline clunk," is weaseling out on your responsibilities to the customer.

Other owners report that the front suspension, steering components, computer modules, and catalytic converter can wear out within as little as 60,000 km. There have also been lots of complaints about electrical, exhaust, cooling, and fuel system bugs; inadequate heating/defrosting; and axle seals wearing out every 12–18 months. Many owners write that they've had to endure a rotten-egg smell coming from the exhaust. Body hardware is fragile, and fit and finish is the pits. One 1994 Safari owner, frustrated by chronic water leaks, replaced the rear hatch moulding three times without any improvement. Squeaks and rattles are legion and hard to locate. Sliding door handles often break off and the sliding door frequently jams in cold temperatures. The hatch release for the Dutch doors occasionally doesn't work and the driver's side seat vinyl lining tears apart. Premature paint peeling/delamination and surface rust are fairly common.

Dealer service bulletins: **1996**—Engine noise (install new valve stem oil seal). • Hard start or no-start. • Intermittent hesitation, low power, and rough running. • Lack of power and early, late, or erratic shifts. • Transmission fluid leak from pump body. • 3–2 part throttle downshift flare and delayed transmission engagement. • Clutch chatter or shudder. • Driveline clunk noises. • Steering column snap or clicking noise. • Fuel tank lines may have insufficient torque. • Insufficient AC performance and/or noisy clutch. • AC odours. • No heat from air ducts. • Abnormal condensation in taillights and other lights. **1997**—Bench seat squeak or squawk noise. • Driveline clunk noises. • Engine noise. • Rough engine performance. • Insufficient heat to driver's side, and cold air comes from the left-side kick panel. • Low-voltage reading or dim lights at idle.

Safety summary/Recalls: **1990–91**—There were 31 complaints of dashboard fires. **1995**—Fire ignited under the dashboard. • Headlight switch shorted out, resulting in a fire. • Driver-side airbag failed to deploy. • Vehicle continues to accelerate after foot is removed from accelerator. • Frequent stalling. • Erratic engine performance due to blocked catalytic converter. • Engine leaks oil. • Loose fan belts cause belt to come off, causing loss of power steering and brakes. • Frequent brake failures. • Loss of braking when going over a bump. • ABS failed to engage. • ABS suddenly engaged for no reason and then wouldn't disengage. • ABS hesitates when applied. • Defective brake calipers. • Power brake hose fell off, causing loss of power brakes. • Front brake hoses collapsed, causing sudden brake failure. • Premature front brake wear. •Power-steering lockup. • Power-steering hose fell off, causing loss of power steering. • Power-steering fluid leaks. • Vehicle jerks to one side when braking. • Front wheels lock up when turning the steering wheel to the right from a stop while in gear. • Steering stuck when turning. • Seatbelts fail to work properly. • Shoulder belt failed to restrain driver in collision. • When the middle seat is removed, the seatbelt that is attached permanently to the headliner swings freely in the passenger compartment. • There is no room to latch the seatbelt. • Passenger-side seatbelt tightens up with every movement. • Seatbelt locks up unexpectedly. • Bench seatbelts are too short and cannot be adjusted. • Fresh-air ventilation system allows fumes from other vehicles to enter interior compartment. • With jack almost fully extended, wheel doesn't lift off of ground. • Spare tire not safe for driving over 35 mph. • Design of horn makes it difficult to use. • Horn buttons require excessive pressure to activate them. • Driver-side window failure. • Sliding door suddenly fell off. • Faulty door hinges allow the door to fall off. • Sliding door rattles. • Front passenger door won't close. • Passenger-side door glass fell out. • Rear hatch latch release failed. • Rear hatch hydraulic rods are too weak to support hatch. • Front passenger's seat reclining mechanism failed. • Poor traction. • Parked in gear and rolled downhill. • Transmission failures. • Left rear axle seal leaks, causing lubricant to burn on brake lining. • Excessive rear-end axle noise. • Sudden wheel bearing failure. • AC clutch fell apart. • Alternator bearing failure. • Windshield wipers fail intermittently. **Recalls**: **1985**—The steering gear may crack or break free, causing sudden steering loss. **1985–91**—The seatback may recline suddenly. **1989**—GM will install and relocate a new AC line. **1990**—GM will repair the fuel return line fitting crimp for free. **1995**—Possible separation of control arm from the frame. • Vehicles with the L35 engine may have fuel line leakage. **1996–97**—Outboard seatbelt webbing on right rear bucket seat can separate during a crash. **1996–98**—Defective child safety seat attaching belts. **1997**—Defective right-hand rear bucket seatbelt webbing.

Secret Warranties/Service Tips

1985–93—Vehicles equipped with a Hydramatic 4L60 transmission that buzzes when the car is in Reverse or idle may need a new oil pressure regulator valve. • A power-steering hiss can be silenced by replacing the power-steering valve assembly. **1988–92**—Hydramatic 4L60/700R4 automatic transmission may click or whine in third or fourth gear. There may also be a rattling noise coming from the rear of the transmission. Correct by installing five new fibre plates in the Low and Reverse clutch. The new plates have a different groove configuration that prevents third and fourth gear vibration. **1989–91**—THM 700-R4 automatic transmissions may exhibit a no-Reverse or delayed Reverse condition in cold weather. This problem can be corrected by replacing the piston outer seal with a long lip design. If this design is already being used or there's no improvement, change the Reverse input clutch housing. • A binding sliding door requires the replacement of the centre track rolling bracket, the centre track assembly, and the lower track striker/bumper assembly. **1992–94**—The PCV hose may freeze, causing oil starvation to the engine and leading to engine failure. Alaska owners are eligible for free higher-flow calibrated PCVs under the emissions warranty. All other cold-weather operators are supposed to be told of the problem and given free servicing on a case-by-case basis. **1993–94**—A faulty speedometer, inability to shift down into second gear, or a transmission stuck in second gear may all be corrected by replacing the C240 connector.

• Poor cold starting can be traced to a defective fuel pump relay. **1993–95**—Malfunctioning gauges and driveability problems may be the result of a short circuit caused by the C110 connector wire rubbing against the AC accumulator pipe. • GM says that a chronic driveline clunk can't be silenced and is a characteristic of its late-model vehicles. **1996–97**—Excessive engine noise can be curtailed by installing an upgraded valve stem oil seal. **1996–98**—Rough engine performance may be caused by a water-contaminated oxygen sensor. **All models/years**—Defective catalytic converters that cause a rotten-egg smell will be replaced free of charge under the vehicle's emissions warranty. • Paint delamination, peeling, or fading (see pages 76, 113).

Astro, Safari Profile

	1991	1992	1993	1994	1995	1996	1997	1998
Cost Price ($)								
Cargo	15,650	16,798	17,638	18,198	22,640	23,475	25,110	25,110
CS/base	17,598	18,500	19,528	19,998	23,295	25,285	26,920	26,920
Used Values ($)								
Cargo ↑	3,500	4,000	4,500	5,500	7,500	8,500	10,000	12,000
Cargo ↓	3,000	3,500	4,000	5,000	7,000	7,500	9,000	11,500
CS/base ↑	5,500	6,500	8,000	9,000	11,000	12,500	14,000	16,500
CS/base ↓	5,000	6,000	7,000	8,500	10,500	11,500	13,000	15,500
Extended Warranty	Y	Y	Y	Y	Y	Y	N	N
Secret Warranty	Y	Y	Y	Y	Y	Y	Y	Y
Reliability	❶	❷	❷	❷	❷	③	③	④
Air conditioning	❷	❷	❷	❷	❷	③	④	④
Automatic transmission	③	③	③	④	④	④	⑤	⑤
Body integrity	❶	❶	❶	❶	❶	❶	❷	❷
Braking system	❶	❶	❶	❷	❷	❷	❷	③
Electrical system	❶	❶	❶	❶	❶	❶	③	③
Engines	❶	❷	③	④	④	④	⑤	⑤
Exhaust/Converter	❷	③	③	③	④	④	④	④
Fuel system	④	④	④	④	④	④	④	④
Ignition system	❶	❶	❷	❷	❷	③	④	④
Rust/Paint	❶	❶	❶	❶	❶	❶	❷	❷
Steering	❷	③	③	③	③	④	④	④
Suspension	❷	❷	❷	❷	❷	③	③	④
Crash Safety	—	❶	❶	❶	—	③	③	③

HONDA

Odyssey

Rating: Above Average (1995–98). An Accord masquerading as a minivan. The only reason the Odyssey doesn't get a recommended rating is its small engine and interior. Consider these smartly styled, fuel-efficient vehicles as full-sized, urban station wagons rather than highway-hauling minivans. **Maintenance/Repair costs**: Higher than average, but any garage can repair these minivans. **Parts**: Limited supply and costly. There aren't many reasonably priced three-year-old Odysseys on the market. If you can't afford to wait, consider the Mercury Villager, Nissan Axxess and Quest, or the Toyota Sienna.

Strengths and weaknesses: One can sum up the strengths and weaknesses of the Odyssey (and its American twin through the 1998 model, the Isuzu Oasis) in three words: performance, performance, and performance. You get carlike performance and handling, responsive steering, and a comfortable ride, offset by slow-as-molasses-in-January acceleration with a full load, a raucous engine, and limited passenger/cargo space due to the narrow body.

The Odyssey has proven to be more reliable and gives better handling than the American rear-drive, truck-inspired minivans. But its high price on the used-car market, weak 2.2L 4-cylinder engine, and small dimensions can't compete with Chrysler and Ford's newest versions. About the same size as the Mercury Villager/Nissan Quest twins, the Odyssey has four sedan-type doors, second-row bench seating, and a third seat that folds flat. With its four sedan-style doors and relatively compact size, this car is more like a tall station wagon.

Like the Accord, component quality and assembly are first class. Nevertheless, potential problem areas are the front brakes (premature wear and noise) and trim and accessory items that come loose, break away, or malfunction. The timing chain may also have to be replaced frequently.

Dealer service bulletins: **1995**—The front door panels may creak. • Speedometer buzzing. • Instrument panel buzz when the brake pedal is depressed. • Fuel gauge may read less than full. • The temperature control lever may be hard to move. • A product update warns owners that the 1995 models may have loose front suspension nuts, causing a pulling to the side, or excessive noise when braking. • Seatbelts that are slow to retract. • Steering wheel shimmy. **1996**—Seatbelts that are slow to retract. • Warped wheel covers. • Loose cigarette lighter socket.

Safety summary/Recalls: Cannot secure a child safety seat with the vehicle's seatbelts; too much play in the lapbelt. **1995**—Airbag failed to deploy. • Front two wheels separated from car while driving. • When vehicle is going over a hill, the cruise control has a tendency to overshoot the set speed by almost 10 km/h. • AC doesn't cool the vehicle sufficiently and it puts out a nasty odour. • Defroster can't clear up fogged windows. • Due to its design, the muffler can suddenly fly off while vehicle is underway. • Far rear restraint retracted so tightly it had to be cut to free the passenger. • Rear middle seat seatbelt slips and won't stay snug around the lap. **1996**—Airbags failed to deploy. • While backing out the vehicle, occupant turned AC on, and vehicle shot forward with no brakes. • Cruise control failed while vehicle was going uphill. **1997**—Sudden acceleration upon brake application. • Minivan was put in Drive, and AC was turned on; vehicle suddenly accelerated, brakes failed, and the minivan hit a brick wall. • Both front airbags failed to deploy in a head-on collision. **1999**—Sliding door is difficult to close. • Design of the gear shifter interferes with the radio controls.

Secret Warranties/Service Tips

1995–96—Warped wheel covers are caused primarily by bent tabs. **1995–97**—In a settlement with the U.S. Environmental Protection Agency, Honda paid fines totaling $17.1 million (U.S.) and extended its emissions warranty on 1.6 million 1995–97 models to 14 years/150,000 miles. This means that costly engine components and exhaust system parts like catalytic converters will be replaced free of charge, as long as the 14-year/150,000 mile limit hasn't been exceeded. Additionally, the automaker will provide a full engine check and emissions-related repairs at 50,000–75,000 miles and will give free tune-ups at 75,000–150,000 miles. The settlement was first reported on page 6 of the June 15, 1998, edition of *Automotive News*. Canadian owners may wish to use this settlement as leverage for free repairs in Canada or use the full-terms of the settlement when visiting the States. One thing is certain, neither Transport Canada nor Environment Canada is sufficiently enthused to render any assistance.

Odyssey Profile

	1995	1996	1997	1998
Cost Price ($)				
Base	28,995	28,795	28,995	29,800
Used Values ($)				
Base ↑	14,500	17,500	20,500	23,000
Base ↓	13,500	17,000	19,500	22,500
Extended Warranty	N	N	N	N
Secret Warranty	N	Y	Y	N
Reliability	⑤	⑤	⑤	⑤
Crash Safety	④	④	④	—

MAZDA

MPV

Rating: Average (1996–98); Below Average (1988–95). The MPV is a good performer hobbled by poor reliability and servicing. The 1997–98 models have lots of potential, if they don't fall apart as they age. **Maintenance/Repair costs**: Higher than average, but any garage can repair these minivans. **Parts**: Limited supply and costly. There are plenty of reasonably priced two- and three-year-old models that have just come off lease on the market. Be wary of earlier high-mileage versions, though; they tend to deteriorate fairly rapidly after the first few years. Other minivans you may wish to consider: a late-model GM Trans Sport, Mercury Villager, Nissan Quest and Axxess, or Toyota's Previa and Sienna.

Strengths and weaknesses: Handling and overall highway performance put the MPV in the top third of the minivan pack, but the poor winter handling, below-average reliability of early models that are no longer under warranty, a "take it or leave it" attitude when handling warranty claims, mediocre servicing, and high fuel and parts costs make pre-1996 versions below-average buys when it comes to overall operating costs.

The 5-speed manual transmission shifts easily and has well-spaced gears, but it's relatively rare. The automatic performs fairly well but sometimes hesitates before going into gear at about 15 mph and again at 35 mph. Steering is crisp and predictable. Rear-drive setup makes for easy load carrying and trailer towing. The base 2.6L 16-valve 121-hp 4-cylinder engine is a dog, especially when hooked up to the automatic 4-speed transmission that robs it of what little power it has. The 3.0L 6-cylinder engine, on the other hand, delivers snappy acceleration with the front-drive, and impressive acceleration with the 4X4.

Overheating and head gasket failures are commonplace with the 4-banger, and the temperature gauge warns you only when it's too late. Some cases of chronic engine knocking in cold weather with the 3.0L have been fixed by installing tighter-fitting, Teflon-coated pistons. Valve lifter problems are also common with this engine. Winter driving is compromised by the MPV's light rear end and mediocre traction, and low ground clearance means that off-road excursions shouldn't be too adventurous.

Owners report that the electronic computer module (ECU), automatic transmission driveshaft, upper shock mounts, front 4X4 drive axles and lash adjusters, AC core, and radiator fail within the first three years. Cold temperatures tend to "fry" the automatic window motor, and the paint is easily chipped and flakes off early, especially around the hood, tailgate, and front fenders. Premature brake caliper and rotor wear and excessive vibration/pulsation are chronic problem areas (repairs are needed about every 12,000 km).

Dealer service bulletins: **1994**—Rough idle on warm restart. • Tips for troubleshooting excessive engine noise. • Loose, rattling outer door handles. • Tips for troubleshooting premature automatic transmission planetary failures. • Soft shifts • Gear slips under heavy throttle. • Brake pulsation repair tips. • Steering column squeaks. • Slightly off-centre steering column. **1995**—Brake pulsation repair. • Loose, rattling outer door handles. • Slightly off-centre steering wheel. • Tips for troubleshooting excessive engine noise. **1996–97**—Brake pulsation repair tips.

Safety summary/Recalls: **1991–93**—Liquid-vapour separator may fail, causing fuel leaks. • The front brake calipers may suddenly grab and then release, losing their efficacy, and the rear brakes may occasionally lock up in emergency situations. • Some models have head restraints and rear seatbelts that are too slack. • Lots of buffeting about by strong winds. **Recalls**: **1989**—Rear brake shoes may fail. **1989–91**—Takata seatbelt replacement. **1990–91**—Rear brakes may be too aggressive when braking at low speeds, causing the rear wheels to lock up with a possible loss of vehicle control. The shoe linings on the rear brakes can change over time and increase friction, causing the rear-wheel ABS to activate prematurely.

Secret Warranties/Service Tips

1988–91—Hard shifting after cold-weather starts can be corrected by installing upgraded synchronizer rings and clutch hub assemblies. **1989–91**—Cold engine piston slapping requires replacement pistons to fix the problem. **1989–93**—AC refrigerant leaks may be caused by a failed O-ring at the block fittings. **1989–94**—Leaking fuel vapour separator that allows fuel vapours to enter the passenger compartment will be fixed up to 11 years from date of purchase. **1990–92**—If parking on an incline makes your MPV impossible to start, install an upgraded inhibitor switch and manual plate. **1992–93**—DSB #007/94 describes Mazda's upgraded head cylinder gasket. **1992–94**—A rough idle following a warm restart can be corrected by installing an upgraded two-stage fuel regulator. **All models/years**—DSB #006-94 looks into all the causes and remedies for excessive brake vibrations, and DSB #11-14-95 gives an excellent diagnostic flow chart for troubleshooting excessive engine noise. • Serious paint peeling and delaminating will be fully covered for up to six years under a Mazda secret warranty, say owners.

MPV Profile

	1991	1992	1993	1994	1995	1996	1997	1998
Cost Price ($)								
Base 4X2	16,900	18,495	19,345	20,615	21,780	27,330	—	—
LX 4X2	17,000	18,895	23,745	24,975	26,720	30,900	27,845	25,199
LX 4X4	24,950	26,095	26,345	28,375	28,675	34,350	35,335	32,355
Used Values ($)								
Base 4X2 ↑	5,000	6,000	8,000	10,000	12,000	14,000	—	—
Base 4X2 ↓	4,500	5,500	7,500	9,500	11,000	13,000	—	—
LX 4X2 ↑	6,000	7,000	9,500	12,000	14,000	16,000	18,000	20,500
LX 4X2 ↓	5,500	6,500	9,000	11,500	13,500	15,500	17,500	20,000
LX 4X4 ↑	7,500	8,000	10,000	12,000	15,500	17,500	19,500	22,000
LX 4X4 ↓	6,000	7,500	9,500	11,500	14,500	16,500	19,000	21,500
Extended Warranty	Y	Y	Y	Y	Y	N	N	N
Secret Warranty	N	N	N	N	N	N	N	N
Reliability	❷	❷	❷	❷	❷	③	③	③
Air conditioning	❷	❷	❷	③	③	③	④	④
Automatic transmission	❷	❷	❷	③	③	③	④	④
Body integrity	❷	❷	❷	❷	❷	❷	❷	❷
Braking system	❶	❶	❶	❷	❷	❷	③	③
Electrical system	❶	❶	❶	❷	❷	❷	❷	③
Engines	❷	❷	❷	❷	③	③	④	④
Exhaust/Converter	❷	❷	③	③	④	④	⑤	⑤
Fuel system	③	③	③	④	④	④	⑤	⑤

Ignition system	③	③	③	④	④	④	⑤	⑤
Rust/Paint	❷	❷	❷	③	③	③	③	④
Steering	❷	❷	③	③	④	⑤	⑤	⑤
Suspension	❷	❷	❷	❷	③	④	④	④
Crash Safety	—	—	—	—	—	④	④	④

NISSAN

Axxess

Rating: Above Average (1993–95); Average (1991–92); Not Recommended (1987–90). **Maintenance/Repair costs**: Below average. Any garage can repair these minivans. **Parts**: Reasonably priced and not hard to find.

Strengths and weaknesses: The Axxess gives you car-like handling and ride comfort in a minivan that's about the size of the higher-priced Honda Odyssey. As with most minivans equipped with a full load and a small engine, acceleration isn't confidence-inspiring on the highway; however, the Axxess is ideal for city-to-suburb commuting. What further sets the Axxess apart, however, is its reasonable price and better-than-average reliability. Unfortunately, 1995 was its last model year, so you may have trouble finding one. But it's worth the wait.

Expect some electrical problems, premature wear of the front disc brakes, manual transmission malfunctions, and paint defects. Owners have also reported that the resonator (located just behind the muffler) fails around the three-year mark and costs about $250 to replace.

Safety summary/Recalls: **Recalls**: **1987–90**—Nissan will install, at no charge, a new cooling fan and radiator to prevent an engine fire from occurring.

Secret Warranties/Service Tips

1990—Improper automatic transmission shift timing may require the rerouting of the transmission breather hose. • A noisy rear suspension can be fixed by installing revised rear shock absorbers and coil springs.

Axxess Profile

	1990	1991	1992	1993	1994	1995
Cost Price ($)						
XE	17,490	16,090	16,290	17,190	17,490	19,090
SE	—	—	—	19,677	18,990	20,590
Used Values ($)						
XE ↑	4,000	4,500	5,000	6,000	8,000	10,000
XE ↓	3,500	4,000	4,500	5,500	7,500	9,000
SE ↑	—	—	—	6,500	8,500	10,500
SE ↓	—	—	—	6,000	8,000	10,000

Extended Warranty	N	N	N	N	N	N
Secret Warranty	N	N	N	N	N	N
Reliability	❶	③	③	④	⑤	⑤
Air conditioning	③	④	⑤	⑤	⑤	⑤
Body integrity	❶	❷	❷	❷	③	④
Braking system	❷	❷	❷	❷	❷	③
Electrical system	❷	❷	❷	❷	③	④
Engines	③	③	③	⑤	⑤	⑤
Exhaust/Converter	③	③	③	④	⑤	⑤
Fuel system	❷	❷	❷	③	④	④
Ignition system	❷	③	③	④	④	④
Manual transmission	❷	③	③	⑤	⑤	⑤
- automatic	③	③	③	③	④	④
Rust/Paint	❶	❷	❷	③	③	④
Steering	③	③	③	④	④	④
Suspension	❷	❷	❷	③	③	⑤
Crash Safety	❶	—	—	—	—	—

TOYOTA

LE, Previa, Sienna

Rating: **LE:** Not Recommended (1984–90). **Previa:** Above Average (1991–97); the Previa has high-priced reliability and mediocre road performance. **Sienna:** Above Average (1998). Hopefully the 1999 Sienna has corrected many of the safety-related complaints recorded by the government relating to poor headlight illumination, brake malfunctions, and faulty sliding doors. **Maintenance/Repair costs**: Higher than average, and only Toyota dealers can repair these minivans, particularly when it comes to troubleshooting the supercharged 2.4L engine and All-Track. **Parts**: Limited supply, but reasonably priced. There are plenty of reasonably priced two- and three-year-old Previas that have just come off lease on the market. Other minivans you may wish to consider: the Mercury Villager, or Nissan Quest. If you have the patience, the new Honda Odyssey is well worth a two-year wait.

Strengths and weaknesses: When you look at Toyota's used minivan offerings, you're faced with the choice of buying an early model poorly performing LE Van, the more refined, though recently discontinued, Previa, or the hard-to-come-by 1998 Sienna.

The LE uses a conventional mechanical layout borrowed from the Toyota truck line. This means that, with its short wheelbase, it has an unusually high centre of gravity and a tendency to tip precariously in tight turns. Add to this the brakes' tendency to lock on hard application and you have a recipe for disaster. The engine's placement under the front seats takes first prize for poor design and makes routine maintenance an all-day affair. Furthermore, the only way to go from the front of the vehicle to the rear is to get out and get back in. The LE is also extremely vulnerable to side winds flinging it about, and the short wheelbase accentuates the discomfort experienced when going over bumpy roads.

LE reliability isn't very good either, with owners reporting lots of premature fuel pump and air conditioning failures, faulty engine oil pressure sensors and cruise control mechanisms, excessive front brake vibrations and pad and disc wear, and poorly engineered, rust-prone steering components.

The redesigned 1991 Previa's performance and reliability are so much improved over its predecessor that it almost seems like a different vehicle. Roomier and rendered more stable thanks to its longer wheelbase, equipped with a new 2.4L engine (supercharged as of the 1994 model year) and loaded with standard safety and convenience features, 1991–97 Previas are almost as driver-friendly as the Chrysler and Mazda competition. Still, it can't match Chrysler for responsive handling and a comfortable ride, and Toyota's small engine is overworked and doesn't hesitate to tell you so.

Previa owners have learned to live with excessive engine noise, poor fuel economy, premature front brake wear and excessive brake vibration and pulsation, electrical glitches, AC malfunctions, and fit and finish blemishes. Four-wheel drive models with automatic transmissions steal lots of power from the 4-cylinder powerplant.

Sienna

Toyota's new Camry-based front-drive minivan replaced the Previa for the 1998 model year. It's built in the same high-quality Kentucky assembly plant as the Camry and comes with lots of safety and convenience features, including side airbags, anti-lock brakes, and a low-tire-pressure warning system.

Sienna abandons the Previa's futuristic look in favour of a more conservative Chevrolet Venture styling. It seats seven, offers a power sliding door (an additional door was added with the 1999 version) with optional remote controls, and has a V6 powerplant (an engine the Previa sorely needed). As with most minivans and vans, you can save about $1,000 by buying the Sienna's cargo version, but you won't get as many features.

Some of the Sienna's strong points: standard ABS and side airbags (LE, XLE); a smooth-running V6 engine; a comfortable, stable ride; a fourth door; a quiet interior; easy entry/exit; and better-than-average fit and finish reliability. Its weak areas: V6 performance is compromised by AC and automatic transmission power drain; it lacks the trailer-towing brawn of rear-drive minivans; although the rear seats fold flat to accommodate the width of a 4' X 8' board, the tailgate won't close, plus the heavy seats are difficult to reinstall (a two-person job, and the centre seat barely fits through the door). There's also no traction control, fuel economy (premium fuel) isn't impressive, the low-mounted radio is hard to reach, and third-row seats lack a fore/aft adjustment to increase cargo space.

Reliability problems include electrical shorts, brake noise and malfunctions, sliding door malfunctions, and various body glitches—not at all unusual for a first-year minivan.

Dealer service bulletins: **1995–96**—Tips for preventing AC evaporator odours. • Tips for troubleshooting excessive brake vibration or pulsation. **1997**—Wind noise repair kit. • CD player does not accept or ejects CDs.

Safety summary/Recalls: Wind buffeting makes the Previa wander, and rear seat head restraints block visibility. **1996**—Defective brake booster. • Brake master cylinder failure. • Multiple ABS failures. • Poor, noisy AC compressor performance. • Windshield wipers suddenly stopped working. • Middle right bench seat lapbelt is impossible to adjust due to its poor design. • Speed sensor failure. **1998**—Sudden acceleration, due to a defective throttle cable; vehicle hit a wall. • Brakes made a loud grinding noise, pedal went to the floor, and vehicle failed to stop. • Extremely loud groaning or grinding noise when braking. • When vehicle is turning left, steering wheel has to be manually returned to the straight-ahead position or vehicle will continue turning. • Many owners have complained the vehicle pulls sharply to one side or another when driving at moderate speeds. • Shape and design of the Sienna creates severe blind spots. • Headlights give poor illumination. • Windshields are distorted, exhibiting a "melted" image. • Rear door doesn't shut tightly. • Sliding door won't latch in cold weather. • Design of the inside sliding door lock release makes it difficult to operate, requiring excessive force to activate. • Poor visibility due to the tinted window design. • Headrests and third-row back seats are loose and vibrate. • Shoulder belts in the middle row lock up instantly when first put on and stay locked up, binding the passenger. **1999**—While being driven, vehicle pulls excessively to the right despite many visits to the dealer to correct the problem. **Recalls**: **1991**—Dealers will replace faulty electrical components in Fujitsu Ten radios to eliminate a fire hazard. • Premature windshield wiper failure. **1997**—Oil leakage will result in a sufficiently low oil supply to cause bearing damage to the front differential unit. This can lead to eventual seizure of the unit, increasing the risk of a crash. Dealers will install a modified air breather plug to prevent such oil loss.

Secret Warranties/Service Tips

1988–92 LE Van—In its U.S. memo to dealers, No. 89-11 issued 2/21/89, Toyota says that it will pay for the replacement of the oxygen sensor for up to 80,000 miles on all of its trucks and vans. • Transmission extension seal leaks require the installation of an upgraded seal (#24201470). **1990 Previa**—If the sliding door makes a scraping noise as it travels, it signals the need to replace the centre rail with an improved part (#68303-28020).

1991—Under a Special Service Program, Toyota has replaced, at no charge, the radiator, fan, and fan shroud with improved parts. • Under Special Service Program LO5, Toyota has also replaced the cylinder head core plugs and gaskets with improved components. If your minivan requires these repairs, ask Toyota for a partial refund. **1992–93**—Panasonic CD players that skip or won't play were upgraded as of June 1993. **1994–96**—Excessive brake vibration or pulsation can be corrected by installing upgraded front brake pads. **All models/years**—Owner feedback confirms that front brake pads and discs will be replaced free of charge if they wear out before 2 years/40,000 km. Improved disc brake pad kits are described in DSB #BR94-004. Brake pulsation/vibration, another generic Toyota problem, is fully addressed in DSB #BR94-002, "Cause and Repair of Vibration and Pulsation."

Previa, Sienna Profile

	1991	1992	1993	1994	1995	1996	1997	1998
Cost Price ($)								
Previa	18,918	20,778	21,448	24,748	28,578	35,908	36,998	—
Sienna CE 3d	—	—	—	—	—	—	—	25,628
Sienna CE 4d	—	—	—	—	—	—	—	26,808
Sienna LE 4d	—	—	—	—	—	—	—	29,558
Used Values ($)								
Previa ↑	7,500	9,000	11,000	12,500	16,00	18,500	20,000	—
Previa ↓	7,000	8,000	10,500	11,500	15,000	17,500	19,500	—
Sienna CE 3d ↑	—	—	—	—	—	—	—	22,000
Sienna CE 3d ↓	—	—	—	—	—	—	—	21,000
Sienna CE 4d ↑	—	—	—	—	—	—	—	22,500
Sienna CE 4d ↓	—	—	—	—	—	—	—	21,500
Sienna LE 4d ↑	—	—	—	—	—	—	—	24,500
Sienna LE 4d ↓	—	—	—	—	—	—	—	23,500
Extended Warranty	Y	Y	Y	Y	Y	Y	Y	N
Secret Warranty	N	N	N	N	N	N	N	N
Reliability	③	③	④	⑤	⑤	⑤	⑤	⑤
Air conditioning	❷	③	③	③	④	⑤	⑤	⑤
Automatic transmission	③	③	③	④	⑤	⑤	⑤	⑤
Body integrity	③	③	④	⑤	⑤	⑤	⑤	④
Braking system	❶	❷	❷	❷	❷	❷	③	③
Electrical system	❷	③	③	③	❶	❶	❷	③
Engines	③	③	④	⑤	⑤	⑤	⑤	⑤
Exhaust/Converter	③	③	④	④	⑤	⑤	⑤	⑤
Fuel system	③	③	④	⑤	⑤	⑤	⑤	⑤
Ignition system	③	③	③	④	⑤	⑤	⑤	⑤
Rust/Paint	④	④	④	⑤	⑤	⑤	⑤	⑤
Steering	④	④	③	④	④	④	⑤	⑤
Suspension	③	③	③	④	④	④	⑤	⑤
Crash Safety	—	—	⑤	❶	④	④	④	⑤

Appendix I
25 BEST INTERNET GRIPE SITES

Websites come and go. Please email *lemonaid@earthlink.net* if you find a site has closed down or moved. Also let me know if you have discovered a website that should be included in next year's *Lemon-Aid* guide.

1. Lemon Org (*www.lemon.org/*)
This website, set up in the States as a non-profit service, helps owners of lemon vehicles register their complaints and correspond with others who have similar problem vehicles.

2. Chrysler Owner Review Committee (email *RAR17@chrysler.com*)
If you're not satisfied by the response you get, phone, fax, or email Mr. Larry Latta, Vice President, Sales and Marketing; or Mr. Lou Spadotto, National Service Manager, Parts, Service, and Engineering, at tel. 519-973-2300, fax 519-561-7005, and email *RAR17@chrysler.com.* They use the Review Committee to take a hard look at all claims, including those that were previously rejected. Three suggestions if you plan to contact the Review Committee: send Chrysler copies of all your repair bills or independent garage estimates, don't accept a refusal based on the fact that you're not the first owner, and finally, don't let Chrysler turn your claim down because the repairs were carried out by an independent repair facility. If you wish to contact CLOG (B.C.), call Patricia Wong at tel. 604-657-2298, fax 604-255-1831, and email *underdog@Lynx.bc.ca.*

3. Chrysler Products' Problem Web Page
(*www.wam.umd.edu/~gluckman/Chrysler/*)
This page was designed to be a resource for Chrysler owners who have had problems in dealing with Chrysler on such issues as peeling paint, transmission failure, the Chrysler-installed Bendix-10 ABS, and other maladies.

4. Ford Paint Delamination/Peeling (*www.ihs2000.com/~peel/*)
Everything you should know about the cause and treatment of Ford paint delamination. It has useful links to other sites and tips on dealing with Ford and GM paint problems.

5. GM Paint Delamination/Peeling
Unfortunately, there's no longer a single website that relates to the above GM problem. (One awesome site existed for two years and then was taken down. A GM payoff? Who knows?) Nevertheless, there are many sites where the problem is discussed. Simply access

Alta Vista or any other search engine and type in "GM paint delamination," or "GM peeling paint."

6. Ford Contour/Mystique Gripe Site (*www.contour.org/FAQ/*)
This website provides a comprehensive listing of major problems affecting the Ford Contour and Mercury Mystique.

7. Dead Ford Page (*www.mindspring.com/~ics_mak/deadford.html*)
An amusing and informative gathering place for Ford car, truck, sport-utility, minivan, and van owners who discuss common problems and solutions. A fun site just to see the FORD VENT page, where an animated character walks across the screen, drops his shorts, and takes a leak up against a thinly disguised Ford logo.

8. Ford Taurus, Sable Automatic Transmission Victims
(*members.aol.com/MKBradley/index.html*)
A great site for learning about Ford's biodegradable automatic transmissions (1991–97).

9. Ford Windstar Engine Head Gasket/Automatic Transmission Site
(*www.fordwindstar.com/*)
This site was constructed by a 1995 Windstar owner who collected complaints from three dozen other Windstar owners who experienced similar powertrain problems.

10. Jeep Paint Delamination/Peeling
(*www.goofball.com/keane/badpaint/*)
Jeep paint and other body defects are covered. Useful links to other sites, including a place where Chrysler's 1990 paint defect service bulletin can be downloaded.

11. The Unofficial BMW Lemon Site (*www.bmwlemon.com/*)
Everything you wanted to know, or maybe didn't want to know, about factory-related defects affecting the 3-series, 5-series, and 7-series, V8-equipped models. Plus a useful link to Dodge Dakota lemon sites (go figure).

12. Nissan Gripe Site (*129.22.253.156*)
For dissatisfied owners of most Nissan vehicles, this site lists common defects and is particularly helpful in providing useful links to groups and government agencies. It also covers in detail what it calls Nissan's "silent recall" of defective engines.

13. Sport-Utility Gripes, Blasts, Whines, and Guilt Trips (The Roadhog Info Trough: *www.suv.org*; Sport Utility Vehicle Anti-Fan Club: *www.howard.net/ban-suvs.htm*)

These anti-SUV sites believe that sport-utilities and trucks are safety and environmental hazards, in addition to being just plain annoying. You'll find lots of important info and links on the environmental and safety downsides of sport-utility, truck, and van ownership, but keep in mind that much of it is one-sided, and some info is outdated. The Friends of the Earth's Roadhog site is a hoot to browse if only because of its Internet survey list of 22 questions and responses that show most respondents don't care much about those policies our Corolla-and-Prizm-hugging activists hold dear.

14. "Lemon Aid or How to Get Carmakers to Call You and Beg You for Mercy!" (*www.saabnet.com/aas/1997.W26/1344522661.26426.html*)
Although not affiliated with the *Lemon-Aid* guides, this site contains a hilarious listing of tactics to use in getting auto manufacturers to return your calls, listen to your complaints, and give you compensation.

15. Canadian Law Resources on the Internet
(*mindlink.net/ drew_jackson/mdj.html*)
If you have to take a dealer or automaker to court and need a lawyer or additional jurisprudence, this is an excellent site to find both. Lots of case summaries, articles, and links to Canadian lawyers and experts. There are a number of discussion groups where lawyers will field your questions for (gasp!) free.

16. *Perez-Trujillo v. Volvo Car Corp.*
(*www.law.emory.edu/1circuit/ mar98/97-1792.01a.html*)
This U.S.-initiated lawsuit provides an interesting, though lengthy, dissertation on the safety hazards that airbags pose and why automakers are ultimately responsible for the injuries and deaths caused by their deployment.

17. Lemon-Aid Feedback (*www.lemonaidcars.com*)
The official site of the *Lemon-Aid* guides, with monthly updates and links to major lemon pages. Comments and critiques are welcome, particularly if you have an experience to relate that can help other *Lemon-Aid* readers.

18. U.S. NHTSA (*www.nhtsa.dot.gov/cars/problems/*)
Run by the Big Daddy of federal government auto safety regulators, the National Highway Traffic Safety Administration, this site has a comprehensive database covering owner complaints, recall campaigns, defect investigations initiated by the department, and automaker service bulletins that may not be found in the Alldata database. Best of all, this data is easily accessed by typing in your vehicle's year, make, and model.

19. Transport Canada (*www.tc.gc.ca/roadsafety/Recalls/search_e.asp*)
Cybersurfers can access this recall database for 1970–2000 model vehicles, but unlike the NHTSA's website, owner complaints aren't listed, defect investigations aren't disclosed, and service bulletin summaries aren't provided. And you can't complain on-line.

Buyers wishing to know if a vehicle can be imported into Canada from Europe or the States can get a list of admissible vehicles (those that conform to Canadian federal safety and pollution regulations) at *www.tc.gc.ca/roadsafety/rsimp_e.htm#US*, or can call the Registrar of Imported Vehicles at 1-800-511-7755.

20. Insurance Institute for Highway Safety (*www.hwysafety.org/*)
A dazzling site that's long on crash photos and graphs that show which vehicles are the most crashworthy.

21. Alldata Service Bulletins (*www.alldata.com/consumer/tsb/yr.html*)
Automotive recalls and technical service bulletins are listed by year, make, model, and engine option. Select a year, and then a manufacturer, to see a summary list of recalls and technical service bulletins for your car or truck.

22. Automobile News Groups
These news groups are compilations of email raves and gripes and cover all makes and models. They fall into four distinct areas: *rec.autos.makers.chrysler* (you can add any automaker's name at the end); *rec.autos.tech*; *rec.autos.driving*; and *rec.autos.misc.*

The following news group bulletin board is particularly helpful to owners with minivan and van problems: *www.he.net/~brumley/family/vanboard.html.*

23. The Auto Channel (*www.theautochannel.com/*)
This website gives you comprehensive information useful in choosing a new or used vehicle, filing a claim for compensation, or linking up with other owners. Lots of background info on ABS defects and paint delamination/peeling, with an update as to where the paint class actions are before the courts.

24. CarCalculator, "Why leasing can be a sucker's bet."
(*www.carcalculator.com/*)
This is an excellent site to learn about the dark side of leasing from leasing expert and consumer advocate Bob Lo Presti.

25. Automobile Protection Association (*www.apa.ca/apa-services.html*)
APA is a Canadian non-profit auto industry watchdog. For $25, plus provincial and federal sales taxes, APA will mail or fax you the dealer's cost or invoice price for any vehicle you are considering buying. It will also provide any information on rebates and interest rate promotions.

Appendix II
SURVEY AND BULLETIN SEARCH

Rate your vehicle

The information found in this book has been garnered from motorists' responses to surveys, like the one on the following page. Your answers help us protect others from sales scams and bad products. Include any photographs (especially of paint defects), diagrams, contracts, or work orders that expose a defect or dishonest practice. If you order a bulletin summary, your survey comments will also help me to zero in on bulletins that may be useful to you.

I pull the computer bulletin summaries myself (it keeps me in touch with readers like you), and usually have them faxed back within a few days. Sorry, I can only process Visa cards.

Phil Edmonston
Email *lemonaid@earthlink.net*

Cut Repair Costs • Fight Fraud and Incompetence with ...
***Lemon-Aid*'s $15 Bulletin Summary**

If you plan to buy a used vehicle, have just bought one, or want to know what to check before the warranty expires, you'll need a bulletin summary computer printout. Nothing gives you a stronger argument with a mechanic than pulling out a confidential bulletin (DSB) that says a failure is factory related or is covered by an extended warranty.

Fill out the request on the next page. You'll receive an exhaustive summary of the DSBs (published in the U.S.) that concern your 1982–98 vehicle. (It's too early for 1999 models.)

Bulletins show repair shortcuts, labour time, lists of upgraded parts, probable defects, recall campaigns, and secret warranties. As well, we'll list your vehicle's present wholesale and retail value.

Order by fax or mail. There's no difference in price, although faxing is much faster. Expect 5–10 faxed pages. You may email your request, but be sure to give a fax number or mailing address because I can only fax or mail the summary.

- **FAX** (Visa only): 954-563-2448 (24 hours a day). Replies by fax should take only a few days—unless I'm on vacation or way behind in my work. You may fax your request and have the bulletin summary mailed. That takes about a week.

- **MAIL**: Make your cheque payable to "DSB" (2805 E. Oakland Park Blvd., PMB 211, Ft. Lauderdale, FL, 33306–1813, U.S.A.). Allow a total of 2–3 weeks for reception and delivery.

Lemon-Aid Survey/DSB Summary Request

❑ I don't need a DSB summary; my survey comments are below.

❑ Please fax ❑ mail ❑ me a DSB summary for my vehicle (CDN$15 fee enclosed) and an estimate of its worth. (From the DSB summary, individual bulletins may then be ordered for an additional CDN$5 per bulletin.)

Fax #: ____________________

VISA (ONLY) #: ____

(It will be used once and then destroyed.)

Signature: ____

Name: ____

Address: ____

____

MY VEHICLE'S PROFILE

Make: ____________________ Model: ______________

Year: _________ Engine (litres): _____ Mileage: ___________

GENERAL COMMENTS

(Include a photo, diagram, or bill.)

Safety: ____

Reliability: ____

____

____

Performance: ____

____

____

JOY OF OWNERSHIP

(Recommended ⑤, Above Average ④, Average ③, Below Average ❷, Not Recommended ❶)

Overall Reliability

Air conditioning	Ignition system
Automatic transmission	Rust/Paint
Body integrity	Steering
Braking system	Suspension
Electrical system	**Dealer service**
Engine	**Maintenance**
Exhaust/Converter	**Parts availability**
Fuel system	**Repair costs**